An Invitation to Biomathematics

An Invitation to Biomathematics

Raina S. Robeva, James R. Kirkwood,
Robin L. Davies, Leon S. Farhy, Michael L. Johnson,
Boris P. Kovatchev, and Marty Straume

Sweet Briar College
Sweet Briar, VA

University of Virginia
Charlottesville, VA

AMSTERDAM • BOSTON • HEIDELBERG • LONDON
NEW YORK • OXFORD • PARIS • SAN DIEGO
SAN FRANCISCO • SINGAPORE • SYDNEY • TOKYO
Academic Press is an imprint of Elsevier

ELSEVIER

Academic Press is an imprint of Elsevier
30 Corporate Drive, Suite 400, Burlington, MA 01803, USA
525 B Street, Suite 1900, San Diego, California 92101-4495, USA
84 Theobald's Road, London WC1X 8RR, UK

This book is printed on acid-free paper. ∞

Library of Congress Cataloging-in-Publication Data
2007926492

British Library Cataloguing in Publication Data
A catalogue record for this book is available from the British Library

ISBN: 978-0-12-088771-2

For all information on all Elsevier Academic Press publications
visit our Web site at www.books.elsevier.com

Cover Art: Nia Kovatcheva

Working together to grow
libraries in developing countries

www.elsevier.com | www.bookaid.org | www.sabre.org

ELSEVIER   BOOK AID International   Sabre Foundation

Transferred to Digital Printing in 2013

PREFACE

In the not so distant past, the sciences of biology, chemistry, and physics were seen as more or less separate disciplines. Within the last half-century, however, the lines between the sciences have become blurred, to the benefit of each. Somewhat more recently, the methods of mathematics and computer science have emerged as necessary tools to model biological phenomena, understand patterns, and crunch huge amounts of data such as those generated by the human genome project. Today, virtually any advance in the life sciences requires a sophisticated mathematical approach. Characterization of biological systems has reached an unparalleled level of detail, and modeling of biological systems is evolving into an important partner of experimental work. As a result, there is a rapidly increasing demand for people with training in the field of biomathematics.

Training at the interface of mathematics and biology has been initiated in a number of institutions, including Rutgers University, the University of California at Los Angeles, North Carolina State University, the University of Utah, and many others. In 2001, a National Research Council panel found that "undergraduate biology education needs a more rigorous curriculum including thought provoking lab exercises and independent research projects." To improve quantitative skills, faculty members should include more concepts from mathematics and the physical sciences in biology classes. Ideally, the report says, "the entire curriculum would be revamped."[1] As the demand for academic programs that facilitate interdisciplinary ways of thinking and problem solving grows, many of the challenges for creating strong undergraduate programs in mathematical biology have become apparent. The report *Math & Bio 2010: Linking Undergraduate Disciplines* summarizes the results of the project *Meeting the Challenges: Education Across the Biological, Mathematical, and Computer Sciences*[2] and emphasizes that interdisciplinary programs should begin as early as the first year of college education, if not in high school. In one of the articles, an editorial reprinted from the journal *Science* and used in the report, the author Louis Gross specifically underscores the importance of finding ways to "teach entry-level quantitative courses that entice life science students through meaningful applications of diverse mathematics to biology, not just calculus, with a few simple biological examples."[3]

The book that you are about to read, our *An Invitation to Biomathematics*, was conceived and written with this exact goal in mind. This book is meant to provide a glimpse into the diverse world of mathematical biology and to invite you to experience, through a selection of topics and projects, the fascinating advancements made possible by the union of biology, mathematics, and computer science. The laboratory manual component of the text provides venues for hands-on exploration of the ever-present cycle of model development, model validation, and model refinement that is inherent in contemporary biomedical research. The textbook aims to provide exposure to some classical concepts, as well as new and ongoing research, and is not meant to be encyclopedic. We have tried to keep this volume relatively small, as we see this text used as a first reading in biomathematics, or as a textbook for a one-semester introductory course in mathematical biology. It is our hope that after reading this *Invitation* you will be inspired to embark on a more structured biomathematical journey. We suggest considering a classical textbook, such as Murray's *Mathematical Biology*, to gain a systematic introduction to the field in general. We also encourage you to delve deeper into some of the more specialized topics that we have introduced, or to take additional courses in mathematical biology.

The textbook is divided into two parts. In Part I, we present some classical problems, such as population growth, predator–prey interactions, epidemic models, and population genetics. While these have been examined in many places, our main purpose is to introduce some core concepts and ideas in order to apply them to topics of modern research presented in Part II. Because we also felt that these topics are likely to be covered in any entry-level course in mathematical biology, we hope that this organization will appeal to college and university faculty teaching such courses. A possible scenario for a one-semester course will be to cover all topics from Part I with a choice of selected topics from Part II that is, essentially, modular in nature. The diagram in Figure 1 outlines the chapter connectivity. Table 1 presents brief chapter descriptions by biological and mathematical affiliation.

A committed reader who has had the equivalent of one semester-long course in each of the disciplines of calculus, general biology, and statistics should be able to follow Chapters 1 to 10. With these prerequisites, we believe that the book can be read, understood, and appreciated by a wide audience of readers. Although Chapters 11 and 12 also comply with those general prerequisites, a quality understanding of the fundamental concepts covered there may require a somewhat higher level of general academic maturity and motivation. Thus, although Chapters 11 and 12 can be considered optional in essence, we would like to encourage the readers to explore them to the extent and level of detail determined by their individual comfort level.

Our rule while writing this book was that the biology problem should lead the mathematics, and that we only present the mathematics on a need-to-know basis and in the amount and level of rigor necessary. As a consequence, very few mathematical theorems are proved or even discussed in the text. We limited ourselves to the minimal

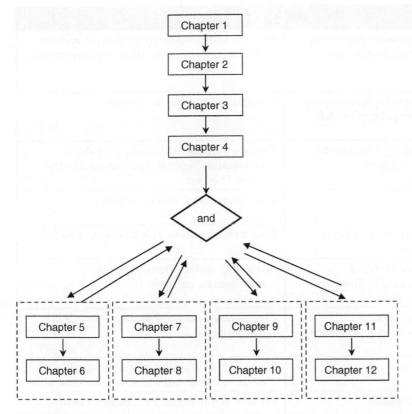

FIGURE I.

mathematical terminology necessary for understanding, formulating, and solving the problem, relying on the reader's intuition for the rest. We felt that in the interest of showing how the tools of mathematics and biology can blend together and work as one when needed, we should resist the urge for possible generalizations (an urge that is almost second nature for those of us trained in the field of theoretical mathematics). The choice not to explore many of the possible exciting mathematical venues that stem from some of the topics and projects was deliberate, and we apologize to those readers who wish we had included them.

We would like to thank all of our students at Sweet Briar College and the University of Virginia, especially Jennifer McDonaugh, Jamie Jensen, and Suzanne Harvey, for providing valuable comments and opinion throughout the development and classroom testing of the textbook and laboratory manual manuscripts. We also thank our colleagues Drs. Marc Breton, Jeff Graham, Stan Grove, David Housman, Eric Marland, Pamela Ryan, Philip Ryan, Karen Ricciardi, and Bonnie Schulman for their feedback on selected chapters and/or laboratory projects, and Anna Kovatcheva for collecting the data used in Exercise 1 of Chapter 4. We appreciate the help of Dr. Stefan Robev and of Ryan King, who carefully proofread the entire first draft of the manuscript, and of Jane Carlson, who assisted with its early technical editing. We are also indebted to all

Chapter	Biological topics	Biological subtopics	Mathematical topics
1	Ecology, Conservation Biology, Toxicology	Population growth, harvesting model, drug dosage model	Discrete and continuous dynamical systems, difference equations, differential equations
2	Ecology, Microbiology, Epidemiology	Epidemic model, predator-prey model, competition model	Continuous dynamical systems
3	Genetics	Hardy-Weinberg law, genetic selection, polygenic inheritance	Discrete dynamical systems, probability histograms, Normal distribution, Central Limit Theorem
4	Genetics, Biostatistics	Heritability	Probability distributions, statistics
5	Physiology, Endocrinology	Blood glucose levels, glucose homeostasis	Data transformation, risk function, statistics
6	Physiology, Microbiology	Development, bacterial infections, cardiac function, premature birth	Probability distributions, statistics, approximate entropy
7	Biochemistry, Physiology, Cell Biology	Hemoglobin function, cooperativity, conformational change	Continuous dynamical systems, probability distributions
8	Biochemistry, Physiology	Ligand binding	Numerical solutions of algebraic equations, iterative computational strategies, time series
9	Endocrinology, Physiology, Cell Biology	Hormone pulsatility	Periodic components, FFT, pulse-detection algorithms
10	Endocrinology, Physiology, Cell Biology	Hormone networks	Continuous dynamical systems with delays
11	Physiology, Cell Biology, Molecular Biology	Circadian rhythms	Confounded time series, rhythm analysis
12	Physiology, Cell Biology, Molecular Biology	Gene chips, molecular biology of circadian rhythms	Data normalization, clustering strategies, time series, rhythm analysis

TABLE 1.
Chapter topics by biological and mathematical affiliation

of our editors at Academic Press/Elsevier: Chuck Crumley, David Cella, Kelly Sonnack, Nancy Maragioglio, Luna Han, and Sally Cheney, for their encouragement and assistance throughout. Our deep gratitude goes to Tom Loftus who put many hours into editing the final draft of the manuscript for style and language consistency. Finally, we appreciate the support of the National Science Foundation under the Department of Undergraduate Education awards 0126740 and 0304930, and the support of the National Institutes of Health under NIDDK awards R25 DK064122, R01 DK51562, and R21 DK72095.

You are now invited to turn the page and begin your exploration of biomathematics.

The Authors
July 20, 2007

1. Morgan, A., for the Committee on Undergraduate Biology Education. (2002). *BIO2010: Transforming Undergraduate Education for Future Research Biologists.* Washington, DC: The National Academy Press.
2. Steen, L. (editor). (2005). *Math & Bio 2010—Linking Undergraduate Disciplines.* Washington, DC: The Mathematical Association of America.
3. Gross, L. G. (2000). Education for a Biocomplex Future. *Science*, Vol. 288. no. 5467, p. 807. The author, Louis Gross is a Professor of Ecology and Evolutionary Biology and Mathematics, University of Tennessee, Knoxville, and Past-President of the Society for Mathematical Biology.

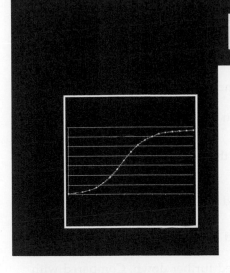

Solutions

Chapter 1

Exercise 1-2.

Because time is measured in decades elapsed since 1800, the value of t for the year 3000 is $t = 120$. Substituting this value for t in Eq (1–7) gives $P(t) = 5.3e^{(0.297)(120)} = 1.594 \times 10^{16}$. This prediction cannot be realistic, because if we divide the earth's land surface (estimated to be approximately 150, 000, 000 km^2) by this number, it would allow less than 0.01 squared meters per person.

Exercise 1-4.

In this case, socioeconomic, cultural, and behavioral factors could play a role in addition to environmental factors and availability of resources.

Exercise 1-6.

a) When $P(0) = 0$, $P(t) = 0$ for all values of t.

b) When $0 < P(0) < K$, $P(t)$ is increasing toward its carrying capacity K and K is a horizontal asymptote for the graph of $P(t)$.

c) When $P(0) = K$, $P(t) = K$ for all values of t.

d) When $P(0) > K$, $P(t)$ is decreasing toward its carrying capacity K and K is a horizontal asymptote for the graph of $P(t)$.

Exercise 1-8.

a) The equilibrium state 0 is unstable, while K is a stable equilibrium.

b) The equilibrium points are p_1, p_2, and p_3: p_1 and p_3 are stable equilibria, while p_2 is unstable. Sample trajectories for the given initial conditions are given in Figure 1-2.

c) The equilibrium point p_1 is neither stable nor unstable. If $P(0) > p_1$, $P(t)$ will be increasing, because $dP/dt > 0$, and thus is moving further away from p_1. If $P(0) < p_1$, $P(t)$ will also be increasing, because $dP/dt > 0$, and thus approaching the level p_1.

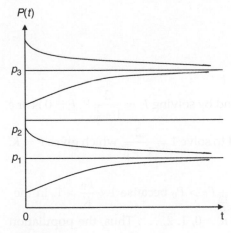

FIGURE 1-2.
Time trajectories for Exercise 1-8, part b.

Exercise 1-10.

a) The equilibrium states are 0 and 400: 0 is unstable and 400 is stable.

b) 82 is an unstable equilibrium and 318 is a stable equilibrium.

c) From the graph in Figure 1-18 in the text, observe that

     i) When $P(t_0) = 50$, $\dfrac{dP}{dt} < 0$ and the population will decrease;

     ii) When $P(t_0) = 150$, $\dfrac{dP}{dt} > 0$ and the population will increase;

         When $P(t_0) = 400$, $\dfrac{dP}{dt} < 0$ and the population will decrease.

d) The original graph will be shifted further down. Compared with the graph from part (a), it will be shifted downwards by the amount of cattle sold per year.

e) To answer this question, we need to find the distance h as shown in Figure 1-5, such that the graph only touches the horizontal axis. As long as the number of cattle sold per year is strictly smaller than h, the graph will be crossing the horizontal axis in two points, thus guaranteeing the existence of stable equilibria. To find h, notice that it is equal to the vertical coordinate of the maximal value of the function $f(P)$. To find this maximum, we have to solve $f'(P) = 0$; that is $\dfrac{d^2P}{dt^2} = 0$. Computing this derivative gives

$\dfrac{d^2P}{dt^2} = 0.004P - 8 = 0$. Thus, the maximal value occurs at $P = 200$, and $h = f(200) = 0.002(400 - 200)200 = 80$. Any number of cattle sold per year which is less than this number guarantees the existence of a stable equilibrium. Therefore, if 79 or fewer cattle are sold per year, a stable equilibrium will exist.

f) 80.

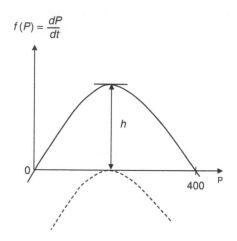

$f(P) = \dfrac{dP}{dt}$

FIGURE 1-5.
Determining the maximal number of cattle sold per year in the context of Exercise 1-10, for which a stable equilibrium exists.

Exercise 1-12.

a) The equilibrium states are found by solving $P = \dfrac{2}{1 + \frac{P}{K}} P$. $P = 0$ is one solution. When $P \neq 0$, we need to solve $1 = \dfrac{2}{1 + \frac{P}{K}}$, which gives $P = K$.

b) When $0 < P(0) < K$, $P_1 = \dfrac{2}{1 + \frac{P_0}{K}} P_0 > P_0$ because $1 + \dfrac{P_0}{K} > 1$. In the same way, $P_{n+1} > P_n$ for any $n = 0, 1, 2, \ldots$. Thus, the population size will be increasing.

c) Denote the limit $P = \lim P_n$, $n \to \infty$. Taking limits on both sides of Eq. (1-27) gives: $P = \lim P_n = \lim \dfrac{2}{1 + \frac{P_n}{K}} P_n = \dfrac{2}{1 + \frac{\lim P_n}{K}} \lim P_n = \dfrac{2}{1 + \frac{P}{K}} P$.

Because $0 < P(0)$ and because the population size increases with time, $P = \lim P_n \neq 0$. This, combined with the above calculations, implies that the limit P satisfies the equation $1 = \dfrac{2}{1 + \frac{P}{K}}$. Solving this equation yields $P = K$.

Exercise 1-14.

The sign of the term $\left(1 - \dfrac{P(t - D)}{K}\right)$ may change over time, thus causing the sign of the derivative $\dfrac{dP}{dt}$ to change. As a result, oscillations in the solution trajectories may develop.

a) To see this, assume that the population size is increasing toward its carrying capacity K over the interval $(t_1 - D, t_1)$, and that at the moment t_1 the population reaches its carrying capacity (that is, $P(t_1) = K$). Then, because $P(t_1 - D) < K$, $1 - \dfrac{P(t_1 - D)}{K} > 0$, causing $\dfrac{dP}{dt} > 0$. At t_1, the population will be increasing and the time trajectory will thus cross over the carrying capacity K, as in Figure 1-22 of the text.

b) The population $P(t)$ will continue to increase in the interval $(t_1, t_1 + D)$, because for any $t_1 < t < t_1 + D, 1 - \dfrac{P(t - D)}{K} > 0$, and thus $\dfrac{dP}{dt} > 0$.

c) When $t = t_1 + D, 1 - \dfrac{P(t - D)}{K} = 1 - \dfrac{P(t_1)}{K} = 0, \dfrac{dP}{dt} = 0$ and thus the population $P(t)$ stops growing at $t = t_1 + D$. Also, because $P(t_1) = K$ and $P(t)$ was shown to be increasing over the time interval $(t_1, t_1 + D)$, this shows that $P(t_1 + D) > K$.

When $t > t_1 + D$, but t is still close to $t_1 + D, 1 - \dfrac{P(t - D)}{K} < 0$, causing $\dfrac{dP}{dt} < 0$. For time instances slightly larger than $t_1 + D$, the population size will be decreasing. If t_2 is the first time after t_1 at which $P(t_2) = K, P(t)$ will be decreasing over the interval $(t_1 + D, t_2 + D)$.

Exercise 1-16.

From the graph in Figure 1-24, the half life can be determined as the value of t at which the concentration $C(t) = 2$. Estimated from the graph, this value appears to be close to 3.5. Analytically, using the given equation, we obtain $\tau = \dfrac{\ln(2)}{0.2} = 3.4657$.

Exercise 1-18.

Thirty minutes after the last dose was given, the concentration will be

$$Ce^{-(10.5)r} + Ce^{-(8.5)r} + Ce^{-(6.5)r} + Ce^{-(4.5)r} + Ce^{-(2.5)r} + Ce^{-(0.5)r}.$$

Exercise 1-20.

From Eq. (1-42) in the text, with $C = 10$ μg/ml, $r = 0.1$ hours^{-1},
$T = 8$ hours, $R = \dfrac{C}{e^{Tr} - 1} = \dfrac{10}{e^{8(0.1)} - 1} = 8.15966$ μg/ml.

Exercise 1-22.

Choose the initial dose to be just under $C_0 = MTC$. Subsequent doses will all be equal to $C = MTC - MEC$. The time between all doses should be calculated as $T = \dfrac{1}{r} \ln \dfrac{MTC}{MEC}$, where r is the elimination constant for the drug.

Exercise 1-24.

The following values would be appropriate:

STARTTIME = 0

STOPTIME = 7

DT = 0.05

Chapter 2

Exercise 2-2.

a) Recall that when $N - \dfrac{\beta}{\alpha} > 0$, the rate of change $\dfrac{dI}{dt}$ is given by the logistic Eq. (2-3) with $K = N - \dfrac{\beta}{\alpha}$. It is known from Chapter 1 then that when $I(0) < N - \dfrac{\beta}{\alpha}$, the function $I(t)$ is increasing toward $K = N - \dfrac{\beta}{\alpha}$. That is, $\lim\limits_{t \to \infty} I(t) = N - \dfrac{\beta}{\alpha}$, and $\lim\limits_{t \to \infty} S(t) = \lim\limits_{t \to \infty} (N - I(t)) = N - \lim\limits_{t \to \infty} I(t) = N - \left(N - \dfrac{\beta}{\alpha}\right) = \dfrac{\beta}{\alpha}$. Because $I(t) + S(t) = N$, the fact that $I(t)$ increasing implies that $S(t)$ is decreasing.

b) When $I(0) > N - \dfrac{\beta}{\alpha} > 0$, we know that the solution of the logistic Eq. (2-3) has a time trajectory $I(t)$ that decreases towards $K = N - \dfrac{\beta}{\alpha}$. Therefore, as in part a), $\lim\limits_{t \to \infty} I(t) = N - \dfrac{\beta}{\alpha}$.

Because $S(t) + I(t) = N$, this is equivalent to $\dfrac{\beta}{\alpha} = N - \lim\limits_{t \to \infty} I(t)$

$= \lim\limits_{t \to \infty} (N - I(t)) = \lim\limits_{t \to \infty} S(t)$, showing that $\lim\limits_{t \to \infty} S(t) = \dfrac{\beta}{\alpha}$.

c) When $I(0) = N - \dfrac{\beta}{\alpha}, \dfrac{dI}{dt}\big|_{t=0} = \alpha I(0) \left(N - \dfrac{\beta}{\alpha} - I(0) \right) = 0$ and the level of infectives would remain at $I(t) = N - \dfrac{\beta}{\alpha}$ for all values of $t > 0$.

Exercise 2-4.

a) Heuristically, the size of the susceptible group should be decreasing, because the model assumes complete immunity after recovery. Thus, once an infective leaves the susceptible group, there is no way for this infective to return to this group, and the number of susceptibles $S(t)$ should be decreasing. Mathematically, the same can be verified by examining the sign of the derivative $\dfrac{dS}{dt}$.

Because $\dfrac{dS}{dt} = -\alpha SI < 0$, the functions $S(t)$ is decreasing.

b) The size of the recovered group is initially zero and should be increasing with the number of individuals who have recovered from the infection. Once in R, no individual can leave the group, because according to the model assumptions, recovery confers permanent immunity. Mathematically, because the derivative $\dfrac{dR}{dt} = \beta I > 0$, the function $R(t)$ is increasing.

c) Because $S(t) + I(t) + R(t) = N$ for all values of t, $\dfrac{dS}{dt} + \dfrac{dI}{dt} + \dfrac{dR}{dt} = 0$.

Thus $-\alpha SI + \dfrac{dI}{dt} + \beta I = 0$, and $\dfrac{dI}{dt} = \alpha SI - \beta I$.

Exercise 2-6.

Because $\dfrac{dS}{dR} = \dfrac{dS/dt}{dR/dt} = -\dfrac{\alpha}{\beta} S$, we obtain $\dfrac{dS}{S} = -\dfrac{\alpha}{\beta} dR$. Integrating both sides yields $\displaystyle\int \dfrac{dS}{S} = -\dfrac{\alpha}{\beta} \int dR$

$\ln(S) = -\frac{\alpha}{\beta}R + C$, where C is a constant of integration. Thus,

$S(t) = e^{-\frac{\alpha}{\beta}R(t)+C} = e^C e^{-\frac{\alpha}{\beta}R(t)}$. For $t = 0$, because $R(0) = 0$, we obtain $S(0) = e^C e^{-\frac{\alpha}{\beta}R(0)} = e^C$, which proves that $S(t) = S(0)e^{-\frac{\alpha}{\beta}R(t)}$.

Exercise 2-8.

When $\frac{dR}{dt} > 0$ for all values of t, mathematically, this means that the function $R(t)$ is increasing for all values of t. There are two possibilities. Either $\lim_{t \to \infty} R(t) = \infty$ or $\lim_{t \to \infty} R(t) = L < \infty$. However, in the latter case, L would be a horizontal asymptote for $R(t)$, which would imply $\lim_{t \to \infty} \frac{dR}{dt} = 0$. Because we know that $\lim_{t \to \infty} \frac{dR}{dt} > 0$, this leaves us with the first possibility, namely $\lim_{t \to \infty} R(t) = \infty$.

Exercise 2-10.

Possible scenarios may include:

1. The owl population goes extinct and the vole population grows undisturbed in the absence of predators;

2. The vole population goes extinct because of a lack of resources, and the owl populations follows because they have no prey to feed on;

3. The vole and owl populations reach equilibrium at levels that allow for the two populations to coexist;

4. The vole and owl populations coexist but neither population is in equilibrium. The balance is maintained by sustained oscillations in the population sizes of predator and prey similar to those depicted in Figures 2-15, 2-16, and 2-17.

Exercise 2-12.

a) The equilibrium states are the solutions of the system of equations

$$\frac{dV}{dt} = f(V, O) = \alpha V - \beta V^2 - \gamma OV = V(\alpha - \beta V - \gamma O) = 0$$

$$\frac{dO}{dt} = g(V, O) = -\delta O + \varepsilon OV = O(-\delta + \varepsilon V) = 0.$$

Notice that when $O = 0$, the equation $\frac{dO}{dt} = 0$ is satisfied, regardless of the value of V. In the same way, when $V = \frac{\delta}{\varepsilon}$, the equation $\dfrac{dO}{dt} = 0$ is satisfied, regardless of the value of O.

Substituting $O = 0$ in the equation $\dfrac{dV}{dt} = 0$ gives $V(\alpha - \beta V) = 0$,

which implies $V = 0$ and $V = \dfrac{\alpha}{\beta}$. Thus $(0,0)$ and $(\dfrac{\alpha}{\beta}, 0)$ are

equilibrium states. Substituting $V = \dfrac{\delta}{\varepsilon}$ in the equation $\dfrac{dV}{dt} = 0$ gives

$\dfrac{\delta}{\varepsilon}\left(\alpha - \beta\dfrac{\delta}{\varepsilon} - \gamma O\right) = 0$. Solving this equation for O implies

$O = \dfrac{\alpha}{\gamma} - \dfrac{\beta\delta}{\gamma\varepsilon}$, and establishes that $\left(\dfrac{\delta}{\varepsilon}, \dfrac{\alpha}{\gamma} - \dfrac{\beta\delta}{\gamma\varepsilon}\right)$ is also an equilibrium state.

b) Computing the partial derivative for the functions f and g above gives:

$\dfrac{\partial f}{\partial V} = \alpha - 2\beta V - \gamma O$ $\qquad$ $\dfrac{\partial f}{\partial V}\left(\dfrac{\delta}{\varepsilon}, \dfrac{\alpha}{\gamma} - \dfrac{\beta\delta}{\gamma\varepsilon}\right) = \alpha - 2\beta\dfrac{\delta}{\varepsilon} - \gamma\left(\dfrac{\alpha}{\gamma} - \dfrac{\beta\delta}{\gamma\varepsilon}\right) = \dfrac{\beta\delta}{-\varepsilon}$

$\dfrac{\partial f}{\partial O} = -\gamma V$ $\qquad$ $\dfrac{\partial f}{\partial O}\left(\dfrac{\delta}{\varepsilon}, \dfrac{\alpha}{\gamma} - \dfrac{\beta\delta}{\gamma\varepsilon}\right) = -\gamma\dfrac{\delta}{\varepsilon} = -\dfrac{\gamma\delta}{\varepsilon}$

$\dfrac{\partial g}{\partial V} = \varepsilon O$ $\qquad$ $\dfrac{\partial g}{\partial V}\left(\dfrac{\delta}{\varepsilon}, \dfrac{\alpha}{\gamma} - \dfrac{\beta\delta}{\gamma\varepsilon}\right) = \varepsilon\left(\dfrac{\alpha}{\gamma} - \dfrac{\beta\delta}{\gamma\varepsilon}\right) = \dfrac{\alpha\varepsilon - \beta\delta}{\gamma}$

$\dfrac{\partial g}{\partial O} = -\delta + \varepsilon V$ $\qquad$ $\dfrac{\partial g}{\partial O}\left(\dfrac{\delta}{\varepsilon}, \dfrac{\alpha}{\gamma} - \dfrac{\beta\delta}{\gamma\varepsilon}\right) = -\delta + \varepsilon\dfrac{\delta}{\varepsilon} = 0$

Exercise 2-14.

a) We begin by writing the general form of the equation describing the change in the population N: $\dfrac{dN}{dt} = r(N,0)N$. Because the resources are limited, the per capita rate of growth $r(N,0)$ will have to be decreasing as N is increases. Following arguments similar to those in Chapter 1, the following logistic equation may be used to model the situation: $\dfrac{dN}{dt} = a\left(1 - \dfrac{N}{K}\right)N$, where K is the carrying capacity for the population N and a is its inherent per capita rate of growth.

b) As in part a), if $N = 0$, we can use the logistic equation $\dfrac{dP}{dt} = c\left(1 - \dfrac{P}{M}\right)P$ to describe the change in the size of population P.

Here M is the carrying capacity for this population and c is its inherent per capita rate of growth.

c) The equations from parts a) and b) are logistic equations describing the growth of a single population, as introduced in Chapter 1.

d) For part a), K is the carrying capacity of population N, and a is its inherent per capita rate of growth. For part b), M is the carrying capacity of population P, and c is its inherent per capita rate of growth.

Exercise 2-16.
....

To determine the equilibrium states of the model described by Eqs. (2-15) in the text, we need to determine the values of N and P for which

$$\frac{dN}{dt} = a\left(1 - \frac{N+bP}{K}\right)N = 0$$

$$\frac{dP}{dt} = c\left(1 - \frac{P+gN}{M}\right)P = 0.$$

This leads to the following 4 possibilities:

1. $N = 0$ and $P = 0$;

2. $N = 0$ and $1 - \frac{P+gN}{M} = 0$ (that is, $1 - \frac{P}{M} = 0$), giving the equilibrium state $N = 0$ and $P = M$;

3. $P = 0$ and $1 - \frac{N+bP}{K} = 0$ (that is $1 - \frac{N}{K} = 0$), giving the equilibrium state $N = K$ and $P = 0$;

4. $1 - \frac{N+bP}{K} = 0$ and $1 - \frac{P+gN}{M} = 0$, which yields the system of equations

$$K - N - bP = 0$$
$$M - P - gN = 0.$$

Solving for N and P gives the fourth equilibrium state $N = \frac{K-bM}{1-bg}$ and $P = \frac{M-gK}{1-bg}$.

Exercise 2-18.
....

In the model from Eq. (2-15), let $g = 0.51$, $K = 2000$, and $M = 1000$. Prove that for these values of the parameters, the equilibrium state $(K, 0)$ is asymptotically stable, regardless of the values of $a > 0$, $b > 0$, and $c > 0$.

As in Exercise 2-17, denote

$$\frac{dN}{dt} = r(N,P)N = a\left(1 - \frac{N+bP}{K}\right)N = f(N,P)$$

$$\frac{dP}{dt} = k(N,P)P = c\left(1 - \frac{P+gN}{M}\right)P = g(N,P),$$

and compute

$$\frac{\partial f}{\partial N} = a - \frac{2a}{K}N - \frac{ab}{K}P \qquad \frac{\partial f}{\partial N}(K,0) = -a$$

$$\frac{\partial f}{\partial P} = -\frac{ab}{K}N \qquad \frac{\partial f}{\partial P}(K,0) = -ab$$

$$\frac{\partial g}{\partial N} = -\frac{cg}{M}P \qquad \frac{\partial g}{\partial N}(K,0) = 0$$

$$\frac{\partial g}{\partial P} = c - \frac{2c}{M}P - \frac{cg}{M}N \qquad \frac{\partial g}{\partial P}(K,0) = c - 2cg.$$

Then, $J = \begin{pmatrix} -a & -ab \\ 0 & c-2cg \end{pmatrix}$, and $\quad \det \ (J) = -a(c - 2cg) = ac(2g - 1)$,

$\mathrm{trace}(J) = -a + (c - 2cg) = -a - c(2g - 1)$. Because for $g > 1/2$, $\det(J) > 0$ and $\mathrm{trace} \ (J) < 0$ (independent of the values of a, b, and c), the equilibrium state is asymptotically stable for $g = 0.51$.

Chapter 3

Exercise 3-2.

In this case, there are 3 sites that we can assume are independent because they are on different chromosomes. The possibilities for the "**A**" site are **AA**, **Aa**, **aA**, and **aa**; the possibilities for the "**B**" site are **BB**, **Bb**, **bB**, and **bb**; and the possibilities for the "**C**" site are **CC**, **Cc**, **cC**, and **cc**.

a) $(\frac{3}{4})(\frac{3}{4})(\frac{3}{4}) = 27/64$

b) $(\frac{3}{4})(\frac{3}{4})(\frac{1}{4}) = 9/64$

c) $(\frac{3}{4}) \ (\frac{1}{4})(\frac{1}{4}) = 3/64$

d) $(\frac{1}{4})(\frac{1}{4})(\frac{1}{4}) = 1/64$

e) In addition to the phenotypes from parts a)-d) the following are possible: the **A** and **C** sites are dominant and the **B** site is recessive (with probability 9/64); the **B** and **C** sites are dominant and the

A site is recessive (with probability 9/64); the **B** site is dominant and the **A** and **C** sites are recessive (with probability 3/64); the **C** site is dominant and the **A** and **B** sites are recessive (with probability 3/64).

Exercise 3-4.

Because females have two X chromosomes, the possible female genotypes are: **CC**, **Cc** and **cc** with frequencies p^2, $2pq$, and q^2, respectively. Because males have one X and one Y chromosome, and because the gene **C** is found on the human X chromosome only, the possible male genotypes will be **C** and **c**, with frequencies p and q.

Exercise 3-6.

With $m = 6$ genes, there are $N = 2m = 12$ alleles and $2^{12} = 4096$ allele combinations. Let k denote the number of contributing alleles.

a) $\binom{12}{12} \Big/ \left(\frac{1}{2^{12}}\right) = \frac{1}{4096} = 0.000244$

b) $\binom{12}{5} \Big/ \left(\frac{1}{2^{12}}\right) = \frac{12!}{5!7!} \cdot \frac{1}{4096} = 0.193359$

c) $\binom{12}{6} \Big/ \left(\frac{1}{2^{12}}\right) = \frac{12!}{6!6!} \cdot \frac{1}{4096} = 0.225586$

d) $\binom{12}{2} \Big/ \left(\frac{1}{2^{12}}\right) = \frac{12!}{2!10!} \cdot \frac{1}{4096} = 0.016113$

Chapter 4

Exercise 4-2.

HYPOTHESES

(1) The right hand of a person has a larger span;

(2) The spans of left and right hands are correlated across subjects;

(3) The dominant hand of a person has a larger span;

(4) The spans of dominant and nondominant hands are correlated across subjects.

Study design and variables: This study has a repeated measures design—each participant contributes two measurements of her left and right hands. These measurements are also paired differently—dominant versus nondominant hand. Thus, paired tests will be needed to address hypotheses 1 and 3. These tests will use one-tail significance level because hypotheses 1 and 3 are *directional*. Hypotheses 2 and 4 will be addressed by directly correlating the measurements of the two hands of all participants:

Hypothesis 1: Table 4-2 presents the average span of the left versus right hand of the participants. A paired *t* tests asserts that the observed distance is significant at $p < .05$.

Hypothesis 2: The measurement of left and right hands of the participants are highly correlated ($r = 0.87$; $p < .001$). This confirms that when a person has one hand which is larger than average, it should be expected that this difference will hold for the other hand.

Hypothesis 3: Table 4-2 presents the average span of the dominant versus nondominant hand of the participants. A paired *t* tests asserts that the observed distance is highly significant.

Hypothesis 4: The measurement of dominant and nondominant hands of the participants are highly correlated ($r = 0.91$; $p < .001$). This correlation is apparently (although not statistically significantly) stronger than the correlation between left and right hands of the participants.

FUTURE RESEARCH

It appears that the conclusions derived from the tests of dominant versus nondominant hand are similar to, but stronger than, the conclusions derived for right versus left hand, both in terms of differences in span and in terms of correlations. Because for most participants in this study the dominant hand was also the right hand, we can speculate that the observed differences in the hand span are caused by the dominant hand being larger. This speculation is reinforced by a close examination of the data of the two participants who reported a dominant left hand—for both, their left hands had larger spans than their right hands. Thus,

	Test of Hypothesis 1		Test of Hypothesis 3	
	Right Hand Span	Left Hand Span	Dominant Hand Span	Non-Dominant Hand Span
Mean	19.10 cm	18.73 cm	19.20 cm	18.63 cm
Difference (SE)	0.37 (0.19)		0.57 (0.15)	
Paired *t* test; P	$t = 1.94$; 1-tail $p = .036$		$t = 3.77$; 1-tail $p = .001$	

TABLE 4-2.
Results for Exercise 4-2.

further research is needed to determine that the difference in hand span is entirely accounted for by dominant versus nondominant hand, and the distinction between right and left hand is not important. Such research would require recruiting a case-controlled sample of an equal number of participants who are right-handed and left-handed. If such a sample is well-balanced, we should expect no difference between the span of left versus right hands, and no correlation between left and right hands across the participants. However, we should expect significant differences and correlations between dominant and nondominant hands.

Chapter 5

Exercise 5-2.

First, we need to compute $f(1.8)$, $f(3.9)$, $f(6.25)$, and $f(24.6)$, from the function $f(BG) = 1.774\,[(\ln(BG))^{1.0329} - 1.8707]$. Next, for each of these values, we need to compute the risk function as $r(BG) = 10[f(BG)]^2$.

For example, $f(3.9) = 1.774\,[(\ln(3.9))^{1.0329} - 1.8707] = -0.87$ and $r(3.9) = 10[f(3.9)]^2 = 7.58$. The complete set of values is presented in Table 5-1.

Exercise 5-4.

The results in Table 5-1 of the text illustrate that LBGI and HBGI appear to be capable of capturing the differences between the two data sets in Figure 5-10. However, because the data comes from only 2 subjects (1 with T1DM and 1 with T2DM), the results cannot be used as a validation of the model, as it is likely that some (or all) of the observed phenomena can be attributed to chance. A credible validation should include a large number of T1DM and T2DM subjects to minimize the element of chance caused by unavoidable differences in the patients' BG control.

BG	f(BG)	r(BG)
1.80	−2.29	52.62
3.90	−0.88	7.74
6.25	0.00	0.00
24.60	2.58	66.81

TABLE 5-1.
Results for Exercise 5-2.

Chapter 6

Exercise 6-2.

Consider, for example, the following two series of 10 RR intervals each, given in milliseconds:

$A = \{658.3, 666.7, 659.3, 666.7, 697.7, 652.2, 666.7, 681.8, 705.9, 666.7\}$

$B = \{659.0, 674.2, 659.3, 666.7, 625.0, 652.2, 666.7, 674.2, 631.6, 674.2\}$

Series A has median value of 666.7 and series B has median equal to 663.

The standard deviations of the RR intervals in series A and B are identical, both equal to 17.51 milliseconds, as presented in the table below. However, series A (with transient decelerations) has a much higher R_2 that R_1, while series B (with transient accelerations) has much higher R_1 than R_2. These results are summarized in Table 6-1.

	Standard Deviation	R_1	R_2
Series A	17.51	33.36	272.56
Series B	17.51	257.63	40.37

TABLE 6-1.
Results for Exercise 6-2.

Exercise 6-4.

Similar to the LBGI and HBGI defined in Chapter 5 that provide cumulative measures for the risk for hypoglycemia and hyperglycemia, the measures R_1 and R_2 defined in Eq. (6-2) of the text provide a cumulative measure for the risk associated with accelerations and decelerations for a data set of RR measurements. In fact, the mathematical definition of R_1 and R_2 in Eq. (6-2) based on risk functions is exactly the same as that for the LBGI and HBGI. The difference is in the specific definitions of the risk functions $rd(x)$ and $ra(x)$, as outlined in Exercise 6-3 above.

Exercise 6-6.

Note that the sum in the definition of R_1 is over the entire sequence (n data points) and R_1 is the average risk for RR accelerations for the sequence. In the same way, R_2 is the average risk for RR decelerations. Further, recall that a data point that corresponds to a measurement larger than the median, has $ra(x) = 0$ and $rd(x) > 0$. Conversely, a data point that corresponds to an RR measurement smaller than the median, has $ra(x) > 0$ and $rd(x) = 0$. Thus, for two sequences of equal length, R_1 is larger for the sequence with more accelerations and R_2 is larger for the sequence with more decelerations. Subsequently, the ratio $SA = R_2/R_1$ will grow when there are more decelerations in the RR sequence and will decrease when there are more accelerations. This establishes (1) and (2). To verify (3), notice that because both $ra(x) \geq 0$ and $rd(x) \geq 0$, both R_1 and R_2 are non-negative, and thus $SA = R_2/R_1 \geq 0$. Finally, when the distribution for an RR sequence is perfectly symmetric, for any data point x with a non-zero value for $ra(x)$ there will be another data point y in the sequence with $rd(y) = ra(x)$.

This will cause the values of R_1 and R_2 to be exactly the same, leading to $SA = R_2/R_1 = 1$. This establishes the condition (4).

Exercise 6-8.

The standard deviation for S1 is SD = 0.527. Thus $t = r \cdot SD = (0.2)(0.527) = 0.1054$. Because the standard deviations for the sequences S2 and S1 are the same (since they both have five 0's and five 1's), the tolerance value for S2 is also $t = 0.1054$.

Exercise 6-10.

Calculate R_1, R_2 and the Sample Asymmetry (SA). The SA of RRI series with increased proportions of decelerations will be higher than 1, while the SA of RR series with increased proportions of accelerations will be lower than 1. For example, in the data of Exercise 6-2, for series A the SA=8.02, while for series B the SA=0.16.

Chapter 7

Exercise 7-2.

Applying the law of mass action to the equation $Hb_n + nO_2 \leftrightarrow Hb_n(O_2)_n$, we obtain $[Hb_n(O_2)_n] = k[Hb_n][O_2]^n$, where k is the association constant. Then for the fractional saturation we calculate:

$$\overline{Y} = \frac{[Hb_n(O_2)_n]}{[Hb_n] + [Hb_n(O_2)_n]} = \frac{k[Hb_n][O_2]^n}{[Hb_n] + k[Hb_n][O_2]^n} = \frac{k[O_2]^n}{1 + k[O_2]^n},$$

as in Eq. (7-9) of the text.

Exercise 7-4.

a) Differentiating Eq. (7-10) in the text with respect to the oxygen concentration $[O_2]$ gives

$$\frac{d\overline{Y}}{d[O_2]} = \frac{nk[O_2]^{n-1}(1 + k[O_2]^n) - k[O_2]^n nk[O_2]^{n-1}}{(1 + k[O_2]^n)^2}$$

$$= \frac{nk[O_2]^{n-1} + k[O_2]^n nk[O_2]^{n-1} - k[O_2]^n nk[O_2]^{n-1}}{(1 + k[O_2]^n)^2} = \frac{nk[O_2]^{n-1}}{(1 + k[O_2]^n)^2}.$$

b) When the concentration of $[O_2]$ decreases to zero, we obtain from part a) that the denominator $(1 + k[O_2]^n)^2 \to 1$ when $[O_2] \to 0$.

For the numerator, because $n > 1$ and, thus, $n-1 > 0$, we obtain $nk[O_2]^{n-1} \to 0$, and thus $\lim\limits_{[O_2] \to 0} \dfrac{d\overline{Y}}{d[O_2]} = \lim\limits_{[O_2] \to 0} nk[O_2]^{n-1} = 0.$

Exercise 7-6.

In the context of the model used to obtain Eq. (7-6) in the text, we need to examine myoglobin–oxygen binding. Because myoglobin binds one oxygen molecule, only two binding species will be present in the solution: unbound myoglobin and myoglobin–oxygen complex. Recall that the binding polynomial Ξ is defined as the sum of the concentrations of all the binding species present in the solution. In this case, therefore, we obtain $\Xi = [M] + [MO_2] = [M] + K_a[M][O_2]$ where M denotes myoglobin and K_a is the myoglobin-oxygen association constant. Expressing the units of myoglobin concentration as a fraction of the unoxygenated myoglobin concentration (as was done to obtain the binding polynomials in Eqs. (7-27)), we obtain $\Xi = 1 + K_a[O_2]$.

Further, because there is a single binding site, $\overline{Y} = \overline{N}$, and from Eq. (7-20) we obtain $\overline{Y} = \overline{N} = [O_2] \dfrac{\partial \ln \Xi}{\partial [O_2]} = \dfrac{[O_2]}{\Xi} \cdot \dfrac{\partial \Xi}{\partial [O_2]}.$

because $\dfrac{\partial \Xi}{\partial [O_2]} = K_a$, we obtain $\overline{Y} = \dfrac{K_a[O_2]}{1 + K_a[O_2]}.$

Chapter 8

Exercise 8-2.

For the model $Y = G(a; X) = aX$, the sum of squared residuals measure is

$$SSR(a) = \sum_{i=1}^{n} r_i^2 = \sum_i [Y_i - G(a; X_i)]^2 = \sum_i [Y_i - aX_i]^2.$$

The value for a that minimizes this expression can be found as the solution of the equation:

$$\frac{\partial(SSR)}{\partial a} = -2 \sum_i X_i[Y_i - aX_i] = 0, \text{ or } \sum_i X_i Y_i - a \sum_i X_i^2 = 0.$$

Therefore, for the linear model $Y = G(a; X) = aX$, the least squares estimate for the parameter a is given by the formula

$$a = \frac{\sum\limits_i X_i Y_i}{\sum\limits_i X_i^2}.$$

Exercise 8-4.

Computing the partial derivatives $\dfrac{\partial SSR(r,c)}{\partial r}$ and $\dfrac{\partial SSR(r,c)}{\partial c}$ for the function $SSR(r,c) = \sum_{i=1}[Y_i - G(r,c;X_i)]^2$ gives

$$\frac{\partial SSR(r,c)}{\partial r} = -2\sum_i [Y_i - G(r,c;X_i)]\frac{\partial G(r,c;X_i)}{\partial r}$$

$$\frac{\partial SSR(r,c)}{\partial c} = -2\sum_i [Y_i - G(r,c;X_i)]\frac{\partial G(r,c;X_i)}{\partial c}, \text{ showing that}$$

$$P^T Y^* = -2\begin{bmatrix} \dfrac{\partial SSR(r,c)}{\partial r} \\[2ex] \dfrac{\partial SSR(r,c)}{\partial c} \end{bmatrix}.$$

Next, because ε can be expressed as $\varepsilon = (P^T P)^{-1}(P^T Y^*)$ (see Example 8-5 in the text), $(P^T P)^{-1}(P^T Y^*) = 0$ when $\varepsilon = 0$. This means that either $(P^T P)^{-1} = 0$ or $(P^T Y^*) = 0$. However, $(P^T P)^{-1}$ cannot be the zero matrix because it is invertible. Thus, $(P^T Y^*) = 0$, implying that

$$P^T Y^* = -2\begin{bmatrix} \dfrac{\partial SSR(r,c)}{\partial r} \\[2ex] \dfrac{\partial SSR(r,c)}{\partial c} \end{bmatrix} = 0. \text{ Thus, } \frac{\partial SSR(r,c)}{\partial r} = 0 \text{ and } \frac{\partial SSR(r,c)}{\partial c} = 0$$

when $\varepsilon = 0$. This shows that the Gauss–Newton procedure described in the text produces the least squares values for the parameters r and c.

Exercise 8-6.

(a) For $Y = G(a,b,c,d,e,f;X) = aX^5 + bX^4 + cX^3 + dX^2 + eX + f$, the first-order derivatives with respect to the parameters are

$$\frac{\partial G(a,b,c,d,e,f;X_i)}{\partial a} = X_i^5, \frac{\partial G(a,b,c,d,e,f;X_i)}{\partial b} = X_i^4,$$

$$\frac{\partial G(a,b,c,d,e,f;X_i)}{\partial c} = X_i^3, \frac{\partial G(a,b,c,d,e,f;X_i)}{\partial d} = X_i^2,$$

$$\frac{\partial G(a,b,c,d,e,f;X_i)}{\partial e} = X_i, \text{ and } \frac{\partial G(a,b,c,d,e,f;X_i)}{\partial f} = 1. \text{ Therefore, all}$$

higher order derivatives for the model will be equal to zero, leaving only the first-order terms in Eq. (8-38) in the text, which corresponds to a linear model.

(b) Analogously to part (a), any model $Y = G(a_m, a_{m-1}, \ldots a_1, a_0; X)$ $= a_m X^m + a_{m-1}X^{m-1} + \ldots + a_1 X + a_0$ where the degree of the

polynomial $Y = a_m X^m + a_{m-1} X^{m-1} + \ldots + a_1 X + a_0$ is a fixed positive integer, is a linear model. This holds because
$$\frac{\partial G(a_m, a_{m-1}, \ldots, a_1, a_0; X_i)}{\partial a_k} = X_i^k, \text{ for all } k = 0, 1, \ldots, m, \text{ and all}$$
higher order derivatives with respect to the parameters are zero.

Exercise 8-8.

(a) This follows from direct matrix multiplication that establishes
$$P\varepsilon = Y^*$$

(b) Multiplying both sides of the equation from part a) by the transposed matrix P^T gives $(P^T P)\varepsilon = (P^T Y^*)$. Multiplying both sides by the inverse $(P^T P)^{-1}$ leaves to $(P^T P)^{-1}(P^T P)\varepsilon = (P^T P)^{-1}(P^T Y^*)$ or, in simplified form $\varepsilon = (P^T P)^{-1}(P^T Y^*)$.

Chapter 9

Exercise 9-2.

Notice that $h(t + 5) = \sin\left(\dfrac{2\pi(t + 5)}{5}\right) = \sin\left(\dfrac{2\pi t}{5} + 2\pi\right) = \sin\left(\dfrac{2\pi t}{5}\right)$
$= h(t)$, which shows that $h(t)$ is periodic with period $T = 5$.

In the same way, $r(t + 5) = \cos\left(\dfrac{2\pi(t + 5)}{5}\right) = \cos\left(\dfrac{2\pi t}{5} + 2\pi\right)$
$= \cos\left(\dfrac{2\pi t}{5}\right) = r(t)$, which shows that $r(t)$ is periodic with period $T = 5$.

Chapter 10

Exercise 10-2.

We follow the steps outlined in the hint to compute
$$C'(t) = \lim_{h \to 0} \frac{C(t + h) - C(t)}{h}.$$

(1) $C(t + h) = \displaystyle\int_{-\infty}^{t+h} S(z) e^{-\alpha(t+h-z)} dz = \int_{-\infty}^{t+h} S(z) e^{-\alpha h} e^{-\alpha(t-z)} dz$

$$= e^{-\alpha h} \int_{-\infty}^{t+h} S(z) e^{-\alpha(t-z)} dz.$$

$$(2) \quad C(t+h) - C(t) = e^{-\alpha h} \int_{-\infty}^{t+h} S(z)e^{-\alpha(t-z)}dz - \int_{-\infty}^{t} S(z)e^{-\alpha(t-z)}dz$$

$$= e^{-\alpha h}\int_{-\infty}^{t} S(z)e^{-\alpha(t-z)}dz + e^{-\alpha h}\int_{t}^{t+h} S(z)e^{-\alpha(t-z)}dz - \int_{-\infty}^{t} S(z)e^{-\alpha(t-z)}dz$$

$$= (e^{-\alpha h} - 1)\int_{-\infty}^{t} S(z)e^{-\alpha(t-z)}dz + e^{-\alpha h}\int_{t}^{t+h} S(z)e^{-\alpha(t-z)}dz.$$

$$(3) \quad \frac{C(t+h) - C(t)}{h} = \frac{e^{-\alpha h} - 1}{h}\cdot\int_{-\infty}^{t} S(z)e^{-\alpha(t-z)}dz + \frac{e^{-\alpha h}}{h}\int_{t}^{t+h} S(z)e^{-\alpha(t-z)}dz$$

$$= \frac{e^{-\alpha h} - 1}{h}C(t) + \frac{e^{-\alpha h}}{h}\int_{t}^{t+h} S(z)e^{-\alpha(t-z)}dz.$$

(4) Using that $\lim\limits_{h\to 0}\dfrac{1}{h}\int_{t}^{t+h} S(z)e^{-\alpha(t-z)}dz = S(t)$ and $\lim\limits_{h\to 0}\dfrac{e^{-\alpha h} - 1}{h} = -\alpha,$

we obtain

$$\lim_{h\to 0}\frac{C(t+h) - C(t)}{h} = \lim_{h\to 0}\frac{e^{-\alpha h} - 1}{h}C(t) + \lim_{h\to 0}\frac{e^{-\alpha h}}{h}\int_{t}^{t+h} S(z)e^{-\alpha(t-z)}dz$$

$$= -\alpha C(t) + S(t).$$

Exercise 10-4.

In the context of Eq. (10-3); that is, $C(t) = (C_0 - S/\alpha)e^{-\alpha t} + S/\alpha$, the problem specifics are as follows:

Half-life $t_0 = 12$ minutes. Because

$$t_0 = \frac{\ln 2}{\alpha}, \alpha = \frac{\ln 2}{t_0} = \frac{\ln 2}{12} = 0.0578[\text{min}^{-1}].$$

$S(t) = S = \text{const.}$

$C(0) = 0$

$C(3) = 500 \ \text{ng/ml}$

We seek the value of S for which

$500 = C(3) = (C_0 - S/\alpha)e^{-\alpha t} + S/\alpha$, which after substitution of the above values gives

$$500 = -\frac{S}{0.0578}e^{-3\times0.0578} + \frac{S}{0.0578} = \frac{S}{0.0578}(1 - e^{-3\times0.0578}), \text{ and}$$

$$S = \frac{500(0.0578)}{1 - e^{-3\times0.0578}} = 181.5 \text{ ng/ml/min}.$$

Exercise 10-6.

In this case, Eq. (10-1) becomes

$$\frac{dC_A}{dt} = -\alpha C_A(t) + S_A(t) = -\alpha C_A(t) + S_{A,basal} + aF_{up,(down)}(C_B).$$

Because the maximal and minimal concentrations of Λ are achieved when $\frac{dC_A}{dt} = 0$, this means that the minimal and maximal concentrations are achieved when $-\alpha C_A(t) + S_{A,basal} + aF_{up,(down)}(C_B) = 0$ or, equivalently, $C_A(t) = \frac{S_{A,basal} + aF_{up,(down)}(C_B)}{\alpha}$.

(1) As the maximal value of $F_{up(doun)}$ is 1, $C_A(t) \leq \frac{S_{A,basal} + a}{\alpha}$ for all values of t, and therefore $C_{A,max} = \frac{S_{A,basal} + a}{\alpha}$. From here,
$a = \alpha C_{A,max} - S_{A,basal}$.

(2) As the minimal value of $F_{up(doun)}$ is 0, $C_A(t) \geq \frac{S_{A,basal}}{\alpha}$ for all values of t, and therefore $C_{A,min} = \frac{S_{A,basal}}{\alpha}$. From here, $S_{A,basal} = \alpha C_{A,min}$.

Exercise 10-8.

(a) For simplicity, denote $C_A(t) = x(t) = x$, $C_B(t) = y(t) = y$, and

$$F(y(t-D_B)) = a\frac{1}{(C_B(t-D_B)/T_B)^{n_B} + 1} = a\frac{1}{(y(t-D_B)/T_B)^{n_B} + 1}$$

$$G(x) = b\frac{(C_A(t)/T_A)^{n_A}}{(C_A(t)/T_A)^{n_A} + 1} = b\frac{(x/T_A)^{n_A}}{(x/T_A)^{n_A} + 1}.$$

Notice that $F(y)$ is a down-regulatory Hill function and is, therefore, monotone decreasing as a function of y. The function $G(x)$ is and up-regulatory Hill function and is, therefore, monotone increasing as a function of x. Also, both functions are strictly positive; that is $F(y) > 0$ for all values of y and $G(x) > 0$ for all values of x.

(1) With this notation the system from Eqs. (10-14) becomes

$$\frac{dx}{dt} = x' = -\alpha x + F(y(t-D_B))$$

$$\frac{dy}{dt} = y' = -\beta y + G(x)$$

(2) The equilibrium states of this system of equations are the pairs of time-independent constants x and y satisfying the system:

$$-\alpha x + F(y) = 0$$
$$-\beta y + G(x) = 0$$

or, equivalently, the system

$$\alpha x = F(y)$$
$$\beta y = G(x).$$

(3) Expressing $y = G(x)/\beta$ from the second equation, and substituting into the first yields $x = F(G(x)/\beta)/\alpha$. The solutions of this equation then will correspond to the intersection points of the graphs of the functions $z = F(G(x)/\beta)/\alpha$ and $z = x$. We next show that there is a unique positive intersection point.

Because the function $G(x)$ is monotone increasing, and the function $F(y)$ is monotone decreasing, the composition $F(G(x)/\beta)/\alpha$ is also (because $\alpha > 0$ *and* $\beta > 0$) monotone decreasing as a function of x. Also, because at $x = 0$, $F(G(0)/\beta)/\alpha > 0$, the graph of the function $z = F(G(x)/\beta)/\alpha$ is above the x – axis for $x = 0$ and decreases with x. As the graph of $z = x$ begins at $z = 0$ for $x = 0$ and increases with x, the two functions will cross at a single point. This proves that the equation $x = F(G(x)/\beta)/\alpha$ has only one positive solution.

Then, because y is calculated as $y = G(x)/\beta$, we obtain that the system of differential equations (14) has an equilibrium state that is unique.

(b) When $D_B = 0$, the function $F(y) = \dfrac{1}{(C_B(t)/T_B)^{n_B} + 1} = \dfrac{1}{(y/T_B)^{n_B} + 1}$.

(4) The Jacobian of the system of differential equations from part (1) above is

$$J = \begin{pmatrix} -\alpha & F'(y) \\ G'(x) & -\beta \end{pmatrix},$$

$tr(J) = -(\alpha + \beta) < 0$, and $\quad$ det $\quad (J) = \alpha\beta - F'(y)G'(x)$. Because $F(y)$ is strictly monotone decreasing, $F'(y) < 0$ for all values of y. Because the functions $G(x)$ is strictly monotone increasing, $G'(x) > 0$. Then $F'(y)G'(x) < 0$ and, thus, $\det(J) = \alpha\beta - F'(y)G'(x) > 0$. This proves that when $D_B = 0$, the single equilibrium state for the system of Eqs. (10-14) is asymptotically stable.

Exercise 10-10.

(a) Passing to a limit for $T_A \to 0$ in the second equation of Eqs. (10-17) gives $0 < \dfrac{b}{\beta} - \varepsilon \leq C_B(t) \leq b/\beta + \varepsilon$ for sufficiently large t. Because these inequalities hold for any sufficiently small $\varepsilon > 0$, this proves that when T_A decreases, the pulsatility of hormone B vanishes.

(b) Passing to a limit for $T_B \to \infty$ in the first equation of Eqs. (10-17) gives (because $\left(\dfrac{b}{\beta T_B}\right)^{n_B} \to 0$) $0 < \dfrac{a}{\alpha} - \varepsilon \leq C_A(t) \leq \dfrac{a}{\alpha} + \varepsilon$ for sufficiently large t. Because these inequalities hold for any sufficiently small $\varepsilon > 0$, this proves that when T_B increases, the pulsatility of hormone A vanishes.

To provide a heuristic explanation for the observed changes in model behavior, we note that if a system change results in a separation of the concentration of one of the hormones from its action threshold, the oscillations will vanish, because the system will be less sensitive to variations in the concentrations of this hormone. In our particular case, the hormone concentrations in the circulation are limited. Therefore, the concentration of A and B will be disassociated from their corresponding action thresholds if either $T_A \to \infty$ or $T_B \to \infty$ and the oscillations will disappear. In addition, under no circumstances can B fully block the release of A. Therefore, when T_A decreases $(T_A \to 0)$, we will see again a separation of the concentration of A from its action threshold. On the other hand, if $T_B \to 0$ the concentration of B may be able to drop below this threshold, because the potential of B to block the secretion of A will increase, which would lead to reduced secretion of A and, hence, a lower secretion of B. Therefore, the restorative process that generates the oscillations in the model would still operate.

Chapter 11

Exercise 11-2.

COSIN2NL: The results for the ARFILTERed time series are close to those for the original time series. In both cases, the period is estimated to be approximately 23.3 ± 0.19 hours. Thus, filtering the time series before the analysis does not appear to be necessary or beneficial.

FFT-NNLS: For this analysis, too, the results for the ARFILTERed time series are close to those for the original time series. In both cases,

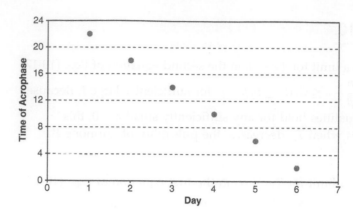

FIGURE 11-1.
A plot depicting the daily time of acrophase for a time series with a period of 20 hours.

four periodic components were identified, with the circadian period estimated at 23.4 ± 0.31 for the original time series and 23.4 ± 0.28 for the AIRFILTERed series. Thus, filtering the time series before the analysis does not appear to be necessary or beneficial.

PHASEREF: In this case, the results for the ARFILTERed time series and for the original time series differ significantly. The mean period values for the ARFILTERed time series ($\pm$ SD, with SEM values in parentheses) presented next to the graphs in Figure 11-29 are a lot closer to the actual simulated values of 24 hours than those next to the graph for the noisy time series in Figure 11-25. This is a consequence of noise confounds creeping in for the original time series, beginning at about 100 hours of x-axis time and manifesting consistently beyond about 200 hours of x-axis time (Figure 11-26). In contrast, for the smoothed time series, there are fewer instances in which values for period estimates are considerably shorter than 24 hours and they do not appear until about 217 hours of x-axis time (Figure 11-30). As a result, the period estimates are more accurate for the ARFILTERed time series, making the use of ARFILTER before PHASEREF beneficial.

Chapter 12

Exercise 12-2.

Let $x_i = (x_{i1}, x_{i2}, \ldots, x_{in})$ and $x_k = (x_{k1}, x_{k2}, \ldots, x_{kn})$ be the i-the and k-th row of the gene expression matrix. We want to show that the dissimilarity measure

$$d(x_i, x_k) = \sqrt{\sum_{j=1}^{n} (x_{ij} - x_{kj})^2}$$

satisfies conditions (1) through (3).

(1) Because $d(x_i, x_k)$ is chosen to be the positive value of the square root, the condition $d(x_i, x_k) \geq 0$ is satisfied.

(2) $d(x_i, x_k) = d(x_k, x_i)$ is obvious, because

$$\sqrt{\sum_{j=1}^{n} (x_{ij} - x_{kj})^2} = \sqrt{\sum_{j=1}^{n} (x_{kj} - x_{ij})^2}.$$

3) Let $x_s = (x_{s1}, x_{s2}, \ldots, x_{sn})$ be any other vector. Then we want to show that

$$d(x_i, x_s) + d(x_s, x_k) = \sqrt{\sum_{j=1}^{n} (x_{ij} - x_{sj})^2} + \sqrt{\sum_{j=1}^{n} (x_{sj} - x_{kj})^2}$$

$$\geq \sqrt{\sum_{j=1}^{n} (x_{ij} - x_{kj})^2} = d(x_i, x_k).$$

We will use the following fact, known as the Cauchy–Schwartz inequality: If $a_1, a_2, \ldots, a_n$ and $b_1, b_2, \ldots, b_n$ are two sets of real numbers, then

$a_1 b_1 + a_2 b_2 + \cdots + a_n b_n \leq \sqrt{a_1^2 + a_2^2 + \cdots + a_n^2} \sqrt{b_1^2 + b_2^2 + \cdots + b_n^2}$, or,

equivalently, $\sum_{j=1}^{n} a_j b_j \leq \sqrt{\sum_{j=1}^{n} a_j} \sqrt{\sum_{j=1}^{n} b_j}$.

We now use this inequality to show that $d(x_i, x_k) \leq d(x_i, x_s) + d(x_s, x_k)$.

$$d^2(x_i, x_k) = \sum_{j=1}^{n} (x_{ij} - x_{kj})^2 = \sum_{j=1}^{n} (x_{ij} - x_{sj} + x_{sj} - x_{kj})^2$$

$$= \sum_{j=1}^{n} [(x_{ij} - x_{sj})^2 + (x_{sj} - x_{kj})^2 + 2(x_{ij} - x_{sj})(x_{sj} - x_{kj})]$$

$$= \sum_{j=1}^{n} (x_{ij} - x_{sj})^2 + \sum_{j=1}^{n} (x_{sj} - x_{kj})^2 + 2 \sum_{j=1}^{n} (x_{ij} - x_{sj})(x_{sj} - x_{kj})$$

$\leq$ (applying Cauchy–Schwartz)

$$\sum_{j=1}^{n} (x_{ij} - x_{sj})^2 + \sum_{j=1}^{n} (x_{sj} - x_{kj})^2 + 2 \sqrt{\sum_{j=1}^{n} (x_{ij} - x_{sj})^2} \sqrt{\sum_{j=1}^{n} (x_{sj} - x_{kj})^2}$$

$$= d^2(x_i, x_s) + d^2(x_s, x_k) + 2d(x_i, x_s)d(x_s, x_k) = [d(x_i, x_s) + d(x_s, x_k)]^2$$

Thus, $d^2(x_i, x_k) \leq [d(x_i, x_s) + d(x_s, x_k)]^2$, which yields $d(x_i, x_k) \leq d(x_i, x_s) + d(x_s, x_k)$.

Exercise 12-4.
....

Let $x_i = (x_{i1}, x_{i2}, \ldots, x_{in})$ and $x_k = (x_{k1}, x_{k2}, \ldots, x_{kn})$ be the i-the and k-th row of the gene expression matrix. We want to show that the dissimilarity measure

$$d(x_i, x_k) = \max_j |x_{ij} - x_{kj}|$$

satisfies conditions (1) through (3).

(1) Because the absolute value is always non-negative and the dissimilarity measure is defined as the maximum of the absolute values, $d(x_i, x_k) \geq 0$ is satisfied.

(2) $d(x_i, x_k) = d(x_k, x_i)$ is obvious, because for any two numbers a and b, $|a - b| = |b - a|$, and thus

$$d(x_i, x_k) = \max_j |x_{ij} - x_{kj}| = \max_j |x_{kj} - x_{ij}| = d(x_k, x_i).$$

3) To prove that $d(x_i, x_k) \leq d(x_i, x_s) + d(x_s, x_k)$, we will use the well-known triangle inequality for real numbers which states that for any real numbers a, and b, $|a| + |b| \geq |a + b|$.

Now,

$$d(x_i, x_s) + d(x_s, x_k) = \max_j |x_{ij} - x_{sj}| + \max_j |x_{sj} - x_{kj}| \geq$$
$$\max_j(|(x_{ij} - x_{sj})| + |(x_{sj} - x_{kj})|) \geq \max_j |x_{ij} - x_{kj}| = d(x_i, x_k),$$

establishing that $d(x_i, x_k) \leq d(x_i, x_s) + d(x_s, x_k)$. The last inequality in the chain of inequalities above follows from the triangle inequality, because

$$|x_{ij} - x_{sj}| + |x_{sj} - x_{kj}| \geq |(x_{ij} - x_{sj}) + (x_{sj} - x_{kj})| \geq |x_{ij} - x_{kj}|.$$

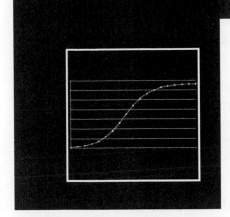

Chapter 1

PROCESSES THAT CHANGE WITH TIME: INTRODUCTION TO DYNAMICAL SYSTEMS

Life belongs to the living, and he who lives must be prepared for changes.

Johann Wolfgang von Goethe (1749–1832)

According to *Encyclopædia Britannica*, a *mathematical model* is defined as "either a physical representation of mathematical concepts or a mathematical representation of reality." Physical mathematical models, such as graphs of curves or surfaces defined by analytic equations or three-dimensional replicas of cylinders, pyramids, and spheres, are used to visualize mathematical terms and concepts. Such models present realistic depictions of abstract mathematical definitions. In contrast, a mathematical representation of reality uses mathematics to describe a phenomenon of nature. There are many mathematical tools that can be used in this process, including statistics, calculus, probability, and differential equations. Different methods may provide insights to different aspects of the problem, and there is often much debate about what approach is preferable. Mathematical models that represent reality are the subject of this text.

Building a good mathematical model is a challenging task that requires a solid understanding of the nature of the system being modeled, as well as the mathematical tools being used to describe it. Because mathematical models are quite diverse, it is difficult to specify a process that would apply to all problems. However, there are fundamental principles that facilitate and guide the creative process. They are:

1. Initially, a model should be simple.

2. It is crucial to test the model under as many conditions as reasonable.

3. If the model seems to be successful in some ways but fails in others, try to modify the model rather than starting over.

In this chapter, we discuss how biological models of one variable change over time. The first model we study is growth of a population. Our initial attempt is based on numerical data. Later, we build the model based on conjectures about "how populations should grow." Both models yield essentially the same result, and although these constructions are successful in the short term, both are flawed because the long-term behavior they predict is unrealistic. We then look at the long-term growth of a yeast culture to build a more believable model.

The first models we construct are of exponential growth. Later in the chapter, we study related models describing exponential decrease in the concentration of drugs in the bloodstream. These exponential growth/decay models are derived from the hypothesis that the time rate of change (i.e., the derivative with respect to time) of a quantity is proportional to the amount present.

We begin with a problem popularized in the late eighteenth century by Thomas Robert Malthus—the growth of human populations.

I. USING DATA TO FORMULATE A MODEL

Contemporary research is hypothesis-driven and is based on experimental evidence. A properly designed experiment can corroborate a hypothesis, prove it false, or produce inconclusive data. An experiment can also suggest new hypotheses that, in turn, will need to be tested. This leads to an ever-repeating cycle of collecting data, formulating hypotheses, designing new experiments to attempt to corroborate them, and collecting new data. It should be emphasized, however, that ultimately the validity of a hypothesis can never be proved. Karl Popper gives the following very instructive example: If somebody sees one, two, or three white swans, he or she may hypothesize, "All swans are white." Each white swan seen corroborates the hypothesis but does not prove it, because the first black swan would invalidate it completely. This demonstrates the necessarily close interdependence between hypothesis and experiment.

In this section, we explore the process of creating mathematical models that describe the growth (or decline) in the size of populations of living organisms. We would like to express the size as a mathematical function of time. Although one model will not work for all species, there are certain fundamental principles that apply almost universally. Our first goal is to identify some of these principles and determine the best way to express them mathematically. We begin by considering U.S. census data for 1800–1860 (U.S. Census Bureau [1993]). Table 1-1 presents the figures for the population of the United States over these 6 decades.

Examining the data plot is always a good idea, as it may suggest certain relationships. Letting $t = 0$ be the year 1800 and one unit of time $= 10$ years, we present the data plot in Figure 1-1. Unfortunately, the conventional plot of the data is not very illuminating. It is evident that the growth is nonlinear, but it is not possible to determine the type of nonlinear dependence by mere observation. There are many mathematical functions that exhibit similar growth patterns. For example, if $P(t)$ represents the U.S. population as a function of the

Year	U.S. Population (millions)
1800	5.3
1810	7.2
1820	9.6
1830	12.9
1840	17.1
1850	23.2
1860	31.4

TABLE 1-1.
Population of the United States from 1800 to 1860.

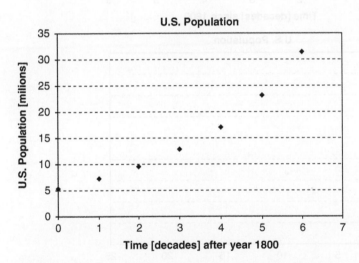

FIGURE 1-1.
Plot of U.S. population versus time. A graph of the data shown in Table 1-1.

time t, the data points in Figure 1-1 may have come from sampling the function $P(t) = at^2$ or $P(t) = at^3$, where $a > 0$ is a constant, or some other power law. It may also be that the data follow an exponential law of increase with the general form $P(t) = ae^{bt}$ where $a > 0$ and $b > 0$ are constants. To determine the specific nonlinear function that provides the best fit for the data, we examine the *change* in U.S. population per decade; that is, the *rate of change*. In our example, they appear to be growing with time—the population change is 1.9 million from 1800 to 1810 but 8.2 million from 1850 to 1860 (more than four times as large). Thus, the rate of population growth increases as the U.S. population increases.

These observations lead to two different ways of plotting the data: (1) The change in population size per decade versus time, and (2) the change in population size per decade versus population size at the beginning of decade. While the graph in Figure 1-2(A) is still not very telling, the one in Figure 1-2(B) is strikingly *linear*. Is this a mere coincidence, or are

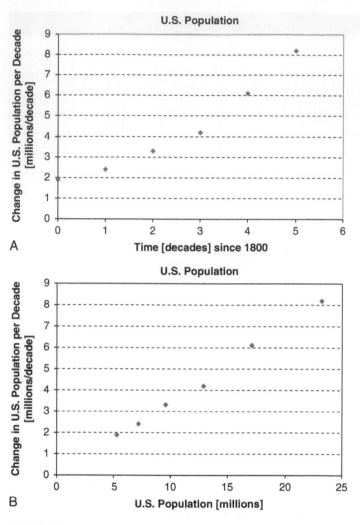

FIGURE 1-2.
Comparison of rate of change versus time to rate of change versus population size. Panel A:
Population rate of change versus time; panel B: Population rate of change versus population size.

we on to something important? Observing the data prompts us to make
the following conjecture:

*There is a linear dependence between the rate of change in population size and
the population size itself.*

We now have a hypothesis. How should we proceed in order to
corroborate or reject it? In general, the process involves the following
major components, presented here in their natural order:

1. *Solicit expert opinion.* In this case, discuss the conjecture with population biologists. If they cannot dismiss the hypothesis right away
   by providing examples that clearly contradict it, it merits further
   investigation.

2. *Describe the conjecture, as well as possible, in quantitative and analytical terms.* This phase may involve statistics, mathematical formulations, and follow-up analyses. Statisticians and mathematicians will usually carry out this phase in close collaboration with biologists. This process often leads to clarifying and refining the hypothesis.

3. *Test the refined hypothesis on several data sets.* Consider the limitations of previous experiments, and design your own new data collection in order to address them. Formulate your refined conjectures.

Each of these steps can sometimes be carried out and thoroughly explored within hours or days. In other cases, it may take much longer. Charles Darwin, for example, took several decades to systematically collect data for his famous *On the Origin of Species by Means of Natural Selection.*

When applying the steps outlined above to the growth of populations, our hypothesis passed the "expert opinion test," but only conditionally. We learned that the rate of growth of populations might, indeed, be proportional to the size of the population, but only during *the initial phases* of their growth. This phase could be characterized as a period during which an abundance of resources allows for unfettered growth. During later phases, the growth of the population might be impeded by competition or a shortage of resources. So our hypothesis had potential and, in fact, it seemed reasonable that the period from 1800 to 1860 was an "initial phase of growth" for the U.S. population. However, the model developed on our general hypothesis had its limitations—not a big surprise, given that it was our first model. We also began to understand some of the rationale for these limitations. We decided, nevertheless, to move on to describing our hypothesis quantitatively and analytically.

Denoting the U.S. population at the end of the n-th decade by P_n (where n can take the integer values $0, 1, 2, 3, \ldots$). We can express the change in population size from the beginning of one decade to the next by $P_n - P_{n-1}$, for $n = 1, 2, 3, \ldots$. The conjectured linear relationship between the rate of change of population and the population size itself then means that the two quantities are proportional. Thus, there is a constant k such that the relationship

$$P_n - P_{n-1} = kP_{n-1} \qquad (1\text{-}1)$$

is satisfied for any value $n = 1, 2, 3, \ldots$. In particular, for $n = 1$, we have $P_1 - P_0 = k\,P_0$; for $n = 2$, $P_2 - P_1 = k\,P_1$; etc. Notice that the constant of proportionality k can be interpreted as the *net per capita rate of change* (also referred to as the *net per capita growth rate*) for the population. The left-hand side of our model represents the change per decade, and the right-hand side expresses this change as a multiple (k) of the population size in the beginning of the decade.

Time (decades) n	U.S. Population (millions) P_n	Change in Population $P_n - P_{n-1}$	$k = (P_n - P_{n-1})/P_{n-1}$
0	5.3	—	—
1	7.2	1.9	0.358
2	9.6	2.4	0.333
3	12.9	3.3	0.344
4	17.1	4.2	0.326
5	23.2	6.1	0.357
6	31.4	8.2	0.353

TABLE 1-2.
Estimation of k from U.S. population data.

We next estimate the numerical value of k from the data. The calculations are summarized in Table 1-2. Ideally, if all data points (P_{n-1}, $P_n - P_{n-1}$), $n = 1, 2, 3, \ldots$, whose coordinates are given in the second and third columns were perfectly lined up, the values of k calculated as $k = (P_n - P_{n-1})/P_{n-1}$ in the third column would be exactly the same. In reality, because of the noise and small inconsistencies that are always present in the world of experimental data, the values of k vary slightly.

The numerical value chosen for k should be the value that provides the best agreement between the actual population sizes and the values predicted by the model. We could, in principle, test all of them and visually determine the best fit of the predicted data with the actual data. We did this for the smallest value of k ($k = 0.326$), the largest value of k ($k = 0.358$), and the average of all calculated values for k ($k = 0.345$). The results and corresponding graphs are presented in Figure 1-3.

Not surprisingly, the smallest k-value produces predictions that systematically underestimate the population, while the largest value generates overestimates. Using the average of the k-values in Table 1-2, however, gave a very good overall fit. The question of what is meant by "best fit" is certainly nontrivial and will be addressed later in detail. For now, we shall note that the value of $k = 0.344$ provides the best fit with the data—just 0.001 below the average value of k we calculated above.

II. DISCRETE VERSUS CONTINUOUS MODELS

Our model is now $P_n - P_{n-1} = (0.345)P_{n-1}$. One limitation of this model is apparent almost immediately: Our model is *discrete*, that is, it can only be used to describe changes that occur at specific time intervals. The smallest unit it works with is a decade, and, thus, the model is incomplete. For example, it does not allow us to compute the

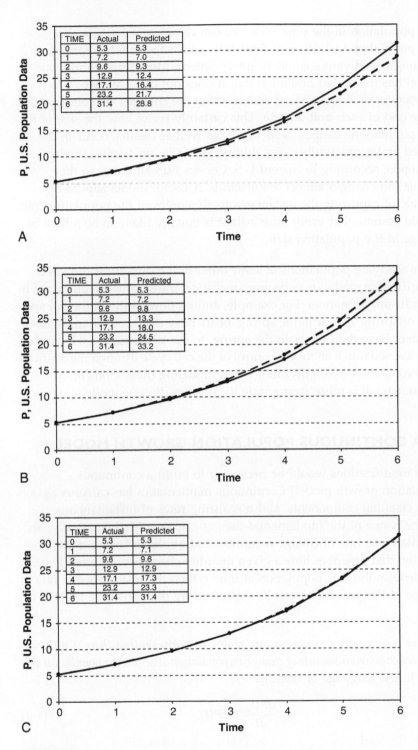

FIGURE 1-3.
Comparison of plots of three different values of k. Panel A: Actual (*solid line*) and predicted (*dashed line*) U.S. population values for $k = 0.326$; panel B: Actual (*solid line*) and predicted (*dashed line*) U.S. population values for $k = 0.358$; panel C: Actual (*solid line*) and predicted (*dashed line*) U.S. population values for $k = 0.345$.

U.S. population in the year 1875. We can calculate values for the U.S. population in 1870 or in 1880, but not for the intermediate years (although such values could be interpolated). More importantly, our model has the added limitation that it does not capture change as it occurs over time and instead assumes that the changes are compounded at the end of each unit of time. This certainly is not how the size of the U.S. population changes. New births, as well as deaths, occur in the United States practically every minute (actually, on average, every 8 seconds, according to current U.S. Census Bureau data), so the population changes almost *continuously*. A useful model should be capable of capturing the instantaneous dynamics of the population and should assume that every time instant is equally likely to be a time of change in the population size.

When studying populations of some other living organisms, however, using discrete models may be more realistic if the organisms reproduce in a synchronized manner. For example, annual flowers die in the fall and their offspring appear in the spring, bears have their cubs in midwinter, and deer have their fawns in the spring. In the laboratory, cell biologists have learned much about the control of the cell cycle through the artificial synchronization of cell division. When modeling these kinds of phenomena, it is more appropriate to consider discrete models.

III. A CONTINUOUS POPULATION GROWTH MODEL

What modifications would be necessary to build a continuous population growth model? Continuous mathematics has calculus as one of its essential components, and measuring rates of instantaneous change is one of the fundamental uses of calculus. Mathematically, an instantaneous rate of change is represented by the *derivative* of the function that describes how a given quantity changes with time. Thus, if $P(t)$ denotes the U.S. population at time t, then the instantaneous rate of change of the population can be expressed by the derivative $dP(t)/dt$ or $P'(t)$.

We are now ready to express our major hypothesis that *there is a linear dependence between the rate of change in population size and the population size itself*. In the language of calculus:

$$\frac{dP(t)}{dt} = rP(t). \qquad (1\text{-}2)$$

The left-hand side of this equation gives the (instantaneous) rate of change for $P(t)$ at time t. The right-hand side expresses this rate as a fraction (r) of the current population size $P(t)$. Notice that this model represents exactly the same hypothesis as before. The only reason Eq. (1-1) looks different from Eq. (1-2) is that they state our hypothesis in two different languages—Eq. (1-1) uses the language of discrete mathematics, whereas Eq. (1-2) uses the language of continuous mathematics.

Equation (1-2) is in the form of a *differential equation*; that is, it contains information about the derivative of the unknown function $P = P(t)$, which we hope to find. Rewriting Eq. (1-2) as $dP/P = rdt$ and integrating, we obtain:

$$\int \frac{dP}{P} = \int r\, dt,$$

so that

$$\ln(P) = rt + C,$$

where C is the constant of integration. Thus:

$$P(t) = e^{\ln(P(t))} = e^{rt+C} = e^{rt}e^{C} = C_1 e^{rt}, \tag{1-3}$$

where $C_1 = e^C$ is a constant.

Usually, we know the *initial population* $P(0)$, and we can thus determine C_1. From Eq. (1-3), using $t = 0$, we obtain $P(0) = C_1 e^{r0} = C_1$, so C_1 is $P(0)$. This gives us the solution of Eq. (1-2) for the unknown function $P(t)$:

$$P(t) = P(0)e^{rt}. \tag{1-4}$$

Equation (1-4) is the fundamental equation of unfettered growth. We want to estimate r from the data in Table 1-1 as we estimated k earlier. Now

$$P(t) = P(0)e^{rt} \quad \text{and} \quad P(t+1) = P(0)e^{r(t+1)},$$

so:

$$\frac{P(t+1)}{P(t)} = \frac{P(0)e^{r(t+1)}}{P(0)e^{rt}} = e^{r}. \tag{1-5}$$

Thus, we can estimate r by:

$$r = \ln\left(\frac{P(t+1)}{P(t)}\right) = \ln\left(P(t+1)\right) - \ln(P(t)). \tag{1-6}$$

Using that $P(0) = 5.3$, $P(1) = 7.2$, and so forth, we give the estimated values of r in column 3 of Table 1-3. If we average the values of r (the method that gave the best estimate in the discrete case), we get $r = 0.297$. We can now estimate the population by using:

$$P(t) = 5.3\, e^{0.297t}, \tag{1-7}$$

where t is the number of decades after 1800. The predicted U.S. population appears in column 4 of Table 1-3.

Time t (decades)	U.S. Population $P(t)$ (millions)	$r = \ln(P(t+1)) - \ln(P(t))$	Predicted U.S. Population [millions] for $r = 0.297$	Relative Error [%] $= \dfrac{\lvert \text{Predicted} - \text{Actual}\rvert}{\text{Actual}} 100$
0	5.3	0.306	5.300	0.000
1	7.2	0.288	7.133	0.931
2	9.6	0.295	9.599	0.010
3	12.9	0.282	12.919	0.147
4	17.1	0.305	17.387	1.678
5	23.2	0.303	23.399	0.858
6	31.4	0.306	31.491	0.290

TABLE 1-3.
Determination of r and evaluation of predicted population values.

As in the discrete case, our method of estimating the value of r was rather primitive. The average value $r = 0.297$ showed a good fit with the census data, but we defer how to find the best value of r until Chapter 8.

One purpose of a mathematical model may be to predict values that cannot be measured directly. In our example, these may be values of the U.S. population for past years for which no U.S. census data are available, or values of the U.S. population for future years. In particular, can we use the discrete and continuous models (1-1) and (1-2) (with our best values of $k = 0.345$ and $r = 0.297$) to predict the U.S. population in the year 3000? Mathematically, this is not a problem. In the discrete case, we rewrite our model $P_n - P_{n-1} = (0.345)\, P_{n-1}$ as $P_n = (1.345)\, P_{n-1}$. Because time is measured in decades beginning with the year 1800, the year 3000 will correspond to $n = 120$, and so we need to find the value of P_{120}. Knowing the U.S. population for $n = 0$ to be 5.3 million, we have $P_0 = 5.3$ and can compute $P_1 = (1.345)\, P_0 = (1.345)\,(5.3) = 7.1$. Having calculated P_1, we can calculate $P_2 = 1.5\, P_1 = (1.345)\,(7.1) = 9.6$, and so on. We would therefore need to calculate 120 consecutive values before we get P_{120}. Alternatively, we could use a computer to get the value of P_{120}. In the continuous case, of course, we just substitute 120 for t into Eq. (1-7). Exercise 1-1 shows that a formula for direct computation of P_{120} can also be calculated for the discrete model.

EXERCISE 1-1

For the model $P_n - P_{n-1} = k\, P_{n-1}$, show that:

(a) $P_n = (1+k)P_{n-1}$

(b) $P_n = (1+k)^n P_0$.

(1-8)

The expression $P_n = (1+k)^n\, P_0$ represents the *analytical solution* for Eq. (1-1). Because we know the net per capita growth rate $k = 0.345$ and the initial population size $P_0 = 5.3$, the solution allows us to compute directly the population P_n for any value of n. For example, when $n = 120$, we can use Eq. (1-8) to compute the model prediction for the U.S. population in the year 3000:

$$P_{120} = (1.345)^{120}(5.3) = 1.48234 \times 10^{16} = 14.8 \text{ quadrillion.}$$

How realistic do you think this prediction is? Why?

EXERCISE 1-2

Use the continuous model from Eq. (1-7) to predict the U.S. population in the year 3000. Is this prediction realistic? Why?

EXERCISE 1-3

The data in Table 1-4 show the initial phase of yeast culture growth over 7 hours (Carlson [1913]; Pearl [1927]). The size of the yeast population is measured in terms of *biomass*. Biomass is simply the weight of living material. For yeast or bacteria, population growth may also be measured by taking advantage of the fact that, as they grow, the medium in which they are growing becomes increasingly turbid. A spectrophotometer is used to determine the amount of light scattered by samples of the culture.

(a) Use this data to determine the best values for k and r for the discrete model (1-1) and the continuous model (1-2).

(b) Use the values determined in part (a) to create a table displaying the actual and predicted values from the discrete and continuous models.

(c) For the value of r determined in (a), plot the predictions of the continuous model, and consider the graph. Based on the graph, do you expect the continuous model to remain accurate in predicting the long-term growth behavior of the culture?

Time (hours)	0	1	2	3	4	5	6	7
Biomass	9.6	18.3	29.0	47.2	71.1	119.1	174.6	257.3

TABLE 1-4.
Growth of yeast population over 7 hours.

IV. THE LOGISTIC MODEL

The exercises from the last section raise some important questions. In particular, the solutions of both the discrete and the continuous models are unbounded functions, and therefore describe *unlimited growth*. In any given environment, however, the factors that support growth—for example, the availability of food or nesting sites—are limited. Any environmental degradation, such as air or water pollution, may also limit population growth. These limiting factors determine the *carrying capacity* of an environment—the maximum number of organisms the environmental system can support. This is the upper limit on a sustainable population.

EXERCISE 1-4

What factors do you think would determine the carrying capacity for human populations?

To illustrate that populations do not grow without limit, Figure 1-4 shows the growth of the same yeast culture from Table 1-4 throughout the entire 18-hour data collection period (Carlson [1913]). Figure 1-4 also contains the solution curve of our continuous model with $r = 0.49$. As anticipated, our model exhibits unlimited growth, while the actual yeast culture appears to approach a maximum population size. One might suppose that the decrease in growth rate is caused by depletion of the yeast's food supply—namely, sugar. However, analysis of the medium showed that sugar was still available (Richards [1928]). Rather, the

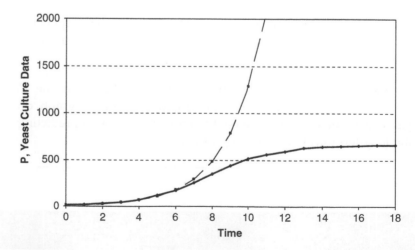

FIGURE 1-4.
Predicted and actual values for yeast population. Comparison between the solution curve of the model $dP/dt = rP(t)$ with $r = 0.49$ (*dashed line*) and the 18-hour yeast growth data (*solid line*), after Carlson (1913).

slowing of growth is caused by the increasing concentration of ethanol in the medium, alcohol being one of the products of anaerobic respiration or fermentation, as any brewer or vintner can attest. The concentration of ethanol rises until it reaches levels toxic to new yeast cells, resulting in the observed decrease in the growth rate.

It is evident from this graph that the ability of the environment to support the growth of the yeast diminishes as the population increases—a reality we must modify our models given by Eqs. (1-1) and (1-2) to reflect. Modification of the continuous model is discussed in detail; the discrete case is left as an exercise.

Given that an environment can sustain only so many organisms, we need to modify the model so the net per capita growth rate r depends on the size of the population. In terms of an equation, we could say:

$$\frac{dP}{dt} = r(P(t))P(t), \tag{1-9}$$

emphasizing that r now is not constant but depends on P.

Specifically, we now assume:

1. The environment can sustain a maximum population of the species, reflecting its carrying capacity, K.

2. The smaller the population, the higher the per capita rate of population growth. In general, as long as the population remains smaller than the carrying capacity K, the population will grow, but the closer to K the population gets, the slower the growth rate will be.

3. If the population ever exceeds K (e.g., by immigration), then the population will diminish and approach K; that is, the net per capita rate of change should be negative for $P > K$.

We want to modify our model to reflect the simplest case—that the environment accommodates zero growth when the population is K and the maximal per capita growth rate when the population is near zero. Suppose the highest per capita growth rate is $a > 0$. If we want to graph per capita growth rate versus population size (see Figure 1-5), we want no growth when the population is K, so $(K, 0)$ must be a point on our graph. We also want the maximum growth rate to be at the hypothetical population of 0, so $(0, a)$ is another point on our graph. Letting $(x_1, y_1) = (K, 0)$ and $(x_2, y_2) = (0, a)$, the slope of the line passing through these two points is

$$m = \frac{y_2 - y_1}{x_2 - x_1} = \frac{a - 0}{0 - K} = -\frac{a}{K}. \tag{1-10}$$

The graph of this line is depicted in Figure 1-5, and its equation is

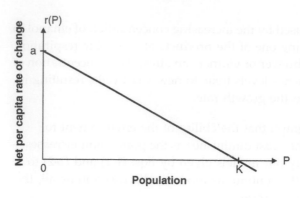

FIGURE 1-5.
Net per capita population growth rate as a function of the population size P. As the population size increases to the carrying capacity K, the net per capita growth rate decreases, in a linear fashion, to 0.

$$r(P(t)) = -\frac{a}{K}P(t) + a = a\left(1 - \frac{P(t)}{K}\right). \tag{1-11}$$

Substituting this into Eq. (1-9) leads to the modified model:

$$\frac{dP}{dt} = a\left(1 - \frac{P(t)}{K}\right)P(t), a > 0, K > 0, \tag{1-12}$$

which is called the *logistic model*. The value $a > 0$, corresponding to the maximal per capita growth rate, is called the population's *inherent per capita growth rate*. The value $K > 0$ represents the *carrying capacity* of the environment.

EXERCISE 1-5

What does the logistic model predict about the population change if:

(a) $P(t) < K$?

(b) $P(t) = K$?

(c) $P(t) > K$?

With enough effort (the details of which are left as an exercise for the readers who enjoy calculus), we can obtain the analytic solution for the model from Eq. (1-12) describing the growth of the population over time:

$$P(t) = \frac{KP(0)}{P(0) + (K - P(0))e^{-akt}}. \tag{1-13}$$

We refer to the graph of the solution of a differential equation as its *time trajectory*, or simply its *trajectory*. As the solution is a function of time, the

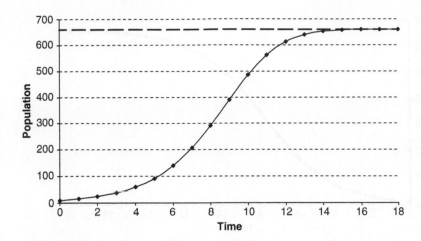

FIGURE 1-6.
A logistic curve. The solution of the logistic equation (1-12) for $P(0)= 5$, $K = 660$, and $a = 0.7$ (*solid line*). The dashed line corresponds to the carrying capacity K.

trajectory describes the evolution of the function quantity in time. The graph of the solution given by Eq. (1-12), for the special case of $P(0)= 5$, $K = 660$, and $a = 0.7$, is shown in Figure 1-6.

EXERCISE 1-6

Consider the solution of the logistic model (1-13). What happens to $P(t)$ as time gets very large ($t \rightarrow \infty$)? Consider the following cases separately:

(a) $P(0) = 0$,

(b) $0 < P(0) < K$,

(c) $P(0) = K$, and

(d) $P(0) > K$.

It is gratifying that the solution (1-13) of our modified model produces the distinctive sigmoidal (S-shaped) curve exhibited by the yeast growth data in Figure 1-4. Comparing the model predictions with the actual data is also encouraging. Using the value of $a = 0.543$ calculated in Exercise 1-3 as the per capita growth rate during the initial growth phase (which is the inherent per capita growth rate for the logistic model) and a value for the carrying capacity $K = 660$, estimated from the data, we obtain the graph in Figure 1-7.

We note that function (1-13) is just one of many different functions exhibiting S-shaped trajectories like the one in Figure 1-7. Such trajectories are often given the generic name "logistic curves," a term

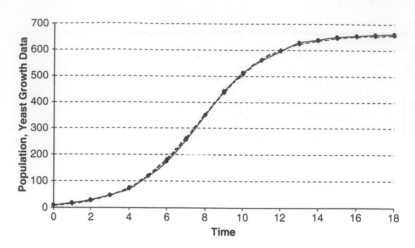

FIGURE 1-7.
Comparison of logistic model and actual yeast population growth. Numerical solution of the logistic model from Eq. (1-12) with $a = 0.543$ and $K = 660$ (*dashed line*) and yeast growth data (*solid line*).

introduced by the Belgian mathematician Pierre-François Verhulst in 1845, and are also referred to as "logistic shapes." In terms of their specific meaning and analytic expressions, however, these curves may be quite different. We have to be careful, therefore, not to assume these functions have the same analytic form as (1-13) simply because their graphs appear similar to the solution of the logistic equation.

V. AN ALTERNATIVE DERIVATION OF THE LOGISTIC MODEL

In the previous section, we derived the logistic model based on the assumption there is a maximum population the environment can sustain, reflecting limited available resources. In this section, we build a model to determine the carrying capacity based on maximum available resources and consumption rates. To keep the model as simple as possible, we assume a single essential resource. We begin by recalling that the net per capita growth rate is not constant but is population-dependent, as shown in Eq. (1-9):

$$\frac{dP}{dt} = r(P(t))P(t).$$

Now, however, we assume that the net per capita growth rate r depends on the amount of resource available, which, in turn, depends on the population size: the higher the population, the lower the resource available. We denote the value of the available resource by $R = R(P)$, and rewrite Eq. (1-9) as:

$$\frac{dP}{dt} = r(R)P(t).$$

We shall model the functions $R = R(P)$ and $r = r(R)$ next, beginning with the function $R(P)$. Assume the resource exists in two forms: free and bound (or consumed) by the population. Let F be the maximum amount of free resource available when the population size $P = 0$. When $P > 0$, the amount of available resource will decrease as P increases. Assuming a fixed per capita rate of consumption $c > 0$, we can write:

$$R = F - cP. \tag{1-14}$$

To model the dependence $r = r(R)$, notice that the net per capita growth rate needs to satisfy the following conditions:

1. The population should be declining when no free resource is available; so when $R = 0$, the net per capita growth rate should be negative: $r(0) < 0$.

2. The population should be growing when the free resource is available. More of the free resource will cause a higher per capita growth rate, so the function $r = r(R)$ should be an increasing function of R.

The simplest mathematical dependency $r = r(R)$ that satisfies conditions 1 and 2 is the line

$$r(R) = mR - n, \tag{1-15}$$

where $m > 0$ represents the rate the free resource affects the per capita net growth rate, and $n > 0$ represents the per capita rate at which the population size will decline when the resource is lacking.

Substituting R from Eq. (1-14) into Eq. (1-15) yields

$$r(R) = m(F - cP) - n$$

and a subsequent substitution into Eq. (1–9) gives the following *resource-based* population growth model:

$$\frac{dP}{dt} = (m(F - cP) - n)P. \tag{1-16}$$

Equation (1-16) can be rewritten as

$$\frac{dP}{dt} = (mF - n)\left[1 - \frac{mc}{mF - n}P\right]P = a\left(1 - \frac{P}{K}\right)P, \tag{1-17}$$

where

$$a = mF - n \quad \text{and} \quad K = \frac{mF - n}{mc}. \tag{1-18}$$

Therefore, this model is the same as the logistic model from Eq. (1-12), with inherent per capita growth rate and carrying capacity as given by Eq. (1-18). Notice that the inherent per capita growth rate *a* corresponds to the special case of Eq. (1-15) when all of the available resource is unbound. The expression for the carrying capacity *K* provides insight to the dependence of this empirical parameter upon available resources and rate of consumption. As should be expected, *K* grows with *F* and declines as the consumption rate *c* increases.

VI. LONG-TERM BEHAVIOR AND EQUILIBRIUM STATES

In the continuous models from Eqs. (1-2) and (1-12), we found a solution that gives *P(t)*. Differential equations, however, are often impossible to solve explicitly. In spite of this, we can still glean essential information about the long-term behavior of the model from these equations. We shall now examine techniques for this type of analysis with the logistic equation.

We begin by recalling that there is dependence between the increase/decrease behavior of a function and the sign of its derivative. Namely, if the derivative is positive over a certain time interval, the function is increasing, while a negative derivative indicates the function is decreasing. When the derivative is zero, the function exhibits no change.

In the logistic model [Eq. (1-12)], the governing differential equation can also be written as:

$$\frac{dP}{dt} = a'(K - P)P, \quad \text{where } a' = \frac{a}{K}. \tag{1-19}$$

What does this differential equation tell us? The derivative is zero at two values of *P*: when $P = 0$ and when $P = K$. When we graph *P* versus *t* (population vs. time), these values divide the graph into two regions— values of *P* larger than the carrying capacity *K* and values of *P* smaller than *K* (see Figure 1-8). Suppose we begin a new culture with a very small quantity of yeast. Because the population is small, $P(t) < K$, then $\frac{dP}{dt}$ is positive, and *P(t)* will increase (see the curve labeled P_1). This does not give the complete information that the solution of the logistic curve gave, but it gives valuable information for very little effort. Similarly, if a huge amount of yeast was introduced (greater than the carrying capacity, so $P(t) > K$), then the derivative is negative and the population will diminish (see the curve labeled P_2).[1] There is an underlying lesson here that is very important: Namely, that we do not have to

_____

1. Notice that none of the arguments determining the long-term behavior of *P(t)* here depends on the actual value of the parameter $a > 0$. We will need this observation in Section VIII, where we discuss discrete analogues of the logistic model (1-12). In contrast with the continuous logistic model, those may exhibit radically different behavior, depending upon the value of $a > 0$.

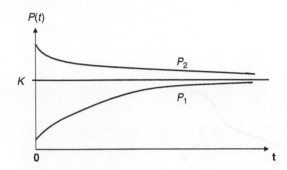

FIGURE 1-8.
Behavior of trajectories beginning at different population sizes. Trajectories that begin at population size values smaller than the carrying capacity K increase toward K, whereas those that begin at population size values larger than K decrease toward K.

explicitly solve the many differential equations that we shall encounter to understand how the solutions evolve.

<hr>

EXERCISE 1-7

What would happen if there were a time when $P(t) = K$?

<hr>

An *equilibrium state* is one in which the quantity in question remains constant over time. These will be the values for which the derivative is zero. In Eq. (1-19), they will be where $P = 0$ and $P = K$.

VII. ANALYZING EQUILIBRIUM STATES

Suppose that we have an equation of the type:

$$\frac{dP}{dt} = f(P). \tag{1-20}$$

This shows that the rate of change depends only on the value of the population and not on when that value is attained. The logistic model from Eqs. (1-12) and (1-19) has this form. The values of P for which $f(P) = 0$ define the equilibrium states, because then $\frac{dP}{dt} = 0$. In this section, we shall show how to classify the equilibrium states as stable or unstable, based on the sign of the function $f(P)$ near each equilibrium state.

If $f(P) > 0$, then the derivative is positive and P will increase. If $f(P) < 0$, then the derivative is negative and P will decrease. A very helpful tool is to graph $y = f(P)$ versus P. From the graph, we can then easily decide where P will increase or decrease. Suppose $\frac{dP}{dt} = f(P)$, and the graph of $f(P)$ is shown in Figure 1-9.

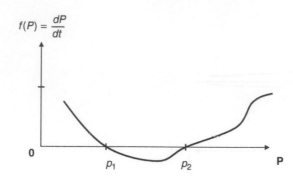

FIGURE 1-9.
Plot of *dP/dt* versus *P.* The graph of *dP/dt* versus *P* helps to visualize how the rate of change of the population depends upon the population size.

The derivative is zero where the graph of $f(P)$ crosses the horizontal axis at points p_1 and p_2. These are equilibrium states. The derivative is positive when the graph of $f(P)$ is above the horizontal axis and negative when it is below. From the graph, $f(P)$ is positive, and the population is growing when P is less than p_1 or greater than p_2. When P is between p_1 and p_2, $f(P)$ is negative.

Let's analyze how P changes if it is near the equilibrium states.

State p_1:

The graph of $f(P)$ for P near p_1 is shown in Figure 1-10.

Suppose P is slightly smaller than p_1. Then $f(P) > 0$, which means $\dfrac{dP}{dt} > 0$, so P is increasing toward p_1. On the other hand, if P is slightly larger than p_1, then $f(P) < 0$, so $\dfrac{dP}{dt} < 0$ and P decreases, again moving toward p_1. In either case, if P is slightly different than p_1, then P moves toward p_1. We refer to a point such as p_1 as a *stable equilibrium point*.

State p_2:

The graph of $f(P)$ for P near p_2 is shown in Figure 1-11.

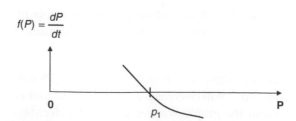

FIGURE 1-10.
Interpreting *dP/dt* versus *P.* When the graph of *dP/dt* versus *P* crosses the horizontal axes at a point p_1 while decreasing near the point of crossing, the value p_1 is a stable equilibrium.

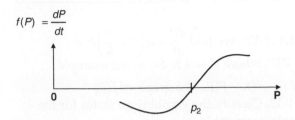

$$f(P) = \frac{dP}{dt}$$

FIGURE 1-11.
Interpreting dP/dt versus P. When the graph of dP/dt versus P crosses the horizontal axes at a point p_2 while increasing near the point of crossing, the value p_2 is an unstable equilibrium.

Now if P is slightly less than p_2, then $f(P) < 0$ and the derivative is negative, so P will decrease and move away from p_2. Similarly, if P is slightly greater than p_2, then $f(P) > 0$, and the derivative is positive, so P will increase and again move away from p_2. In either case, if P is slightly different than p_2, then P will move away from p_2. A point such as p_2 is called an *unstable equilibrium point*.

A physical example of stable and unstable equilibrium points is shown in Figure 1-12. If a roller coaster cart is stopped at the positions indicated, it will remain there. If the cart is at positions 2 or 3 and is nudged gently, it will return to its original position. On the other hand, if the cart is at positions 1 or 4 and is nudged, it will roll down the track and away from the position at which it was balanced.

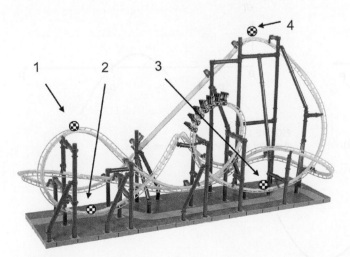

FIGURE 1-12.
A roller coaster model of equilibrium points. Positions 2 and 3 represent stable equilibria while positions 1 and 4 represent unstable equilibria. (Scorpion Roller Coaster Modeling System photograph from www.coasterdynamix.com. Used by permission.)

(a) For the logistic model (1-12), we had $\dfrac{dP}{dt} = a\left(1 - \dfrac{P}{K}\right)P = \dfrac{a}{K}(K - P)P = a'(K - P)P$, where $a' = a/K$. So in the example discussed above, $f(P) = a'(K - P)P$. The graph of $f(P)$ is shown in Figure 1-13. Classify the equilibrium states for the logistic model as stable or unstable.

(b) Suppose $\dfrac{dP}{dt} = f(P)$ and the graph of $f(P)$ is shown in Figure 1-14.

     (i) Locate the equilibrium points, and classify them as stable or unstable.

     (ii) Sketch the trajectory $P(t)$, as in Figure 1-8, for $P(0)$ in the following regions (i.e., $P(0) < p_1$; $p_1 < P(0) < p_2$; $p_2 < P(0) < p_3$; $P(0) > p_3$).

(c) Suppose $\dfrac{dP}{dt} = f(P)$ and the graph of $f(P)$ is shown in Figure 1-15. Describe what happens if P is close to the equilibrium point p_1.

So far, we have only considered questions related to population growth. The techniques described, however, are quite general and can be used to answer a variety of questions related to quantities that change with time, as the following examples illustrate.

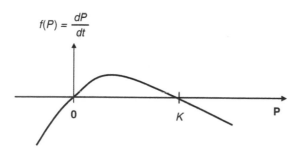

FIGURE 1-13.
The graph of $f(P) = dP/dt$ versus P for the logistic Eq. (1-12).

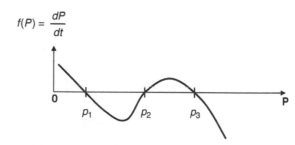

FIGURE 1-14.
A model with three equilibrium states.

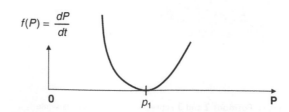

FIGURE 1-15.
A model with one equilibrium state that never decreases.

Example 1-1

Global Temperature Analysis. It has been estimated that the average surface temperature of the Earth has risen 0.8 to 1.0°F in the last century. This increase with time (presented in Figure 1-16 as deviations from long-term averages) is fueling a heated debate over what may be causing the accelerated global warming of the past 2 decades.

The many factors that impact the Earth's temperature can be broadly grouped as external and Earth factors. External factors include solar output, earth–sun geometry, and stellar dust, to name a few. Earth factors include volcanic activity,[2] ocean heat exchange, and the

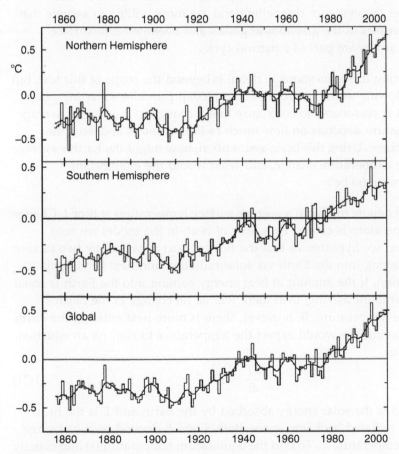

FIGURE 1-16.
Deviations of global temperature from long-term averages. Values for the northern and southern hemispheres are also shown. (Source: http://www.cru.uea.ac.uk/cru/data/temperature/. Used by permission of the Climatic Research Unit, University of East Anglia, Norwich, UK.)

---

2. Volcanic activity can significantly impact temperature when volcanic dust lifted high in the atmosphere blocks solar radiation. For example, climatologists blame increased volcanic activity during the period 1812–1817, especially the volcanic eruption of Mount Tambora in Indonesia in 1815, for the "year with no summer" in New England in 1816 (Lamb [1995]).

atmospheric greenhouse gases, such as carbon dioxide, methane, and nitrous oxide. As solar energy heats the Earth's surface, it in turn emits energy back into space. The greenhouse gases trap some of this energy, and without this natural greenhouse effect the Earth's temperature would be much lower, and life as we know it would not exist. If the concentration of greenhouse gasses increases, however, ever higher average temperatures may produce catastrophic results. According to the U.S. Environmental Protection Agency, during the last few decades the atmospheric concentrations of carbon dioxide have increased nearly 30%; methane concentrations have more than doubled; and nitrous oxide concentrations have risen by about 15% (U.S. Environmental Protection Agency [2000]). Although many scientists attribute these changes to increased pollution caused by industrial activities, others do not consider the evidence compelling and are more willing to assume that the increases in the greenhouse gasses and average Earth surface temperatures are part of a natural cycle.

Examining these questions in detail is beyond the scope of this text, but the following example (from Taubes [2001]) provides a starting point. First, it is reasonable to conjecture that a change in the Earth's average temperature depends on how much radiation enters and exits the atmosphere. Using this basic assumption, how might the Earth's average surface temperature change, and what would the equilibrium temperature(s) be?

Let $T(t)$ denote the Earth's average surface temperature at time t. Change in temperature is caused by a flow of heat. In the model we now consider, we hypothesize that the flow of heat is caused by two factors: heat flowing into the Earth via solar radiation and heat leaving via irradiation. If the amount of heat energy coming into the Earth is equal to the amount leaving, then there will be no change in the Earth's average temperature. If, however, there is more heat entering the Earth than leaving, we would expect the temperature to rise. As an equation, we have:

$$\frac{dT}{dt} = S - E, \tag{1-21}$$

where S is the solar energy absorbed by the Earth and E is the heat energy radiated back into space. Both S and E depend on the average Earth temperature T. To find the equilibrium temperature(s) and classify their stability, we need to know exactly how S and E depend on the Earth's temperature. It stands to reason that as the Earth's temperature increases, the Earth emits more heat (i.e., as T increases, E increases). Perhaps surprisingly, the heat absorbed from the sun, S, also increases as T increases. This is partly because of the decreased reflection of solar energy by smaller ice cover at higher Earth temperatures.[3] Assume that the graphs of $S(T)$ and $E(T)$ are as shown in Figure 1-17.

---

3. See Taubes (2001), p. 50, for more details.

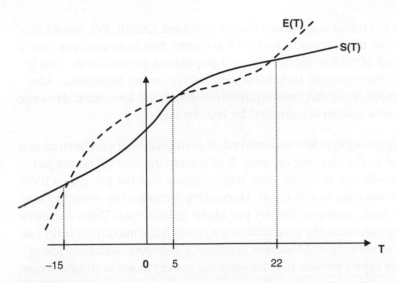

FIGURE 1-17.
Hypothetical graphs of the solar energy absorbed by the Earth, $S(T)$, and the energy radiated by the Earth, $E(T)$, as functions of the average Earth temperature T.

EXERCISE 9

(a) Using Figure 1-17, sketch the graph of $\dfrac{dT}{dt}$ versus the average Earth temperature T.

(b) Use the graph from part (a) to show that this model has two stable equilibrium states.

(c) Tell how the temperature will evolve if at time t_0 we have:

    (i) $T(t_0) = 23°C$,

    (ii) $T(t_0) = 18°C$,

    (iii) $T(t_0) = -16°C$, and

    (iv) $T(t_0) = 4°C$.

(d) Sketch the time trajectories of the temperature $T(t)$ for the four cases in part (c).

Example 1-2

Harvesting a Renewable Resource. As the world's population grows, humans must be increasingly conscious that natural resources are limited. The food supply is one resource that can be exploited in catastrophic ways. For example, overharvesting has led to the collapse of important fisheries in several areas of the world. The question of determining ecologically sustainable approaches to harvesting is therefore of critical importance (Smith and Link [2005]). We would like to determine the maximum level of harvesting that is sustainable over a long period of time without driving a population to extinction. This is known as the *maximum sustainable yield* (MSY) for the population. Our next example illustrates how equilibrium states and long-term dynamic behavior of a system are affected by harvesting.[4]

When a population is left undisturbed, it maintains near equilibrium at a level close to the carrying capacity K of the environment. The net per capita growth rate is nearly zero, which means that the per capita birth and death rates are nearly equal. Harvesting increases the mortality rate, which, in turn, decreases the net per capita growth rate. Thus, excessive harvesting can cause the mortality rate to exceed the maximum birth rate and lead to extinction. Moderate harvesting, however, will only lower the net per capita growth rate, causing the system to settle around a new equilibrium level lower than K.

To illustrate this concept mathematically, assume that a population grows according to the logistic model $\dfrac{dP}{dt} = a\left(1 - \dfrac{P}{K}\right)P$ and that harvesting yield per time unit is proportional to the size of the population. The harvesting will then decrease the rate of change for the population by a factor of bP, where $b > 0$ represents the harvesting effort. The rate of change of the population size accounting for the harvesting will then be $\dfrac{dP}{dt} = a\left(1 - \dfrac{P}{K}\right)P - bP$. The new nonzero equilibrium state for this model is $P = K\left(1 - \dfrac{b}{a}\right)$, which corresponds to harvesting yield $Y(b) = bP = bK\left(1 - \dfrac{b}{a}\right)$. This equilibrium state will be non-negative if $b/a < 1$; that is, if $b < a$. Therefore, if the harvesting effort b is less than the inherent per capita growth rate a, the harvesting effort is sustainable. Conversely, if $b > a$, the population will die out. The yield $Y(b) = bP = bK\left(1 - \dfrac{b}{a}\right)$ achieves its maximum at $b = \dfrac{a}{2}$ (the reader should verify this), which shows that the maximum sustainable yield in this case is

$$\text{MSY} = Y_{\max} = Y\left(\frac{a}{2}\right) = \frac{a}{2}K\left(1 - \frac{a}{2a}\right) = \frac{aK}{4}.$$

In this example, we made the assumption that yield is proportional to population size. This assumption is certainly justified when fishing or hunting is involved. As our next exercise shows, in more controlled environments, the harvesting rate may be independent from the population size. In such cases, a model may have more than one nonzero

---

4. An expanded analysis of these models can be found in Hoppensteadt and Peskin (2002).

equilibrium state. The MSY is then determined as the largest yield that will guarantee the existence of a nonzero equilibrium.

EXERCISE 1-10

Suppose we have a farm that can sustain a herd of 400 head of cattle and that the cattle population is modeled by the logistic equation:

$$\frac{dP}{dt} = 0.002(400 - P)P, \tag{1-22}$$

where $\frac{dP}{dt}$ is the rate of change in the number of head per year. The graph of $0.002(400 - P)P$ versus P is given in Figure 1-18 (*solid line*).

(a) Give the equilibrium state(s) for the model, and classify each as stable or unstable.

If we decide to sell one animal per week (or 52 per year), then the new equation governing the population would be:

$$\frac{dP}{dt} = 0.002(400 - P)P - 52. \tag{1-23}$$

The graph of $0.002(400 - P)P - 52$ versus P is shown in Figure 1-18 (*dashed line*). The equilibrium states are approximately 82 and 318 cattle.

(b) Classify the new equilibrium states as stable or unstable.

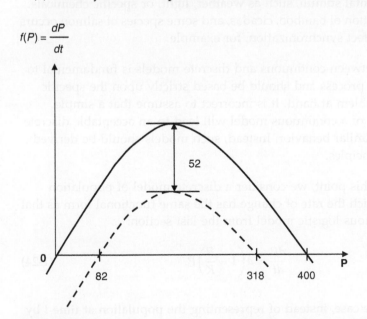

FIGURE 1-18.
Models showing the logistic curves with and without harvesting. The graphs of $0.002(400 - P)P$ (solid line) and $0.002(400 - P)P - 52$ (dashed line) versus P. The graph of $0.002(400 - P)P - 52$ is obtained by shifting downward the graph of $0.002(400 - P)P$ by 52 units.

(c) What happens to the population described by Eq. (1-23) if:

   (i) $P(t_0) = 50$,

   (ii) $P(t_0) = 150$, and

   (iii) $P(t_0) = 400$?

(d) Describe how the graph in Figure 1-18 will change when the harvesting rate is greater than 52 cattle/year.

(e) What is the largest number of cattle/year that could be sold while guaranteeing the existence of a stable equilibrium for the model?

(f) What is the maximal sustainable yield for this model?

VIII. THE VERHULST MODEL FOR DISCRETE POPULATION GROWTH

For modeling purposes, it is appropriate to hypothesize that many species reproduce at a uniform rate. Said another way, there is no preferred time at which reproduction occurs. In these cases, it is often best to use a continuous model. As discussed in Section II, the use of discrete models may be appropriate when the population remains constant throughout intervals of time and then changes with a jump at the end of an interval. Reproduction in such cases may be synchronized to environmental stimuli, such as weather, light, or specific chemicals. The reproduction of bamboo, cicadas, and some species of salmon occurs in almost perfect synchronization, for example.

The choice between continuous and discrete models is fundamental to the modeling process and should be based strictly upon the specific biological problem at hand. It is incorrect to assume that a simple discretization of a continuous model will lead to an acceptable discrete model with similar behavior. Instead, such models should be derived from first principles.

To illustrate this point, we consider a discrete model of population growth in which the rate of change has the same functional form as that of the continuous logistic model from the last section:

$$\frac{dP}{dt} = a\left(1 - \frac{P}{K}\right)P. \tag{1-24}$$

In the discrete case, instead of representing the population at time t by $P(t)$, we denote the population size throughout the n-th generation by P_n. The change from the n-th to the $(n+1)$-st generation is then given by $P_{n+1} - P_n$ and the model is:

$$P_{n+1} - P_n = r(P_n)P_n = a\left(1 - \frac{P_n}{K}\right)P_n. \qquad (1\text{-}25)$$

The model from Eq. (1-25) is sometimes called the Verhulst model, after the Belgian mathematician Pierre Verhulst (1804–1849), who first studied it in 1846. The model given by Eq. (1-25) may also be obtained by modifying the unlimited growth model from Eq. (1-1) to allow net per capita growth rate to vary with population size. In Eq. (1-25), as in Eq. (1-24), K is the carrying capacity for the population; $a > 0$ is the inherent per capita growth rate; and the arguments for choosing $r(P_n) = a\left(1 - \frac{P_n}{K}\right)$ will be the same as for the continuous logistic model developed in Section IV.

Although the Verhulst model has the same equilibrium states as the continuous logistic model, it can exhibit radically different long-term behavior. Recall that an equilibrium state P is one in which the quantity in question remains constant over time. For discrete models such as (1-25), these are the values at which the system exhibits no change (i.e., $P_n = P$, for all $n = 0, 1, 2, \dots$). Equivalently, these are the values for which $P_n = P_{n-1}$ for all values of $n = 1, 2, 3, \dots$.

EXERCISE 1-11

Show that the equilibrium states for Verhulst model [Eq. (1-25)] are $P = 0$ and $P = K$.

In Section VI, we proved that for any value of $a > 0$ and any nonzero initial population size $P(0)$, the logistic model (1-24) exhibits convergence for $t \to \infty$ to its equilibrium state $P = K$. For $P(0) < K$, the population size $P(t)$ is continuously increasing to K when $t \to \infty$ while if $P(0) > K$, the population size $P(t)$ is continuously decreasing to K when $t \to \infty$ (Figure 1-8). The Verhulst model offers cases of considerably more complex long-term behavior—the system could converge to an equilibrium state through oscillations, exhibit lack of convergence because of periodic oscillatory behavior, or be driven to chaos.

To demonstrate this, let $x_n = \frac{P_n}{K}$ so that x_n *is the fraction of the maximum population the environment can sustain*. With this notation, the Verhulst model takes the equivalent form:

$$x_{n+1} - x_n = a(1 - x_n)x_n, \qquad (1\text{-}26)$$

and the carrying capacity of the model in Eq. (1-26) is equal to 1. Equation (1-26) represents the *nondimensional* form of the Verhulst model from Eq. (1-25). This is due to the fact that P_n and K are measured in the same units, so the quantity $x_n = \frac{P_n}{K}$ is nondimensional. This representation has several mathematical advantages. First, it decreases the number of model parameters, as the carrying capacity parameter K in Eq. (1-25) is now scaled to 1. Second, the results obtained for Eq. (1-26) will be independent of the units of measurement.

An interesting aspect of Eq. (1-26) is its sensitivity to the initial value x_0 and different values of a. Figure 1-19 illustrates several different scenarios. The ability of these models to generate oscillating trajectories is particularly interesting.

The mathematics required for understanding and classifying all the trajectories of Eq. (1-26) are not trivial. This may seem surprising, given that our continuous logistic model was simple to understand. It was not until the latter half of the twentieth century that mathematicians began to discover the peculiarities of discrete models, such as the Verhulst model. Decades of interdisciplinary work involving mathematicians, ecologists, biologists, physicists, and computer scientists were necessary before some satisfactory answers were found, and the theory is still far from complete. Equation (1-26) is one of the seeds from which the *mathematical theory of chaos* grew. We refer the reader to Gleick (1987) for the fascinating history behind "discovering chaos" and to Hirsch et al. (2003) for an introduction to the mathematical theory.

To get a heuristic impression of why oscillations occur for the discrete models [Eqs. (1-25) and (1-26)], notice that the net per capita growth rate $r = r(P_n)$ in Eq. (1-25) uses current population size to predict growth during the next generation. As changes in population size occur only at designated, equally spaced time intervals, there is a lag that may cause overshooting or undershooting, similar to the inertial effect in physical systems. Mathematically, the following argument provides quantitative insight. When $ax_n > 1$, Eq. (1-26) implies that $|x_{n+1} - x_n| > |1 - x_n|$; that is, the distance between the current level of the population x_n and the maximum level is smaller than the distance between the current level x_n and the level x_{n+1} of the next generation (see Figure 1-20). Thus, in this case, x_n and x_{n+1} will always be on opposite sides of the maximum level 1, causing oscillatory behavior.

Table 1-5 presents several values from Eq. (1-26) of x_n, for $n \geq 100$ with $x_0 = 0.8$, and for different values of a to demonstrate the dependence of the long-term behavior of the process upon the value of a. For $a = 1.5$, the system oscillates above and below 1 before settling into the equilibrium state of 1. For $a = 2.10$, the system oscillates between two values. As a increases further, the system will oscillate among four values. This is an example of *period doubling*. If a were to continue increasing, the system would be driven to chaos.

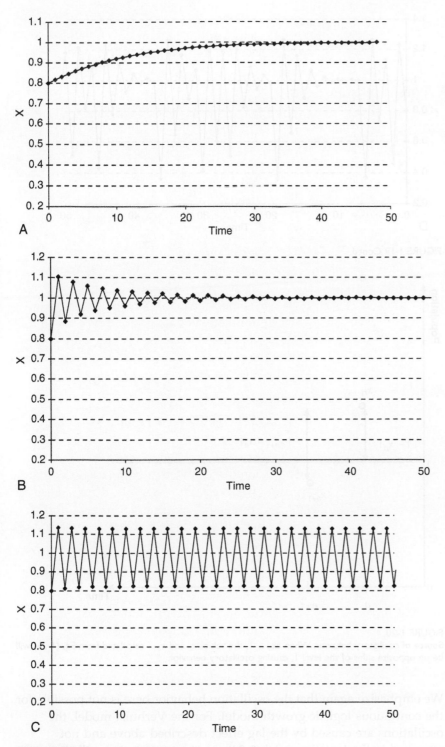

A

B

C

FIGURE 1-19.
Solutions of Eq. (1-26) for initial condition $x_0 = 0.8$ and different values of a. Panel A: $a = 0.1$; panel B: $a = 1.9$; panel C: $a = 2.1$; panel D: $a = 2.7$.

(Continued)

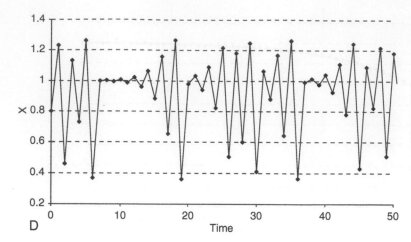

D

FIGURE 1-19 Cont'd.

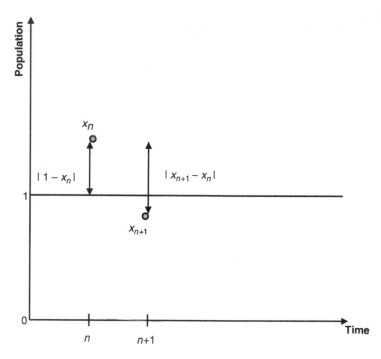

FIGURE 1-20.
Source of oscillatory behavior. When $ax_n > 1$, the values x_n and x_{n+1} calculated from Eq. (1-26) will be on opposite sides of the level 1, causing oscillatory behavior.

We emphasize again that the oscillation behavior here is not possible for the continuous logistic growth model. For the Verhulst model, the oscillations are caused by the lag effect described above and not observed in the logistic model (1-24). As the next section will show, logistic equations are capable of generating oscillations when explicit delay is introduced.

Our next exercise presents a discrete population model with long-term behavior similar to the continuous logistic population model.

a	$x_{100}, x_{101}, x_{102}, \ldots$
1.50	1, 1, 1, 1, ...
2.10	0.823735, 1.12864, 0.823735, 1.12864, 0.823735, 1.12864, 0.823735, 1.12864, ...
2.50	1.225000, 0.535948, 1.157720, 0.701238, 1.225000, 0.535948, 1.157720, 0.701238, ...

TABLE 1-5.
Long-term behavior depends upon the value of parameter a.

EXERCISE 1-12

Consider the following population model:

$$P_{n+1} = \frac{2}{1 + \frac{P_n}{K}} P_n. \qquad (1\text{-}27)$$

where K is the carrying capacity.

(a) Show that $P = 0$ and $P = K$ are the equilibrium states.

(b) Show that if $0 < P(0) < K$, the population will be increasing.

(c) Assuming that $\lim P_n$ exists as $n \to \infty$, show that, under the conditions of part (b), $\lim P_n = K$.

IX. A POPULATION GROWTH MODEL WITH DELAY

Despite substantial improvement over the "unlimited" population growth model $\frac{dP}{dt} = rP(t)$, the logistic growth model (1-12) has one major drawback—replacing r with the factor $r(P(t)) = a(1 - P(t)/K)$ only provides a mechanism for the net per capita growth rate to adjust itself based on *current* population size. The logistic model (1-24) is entirely based on *the present* and disregards, to a large extent, *the past*. In reality, certain delay effects are essential, although this logistic model does not account for them.

Just as we need to appreciate the logistic model's merits, we also need to understand its limitations. In Section IV, we found that the logistic model described yeast and bacterial growth with great accuracy. It may be less successful, however, in describing the growth of populations of more complex organisms. For example, populations of the water flea *Daphnia* have been observed to oscillate when cultures are maintained at 25°C (Pratt [1943]), as shown in Figure 1-21.

To recognize the effect caused by delay, notice, for example, that because the gestational age of newborn human babies is about 9 months, the number of babies born on January 1, 2005 was generally determined 9 months earlier, on April 1, 2004. In order to refine our model, we need

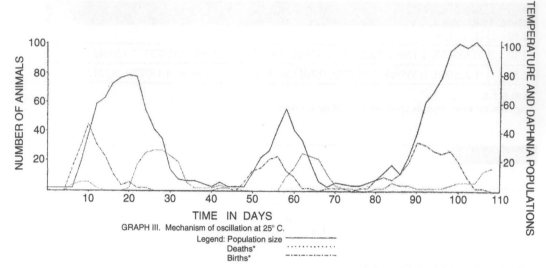

GRAPH III. Mechanism of oscillation at 25° C.

Legend: Population size ————————
Deaths* ·················
Births* ——·——·——·——

* If the actual number of deaths and births occurring on each day is plotted, the resulting curves are too irregular and too low to read with ease. Accordingly, each number was doubled, and the curves smoothed by plotting the points as 3-point moving averages.

FIGURE 1-21.
Oscillations in the size of a water flea (*Daphnia*) population. (From Pratt, D. M. [1943]. *Biological Bulletin* 85, 116–140. Used by permission.)

to offset the dependence $r(t) = r(P(t))$ to account for this time lag. In nonmathematical terms, we say that the *present* value of $r(t)$ is determined by the population size at a specific time in *the past*. The simplest way to model this is to postulate that $r(t) = r(P(t - D))$, where $D > 0$ is the measured delay. In the example above, D will be equal to nine months—a baby's average gestational period. The logistic model (1-12) can now be modified so:

$$\frac{dP}{dt} = a\left(1 - \frac{P(t - D)}{K}\right)P(t).\tag{1-28}$$

Notice that this model preserves our fundamental hypothesis that the rate of change in population size is proportional to the population size.

Exercise 1-13

List the limitations of the model given by Eq. (1-28).

The model from Eq. (1-28) is quite different mathematically from the logistic model. To obtain an exact analytical solution for Eq. (1-28), we need to know the values of the solution $P(t)$ over the whole interval $[0, D]$. In contrast, knowing the value of $P(t)$ at just one point, say $t = 0$, is

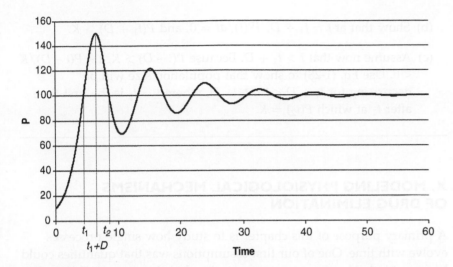

FIGURE 1-22.
A numerical solution of the model described by Eq (1-28). In this case, the delay was $D = 2.0$, and the parameters were $a = 0.61$; $K = 100$.

enough to solve logistic model (1-24) and find the unique solution satisfying the initial condition $P(0) = P_0$. A computer-generated solution of Eq. (1-28) is presented in Figure 1-22.

The pattern of convergence to the equilibrium state $P = K$ through oscillations with decreasing amplitudes (damped oscillations) is similar to one of the solution trajectories observed in the discrete Verhulst model [see Figure 1-19(B)]. The heuristic reason for the oscillations is also essentially the same. The net per capita growth rate used to determine the population's rate of change at time t uses information on the availability of resources based on population size at D units of time earlier. This causes over- or underestimating of the slope while the system adjusts itself, leading to damped oscillations about the carrying capacity K. Exercise 14 provides greater insight into the mathematical properties that allow oscillations to develop in the solution time trajectories.

EXERCISE 1-14

Follow the steps outlined below to show that the sign of the derivative dP/dt in Eq. (1-28) may change over time, causing oscillations.

(a) Assume that at a moment t_1, the population reaches its carrying capacity [that is, $P(t_1) = K$] and for t in the interval $(t_1, t_1 + D)$ $P(t - D) < K$, as in Figure 1-22. Then for $t < t_1 + D, 1 - P(t - D)/K > 0$. Now use Eq. (1-28) to show that for $t < t_1 + D, dP(t)/dt > 0$, and population size is increasing. In particular, population size is still increasing at $t = t_1$, thus overshooting the carrying capacity K.

(b) Show that at $t = t_1 + D$, $dP(t)/dt = 0$, and $P(t_1 + D) > K$.

(c) Assume now that $t > t_1 + D$. Because $P(t - D) > K$, $1 - P(t - D)/K < 0$. Use Eq. (1-28) to show that population size will be decreasing for $t_1 + D < t < t_2 + D$, where $t = t_2$ is the first time after t_1 at which $P(t_2) = K$.

X. MODELING PHYSIOLOGICAL MECHANISMS OF DRUG ELIMINATION

A primary purpose of this chapter is to study how single processes evolve with time. One of our first assumptions was that quantities could be expected to change at a rate proportional to the amount of quantity present. We have demonstrated that this may work well for modeling population growth for relatively short periods of time, but environmental limitations will eventually cause the growth rate to abate. In other situations, however, quantities diminish rather than increase, in proportion to the amount present. While not extremely common for population changes, there are many biological processes that do change in this way. For example, in many organisms, foreign materials are excreted at a rate proportional to their concentration.

To illustrate this phenomenon of *exponential decay*, we shall study how the body eliminates drugs by modeling concentrations of physiologically active substances in the bloodstream. Each drug dose received increases its concentration in the bloodstream, but, simultaneously, the kidneys are working to remove the drug. It has been experimentally determined, in fact, that substances entering the bloodstream are eliminated by the kidneys at a rate proportional to their concentration. Clearly, the factors governing this physiological process are different from the factors that determine population growth but, as we shall see, the underlying mathematical models are very similar.

Ask yourself: Why do you need to take two acetaminophen tablets every 4 to 6 hours when you have a headache? Why is there a warning label that cautions you not to take any more than four doses in a given 24-hour period? And why does your head start to ache again after four hours when the warning suggests you really ought to wait six hours before you take the next dose?

When a physician administers a drug to a patient, he or she has two important aims—ensuring that the dosage is high enough to provide the desired effect while ensuring that it is not so high that the drug becomes toxic. These aims illustrate two critical drug concentrations—the minimum effective concentration (MEC) and the minimum toxic concentration (MTC). Between these limits lies the therapeutic window

(TW)—the range of concentrations over which the drug is both effective and safe (see Figure 1-23). Our goal is to construct a model describing how to achieve and maintain drug concentrations in this range.

Before we can begin to produce a model, we should consider the factors influencing a drug's concentration in the bloodstream. These may be roughly divided into factors controlling entry of the drug into the bloodstream (absorption) and factors controlling its exit (elimination).

To consider absorption, we must first determine how the drug will be introduced into the patient—orally, intravenously, intramuscularly, transdermally (through the skin), or by inhalation. We must consider whether the drug undergoes any physical or chemical changes during administration. For example, is it a solid that must be dissolved in the stomach or small intestine, or will it be introduced as a solution? We must also consider the characteristics of the drug that will control how it will be absorbed into the bloodstream. A fat-soluble molecule can diffuse through intestinal cell membranes, whereas a charged or polar molecule has to rely on transport proteins.

Once the drug has entered the bloodstream, we must consider how it disperses through the body (distribution). For example, does the drug bind to serum proteins, or does it circulate as a free drug? Distribution also involves the drug's movement from the bloodstream to the tissue, organ, or other area of the body where it has its effect—in the case of the acetaminophen, your poor aching head.

While some acetaminophen will exit the bloodstream to reduce your headache, the remainder will be removed by processes grouped under the term *elimination*. Elimination may involve *metabolism*, the term for any chemical reaction the drug undergoes in the body, as well as excretion, which is usually the action of the kidneys.

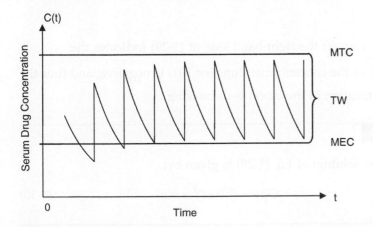

FIGURE 1-23.
Changes in serum drug concentration. The drug concentration changes because of the cumulative effect of multiple doses and the elimination of the drug from the bloodstream. When maintained within the therapeutic window, the drug concentration is both effective and safe.

Metabolism occurs through the action of protein catalysts called enzymes. Metabolism of drugs most often occurs in the liver, although there are some enzymes in the bloodstream as well. Liver enzymes, such as epoxide hydratase and the cytochrome P450 family, catalyze the chemical modification of the drugs, often oxidizing them and making them more amenable to excretion.

In excretion, the kidneys filter the blood and remove drugs as well as metabolic wastes. Drugs may also be excreted by the liver in the bile, a substance necessary for digesting fats. Bile is made by the liver, stored in the gall bladder, and delivered into the small intestine. Drugs may also be removed from the bloodstream by the lungs and then exhaled. (If this sounds unlikely, just think of the "breathalyzer" commonly used for testing persons suspected of driving under the influence of alcohol!) By this point, it should be clear that we need to begin our model construction by making some simplifying assumptions.

The simplest model would be to assume instantaneous entry of drug into the bloodstream, followed by its gradual clearance from the bloodstream. An intravenous injection might well approximate instantaneous entry. It is known from experimental data that the clearance rate of drugs from the bloodstream is generally proportional to the amount present in the bloodstream. Therefore, we are looking at an example of exponential decay—the reverse of our first population model.

If $C(t)$ is the drug concentration at time t, then the fact the drug is eliminated from the bloodstream at a rate proportional to the amount present can be expressed as:

$$\frac{dC(t)}{dt} = -rC(t), \tag{1-29}$$

where $r > 0$.

The negative sign on the right-hand side of (1-29) indicates the derivative $\frac{dC}{dt}$ of the concentration function $C(t)$ is negative, and thus the drug's concentration in the blood is decreasing.

EXERCISE 1-15

Show that the solution of Eq. (1-29) is given by:

$$C(t) = C(0)e^{-rt}. \tag{1-30}$$

Equation (1-29) has one parameter, represented by the elimination rate constant r. Larger values of r correspond to faster elimination. By inspection of Eq. (1-29), it can be seen that the units for r are time^{-1}; for example, hours^{-1}, min^{-1}, or days^{-1}. In Figure 1-24, we plot the drug concentration $C(t)$ versus time with $C(0) = 4\ \mu g/ml$ and for three different values of r: $r = 0.3$, $r = 0.2$, and $r = 0.1$ hours^{-1}.

The elimination parameter r is closely related to the drug's *half-life*, τ, defined as the time necessary to reduce the concentration of the drug in the blood by 50%. In mathematical terms, the half-life τ is the time elapsed since the initial moment $t = 0$ for which $C(\tau) = 0.5C(0)$. Using Eq. (1-30) gives us $C(0)e^{-r\tau} = 0.5C(0)$, or $e^{-r\tau} = 0.5$. Thus $-r\tau = \ln(0.5)$, leading to the following connection between the elimination rate constant r and the drug's half-life: $\tau = \dfrac{\ln(2)}{r}$.

EXERCISE 1-16

From the graph in Figure 1-24, corresponding to $r = 0.2$ hours^{-1}, estimate the drug's half-life; then compute τ from the equation above and compare the two values.

EXERCISE 1-17

The half-life of acetaminophen is 2.5 hours. If a single dose is administered at 12:00 noon, how long will it take for the

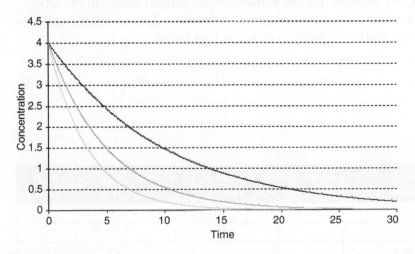

FIGURE 1-24.
Serum drug concentration for different values of r. From top to bottom, the graph corresponds to $r = 0.1$, $r = 0.2$, and $r = 0.3$ hours^{-1}, respectively. Larger values of r signify faster elimination of the drug from the bloodstream.

concentration in the bloodstream to fall below 1% of the initial concentration?

The following example shows how to mathematically model multiple doses.

Example 1-3

A drug whose elimination constant is r hours^{-1} is first administered at 1:00 A.M. at a dosage C µg/ml and in the same dosage every three hours afterwards. What is the drug's concentration at 12:00 noon?

SOLUTION:

By 12:00 noon, we will have given four doses. The total concentration is the sum of the concentrations of each dose. Because the doses were given at different times, the effect at 12:00 noon is different for each dose, as shown in Table 1-6.

The total concentration at 12:00 noon is then equal to:

$$Ce^{-11r} + Ce^{-8r} + Ce^{-5r} + Ce^{-2r} = Ce^{-2r}(e^{-9r} + e^{-6r} + e^{-3r} + 1). \qquad (1\text{-}31)$$

For a more compact form of Eq. (1-31), observe that if $b = e^{-3r}$, then $b^2 = e^{-6r}$, and $b^3 = e^{-9r}$. Thus, the concentration becomes:

$$Ce^{-2r}(b^3 + b^2 + b + 1). \qquad (1\text{-}32)$$

From this example, we can extract a more general result. In the term e^{-3r}, the number 3 arises from the time between dosages. Notice also that the parenthetical expression contains four terms:

$$b^3 + b^2 + b + 1 = b^3 + b^2 + b + b^0, \qquad (1\text{-}33)$$

Time of Dose	Length of Time in Body (hours) by 12:00 Noon	Residual Concentration from the Dose at 12:00 Noon (µg/ml)
1:00 A.M.	11	Ce^{-11r}
4:00 A.M.	8	Ce^{-8r}
7:00 A.M.	5	Ce^{-5r}
10:00 A.M.	2	Ce^{-2r}

TABLE 1-6.
Contribution of multiple doses to serum drug concentration.

and four doses have been given. The value C is the dosage, and the 2 in Ce^{-2r} comes from the fact that it has been 2 hours from the last administered dose.

EXERCISE 1-18

We give a dosage of C µg/ml at 2-hour intervals. The elimination constant is r hours^{-1}. There are six doses given. Give an expression for the concentration 30 minutes after the last dose.

In general, the drug's concentration follows the pattern shown in Figure 1-25.

If the drug is administered in dosages C at intervals of length T, then at the end of the n-th period the concentration is:

$$R_n = C[(e^{-Tr})^n + (e^{-Tr})^{n-1} + (e^{-Tr})^{n-2} + \ldots + e^{-Tr}]. \qquad (1\text{-}34)$$

Thus, R_n is the residual concentration from the first n doses immediately before the next dose is administered. Immediately after the next dose, the concentration rises to:

$$R_n + C = C[(e^{-Tr})^n + (e^{-Tr})^{n-1} + (e^{-Tr})^{n-2} + \ldots + e^{-Tr} + 1] \qquad (1\text{-}35)$$

To simplify expressions (1-34) and (1-35) and write them in closed form (i.e., without using the "and so on" symbol "..."), we need some preliminary mathematics.

An expression of the form:

$$a + ab + ab^2 + ab^3 + \ldots \qquad (1\text{-}36)$$

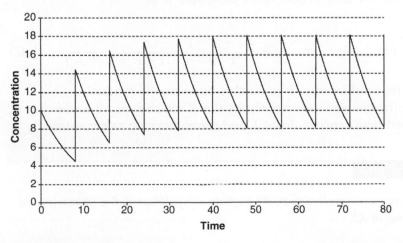

FIGURE 1-25.
Serum drug concentration with regular doses. A dose $C = 10$ µg/ml is administered every $T = 8$ hours. The elimination constant for the drug is $r = 0.1$ hours^{-1}.

is called a *geometric series*. We let S_n denote the sum of the first n terms of such a series, so:

$$S_n = a + ab + ab^2 + ab^3 + \ldots + ab^{n-1}. \tag{1-37}$$

Then:

$$bS_n = ab + ab^2 + ab^3 + \ldots + ab^{n-1} + ab^n \text{ and } S_n - bS_n = (1-b)S_n = a - ab^n. \tag{1-38}$$

If $b \neq 1$, we obtain the following compact formula for S_n:

$$S_n = \frac{a - ab^n}{1 - b}. \tag{1-39}$$

When $|b| < 1$, the limit below can be calculated to be:

$$\lim_{n \to \infty} S_n = \frac{a}{1 - b} \tag{1-40}$$

because when $|b| < 1$, $\lim_{n \to \infty} b^n = 0$.

Applying Eq. (1-40) with $a = Ce^{-Tr}$ and $b = e^{-Tr}$, the residual concentration after n doses from Eq. (1-34) can be written as:

$$\begin{aligned}
R_n &= C([e^{-Tr}]^n + [e^{-Tr}]^{n-1} + [e^{-Tr}]^{n-2} + \ldots + e^{-Tr}) \\
&= Ce^{-Tr}([e^{-Tr}]^{n-1} + [e^{-Tr}]^{n-2} + \ldots 1) = Ce^{-Tr}\frac{1 - (e^{-Tr})^n}{1 - e^{-Tr}}. \tag{1-41}
\end{aligned}$$

What happens to the residual amounts as the number of doses increases? It appears from Figure 1-25 that, after several doses, the residual values stabilize around a value slightly higher than 8 μg/ml. To see if this is true in general, we need to find $\lim R_n$ as n $\to \infty$. Indeed, using Eq. (1-40), the limit is now easily computed to be:

$$R = \lim_{n \to \infty} R_n = \lim_{n \to \infty} Ce^{-Tr}\frac{1 - (e^{-Tr})^n}{1 - e^{-Tr}} = \frac{Ce^{-Tr}}{1 - e^{-Tr}} = \frac{C}{e^{Tr} - 1}. \tag{1-42}$$

Thus, for a sufficiently large number of doses, the residual concentrations stabilize around the value R, which depends on the dose C, the fixed time between doses T, and the elimination rate constant r.

Exercise 1-19

Is R_n larger or smaller than R? Explain why. What is the physiological meaning of R?

Compute the exact value of the limit R for the example in Figure 1-25: $C = 10\ \mu g/ml$, $r = 0.1$ hours^{-1}, $T = 8$ hours.

Knowing the MEC, the MTC, and the drug's half-life (or its elimination rate constant r), we now want to design a therapeutic regimen with maximal benefits. Equal doses C of the drug should be given at equal time intervals T. Once the concentration reaches the MEC, it should remain between the MEC and MTC.

The graph in Figure 1-25 shows that after a few dosages the drug's concentration is almost between R and $R + C$. (In fact, this is not quite correct, but the difference is so small that it is not enough to have an effect on the treatment's safety or effectiveness.) Because one goal is to maintain the concentration between the MEC and MTC, we can determine R and C from the conditions:

$$R = MEC, \qquad R + C = MTC. \qquad (1\text{-}43)$$

Because the MEC and MTC are known for every drug, we can determine the dose C as:

$$C = MTC - MEC. \qquad (1\text{-}44)$$

Using these values for R and C in Eq. (1-42), we obtain:

$$MEC = R = \frac{C}{e^{Tr} - 1} = \frac{MTC - MEC}{e^{Tr} - 1}. \qquad (1\text{-}45)$$

Solve the equation $MEC = \dfrac{MTC - MEC}{e^{Tr} - 1}$ for T to show that $T = \dfrac{1}{r} \ln \dfrac{MTC}{MEC}$.

Requiring all doses to be the same has the obvious disadvantage that a certain build-up period is required before the concentration reaches the MEC. For some drugs, such as certain antidepressants, a slow build-up is necessary to minimize side effects. For many other common drugs, however, the dosage schedule tolerates a larger first dose to achieve the maximal effective concentration as quickly as possible.

If a drug's MEC, MTC, and elimination constant r are known, determine a drug intake schedule that maximizes the drug's therapeutic effect under the following constraints:

1. All doses must be given at equal time intervals.

2. All doses, except possibly for the first dose, must be equal.

3. The MEC must be achieved as quickly as possible.

4. The concentration of the drug should remain between the MEC and MTC at all times.

EXERCISE 1-23

It is not common practice for pharmaceutical companies to make the MEC and MTC figures for their drugs available. Instead, users are provided with recommended doses and time intervals. It may be interesting to consider the following question: Assuming that pharmaceutical companies follow objectives 1–4 from Exercise 1-22 when determining the dose regimens for their drugs and that the half-life is known, could you estimate the drug's MEC and MTC?

XI. USING COMPUTER SOFTWARE FOR SOLVING THE MODELS

For most models developed thus far, we have presented analytical solutions. Knowing the analytical form of a solution allows for direct calculation of the predicted value. For example, knowing that the solution of Eq. (1-2) is given by $P(t) = P(0)e^{rt}$, where $P(0) = 5.3$ and $r = 0.297$, we can calculate that for $t = 2.5$, the model predicts a population size of $P(2.5) = 5.3e^{(0.297)(2.5)} = 11.1$ million for the United States for the year 1825. In the same way, using the solution of the discrete model $p_n = (1 + k)^n p_0$ from Exercise 1-1, we can calculate that if $p_0 = 5.3$ and $k = 0.345$, according to the discrete model (1–1), the U.S. population in 1880 will be $p_8 = (1 + k)^8 p_0 = (1.345)^8(5.3) = 56.8$ million.

It is not always easy to solve a model analytically, and, as the sophistication of the models increases, the mathematics for solving the equations become increasingly more challenging. When it is difficult (or sometimes impossible!) to obtain the actual analytic solution, *numerical solutions* are used instead. A numerical solution does not give us a function as the analytical solution does, but instead provides us with a table of values for the unknown function. For example, a numerical solution for the problem $\dfrac{dP(t)}{dt} = rP(t), P(0) = 5.3$, for $r = 0.297$ is presented in Table 1-7.

The left column contains a list of values for t, and the right column contains the values of the numerical solution $P(t)$ at these points. The

t (DT = 0.5)	$P(t)$
0.0	5.300
0.5	6.148
1.0	7.133
1.5	8.275
2.0	9.599
2.5	11.136
3.0	12.919
3.5	14.987
4.0	17.387
4.5	20.170
5.0	23.399
5.5	27.145
6.0	31.491

TABLE 1-7.
A numerical solution for $dP(t)/dt = rP(t)$.

increment used on the time variable t is often denoted by DT and is 0.5 in this example. Changing the value of DT allows for creating a specific mesh of points at which the value of the function $P(t)$ will be calculated. For DT = 1, the time values for which $P(t)$ will be calculated will be $t = 0,1,2,3$, etc. For DT=0.2, the time values will be $t = 0, 0.2, 0.4, 0.6$, etc. There are a number of software products that can be used to model and analyze dynamical systems and obtain numerical solutions of differential equations, including *MATLAB®*, BERKELEY MADONNA, Stella®, Vensim®, and others. For the rest of the chapter, we refer to the specific syntax of *BERKELEY MADONNA*, although any of the above software packages can be employed instead. A functional version (with some limitations on saving and printing) of *BERKELEY MADONNA* can be downloaded at no charge from the Web site, listed in Internet Resources at the end of this chapter. The remaining chapters of the text do not pertain to particular software, although references to relevant programs are provided at the end of each chapter, where appropriate.

The initial and final values of the time interval over which we would like to know the values of the solution should be specified. In *BERKELEY MADONNA*, they are called STARTTIME and STOPTIME. In the example above, we had the values of $P(t)$ calculated over the interval [0, 6], corresponding to STARTTIME = 0 and STOPTIME = 6.

We now give a basic introduction that will allow you to enter mathematical models in *BERKELEY MADONNA* and obtain their numerical solutions. We shall use the models developed in this chapter as examples.

XII. SOME *BERKELEY MADONNA* SPECIFICS

When you start *BERKELEY MADONNA*, the screen that will contain your model appears. There is even some code that has already been written:

```
METHOD RK4

STARTTIME = 0

STOPTIME = 10

DT = 0.02
```

The first line specifies the numerical method that will be used by the program for computing the numerical solution. You can safely ignore this for now and accept the default algorithm.[5] The remaining lines

---

5. For those readers familiar with the theory of numerical methods for solving ordinary differential equations, we would add that *BERKELEY MADONNA* allows you to choose from a set of built-in algorithms, including Euler's method and two types of Runge–Kutta methods. More details on this and other specifics related to the software can be found in *BERKELEY MADONNA*'s brief documentation accessible under the Help menu.

specify that the time values for which the function values will be calculated begin at $t = 0$, end with $t = 10$, and contain all points in between with increments of DT $= 0.02$.

BERKELEY MADONNA is case-insensitive—*m* and *M* are treated as being exactly the same. It is up to you whether to use all lower-case, all upper-case, or mixed cases. Blank lines do not matter—include as many or as few as you need to make your equations more readable.

To enter the model $\dfrac{dP(t)}{dt} = rP(t), P(0) = 5.3$, begin typing at the end of the code that is already there. Enter the following:

```
d/dt (P) = r* P

init P = 5.3

r = 0.297
```

The first line is, of course, the model itself. Notice that we have completely ignored the fact that $P = P(t)$ is a function that depends on the time variable t—it is understood by default.

We use `init` P to specify the *initial condition* $P(0) = 5.3$. Finally, on the last line, we give the specific value for r. Run the model by clicking the Run button in the upper left corner. The graph of the solution will appear, as in Figure 1-26.

To see the numerical solution as a table of values, click on the Table button found across from the Run button (the icon depicts two squares offset from one another). You should be looking at output similar to Figure 1-27.

To compare the model predictions with the actual U.S. census data, enter the data from Table 1-1 into a text file (separating the two columns by a blank space or tabs), and save the file as *U.S.Pop.txt*. To import this file into *BERKELEY MADONNA*, select File > Import Dataset from the main menu. Navigate to your *U.S.Pop.txt* file and open it. Click OK in the Import Dataset dialog box. The data should now appear on the plot.

EXERCISE 1-24

Select appropriate values for DT, STARTTIME, and STOPTIME to obtain the numerical solution for $P(t)$ that:

(a) Contains the values for $P(t)$ at integer time values from $t = 1$ to $t = 7$; and

(b) Allows you to use the numerical solution to obtain the value $P(2.35)$.

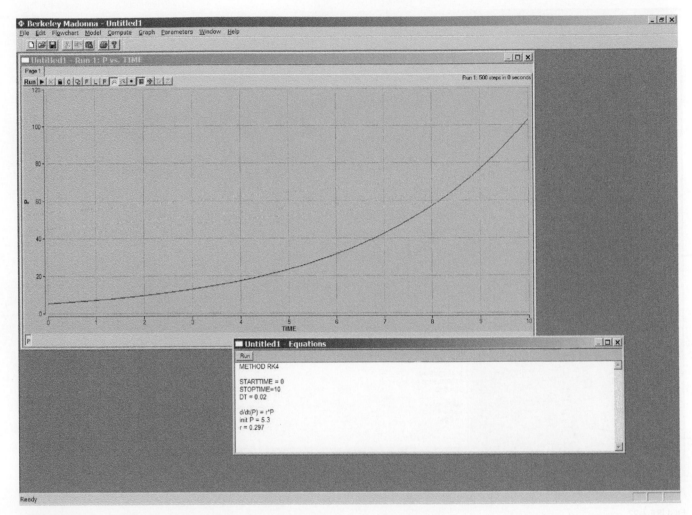

FIGURE 1-26.
BERKELEY MADONNA numerical solution: Graphical output and model.

EXERCISE 1-25

(a) Obtain a numerical solution at the integer values of t from $t = 0$ to
$t = 20$ for $\dfrac{dP(t)}{dt} = a\left(1 - \dfrac{P(t)}{K}\right)P(t)$, $P(0) = 9.6$, $K = 660$, and
$a = 0.608$.

(b) Import the data from Table 1-8, and compare them with the model
prediction.

Solving discrete models in *BERKELEY MADONNA* is very similar. There
are no derivatives involved here, so entering the actual model equation
is done in a slightly different way. For example, to enter our discrete
model $p_n - p_{n-1} = k\,p_{n-1}$, first rewrite it as $p_n = p_{n-1} + k\,p_{n-1} = (1 + k)\,p_{n-1}$.

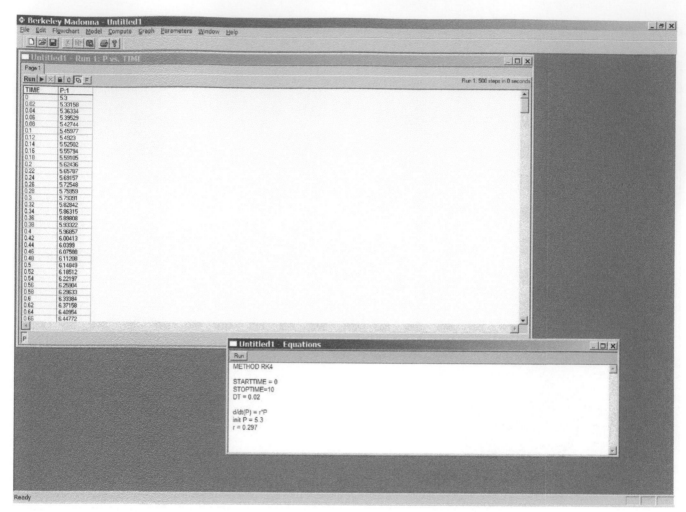

FIGURE 1-27.
BERKELEY MADONNA numerical solution as a table of values.

Time t	Biomass $P(t)$	Time t	Biomass $P(t)$	Time t	Biomass $P(t)$
0	9.6	7	257.3	14	640.8
1	18.3	8	350.7	15	651.1
2	29	9	441	16	655.9
3	47.2	10	513.3	17	659.6
4	71.1	11	559.7	18	661.8
5	119.1	12	594.8	—	—
6	174.6	13	629.4	—	—

TABLE 1-8.
Yeast culture growth data. (Data taken from Carlson [1913] and Pearl [1927].)

Notice that this model is *recursive*; knowing the "present value" p_{n-1}, we can determine the "next value" p_n. This is exactly the syntax used in *BERKELEY MADONNA* to describe the equations.

Enter the following (the first line is the default and we have specified appropriate values for STARTTIME, STOPTIME, and DT):

METHOD RK4

STARTTIME = 0

STOPTIME = 10

DT = 1

init $p = 5.3$

next $p = (1 + k)*p$

$k = 0.345$

Run the model to obtain the graph and a table of values for the population size p (see Figure 1-28). Pressing the Data Points button (an icon with a solid black dot located across from the Run button) will show you the points on the graph that make the numerical solution table for p.

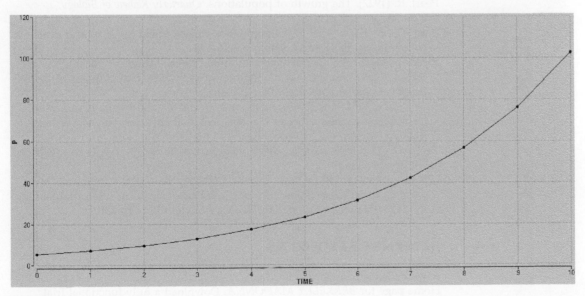

FIGURE 1-28.
BERKELEY MADONNA graphical output depicting the data points representing the numerical solution.

XIII. SUGGESTED BIOLOGY LABORATORY EXERCISES FOR CHAPTER I

1. **Monitor the growth of a bacterial population using visible light spectrophotometry.** This exercise can be accomplished in one day, with demonstrable exponential growth within a few hours. Demonstration of logistic growth will require data points taken over the course of the day or into the evening.

2. **Monitor the growth of a population of *Drosophila melanogaster*.** This is a long-term exercise, requiring periodic counting of flies over a couple of months. To simplify matters, (chiefly to avoid having the experimental organisms fly away while they are being counted), we recommend the use of a flightless mutant strain, such as *apterous*. This exercise can be modified by manipulating the initial population density—start several cultures each with a different number of flies.

REFERENCES

Carlson, T. (1913). Über Geschwindigkeit und Grösse der Hefevermehrung in Würze. *Biochemische Zeitschrift, 57,* 313–334.
Gleick, J. (1987). *Chaos. Making a new science.* New York: Penguin Books.
Hirsch, M., Smale, S., Devaney, R. (2003). *Differential equations, dynamical systems, and an introduction to chaos* (2nd ed.). New York: Academic Press.
Hoppensteadt, F. C., Peskin, C. S. (2002). *Modeling and simulation in medicine and the life sciences* (2nd ed.). New York: Springer-Verlag.
Lamb, H. H. (1995). *Climate, history, and the modern world* (2nd ed.). New York: Routledge.
Pearl, R. (1927). The growth of populations. *Quarterly Review of Biology, 2,* 532–548.
Pratt, D. M. (1943). Analysis of population development in *Daphnia* at different temperatures. *Biological Bulletin, 85,* 116–140.
Richards, O. W. (1928). Potentially unlimited multiplication of yeast with constant environment, and the limiting of growth by changing environment. *Journal of General Physiology, 11,* 525–538.
Smith, T. D., Link, J. S. (2005). Autopsy your dead…and living: A proposal for fisheries science, fisheries management, and fisheries. *Fish and Fisheries, 6,* 73–87.
Taubes, C. H. (2001). *Modeling differential equations in biology.* Upper Saddle River, NJ: Prentice Hall.
U.S. Environmental Protection Agency. (2000, April). *Global warming and our changing climate. Answers to frequently asked questions* (EPA Publication No. 430-F-00-011). Washington, DC: U.S. Government Printing Office.

INTERNET RESOURCES

www.berkeleymadonna.com
Home page for *BERKELEY MADONNA*. Download a fully functional trial version, and investigate the tutorials here.
www.mathworks.com

Home page for the suppliers of *MATLAB*.
http://www.iseesystems.com/index.aspx
Home page for the suppliers of *STELLA*.
http://www.vensim.com/
Home page for the suppliers of *Vensim*.
http://www.census.gov/population/censusdata/table-2.pdf
U.S. Census Bureau data site.

FURTHER READING

Mathematical model. (2006). In *Encyclopædia Britannica*, Retrieved September 21, 2006, from Encyclopædia Britannica Online: http://www.britannica.com/eb/article-9051377.

U.S. Census Bureau. (1993, August 27). *Selected historical decennial census population and housing counts. Population, housing units, area measurements, and density: 1790 to 1990.* Retrieved September 21, 2006, from http://www.census.gov/population/censusdata/table-2.pdf

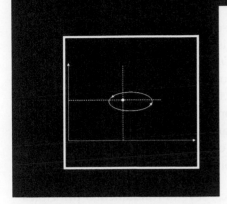

Chapter 2

COMPLEX DYNAMICS EMERGING FROM INTERACTING DYNAMICAL SYSTEMS

There is something fascinating about science. One gets such wholesale returns of conjecture out of such a trifling investment of fact.
Mark Twain (1835–1910)

In the first chapter, we modeled the behavior of several single-population biological systems. Few populations exist in isolation, however, so we now want to expand the range of biological phenomena we can investigate by modeling the interactions of systems with multiple groups. Our basic questions remain the same: What is the long-term behavior of the system, are there equilibrium states, and how do we classify them? Although our models will be necessarily more complex, in every example a single differential equation determines the evolution of each element of the system.

When two or more groups of organisms interact, a variety of relationships and outcomes are possible. Let us begin by exploring some of the possibilities.

When two or more groups of individuals of different species live in the same area, we describe the members of each species as belonging to a population and the group of populations as belonging to a community. Interactions among community members may take many forms. One population may feed upon another, or two may compete with each other to feed upon a third. Competition between populations, in fact, may occur over any valuable aspect of the habitat, including nesting sites or hiding places. Such competitions may be classified as *exploitative* or *resource competition*, where the two species use the same resources, or as *interference competition*, where the two species cause harm to each other.

Interactions between species need not always be harmful. *Mutualism*, which is also sometimes called *symbiosis*, occurs when two species benefit from living in close association with each other. A good example of mutualism occurs when an alga and a fungus combine to form a lichen and together wrest a living from extremely inhospitable environments, such as bare rock faces. Two related associations are *commensalism*, where one species benefits and the other is unharmed, and *amensalism*, where one species suffers but the other is unaffected.

Predations are interactions in which one species benefits while the other is harmed. In addition to the well-known example of *carnivory*, where one meat-eating species hunts and feeds on another animal, if the food organism is a plant, then the relationship is described as *herbivory*. *Parasitism* and *infectious disease* also fall within the definition of predation. In parasitism, one species benefits by living on or inside the body of another. With infectious diseases, a pathogenic microbe (a virus, bacterium, fungus, or protozoan) grows in or upon the body of the host organism, causing varying levels of harm to the host, ranging from annoyance (as with the common cold or athlete's foot) to death (as with AIDS, cholera, or bubonic plague).

Regardless of the difference in specifics, several fundamental principles are followed in the development of most models in which two or more groups interact. In this chapter, our goal will be to understand those principles as they apply to some of the classical epidemic and predator–prey models and emphasize their biological meaning and mathematical formulation.

I. INTRODUCTION TO INFECTIOUS DISEASE

A. Background

Throughout the centuries, more human lives have been lost to infectious diseases than to wars. A well-documented example is the "Spanish" influenza pandemic of 1918–1919, which killed at least 25 million people worldwide. Chillingly, a recent review of mortality figures by Johnson and Mueller (2002) increases the conservative estimate to 50 million. In contrast, deaths from World War I (which was then just ending) are estimated at 8.5 million. Bubonic plague, cholera, rabies, yellow fever, malaria, leprosy, and, most recently, AIDS, are humanity's lethal enemies. Because the cause of such ailments was unknown for many centuries, the fear of contagious diseases is rooted deeply in the human mind. It is hard to fight an "invisible enemy," and the fight against infectious diseases has been slow, dangerous, and heroic.

That some diseases are infectious was recognized long ago, and it was also known that the spread of such diseases could be restricted by isolating the affected persons and places. For thousands of years, isolation was the only effective tactic available in the fight with the invisible enemy.

Unfortunately, we are all familiar with the progression of an infectious illness, from exposure through the miserable symptoms to resolution, either by recovery or death. A mother wipes her infant's runny nose and inadvertently rubs her own nose while on her way to the sink to wash her hands. The viruses on her hand adhere to mucous membrane cells, and a new infection is initiated.

In the beginning, the host does not realize that she is infected. Following infection, the host enters a phase called the *incubation period*, during which the pathogenic microbe begins to multiply. The infected host shows no symptoms at all during this period, which usually lasts from several days to several weeks, but may be as short as a few hours or as long as 10 or 12 years. The length of the incubation period is characteristic of the particular infectious organism, within limits determined by the health of the host and the route of infection. The healthier the host and the longer the path the infecting organism must travel, the longer the incubation period. If the host is healthy, nonspecific defense mechanisms may be able to defeat some of the infecting organisms, thus slowing the progress of the infection. If the path traveled by the infecting organism is longer, it will take more time for the infection to occur because each step in the path (for example, from the mouth, through the gastrointestinal tract, and then into the bloodstream in the case of some *Salmonella* species) will require some amount of time to occur.

The host begins to feel ill during the *prodromal period* or *prodromium*. This is a short period of mild symptoms, which may be difficult to characterize as anything other than "not feeling quite right." The prodromium is rapidly followed by the *period of illness* or *period of invasion*, characterized by the most rapid reproduction of the pathogen. The host develops unmistakable symptoms, which may be quite severe. If the disease is serious and the immune response is weak, delayed, or absent, the host may die during this phase. The peak of this phase is called the *fastigium*.

If all goes well, however, the immune system will begin to bring the infection under control, and the infection will enter the *period of decline*. The host will begin to get better and enter the *convalescent period*, in which the immune system elements (the antibodies and cytotoxic or killer T cells) will be actively targeting and destroying the remaining infectious agents. With many diseases, the host will acquire long-lasting resistance to the disease, such that when she encounters the pathogen again, her immune system will mount such a rapid and strong reaction that an infection will not be established, and the host will not even realize that she has encountered the pathogen again. This resistance to recurring infection is called *immunity*.

For countless years, human beings had no choice but to suffer through the above course of infection and, if they were fortunate, survive. A major breakthrough was made in 1798, when the British physician Edward Jenner (1749–1823) developed a vaccine against smallpox by inoculating people with tissue from individuals infected with cowpox. Even Lord Byron, in his typical ironic manner, praised this great achievement:

> *With it the Doctor paid off an old pox*
> *By borrowing a new one from an ox...*
> Don Juan, *Canto the 1st, CXXIX.*

At about the same time, Daniel Bernoulli (1700–1782) published a mathematical model developed to assess the effect of cowpox inoculation on the spread of smallpox. This work is one of the earliest known scientific efforts to create a quantitative model for assessing the effects of treatment.

B. What is an Epidemic?

Infectious diseases are a fact of life, and ever since humans began to live in large groups, epidemics of infectious diseases have left their mark on human history. To define the term, an *epidemic* occurs when the number of infected individuals in the population increases as time goes by. This definition, however, does not at all convey the human tragedy it entails.

At No. 39 Broadwick Street in the Soho district of London, there is a pub named in honor of a physician, John Snow. A plaque on the wall of that establishment reads, "The Red Granite kerbstone marks the site of the historic BROAD STREET PUMP associated with Dr. John Snow's discovery in 1854 that cholera is conveyed by water." John Snow was a pioneer in the science we now call epidemiology, the study of health and disease in human populations.

In August of 1854, an outbreak of cholera occurred in London. In the period between August 30 and September 9, hundreds of people were stricken and died. John Snow observed that the cases were concentrated in the vicinity of the Broad Street water pump, and he convinced the local government officials to remove the handle from the pump, which they did on September 8. By this time, the peak of the epidemic had already passed (see Table 2-1).

Dr. Snow's action might have seemed irrelevant but for an independent investigation by Rev. Henry Whitehead, a local clergyman. Rev. Whitehead discovered that an infant living at No. 40 Broad Street had been stricken with diarrhea on August 29 and died on September 2. The wash water from the baby's dirty diapers had been dumped in the cesspool at the front of the house, just a few feet from the water pump. At the clergyman's urging, the cesspool was inspected and found to be leaking, contaminating the water from the pump. Further, since the infant's father was stricken with cholera on September 8, John Snow's actions likely prevented a second epidemic caused by drinking water from the Broad Street pump. Snow continued his investigations into the cause of cholera outbreaks and published, in 1855, *On the Mode of Communication of Cholera*, linking cholera to the consumption of water contaminated with fecal matter.

Contaminated drinking water is just one means by which infectious agents are transmitted. Some, such as the smallpox virus, may be

Date	Number of Fatal Attacks	Deaths
August 29	1	1
August 30	8	2
August 31	56	3
September 1	143	70
September 2	116	127
September 3	54	76
September 4	46	71
September 5	36	45
September 6	20	37
September 7	28	32
September 8	12	30
September 9	11	24
September 10	5	18
September 11	5	15
September 12	1	6

TABLE 2-1.
Cases of fatal cholera and cholera deaths in the vicinity of the Broad Street water pump, 1854.
(Excerpted from Snow [1855].)

transmitted by objects used by an infected person. Others, such as the malaria protozoan or the West Nile virus, are transmitted by insects. Still others are communicated by direct contact between infected individuals and susceptible individuals.

The duration, severity, and recurrence of epidemics may also vary widely. Some epidemics, such as the Spanish influenza epidemic of 1918, may develop rapidly, spreading terrible destruction within months and then tapering off (see Figure 2-1). Others may recur with relative regularity, as illustrated by the weekly case notification records of measles in England and Wales before mass vaccination was initiated (see Figure 2-2).

Despite the early work by Daniel Bernoulli, mathematicians were not seriously engaged in the fight against infectious diseases during the eighteenth and nineteenth centuries. In the early twentieth century, the British bacteriologist Ronald Ross (1857–1932) used mathematical modeling in his work with malaria and, more generally, in studying the spread of an infectious disease. Ross received the Nobel Prize in 1902. The now classical works (Kermack and McKendrick [1927; 1932; 1933]) built upon Ross's studies and examined the questions of when a disease will spread and how to find the threshold of an epidemic. In this text, it would be impossible to cover even a fraction of the literature on modeling infectious diseases that is available today.

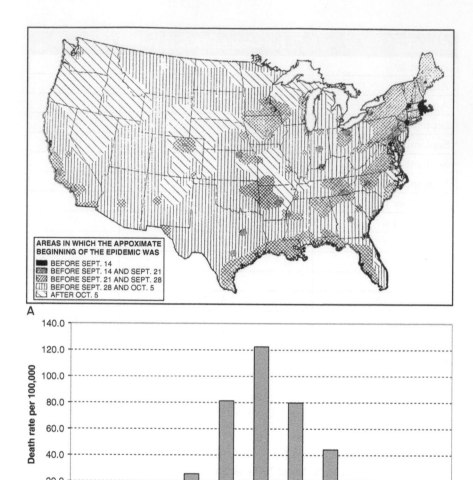

FIGURE 2-1.
The influenza epidemic of 1918 in the United States. Panel A: Territorial spread of the epidemic from September to early October 1918; panel B: Death toll in Richmond, VA, attributed to the influenza and pneumonia. (Panel A is from Crosby, A. W. [1989]. *America's forgotten pandemic: The influenza of 1918*. Cambridge, UK: Cambridge University Press. Used by permission. Data for panel B were obtained from the same source.)

II. THE SPREAD OF AN EPIDEMIC

A. Properties of the Mathematical Models

The models in this chapter will be continuous, and their behavior will be described by coupled differential equations. Typically, there will be two or more interacting groups and one differential equation per group describing how that particular group changes. Coupling means that at

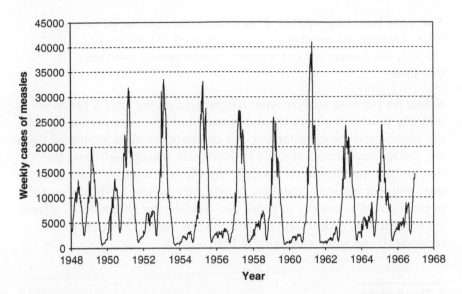

FIGURE 2-2.
Measles in England and Wales. Weekly case notifications of measles in England and Wales before general vaccination was initiated. (Data from http://www.zoo.ufl.edu/bolker/measdata.html.)

least some of the differential equations involve more than one variable. For example, we might have:

$$\frac{dA}{dt} = 2A - 3B$$

$$\frac{dB}{dt} = A + B.$$

This means that the behaviors of the groups A and B are linked.

To provide concrete examples, we shall consider two classical models from mathematical biology—the spread of an epidemic, and predator–prey models. In these examples (and many other problems in biology and chemistry), the dynamics depend on contact between groups. This dependency is manifested by the product of two groups in a differential equation. For example, in the spread of an epidemic, the two groups considered will be the susceptibles (S) and the infectives (I). We shall hypothesize that the susceptibles' rate of infection is proportional to the amount of contact they have with the infectives, and that the flow from S to I occurs at a rate αSI, where $\alpha > 0$ is a constant; that is, at a rate proportional to the product of the group sizes.

Why SI instead of something else? An intuitive way to understand this is to begin with the groups uniformly mixed, then double the number of infectives while keeping the number of susceptibles fixed. We should expect the number of contacts between the groups to then double.

The same reasoning applies if we double the number of susceptibles while keeping the number of infectives fixed. At the end of this chapter, we present a more rigorous mathematical derivation of this idea.

Before we begin to analyze what happens when we have a group of infective people in a population, we need to know several things about the disease and the environment. For example, do those recovered from the disease have immunity, or are they again susceptible? Does the disease have an incubation period? Will the infectives be isolated, by quarantine or natural separation, or will they be spread throughout the population? We consider in detail two models, the SIS model (*S* stands for susceptibles, *I* for infectives) and the SIR model (*R* stands for recovered), and some variations of both.

B. The SIS Model

I. Description of the SIS Model

Our first model will be quite simple. It assumes that the population is divided into two nonintersecting groups—the group of those who have the disease and can infect others (*I*) and the group of those who do not have the disease and can be infected (*S*). Our goal is to build a model that describes the change in the sizes of the susceptible and infective groups with time. We make the following assumptions:

1. The population is fixed and consists of *N* individuals. There are no births or deaths, and no one migrates into or out of the population.

2. There is no incubation period for the disease.

3. The two groups—susceptibles and infectives—are uniformly mixed within the population.

4. Once recovered from the disease, an individual is susceptible again; that is, there is no immunity.

These assumptions may seem unjustifiably restrictive, but our goal is to begin with the simplest model that will allow us to examine some important questions. Also, there are real situations in which these assumptions would be appropriate. Assumption 1, for example, would apply to the SARS epidemic that developed in an apartment building in Hong Kong, when the building was quickly sealed. For us, the SIS model will provide a start, and we shall then proceed to models that remove or relax its restrictive assumptions.

Assume that in a fixed population of size *N*, a small number $I(0)$ of individuals have somehow contracted an infectious disease. As time progresses, the infection may spread in the population, and we want to examine those changes with time. We denote:

$$S(t) = \text{the number of susceptibles at time } t,$$

$$I(t) = \text{the number of infectives at time } t.$$

With this notation, the assumption that the size of the population N remains fixed translates into $S(t) + I(t) = N$, so that $\dfrac{dS}{dt} + \dfrac{dI}{dt} = 0$. We need equations that describe how each of the groups changes. Earlier, we noted that the susceptibles become infected at the rate αIS. The number $\alpha > 0$ is called the *infection rate*. It can be interpreted as the probability that a particular susceptible is infected by a particular infective within a unit of time. We justify this interpretation in Section VIII.

Individuals who recover from the disease rejoin the group of the susceptibles immediately. If we assume that the infectives recover at a constant per capita rate β, then the recovery rate is βI. That is, the flow from I to S occurs at a rate βI. Accounting for outflow and inflow, the rate of change for the size of S is therefore given by:

$$\frac{dS}{dt} = -\alpha SI + \beta I. \tag{2-1}$$

Now, because $\dfrac{dS}{dt} + \dfrac{dI}{dt} = 0$, the rate of change for I will be:

$$\frac{dI}{dt} = \alpha SI - \beta I. \tag{2-2}$$

The mathematical model composed of Eqs. (2-1) and (2-2) is often called the *SIS model*. A pictorial representation of the model is given in Figure 2-3. The rectangles represent the different groups, and the arrows represent the flows between the groups. Each arrow is labeled with the rate of flow between the groups. Such diagrams are often helpful in formulating mathematical models.

We next focus on the meaning of parameter β and the long-term behavior described by the SIS model.

2. Interpreting the Parameter β

The per capita recovery rate β in the SIS model is related to the average length of the infection, $\bar{d}$, in the following way:

$$\bar{d} = \frac{1}{\beta}.$$

Thus, the smaller the value of β, the longer lasting the disease would be on average.

As an optional reading, we next present a mathematical justification for this relation. It requires a certain higher level of calculus proficiency, and its omission will not affect the subsequent sections.

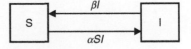

FIGURE 2-3.
Schematic representation of the SIS model. S and I represent the susceptible and infected populations, α is the infection rate, and β is the per capita recovery rate.

Assume that we examine a number of individuals $A_0 = A(0)$ who have contracted the disease at about the same time. Let $A(t)$ be the number of individuals who remain sick after time t. Because the per capita rate of recovery is β, the rate of change of $A(t)$ will be:

$$\frac{dA}{dt} = -\beta A(t).$$

The basic principle that we use is that the mean value of a group of numbers is computed as $\sum_t t p(t)$, where $p(t)$ is the proportion of scores that have the value t.

To compute the average length of the disease, we begin by dividing the interval $[0, \infty)$ into equal (small) subintervals by selecting $0 = t_0 < t_1 < t_2 < \ldots$, with $t_{n+1} - t_n = \Delta t$, for each $n = 0, 1, 2, \ldots$. (To be rigorous, we should consider the interval $[0, L]$ and then take the limit of our answer as $L \to \infty$.)

The number of individuals who recover between t_n and t_{n+1} is $A(t_n) - A(t_{n+1})$ and the approximate length of their infection is t_n. Therefore, the proportion of A_0 that become cured between t_n and t_{n+1} is $\dfrac{A(t_n) - A(t_{n+1})}{A_0}$ and, thus, the mean value of the infection is approximately:

$$\bar{d} \approx \sum_{n=0}^{\infty} t_n \left(\frac{A(t_n) - A(t_{n+1})}{A_0} \right) = \frac{1}{A_0} \sum_{n=0}^{\infty} t_n \left(\frac{A(t_n) - A(t_{n+1})}{\Delta t} \right) \Delta t.$$

Notice now that, as $\Delta t \to 0$, the approximation above improves. Further, because the expression $\left(\dfrac{A(t_n) - A(t_{n+1})}{\Delta t} \right)$ approaches $-\dfrac{dA(t)}{dt}$ and the sum $\dfrac{1}{A_0} \displaystyle\sum_{n=0}^{\infty} t_n \left(\dfrac{A(t_n) - A(t_{n+1})}{\Delta t} \right) \Delta t$ approaches

$\dfrac{1}{A_0} \displaystyle\int_0^{\infty} t \left(-\dfrac{dA(t)}{dt} \right) dt = -\dfrac{1}{A_0} \displaystyle\int_0^{\infty} t\, dA(t)$, we obtain that the average duration of the disease is given by:

$$\bar{d} = -\frac{1}{A_0} \int_0^{\infty} t\, dA(t).$$

To compute the integral above, we integrate by parts to get:

$\displaystyle\int_0^{\infty} t\, dA(t) = tA(t) \Big|_0^{\infty} - \int_0^{\infty} A(t)\, dt = -\int_0^{\infty} A(t)\, dt$. The last equality holds

because $\displaystyle\lim_{t \to \infty} t\, A(t) = 0$ and $0A(0) = 0$ imply $tA(t) \Big|_0^{\infty} = 0$ ($\displaystyle\lim_{t \to \infty} tA(t) = 0$ holds because $A(t)$ decays exponentially).

Now, because $\dfrac{dA}{dt} = -\beta A(t)$, we have $A(t) = -\dfrac{1}{\beta}\dfrac{dA}{dt}$ and:

$$\int_0^\infty A(t)dt = -\frac{1}{\beta}\int_0^\infty dA(t) = -\frac{1}{\beta}A(t)\Big|_0^\infty = \frac{1}{\beta}A(0) = \frac{A_0}{\beta}.$$

Thus, the average duration of the disease is:

$$\bar{d} = -\frac{1}{A_0}\int_0^\infty t\,dA(t) = -\frac{1}{A_0}\left(-\int_0^\infty A(t)dt\right) = \frac{1}{A_0}\left(\int_0^\infty A(t)dt\right) = \frac{1}{A_0}\frac{A_0}{\beta} = \frac{1}{\beta}.$$

3. The Long-Term Evolution of the Disease

As time passes and more susceptibles become infected and more infectives recover, what is the long-term behavior of the disease?

Notice that because of the condition $S(t) + I(t) = N$, Eq. (2-2) of the model could be written as:

$$\frac{dI}{dt} = \alpha I\left(S - \frac{\beta}{\alpha}\right) = \alpha I\left(N - I - \frac{\beta}{\alpha}\right).$$

Next, if $N - \dfrac{\beta}{\alpha} > 0$, we could further rewrite the right-hand side as

$$\alpha I\left(N - I - \frac{\beta}{\alpha}\right) = \alpha I\left(N - \frac{\beta}{\alpha} - I\right) = \alpha\left(N - \frac{\beta}{\alpha}\right)\left(1 - \frac{I}{N - \frac{\beta}{\alpha}}\right)I = r\left(1 - \frac{I}{K}\right)I,$$

where $K = N - \dfrac{\beta}{\alpha}$, and $r = \alpha\left(N - \dfrac{\beta}{\alpha}\right) = \alpha N - \beta > 0$. Thus, Eq. (2-2) takes the form:[1]

$$\frac{dI}{dt} = r\left(1 - \frac{I}{K}\right)I. \tag{2-3}$$

Does this equation look familiar? If $K = N - \dfrac{\beta}{\alpha} > 0$, the rate of change for the group of infectives is given by a logistic equation! This also means that we already *know* the long-term behavior for $I(t)$, because in Chapter 1 we studied the logistic equation in detail. Finally, knowing that $S(t) = N - I(t)$ allows us to derive the long-term behavior of $S(t)$ from that of $I(t)$. We present the results in the next two exercises.

EXERCISE 2-1

Show that for $N - \dfrac{\beta}{\alpha} < 0$ (or $N - \dfrac{\beta}{\alpha} = 0$), the number of infectives $I(t)$ is declining and, thus, there is no epidemic.

---

1. For Eq. (2-3), be sure to properly distinguish between the number 1 and the symbol I that denotes the number of infectives.

EXERCISE 2-2

Let $N - \dfrac{\beta}{\alpha} > 0$.

(a) Show that if $I(0) < N - \dfrac{\beta}{\alpha}$, then the number of infectives is increasing with time, and the number of susceptibles is decreasing with time. Show that $\lim_{t \to \infty} I(t) = N - \dfrac{\beta}{\alpha}$ and thus $\lim_{t \to \infty} S(t) = \dfrac{\beta}{\alpha}$.

(b) Show that if $N > I(0) > N - \dfrac{\beta}{\alpha}$, then the number of infectives is decreasing; the number of susceptibles is increasing; and, once again, $\lim_{t \to \infty} S(t) = \dfrac{\beta}{\alpha}$.

(c) What happens for $I(0) = N - \dfrac{\beta}{\alpha}$?

Figure 2-4 shows typical trajectories for the SIS model. For certain values of the parameters, the trajectories of both $S(t)$ and $I(t)$ stabilize at nonzero levels, meaning that the disease does not die out but remains endemic in the population. Thus, for situations where accounting for the continuing presence of disease is important, the SIS model may offer a good starting point for describing the dynamics of the epidemic.

EXERCISE 2-3

Criticize the SIS model. What assumptions were made to create the model that may not be quite realistic? Suggest improvements and refinements for the model.

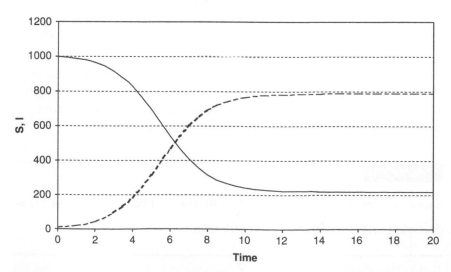

FIGURE 2-4.
Progression of disease in SIS model. Solution trajectories for $S(t)$ (solid line) and $I(t)$ (dashed line) for the SIS model with initial conditions $S(0) = 1000$ and $I(0) = 10$ and parameter values $\alpha = 0.001$, $\beta = 0.22$.

Summary. Before we examine another epidemic model, let us summarize our results with the SIS model:

1. The average lifetime of an infection is $\frac{1}{\beta}$.

2. When $N - \frac{\beta}{\alpha} > 0$, the disease remains endemic in the population.

3. When $N - \frac{\beta}{\alpha} \leq 0$, there is no epidemic.

C. The SIR Model

1. Description of the SIR Model

One of the assumptions we made for the SIS model was that, once recovered, an infective immediately becomes susceptible to the disease. Although there are certain diseases for which this is nearly true (e.g., gonorrhea and syphilis), there are others, such as chicken pox and measles, for which this assumption is not justified. For these diseases, individuals who have recovered from the infection have gained immunity, and the SIS model will not describe such infections accurately.

The SIR model, on the other hand, assumes that, once recovered, the person is immune to the disease and is no longer susceptible. This assumption necessitates a new group being added to the population—the group of recovered, R. The assumptions for the SIR model are:

1. The population is fixed. There are no births or deaths, and no one migrates into or out of the population.

2. There is no incubation period for the disease.

3. As in the SIS model, we assume that the susceptibles and infectives are uniformly mixed.

4. Once a person has recovered, he or she is permanently immune.

Again, these assumptions describe an idealized situation. As with the SIS model, we assume that a small number of members of a population have become sick with an infectious disease. We have three groups:

$S(t) = $ the number of susceptibles at time t;

$I(t) = $ the number of infectives at time t;

$R(t) = $ the number of recovered at time t.

The assumption that the population remains constant is:

$$S(t) + I(t) + R(t) = N,$$

and thus $\dfrac{dS}{dt} + \dfrac{dI}{dt} + \dfrac{dR}{dt} = 0$ for all t.

We need equations describing how each group changes. As with the SIS model, we assume the rate of new infections is given by αSI. In this case, however, there is no flow into S, because the recovered are immune. We then have:

$$\frac{dS}{dt} = -\alpha SI.$$

Individuals join the recovered group after they have been infected. We assume, as before, that the infectives recover at a constant per capita rate β. Then we would have:

$$\frac{dR}{dt} = \beta I.$$

The schematic representation of the SIR model is presented in Figure 2-5:

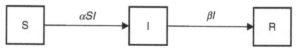

FIGURE 2-5.
Schematic representation of the SIR model. S, I, and R represent susceptible, infected, and recovered, α is the infection rate, and β is the per capita recovery rate.

EXERCISE 2-4

(a) Will $S(t)$ be increasing or decreasing? Give two reasons: one based on physical considerations, the other on knowing the derivative.

(b) Repeat part (a) for $R(t)$.

(c) Show that $\dfrac{dI}{dt} = \alpha SI - \beta I$.

In summary, the SIR model is described by the following system of differential equations:

$$\frac{dS}{dt} = -\alpha SI$$

$$\frac{dI}{dt} = \alpha SI - \beta I \qquad\qquad (2\text{-}4)$$

$$\frac{dR}{dt} = \beta I.$$

Notice that for the SIR model, the parameters α and β have the same meaning as in the SIS model.

2. Does an Epidemic Occur?

Recall that an epidemic occurs if the number of infectives in the population increases. We now examine this question for the SIR model.

For the SIR model described by Eqs. (2-4),

$$\frac{dI}{dt} = \alpha SI - \beta I = I(\alpha S - \beta).$$

Thus, the number of infectives will increase when $\alpha S(t) - \beta > 0$; that is, when $\alpha S(t) > \beta$, and decrease when $\alpha S(t) < \beta$. Because $S(t)$ is largest when $t = 0$, an epidemic will *not* take place if

$$\alpha S(0) - \beta < 0 \text{ or, equivalently, } S(0) < \frac{\beta}{\alpha}.$$

We shall see that this is related to the average number of new infectives that each infective causes (i.e., the number of secondary infections).

We would like to know how many susceptibles are infected by a typical infective. Upon reflection, we might decide that this depends on how many susceptibles are available and how long the infective is available.

EXERCISE 2-5

Are there any other assumptions or factors that should be considered in determining how many susceptibles an infective can infect?

To estimate the average number of secondary infections, consider the following example. Assume that the average number of infections caused by 1 infected individual per unit time is 3 per hour. Then, if the infective remains sick, on average, for 5 hours, he or she would infect

$$(5 \text{ hours}) \times (3 \text{ susceptibles per hour}) = 15 \text{ susceptibles.}$$

In the SIR model, the rate of new infections is given by $\frac{dS}{dt} = -\alpha SI$.

Recall that the outflow from S equals to the inflow to I. Thus, the rate of new infections is given by $\alpha SI = (\alpha S)I$, meaning that the (average) per capita infection rate at time t is αS.

If $I(0) = 1$, we obtain that at $t = 0$, $\frac{dS}{dt} = -\alpha S(0)I(0) = -\alpha S(0)$. Since $S(0)$ is the largest value of $S(t)$, one infected individual can infect, on average, no more than $\alpha S(0)$ susceptibles per unit time. Recall now that the average length of the infection is $\frac{1}{\beta}$. Thus, the average number of secondary infections produced by the single infective among the $S(0)$ susceptibles is at most $\frac{\alpha}{\beta} S(0)$. The number $\frac{\alpha}{\beta} S(0)$ estimates the *basic reproduction number* of the infection defined as the average number of secondary infections that a single infective can produce in a fully susceptible population.

We next determine the long-term behavior of $S(t)$, $R(t)$, and $I(t)$.

3. The Long-Term Evolution of the Disease for the SIR Model

We first show that according to the SIR model not everyone will catch the disease.

EXERCISE 2-6

Use $\dfrac{dS}{dR} = \dfrac{dS/dt}{dR/dt} = -\dfrac{\alpha}{\beta} S$ to show that if $R(0) = 0$, then $S(t) = S(0)e^{-\frac{\alpha}{\beta}R(t)}$.

Let $\lim_{t\to\infty} R(t)$ be denoted by $R(\infty)$. Because $R(t)$ is non-decreasing as a function of t, we obtain:

$$-\frac{\alpha}{\beta}R(t) \geq -\frac{\alpha}{\beta}N.$$

Combined with the result from Exercise 2-6, this yields:

$$S(t) = S(0)e^{-\frac{\alpha}{\beta}R(t)} \geq S(0)e^{-\frac{\alpha}{\beta}N} > 0.$$

Notice that because $S(t)$ is decreasing as time increases, and $R(t)$ is increasing, we know that $S(\infty) = \lim_{t\to\infty} S(t)$ and $R(\infty) = \lim_{t\to\infty} R(t)$ exist. Passing to a limit for t, when $t \to \infty$, we obtain $S(\infty) = \lim_{t\to\infty} S(t) \geq S(0)e^{-\frac{\alpha}{\beta}N} > 0$, showing that in the long run, a fraction of the population will never get infected.

EXERCISE 2-7

We have argued that not everyone will catch the disease. Is this intuitively plausible? If so, explain why. If not, explain some discrepancies between the model and reality.

Next, we show that eventually the number of infectives goes to 0. Notice that because $S(\infty)$ and $R(\infty)$ exist and $S(t) + I(t) + R(t) = N$ for all t, we obtain that:

$$I(\infty) = \lim_{t \to \infty} I(t) = N - S(\infty) - R(\infty)$$

exists as well. There are two possibilities for $I(\infty)$: either $I(\infty) > 0$ or $I(\infty) = 0$. We shall show that $I(\infty) = 0$.

Assume $I(\infty) > 0$.

Because $\dfrac{dR}{dt} = \beta I$, if $I(\infty) > 0$, then we would have

$$\lim_{t \to \infty} \frac{dR}{dt} = \beta I(\infty) > 0,$$

and then $R(t)$ would have to go to infinity (see Exercise 2-8, below). This is impossible, because $R(t) \le N$ for all values of t. Thus, it is impossible for $I(\infty) > 0$, and this implies that the alternative $I(\infty) = 0$ holds. That is, the disease dies out.

Combined with our earlier result that $S(\infty) > 0$, this means that the disease dies out because all infectives have been removed from the population, and not because all susceptibles have been infected.

EXERCISE 2-8

If $\lim_{t \to \infty} \dfrac{dR}{dt} > 0$, why does this mean $R(t)$ would have to go to infinity?

It is unusual to find actual populations as isolated as hypothesized in the SIR model. One famous example occurred at an English boarding school in 1978, as described in The Communicable Disease Surveillance Centre Report (1978). We present this example in Figure 2-6.

Summary. To summarize the results of the SIR model:

1. We have an epidemic if, and only if, $S(0) > \dfrac{\beta}{\alpha}$;

2. The average lifetime of an infection is $\dfrac{1}{\beta}$;

3. Under optimal conditions, the average number of secondary infections one infective can produce in a fully susceptible population is $\dfrac{\alpha}{\beta} S(0)$;

4. The disease dies out, and not all susceptibles will catch the disease.

In the SIR model, it is impossible to solve explicitly for $S(t)$, $I(t)$, and $R(t)$. This is typical for coupled systems of nonlinear differential equations. Furthermore, what is often most important to know is how one group responds to a change in the other groups. When only two groups are

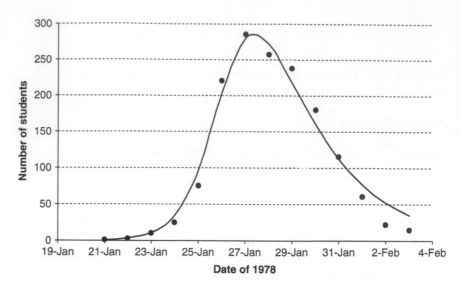

FIGURE 2-6.
Data from an influenza epidemic in a boarding school in England. The solid line represents the solution for the infectives (I) in a SIR model with $S_0 = 762$, $I_0 = 1$, $N = 763$, $\beta = 0.0022$, and $\alpha = 0.455$. (From a report by the Communicable Disease Surveillance Centre and the Communicable Disease [Scotland] Unit in the March 4 issue of the *British Medical Journal*, 1978, p. 587, reproduced with permission from the BMJ Publishing Group.)

involved, a *phase diagram* can be very helpful, and we describe this concept next.

III. PHASE PLANE ANALYSIS

A. The Phase Plane

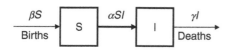

FIGURE 2-7.
Schematic representation of the model represented by Eq. (2-5). S, I, and α are as above, and β and γ represent the per capita rates of birth and death, respectively.

To demonstrate the phase plane technique, we consider a slightly modified SIS model with specific values for the constants. In this version, we allow for births to the susceptibles and consider the infectives to be removed by dying, meaning the population is no longer fixed. The schematic representation of the model is presented in Figure 2-7, and the differential equations are:

$$\frac{dS}{dt} = -\alpha SI + \beta S$$

$$\frac{dI}{dt} = \alpha SI - \gamma I, \tag{2-5}$$

where $\beta > 0$ is the per capita birth rate and $\gamma > 0$ is the per capita death rate (corresponding to the recovery rate of the earlier model).

Before proceeding further, we note an important mathematical fact and establish some vocabulary. The mathematical fact is this:

Theorem. *Suppose x and y are functions of t and* $\frac{dx}{dt} = f(x, y), \frac{dy}{dt} = g(x, y),$ *where f(x,y) and g(x,y) have continuous partial derivatives. Then for any point (x_0, y_0) there is a unique solution $(x(t), y(t))$ to the above system of equations with $x(0) = x_0$ and $y(0) = y_0$.*

This theorem is not nearly as complicated as the mathematical symbolism might make it appear. It simply says the rate of change of each variable depends only on the variables and not time, while the conclusion states that if the initial values of each variable are given, then there is exactly one way the process can evolve. All of the models that we consider will fulfill the hypotheses of this theorem. Such theorems in mathematics are referred to as *existence and uniqueness* theorems.

To gain some perspective on what the theorem means, we go back to the example and let S and I play the roles of x and y, respectively. The values of $\frac{dS}{dt}$ and $\frac{dI}{dt}$ are determined by S and I, rather than having t appear explicitly in the equations. This means that the model from Eq. (2-5) fits the hypotheses of the theorem.

We construct a Cartesian coordinate system, where the axes are the variables of interest—in our case, S and I.

Suppose at some time, usually $t = 0$, we know the values of S and I in the equations for $\frac{dS}{dt}$ and $\frac{dI}{dt}$ given above. Then we have a point $(S(0), I(0))$ in the (S,I) plane (also called the *phase plane*). We can also compute what $\frac{dS}{dt}$ and $\frac{dI}{dt}$ are at that time, so we know the direction of travel. For example, if the coordinates are as shown in Figure 2-8, and if $\frac{dS}{dt} > 0$ and $\frac{dI}{dt} < 0$, then S would be increasing and I would be decreasing. In this case, at this instant of time, we would associate an arrow that points down and to the right. Thus, for each time t, we have a point, $(S(t), I(t))$ in the (S,I) plane and a direction of travel. As t continues to change, we have a curve called a *trajectory* traced out in the (S,I) plane. A trajectory constructed as we have just described defines a solution to the system of equations. The theorem says:

1. Given any initial point in the (S,I) plane, there is a trajectory (solution) that starts at that point; and

2. Different trajectories never intersect.

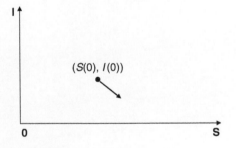

FIGURE 2-8.
Using the phase plane to determine direction of a trajectory. The arrow indicates direction of travel at a point where and at $t = 0$.

B. Constructing the Phase Plane

Consider now the following specific example for the model defined by Eqs. (2-5):

$$\frac{dS}{dt} = -0.002SI + 0.1S = S(-0.002I + 0.1)$$

$$\frac{dI}{dt} = 0.002SI - 0.4I = I(0.002S - 0.4). \tag{2-6}$$

In a phase plane diagram, we have a collection of arrows that describe how the trajectories travel. To construct a phase plane, we begin by finding the *null clines* for S and I; that is, where $\frac{dS}{dt} = 0$ and $\frac{dI}{dt} = 0$. For the first equation in our example,

$$\frac{dS}{dt} = S(-0.002I + 0.1) = 0;$$

the null clines are $S = 0$ or $I = \frac{-0.1}{-0.002} = 50$. If $\frac{dS}{dt} = 0$, there is no horizontal movement, so the arrows will be vertical. Likewise,

$$\frac{dI}{dt} = I(0.002S - 0.4) = 0$$

implies $I = 0$ or $S = \frac{0.4}{0.002} = 200$. Along the null clines for I, the arrows will be horizontal.

In Figure 2-9, we have sketched the null clines and arrows indicating horizontal or vertical travel. Along the line $S = 200$, the first equation from Eq. (2-6) implies that $\frac{dS}{dt} < 0$ when $I > 50$. Thus, above the line $I = 50$, the arrows will point to the left. For $I < 50$, the same equation shows that $\frac{dS}{dt} > 0$, which means that below the line $I = 50$ the arrows will point to the right. Similarly, along the line $I = 50$, we get $\frac{dI}{dt} < 0$ for $S < 200$ and $\frac{dI}{dt} > 0$ for $S > 200$. Therefore, along the line $I = 50$, the

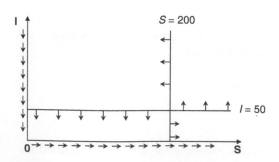

FIGURE 2-9.
Direction of travel near the null clines for the example described by Eq. (2-6).

arrows point down for $S < 200$ and up for $S > 200$. When $S = 0$, Eq. (2-6) implies that $\dfrac{dI}{dt} < 0$, and the arrows along the vertical axis point down. If $I = 0$, we get $\dfrac{dS}{dt} > 0$, and the arrows along the horizontal axis point right.

Points at which two null clines intersect have the property that both $\dfrac{dS}{dt} = 0$ and $\dfrac{dI}{dt} = 0$ at those points. This indicates that there is no motion at those points, and the trajectory will not move from there. Such points are called *equilibrium points* or *equilibrium states*. In this model, the equilibrium points are $(0,0)$ and $(200,50)$.

Because S and I represent members of a population, we are only interested in the part of the phase plane where $S \geq 0$ and $I \geq 0$. The null clines divide this region into four parts. We determine the direction of travel in each region by checking the sign of $\dfrac{dS}{dt}$ and $\dfrac{dI}{dt}$ at one point in that region. For example, in the region where $0 < I < 50$ and $0 < S < 200$, we could take $(10,20)$ as a test point. Then:

$$\left.\frac{dS}{dt}\right|_{(10,20)} = 10(-0.002(20) + 0.1) > 0$$

$$\left.\frac{dI}{dt}\right|_{(10,20)} = 20(0.002(10) - 0.4) < 0$$

indicates a direction to the right and down. We leave it to the reader to check that the signs of the derivatives for the other regions are as shown in Figure 2-10(A), meaning the directions are as shown in Figure 2-10(B).

The phase plane contains a wealth of information about how the process evolves in time, but with what we have done so far, we have an important unanswered question: If you study Figure 2-10(B), it seems that the equilibrium point $(200,50)$ has particular importance (and it does). The question is, how do the trajectories for our example behave regarding this point? Different behaviors, satisfying the conditions from Figure 2-10, are possible for the trajectories. For instance, they could spiral away from the equilibrium point [Figure 2-11(A)], spiral in [Figure 2-11(B)], or orbit around [Figure 2-11(C)].

In the next section, we describe how to determine the behavior of a trajectory near an equilibrium point.

IV. STABILITY OF EQUILIBRIUM POINTS

A. Equilibrium Points

Many processes in nature come to equilibrium, and many do not. For those that have a selection of equilibrium states, some of those states

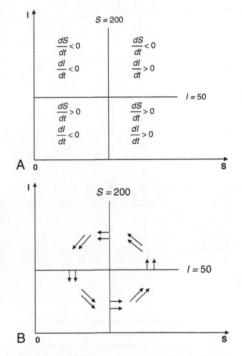

FIGURE 2-10.
Phase plane for Eq. (2-6). Panel A: Regions in the phase plane where the derivatives *dS/dt* and *dI/dt* for Eq. (2-6) do not change sign; panel B: The directions of movement in these regions.

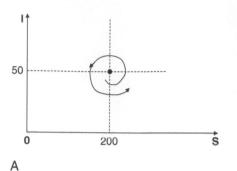

A

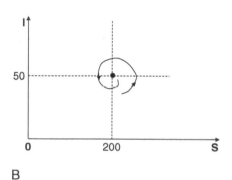

B

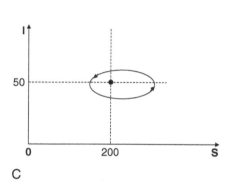

C

FIGURE 2-11.
Possible behaviors of the phase trajectories corresponding to the directional diagrams in Figure 2-10. Panel A: A trajectory that spirals away from the equilibrium point (200,50); panel B: A trajectory that spirals into the equilibrium point; panel C: A trajectory that orbits around the equilibrium point.

may never be observed in nature because they are unstable. We repeat an intuitive example by revisiting Figure 1-12 in Section VII of Chapter 1. Observe that points 1, 2, 3, and 4 are equilibrium points, of which 1 and 4 are unstable, and 2 and 3 are stable.

In a two-dimensional system, we can give a similar heuristic definition for stability, and the phase plane can sometimes be used to classify equilibrium states. For stable equilibrium points, all trajectories that initiate sufficiently close to those point "remain close" for any $t > 0$. This is not the case for unstable equilibrium points.

B. Finding the Equilibrium Points of a Two-Component System

Recall that for a one component system, we had:

$$\frac{dx}{dt} = f(x),$$

and found that the steady-states are values of x_0, where $f(x_0) = 0$.

In an analogous two-component system, if the components are x and y, the equations are

$$\frac{dx}{dt} = f(x,y),$$

$$\frac{dy}{dt} = g(x,y),$$

and the equilibrium states are points (x_0, y_0), where $f(x_0, y_0) = 0$ and $g(x_0, y_0) = 0$.

Example 2-1

If $x \geq 0$ and $y \geq 0$, find the equilibrium states for the system:

$$\frac{dx}{dt} = f(x,y) = y^2 - y + x - 1$$

$$\frac{dy}{dt} = g(x,y) = y - x.$$

SOLUTION:

We need to find the point (x_0, y_0) where $f(x_0, y_0) = 0$ and $g(x_0, y_0) = 0$ Thus, we need to solve the system of equations:

$$y^2 - y + x - 1 = 0$$
$$y - x = 0.$$

The second equations yields $y = x$. After substituting in the first equation, we obtain $x^2 - x + x - 1 = 0$, (i.e. $x^2 - 1 = 0$,). This equation has $x = 1$ and $x = -1$ as solutions. Since we have the restriction that

$x \geq 0$ and $y \geq 0$, the system therefore has $(x_0, y_0) = (1,1)$ as its only equilibrium state.

Next, we examine some analytic methods for determining the stability of equilibrium points.

C. Some Necessary Mathematical Background

In order to determine the stability of the equilibrium point for a system with two groups, we need to compute the partial derivatives of $f(x,y)$ and $g(x,y)$.

In computing the partial derivative of, say:

$$f(x,y) = x^2y^3 + 3x + 4y, \tag{2-7}$$

with respect to x, we treat y as if it were a constant and differentiate the function as if x were the only variable. The symbol for the partial derivative of f with respect to x is $\frac{\partial f}{\partial x}$. So, for Eq. (2-7):

$$\frac{\partial f}{\partial x} = 2xy^3 + 3.$$

Similarly, the partial derivative of f with respect to y is denoted by $\frac{\partial f}{\partial y}$. For Eq. (2-7), this partial derivative is:

$$\frac{\partial f}{\partial y} = 3x^2y^2 + 4.$$

Example 2-2

Let $f(x,y) = 2x^3y - 3xy$. Find $\frac{\partial f}{\partial x}$ and $\frac{\partial f}{\partial y}$ at the point (1,2).

SOLUTION:

We have:

$$\frac{\partial f}{\partial x} = 6x^2y - 3y,$$

so that at the point (1,2):

$$\frac{\partial f}{\partial x}(1,2) = 6(1^2)(2) - 3(2) = 6.$$

Likewise:

$$\frac{\partial f}{\partial y} = 2x^3 - 3x \text{ and } \frac{\partial f}{\partial y}(1,2) = 2(1^3) - 3(1) = -1.$$

The other items we need to compute to classify equilibrium points involve 2×2 matrices. For a 2×2 matrix:

$$J = \begin{pmatrix} a & b \\ c & d \end{pmatrix}$$

we have the *determinant* of J defined as $\det(J) = ad - bc$ and the *trace* of A, defined as $\text{trace}(J) = a + d$.

Example 2-3

Find the determinant and trace of:

$$J = \begin{pmatrix} 1 & 3 \\ 4 & -6 \end{pmatrix}.$$

SOLUTION:

We have:

$$\det(J) = 1(-6) - (3)(4) = -18,$$

$$\text{trace}(J) = 1 + (-6) = -5.$$

D. Stability of a Two-Component System

The setting is the same as before; namely, we have:

$$\frac{dx}{dt} = f(x, y),$$

$$\frac{dy}{dt} = g(x, y). \tag{2-8}$$

To determine the stability of the equilibrium point (x_0, y_0), we form the matrix J, called the *Jacobian*, where:

$$J = J(x_0, y_0) = \begin{pmatrix} \dfrac{\partial f}{\partial x}(x_0, y_0) & \dfrac{\partial f}{\partial y}(x_0, y_0) \\ \dfrac{\partial g}{\partial x}(x_0, y_0) & \dfrac{\partial g}{\partial y}(x_0, y_0) \end{pmatrix}$$

Definition. An equilibrium point (x_0, y_0) of the system of Eqs. (2-8) is called *stable* if for any region in the plane U that contains (x_0, y_0), there exists a smaller region V contained in U, such that all trajectories that initiate from V remain in U for all $t > 0$. An equilibrium point that is not stable is called *unstable*.

A subclass of stable points is of special importance.

Definition. An equilibrium point (x_0, y_0) of the system of Eq. (2-8) is called *asymptotically stable* when *all* trajectories that start in some region that contains the point (x_0, y_0) converge to the point (x_0, y_0) as t becomes large. A stable equilibrium point (x_0, y_0) that is not asymptotically stable is called *neutrally stable*.

Asymptotically stable equilibrium points are of particular importance, as they are most likely to be seen in natural systems. For this type of equilibria, the long-term behavior of the solution trajectories is insensitive to small changes in the initial values. Such equilibrium points are useful for representing the dynamics of many systems in biology, ecology, or medicine by allowing for "normal" variability of the initial conditions without affecting the long-term evolution of the system. The following theorem presents a criterion that allows us to determine whether a given equilibrium point is asymptotically stable.

Theorem. *With the notation above, the equilibrium point (x_0, y_0) is asymptotically stable if* $\det(J) > 0$, *and* $\mathrm{trace}(J) < 0$.

Example 2-4

Consider the spread of an epidemic represented by the diagram in Figure 2-12. This model differs from that defined by Eq. (2-5) in that the incoming flow to the group of susceptibles is constant, and is not dependent upon the size of the population.

We may think of parameter β as the rate of immigration into the group of susceptibles (e.g., periodically a plane full of immigrants free of the disease arrives at a village where the infectious disease is spreading).

The differential equations describing this model are:

$$\frac{dS}{dt} = -\alpha IS + \beta$$

$$\frac{dI}{dt} = \alpha IS - \gamma I. \tag{2-9}$$

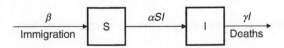

FIGURE 2-12.
Block-diagram representing a model for spread of an infectious disease in a system that allows for immigration at a constant rate β. At any time t, deaths occur at a fixed per capita rate γ and leave the system.

We consider the following special case:

$$\frac{dS}{dt} = -0.00001IS + 0.2$$

$$\frac{dI}{dt} = 0.00001IS - 0.02I.$$

We want to determine the equilibrium points and whether they are asymptotically stable. It is easy to find that this system of equations has one equilibrium point, $S = 2000$ and $I = 10$. Converting to the earlier notations, we can write:

$$f(S, I) = -0.00001IS + 0.2$$

$$g(S, I) = 0.00001IS - 0.02I,$$

and thus:

$$\frac{\partial f}{\partial S} = -0.00001I \qquad \frac{\partial f}{\partial I} = -0.00001S$$

$$\frac{\partial g}{\partial S} = 0.00001I \qquad \frac{\partial g}{\partial I} = 0.00001S - 0.02.$$

The Jacobian associated with the equilibrium state $(S,I) = (2000,10)$ is:

$$J = \begin{pmatrix} \dfrac{\partial f}{\partial S}(2000, 10) & \dfrac{\partial f}{\partial I}(2000, 10) \\ \dfrac{\partial g}{\partial S}(2000, 10) & \dfrac{\partial g}{\partial I}(2000, 10) \end{pmatrix} = \begin{pmatrix} -0.0001 & -0.02 \\ 0.0001 & 0 \end{pmatrix}.$$

Computing the trace and the determinant of J gives trace (J) $= -0.0001 < 0$ and det $(J) = 0.000002 > 0$. According to the criteria given in the previous theorem, the equilibrium point is asymptotically stable.

To summarize, we have now made a certain amount of progress in analyzing phenomena that are governed by the system of differential equations given by Eq. (2-8), where $f(x,y)$ and $g(x,y)$ have continuous partial derivatives. Namely, we know:

1. Given an initial point (x_0, y_0) there is exactly one solution to the system with initial point (x_0, y_0).

2. We can find the equilibrium points of the system.

3. We can distinguish between types of equilibrium points according to their stability.

When we study predator–prey models, we shall return to the idea of stability for two-component systems and give additional descriptions of the equilibrium states.

V. EPIDEMIC MODELS WITH DELAY AND MODELS WITH INTERMEDIATE GROUPS

A. Models with Delay

As already noted, one of the advantages of the SIR model is that it allows for immunity. However, the model is built on the assumption of permanent immunity upon recovery, and this assumption, although reasonable for certain diseases, is not accurate for others. For certain diseases, such as brucellosis or the sexually transmitted diseases gonorrhea, chlamydia, and syphilis, only limited immunity is conferred by having had the disease. This immunity is lost after some time interval has elapsed, and the length of this interval varies, depending on the disease. We shall model this situation using an SIR model with delay. As a first approximation, we assume that the length of this temporary immunity is constant—those who have recovered from the disease lose immunity after a fixed time $D > 0$.

Suppose, for example, that when an individual recovers he or she maintains immunity for 60 days. Consider how S is changing at a time t: as before, S is decreasing in size because of individuals who are becoming infected, and the rate at which this happens is proportional to $S(t)I(t)$. Now, however, S is also increasing because of the individuals who have recovered and have subsequently lost their immunity. The rate at which this happens is exactly the rate at which individuals moved from the infected to the recovered group 60 days before t, and this is $\beta I(t - 60)$. The block diagram representing this model is depicted in Figure 2-13. The delay of length D is represented by a triangle on the arrow from R to S.

The following equations mathematically describe these modifications to the SIR model:

$$\frac{dS}{dt} = -\alpha S(t)I(t) + \beta I(t - D)$$

$$\frac{dI}{dt} = \alpha S(t)I(t) - \beta I(t) \qquad (2\text{-}10)$$

$$\frac{dR}{dt} = \beta I(t) - \beta I(t - D).$$

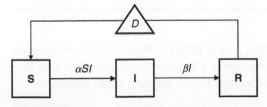

FIGURE 2-13.
Schematic representation of an SIR model with temporary immunity. The recovered individuals remain in R for a fixed time D that corresponds to temporary immunity. After time D, the recovered individuals are again susceptible to the infection.

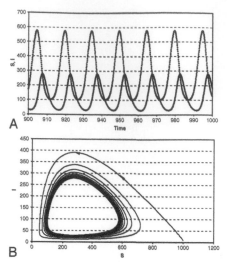

FIGURE 2-14.
Numerical solution of the SIR model with delay described by Eq. (2-10) that exhibits a robust periodicity for large values of t for $\alpha = 0.002$, $\beta = 0.55$, $D = 10$, and $I(t) = I(0)$ for t in the interval $[-D, 0]$. Panel A shows the time trajectories for S and I for large t while panel B depicts the (S, I)-phase trajectory.

An interesting feature of this model is that for certain values of the parameters (and for certain initial conditions) convergence to a steady periodic cycle is possible. This is demonstrated in Figure 2-14 by presenting the trajectories of $I(t)$ and $S(t)$ and a phase diagram of I versus S. This behavior of the solution may be interpreted in the following way. For many infectious diseases, periods of acute epidemic outbreaks may be separated by relatively quiet periods, such as in the measles data depicted in Figure 2-2. A mathematical model capable of describing such events should have oscillating trajectories. Note that in simulations involving delay, initial conditions for some of the functions must be provided for an entire interval of length $D > 0$.

B. Epidemic Models with an Intermediate State

For the SIR and the SIS models, we assumed that an infected person becomes instantaneously infectious. For many infectious diseases (such as measles and AIDS), a latent or incubation period exists, during which the person is infected but is not yet infectious. One way to modify the SIR model to incorporate this scenario is to introduce a new group E for individuals who are infected but not yet infectious. When an individual is infected, he or she is moved to group E and remains there until he or she becomes infectious.

If the incubation period does not vary significantly, we may assume that all individuals from group E move to the group of infectious I after a time period D, where D represents the approximate value of the incubation or latent period. This will result in a model with delay. For some diseases, this assumption will be close to the truth. For example, the incubation period for measles varies from 8 to 13 days, with the bulk of cases being close to 9 or 10 days. On the other hand, this assumption will clearly be unjustified if we consider a disease where the length of delay varies widely. For example, in AIDS the incubation periods range from months to decades.

Where the duration of the incubation period varies significantly, we may assume that a certain proportion of the exposed group becomes infectious over a unit time interval. This is similar to the assumption in the SIR model regarding the flow of infectives into group R. We leave the detailed mathematical description of these models as an exercise.

EXERCISE 2-9

Modify the SIR model by adding a new group E that consists of infected individuals who are not yet infectious. Give the block diagram and the mathematical equations. Consider the following two cases separately:

(a) All infectives spend a fixed time $D > 0$ in group E and are moved to the group of infectious I after that.

(b) Over a unit interval, a certain portion (say γ) of group E moves to the group of infectious I.

VI. PREDATOR–PREY INTERACTIONS

We now investigate what may happen to the sizes of populations when one species preys upon another. One might conjecture several different possible outcomes. The predator may eliminate the prey, and, unless it finds something else to eat, the predator would then follow its prey into extinction. The prey may evade the predator, and the predator would then starve. Finally, predator and prey may exist in a balance, with each population exerting some control over the other in a manner that maintains both populations.

As an example, let us consider the laboratory experiments conducted by the Russian microbiologist Georgii Frantsevich Gause (1910–1986). When Gause combined populations of *Paramecium caudatum*, a ciliated protozoan, and *Didinium nasutum*, a predatory protozoan, in a sediment-free medium, the *Didinium* ate all of the *Paramecium* and then starved to death. However, when he combined the two protozoans in a medium with sediment, the *Paramecium* were able to hide from the *Didinium* and the *Didinium* starved. Finally, if Gause periodically added *Paramecium* to the sediment-free medium, he was able to maintain populations of both predator and prey that would continue rising and falling for a few cycles (Gause [1934]). The last experiment was of particular interest because it indicated that predator and prey populations could coexist in a laboratory setting under careful control. A later experiment by Luckinbill (1973) demonstrated that by increasing the viscosity of the medium (which decreased the encounter rate between predator and prey), it is possible to obtain prolonged coexistence without the periodic "immigrations" of *Paramecium*. Figure 2-15 shows the population levels of *Paramecium aurelia* and *D. nasutum* in a water medium with methylcellulose added for increased viscosity. Here again, it is clear that coexistence involves cycles in the population sizes of both predator and prey.

The sustained oscillations observed in the population levels should not be surprising at a heuristic level. High prey levels naturally stimulate the growth of the predator population, which, in turn, causes a decline in the prey population and a subsequent decrease in the predator population due to the shortage of food resources. The low level of predators then allows for the prey population to increase, and the cycle repeats itself. We next present two more examples illustrating that coexistence of predator and prey populations involves cycles in the population numbers.

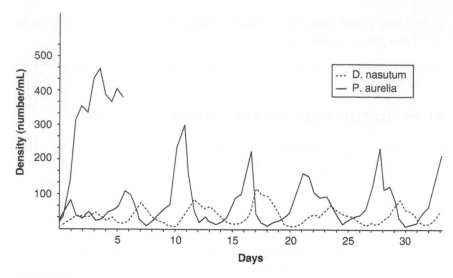

FIGURE 2-15.
Sustained oscillations of predator–prey populations. The plot in the far left is the control showing the increase of *Paramecium*, grown in the experimental medium, in the absence of *Didinium*. (From Luckinbill, L. S. [1973]. Coexistence in laboratory populations of *Paramecium aurelia* and its predator *Didinium nasutum*. *Ecology, 54*, 1320–1327. Used by permission.)

Utida (1957) studied bean weevil beetle (*Callosobruchus chinensis*) and wasp (*Heterospilus prosopidis*) populations. The wasps are parasites, laying their eggs in the weevil larvae. When a wasp egg hatches, the wasp larva eats the host weevil larva and destroys it. The number of beetle larvae therefore affects the number of wasp eggs that hatch and develop into adult wasps in the following generation. During the experiment, food resources for the beetles were not a constraint. Utida observed robust oscillations in the levels of the beetle and wasp populations, presented in Figure 2-16, with predator cycles lagging behind those of the prey.

In a nonlaboratory setting, similar interactions can be inferred from the Hudson Bay Company records of Canadian lynx and snowshoe hare populations. Although lynx also eat mice, voles, squirrels, and carrion, their preference for hare is so strong that many lynx starve to death when the hare population is low. The data in Figure 2-17 represent the fur catches of the company from the 1840s through the 1930s and are some of the few long-term records available in the literature. As it is reasonable to assume that the data represent a fixed fraction of the actual populations, the records can be viewed as a scaled plot of these populations.[2]

---

2. It is often argued that the data set depicted in Figure 2-17 is, in reality, composed of several different data sets collected between 1821 and 1939 (most ending in 1934) from different geographical regions and using different techniques (see, for example, Stenseth et al. 1997). Thus, the assumption that the fur catches represent a constant fraction of the populations may not be entirely accurate and should be made with caution.

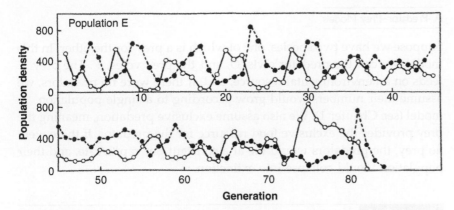

FIGURE 2-16.
Fluctuations in the levels of the azuki been weevil, *Callosobruchus chinensis*, (solid line) and wasp, *Heterospilus prosopidis*, (dashed line) populations. (From Utida, S. [1957]. Cyclic fluctuations of population density intrinsic to the host–parasite system. *Ecology, 38,* 442–449. Used by permission.)

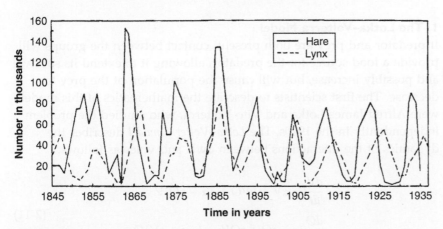

FIGURE 2-17.
Fur catch records for Canadian lynx and snowshoe hare from the Hudson Bay Company. (From *Fundamentals of ecology,* 3rd ed., by ODUM. [1971]. Reprinted with permission of Brooks/Cole, a division of Thomson Learning: www.thomsonrights.com. Fax: (800) 730-2215.)

The cyclic behavior of the predator–prey population levels requires that a delicate balance be maintained between the dynamic rates of change of the interacting populations. Notice, for example, how close the population levels in Figures 2-15, 2-16, and 2-17 come to the horizontal axis at times. If the predator population were driven to extinction, the prey would develop undisturbed. Alternatively, if the prey population were destroyed, the predators would also die off. Although numerous ecological and biological factors need to be simultaneously present to maintain balance, the system should also be robust enough to maintain its dynamics in the presence of common environmental noise. In what follows, we shall consider some mathematical models capable of generating oscillatory behavior and examine conditions under which the system is able to preserve this behavior in the long run.

A. Predator–Prey Models

Suppose we have two species, one of which is a prey for the other. In this model, we call the predator owls, O, and the prey voles, V. The prey feeds on an environmental resource, and, if there were no predators, we assume their numbers would grow according to a single population model (see Chapter 1). We also assume exclusive predation, meaning the prey provides the exclusive food resource for the predator. If there were no prey, the predators would die at a constant per capita rate, and their population would collapse into extinction.

Exercise 2-10

Under the conditions described above, what are some ways predator and prey populations could evolve?

1. The Lotka–Volterra Model

If predator and prey are both present, contact between the groups will provide a food source for the predator, allowing it to extend its survival and possibly increase, but will cause the population of the prey to decrease. The first scientists to describe the mathematics of this model were Alfred James Lotka and Vito Volterra, who studied the problem independently in the 1920s. The *Lotka–Volterra model* describes the dynamics of the interactions between owls and voles as follows:

$$\frac{dV}{dt} = \alpha V - \gamma VO = (\alpha - \gamma O)V$$

$$\frac{dO}{dt} = -\delta O + \varepsilon OV = (-\delta + \varepsilon V)O,$$

(2-11)

where α, γ, δ, and ε are positive constants. Notice that $(\alpha - \gamma O)$ has the meaning of the net per capita growth rate for the vole population and that $(-\delta + \varepsilon V)$ is the net per capita growth rate for the owl population. Thus, the parameters α, γ, δ, and ε have the following biological meanings:

α = net per capita growth rate for the voles in the absence of owls.

δ = per capita death rate of the owls if there is no food (i.e., voles).

γO = per capita death rate of the voles because of predation. The parameter γ represents the effectiveness of the owls as hunters.

εV = per capita birth rate for the owls because of the availability of food resources (the voles). The parameter ε represents the per capita rate at which the increase of the vole population contributes to increasing the owls' per capita birth rate.

According to the Lotka–Volterra model, both predator and prey population levels oscillate with time, with predators lagging behind prey. A typical time–trajectory plot and the corresponding phase plot are presented in Figure 2-18. The closed phase trajectories indicate strictly periodic cycles that repeat indefinitely. The trajectories do not converge to an equilibrium, and no trajectory drifts off to infinity. Instead, any change in the initial conditions gives rise to a new closed phase trajectory [see Figure 2-18(C) and (D)].

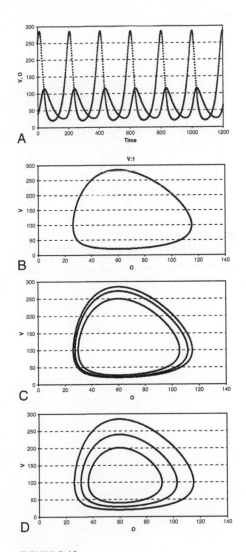

EXERCISE 2-11

(a) Determine the null clines for V and O in the Lotka–Volterra model from Eq. (2-11).

(b) How many equilibrium states does the Lotka–Volterra model have? List them.

(c) Draw a V versus O phase diagram indicating the directional fields in each of the regions between the null clines (see, for example, Figure 2-10).

(d) Based on your answers to (a)–(c) above and Figure 2-18, would you characterize the nontrivial equilibrium state of the Lotka–Volterra model as stable, unstable, or neutrally stable?

2. A Predator–Prey Model with Limited Growth

One weakness of the Lotka–Volterra model is that, in the absence of predators, it allows voles to multiply exponentially. A modified version of the model (2-11) that hypothesizes logistic growth for the voles with carrying capacity K would be:

$$\frac{dV}{dt} = \alpha\left(1 - \frac{V}{K}\right)V - \gamma OV = \alpha V - \beta V^2 - \gamma OV$$

$$\frac{dO}{dt} = -\delta O + \varepsilon OV,$$

(2-12)

where $\beta = \alpha/K$ and α now represents the inherent per capita net growth rate for the voles in the absence of owls. We again assume owls eat only voles, and, if there were no voles, owls would die at a constant per capita rate.

We describe the analysis of the phase plane, leaving the details as an exercise. The null clines for V, where $\frac{dV}{dt} = 0$, are:

$$V = 0 \text{ and } O = \frac{\alpha}{\gamma} - \frac{\beta}{\gamma}V.$$

FIGURE 2-18.
Typical time plots and phase plots for the Lotka–Volterra model. Parameter values: $\alpha = 0.06$, $\delta = 0.02$, $\gamma = 0.001$, $\varepsilon = 0.0002$. Panel A: Time trajectories for the vole (solid line) and owl (dashed line) populations with initial population sizes: $O_0 = 40$, $V_0 = 250$; panel B: Phase trajectory with initial population sizes $O_0 = 40$, $V_0 = 250$; panel C: Phase trajectories for initial population sizes $V_0 = 250$ and, from outside in, $O_0 = 40$, $O_0 = 60$, and $O_0 = 80$; panel D: Phase trajectories for initial population sizes $O_0 = 40$ and, from outside in, $V_0 = 250$, $V_0 = 200$, and $V_0 = 150$.

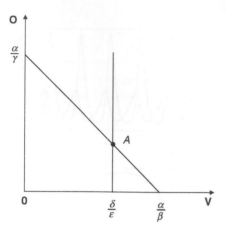

FIGURE 2-19.
Null clines for the model described by Eq. (2-12).

The null clines for O are:

$$O = 0, \text{ and } V = \frac{\delta}{\varepsilon}.$$

The null clines are shown in Figure 2-19.

They intersect inside the region $\{(V, O) : V > 0, O > 0\}$ only when $\frac{\delta}{\varepsilon} < \frac{\alpha}{\beta}$. We prove below that when this condition is satisfied, and thus the null clines intersect, the intersection point A is a stable equilibrium for model (2-12).

Before we do this, we want to discuss the biological meaning of the condition $\frac{\delta}{\varepsilon} < \frac{\alpha}{\beta}$ because our mathematical arguments suggest that when it is not satisfied, the model may exhibit radically different long-term behavior.

First, recall that $\beta = \frac{\alpha}{K}$, and so $\frac{\alpha}{\beta} = K$—the carrying capacity for the vole population in isolation (i.e., in the absence of owls). The condition $\frac{\delta}{\varepsilon} < \frac{\alpha}{\beta}$ now takes the form $\frac{\delta}{\varepsilon} < K$, or $\delta < \varepsilon K$, equivalently. Recall that δ is the owls' per capita death rate when no voles are present and that $\varepsilon V(t)$ is the owls' birth rate due to the presence of voles. Because $V(t)$ can never exceed its carrying capacity K (assuming no immigration), we have $V(t) < K$. Thus, the term εK represents the owls' maximal per capita growth rate controlled by food resources.

The condition $\delta < \varepsilon K$ then requires that the owls' per capita death rate caused by a lack of food be lower than the owls' maximal per capita birth rate caused by the presence of food. We shall see below that under this condition, the vole and owl populations both stabilize around nonzero equilibrium values. When this condition is not satisfied and $\delta > \varepsilon K$, this means [because $V(t) < K$] that at any moment t, $\delta > \varepsilon V(t)$, i.e., the owls' per capita death rate exceeds their per capita birth rate. The owl population will then die out. We shall see below that this is exactly what the model predicts.

Proceeding with the mathematical analyses of the equilibrium states, our previous notation yields:

$$\frac{dV}{dt} = f(V, O) = \alpha V - \beta V^2 - \gamma OV$$

$$\frac{dO}{dt} = g(V, O) = -\delta O + \varepsilon OV. \tag{2-13}$$

We calculate the partial derivatives for the functions $f(V, O)$ and $g(V, O)$ to get:

$$\frac{\partial f}{\partial V} = \alpha - 2\beta V - \gamma O \qquad \frac{\partial f}{dO} = -\gamma V$$

$$\frac{\partial g}{\partial V} = \varepsilon O \qquad \frac{\partial g}{\partial O} = -\delta + \varepsilon V.$$

EXERCISE 2-12

(a) Show that the equilibrium states for the model defined by Eq. (2-13) are $(0,0)$, $\left(\frac{\alpha}{\beta}, 0\right)$, and $\left(\frac{\delta}{\varepsilon}, \frac{\alpha}{\gamma} - \frac{\beta\delta}{\gamma\varepsilon}\right)$.

(b) Show that for the equilibrium point $A = (x_0, y_0) = \left(\frac{\delta}{\varepsilon}, \frac{\alpha}{\gamma} - \frac{\beta\delta}{\gamma\varepsilon}\right)$, we have:

$$\frac{\partial f}{\partial V}(x_0, y_0) = -\frac{\beta\delta}{\varepsilon} \qquad \frac{\partial f}{dO}(x_0, y_0) = -\frac{\gamma\delta}{\varepsilon}$$

$$\frac{\partial g}{\partial V}(x_0, y_0) = \frac{\alpha\varepsilon - \beta\delta}{\gamma} \qquad \frac{\partial g}{\partial O}(x_0, y_0) = 0.$$

Using the results obtained in Exercise 2-12, we form the Jacobian J and use it to analyze the stability of the equilibrium point $A = \left(\frac{\delta}{\varepsilon}, \frac{\alpha}{\gamma} - \frac{\beta\delta}{\gamma\varepsilon}\right)$ (see Figure 2-19). Calculating the determinant and the trace of this matrix, we obtain: $det(J) = \delta\alpha - \frac{\beta\delta^2}{\varepsilon} < \delta\beta\left(\frac{\alpha}{\beta} - \frac{\delta}{\varepsilon}\right)$ and trace $(J) = -\frac{\beta\delta}{\varepsilon}$.

Now we see that trace$(J) < 0$ and $det(J) > 0$ if $\frac{\alpha}{\beta} > \frac{\delta}{\varepsilon}$, which is necessary for the null clines in question to intersect. This shows that the equilibrium point $\left(\frac{\delta}{\varepsilon}, \frac{\alpha}{\gamma} - \frac{\beta\delta}{\gamma\varepsilon}\right)$ for the model defined by Eq. (2-12) is asymptotically stable. The point of using arbitrary constants was to show that as long as $\frac{\delta}{\varepsilon} < \frac{\alpha}{\beta}$, this equilibrium point in the modified Lotka–Volterra model is always asymptotically stable, regardless of the specific parameter values.

3. More on Classifying the Equilibrium States

We now revisit the classification of equilibrium states for the model defined by Eq. (2-8).

Recall that a point (x_0, y_0) is an equilibrium point if and only if $f(x_0, y_0) = 0$ and $g(x_0, y_0) = 0$. An asymptotically stable equilibrium point is one that attracts solutions any time a trajectory passes sufficiently close. Unstable equilibrium points may attract some trajectories, but not all. Among unstable equilibrium points is a special class called *repellers*. Any trajectory coming close to a repeller is forced away from it. Figure 2-20(B) shows a repeller (for simplicity, the trajectories are shown as straight lines).

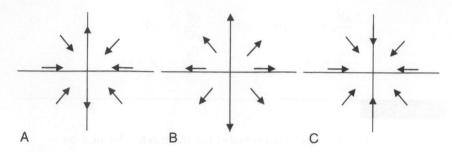

FIGURE 2-20.
Stable and unstable equilibria. Panel A depicts an unstable equilibrium, since the trajectories along the vertical axis move away from the equilibrium. Panel B represents a repeller, because all trajectories close to the equilibrium are forced away from it. Panel C depicts an asymptotically stable equilibrium, attracting all trajectories.

An analytic condition, similar to the criterion for asymptotic stability, is available to determine whether an equilibrium point is a repeller. As before, we use the Jacobian matrix:

$$J = \begin{pmatrix} \dfrac{\partial f}{\partial x}(x_0, y_0) & \dfrac{\partial f}{\partial y}(x_0, y_0) \\ \dfrac{\partial g}{\partial x}(x_0, y_0) & \dfrac{\partial g}{\partial y}(x_0, y_0) \end{pmatrix}.$$

The criterion, the first part of which we presented earlier, is as follows:

Theorem: *If $\det(J) > 0$, and $\mathrm{trace}(J) < 0$, then (x_0, y_0) is asymptotically stable. If $\det(J) > 0$ and $\mathrm{trace}(J) > 0$, then (x_0, y_0) is a repeller.*

The next predator–prey model we examine will exhibit cyclic or periodic behavior. Although there will be equilibrium states, the trajectories do not converge toward them. There is one more idea we need before we can proceed.

Definition. A set U in the (x,y) plane is called a basin of attraction for the system $\dfrac{dx}{dt} = f(x,y), \dfrac{dy}{dt} = g(x,y)$, if whenever a trajectory is in U at time t_0, it remains in U for all $t > t_0$. In other words, if a trajectory ever enters U, it stays there forever.

The main mathematical tool for our next example is:

Theorem (Poincaré–Bendixson). *Suppose $\dfrac{dx}{dt} = f(x,y), \dfrac{dy}{dt} = g(x,y)$, has U as a basin of attraction. Suppose that there is exactly one equilibrium point in U and that point is repelling. Then the system has a periodic solution that remains in U.*[3]

---

3. We call a solution $(x(t), y(t))$ periodic if is not an equilibrium state and there exists $T > 0$ such that $(x(t+T), y(t+T)) = ((x(t), y(t))$ for all t.

The image to have for this dynamic is suggested in Figure 2-21. $P(x,y)$ denotes the periodic solution. A trajectory that begins inside P will spiral outward toward P, and one that is outside P, but within the basin of attraction, will spiral inward toward P. Notice that the theorem says nothing about how to find the periodic solution.

4. Another Revised Predator–Prey Model

We next present an example showing an application of Poincaré–Bendixson's criterion. This example comes from *Mathematics for Dynamic Modeling* by Edward Beltrami (1987), and illustrates how the type of equilibrium may change because of changes in the values of the model parameters.

Again, owls (O) are the predators and voles (V) the prey. The assumption for the voles' growth in the absence of predators is the same as before (i.e., it follows the logistic equation). We change the assumption regarding how owls devour voles, so that instead of being proportional to VO, it is proportional to $\dfrac{VO}{1+V}$. The idea is that an owl can eat only so many voles before becoming sated. The owl then needs some time to digest before being ready to eat again. So, we have:

$$\frac{dV}{dt} = \alpha V - \beta V^2 - \gamma \frac{VO}{1+V}.$$

We also change the assumptions on the owls' rate of change to the following:

$$\frac{dO}{dt} = \delta O\left(1 - \frac{\varepsilon O}{V}\right).$$

Note that if V were a constant, the above equation would be a logistic equation for the owls' population growth with a stable equilibrium equal to V/ε. This equation reflects the fact that the carrying capacity of the predator population is proportional to the number of prey. If $\delta > 0$ and $\dfrac{\varepsilon O}{V} > 1$, the owls will have a negative per capita growth rate and will be dying out.

Following Beltrami and Taubes' *Differential Equations Modeling in Biology* (Taubes [2001]) and using their choice of constants to facilitate the computations, we consider the special case:

$$\frac{dV}{dt} = \frac{2}{3}V - \frac{V^2}{6} - \frac{VO}{1+V} = f(V,O)$$

$$\frac{dO}{dt} = \delta O\left(1 - \frac{O}{V}\right) = g(V,O). \tag{2-14}$$

As we see next, the stability of this system will depend on the value of δ.

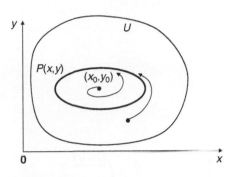

FIGURE 2-21.
Graphical representation of the Poincaré–Bendixson's criterion. U is a basin of attraction and (x_0, y_0) is a unique equilibrium point inside U that is a repeller. Any solution trajectories close to (x_0, y_0) are pushed away from it while forced to remain in the basin of attraction U. Poincaré–Bendixson's result asserts the existence of a periodic solution, that is, the existence of a closed cycle $P(x,y)$ in the phase diagram.

To find the null clines, we have that:

$$\frac{dV}{dt} = V\left(\frac{2}{3} - \frac{V}{6} - \frac{O}{1+V}\right) = 0 \text{ if } V = 0, \text{ or } \frac{2}{3} - \frac{V}{6} = \frac{O}{1+V}.$$

Solving the last equation gives:

$$\left(\frac{2}{3} - \frac{V}{6}\right)(1+V) = \frac{2}{3}\left(1 - \frac{V}{4}\right)(1+V) = O.$$

Thus, the first two null clines are $V = 0$ and $O = \frac{2}{3}\left(1 - \frac{V}{4}\right)(1+V)$.

From the second equation, we obtain $\frac{dO}{dt} = 0$ if $O = 0$ or $1 - \frac{V}{O} = 0$ which

implies $O = V$. Therefore, the other two null clines are $O = 0$ and $O = V$.

The null clines and directions of movement are shown in Figure 2-22.

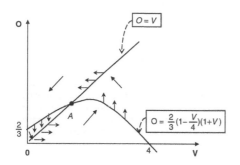

FIGURE 2-22.
Null clines and directions of movement for the model described by Eq. (2-14).

EXERCISE 2-13

Verify that the directions of movement along the null clines $O = V$ and $O = \frac{2}{3}\left(1 - \frac{V}{4}\right)(1+V)$ are as shown in Figure 2-22.

We are interested in the equilibrium point A—the intersection point of the straight line $O = V$ and the parabola $O = \frac{2}{3}\left(1 - \frac{V}{4}\right)(1+V)$. To determine the coordinates of the point A, we observe that the value where:

$$\frac{2}{3}\left(1 - \frac{V}{4}\right)(1+V) = V$$

is $V = 1$. So, in this example, the intersection point A is $(1,1)$.

We calculate:

$$\frac{\partial f}{\partial V} = \frac{2}{3} - \frac{1}{3}V - \frac{O}{(1+V)^2} \qquad \frac{\partial f}{\partial V}(1,1) = \frac{1}{12}$$

$$\frac{\partial f}{\partial O} = -\frac{V}{1+V} \qquad \frac{\partial f}{\partial O}(1,1) = -\frac{1}{2}$$

$$\frac{\partial g}{\partial V} = \delta\frac{O^2}{V^2} \qquad \frac{\partial g}{\partial V}(1,1) = \delta$$

$$\frac{\partial g}{\partial O} = \delta - \frac{2\delta O}{V} \qquad \frac{\partial g}{\partial O}(1,1) = -\delta.$$

The stability of point A is related to the matrix:

$$J = \begin{pmatrix} \dfrac{1}{12} & -\dfrac{1}{2} \\ \delta & -\delta \end{pmatrix}.$$

Now,

$$det(J) = \frac{5}{12}\delta \quad \text{and} \quad \text{trace }(J) = \frac{1}{12} - \delta.$$

Because $det(J) > 0$ for all values of δ, we obtain that $(1,1)$ is an unstable equilibrium point if $0 < \delta < \dfrac{1}{12}$ and stable if $\delta > \dfrac{1}{12}$. The unstable point is also repelling.

Next, we show that the square $0 < V < 4, 0 < O < 4$ is a basin of attraction.

Note that at $V = 4, \dfrac{dV}{dt} \le 0$, so a trajectory cannot cross that boundary going to the right. Similarly, at $V = 0, \dfrac{dV}{dt} = 0$, so no line can cross that boundary going to the left. In the same way, $\dfrac{dO}{dt} \le 0$ if $O = 4$ and $0 \le V \le 4$, and $\dfrac{dO}{dt} = 0$ if $O = 0$, so no solution can cross out of the square along those boundaries. Thus, if $0 < \delta < \dfrac{1}{12}$, we see that $(1,1)$ is a repelling point in the basin of attraction $0 < V < 4, 0 < O < 4$, and, since there are no other equilibrium points in this basin, according to the Poincaré–Bendixson theorem, we have a periodic solution.

VII. A MODEL OF COMPETITIVE INTERACTION

We present a final model describing two species competing for the same crucial resource. It seems obvious that if one species has a decided advantage, the other species will become extinct. However, in some situations, both species can coexist. Figure 2-23 gives the results of two types of yeast, *Sacharromyces cerevisiae* and *Saccharomyces kefir*, competing for a food supply. The experiment, conducted by G. F. Gause in 1932, shows mutual long-term coexistence with diminished saturation levels for both populations.

A well-designed model should be capable of capturing some of these possible long-term behaviors. As always, we begin with a few general assumptions:

1. The environment provides a limited food resource.

2. There are only two species competing for this resource.

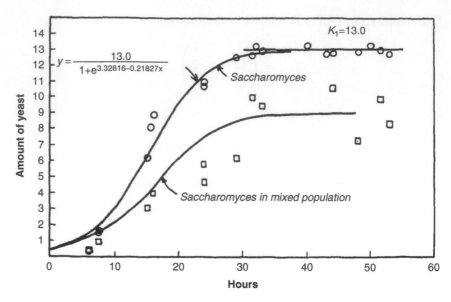

FIGURE 2-23.
Saccharomyces cerevisiae growing alone and in a mixed culture with a population of
Schizosaccharomyces kefir. (From Gause, G. F. [1932]. Experimental studies on the struggle for
existence. I. Mixed population of two species of yeast. *Journal of Experimental Biology, 9,* 389–402.
Used by permission.)

3. Both species use this resource as their only food supply.

We denote the two species by N and P and let $N(t)$ and $P(t)$ denote the
sizes of their populations at time t. As $N(t)$ and $P(t)$ change with time, we
begin with:

$$\frac{dN}{dt} = r(N,P)N$$

$$\frac{dP}{dt} = k(N,P)P.$$

Here $r(N,P)$ and $k(N,P)$ are the net per capita growth rates of N and P,
respectively. Those rates will depend on the sizes of the two
populations, as emphasized by the chosen notation. Our task will be to
find exact forms for $r(N,P)$ and $k(N,P)$, incorporating the assumptions of
our model.

EXERCISE 2-14

(a) Assuming food resources are limited, give a mathematical model
describing the growth of population N in the absence of type P
species ($P = 0$).

(b) Repeat, describing the growth of P in the absence of type N species.

(c) Do the equations you obtained look familiar? Compare them with Eq. (1-12) in Chapter 1.

(d) List the parameters of your models. Explain their biological meaning.

In Exercise 2-14, you should have found that each population grows to its carrying capacity in a logistic fashion when undisturbed by competitors. In the following exercise, we ask you to consider several preliminary questions and examine a model describing a competitive interaction.

EXERCISE 2-15

(a) Let $P \neq 0$. With $r(N,P)$ denoting the per capita growth rate of N, how do you expect $r(N,P)$ to change when P increases? Why?

(b) Let $N \neq 0$. With $k(N,P)$ denoting the per capita growth rate of P, how do you expect $k(N,P)$ to change when N increases? Why?

(c) The following model may be used to describe competition between N and P, where K and M are the carrying capacities for the populations N and P respectively:

$$\frac{dN}{dt} = r(N,P)N = a\left(1 - \frac{N+bP}{K}\right)N$$

(2-15)

$$\frac{dP}{dt} = k(N,P)P = c\left(1 - \frac{P+gN}{M}\right)P$$

where $r(N,P) = a\left(1 - \frac{N+bP}{K}\right)$ and $k(N,P) = c\left(1 - \frac{P+gN}{M}\right)$. Notice

that when $P = 0$, the first equation becomes $\frac{dN}{dt} = a\left(1 - \frac{N}{K}\right)N$ (i.e.,

a logistic equation for N with carrying capacity K). The parameter b measures the competitive effect of P on N, and the parameter g measures the competitive effect of N on P. Describe the meaning of the terms $N + bP$ in $r(N,P)$ and $P + gN$ in $k(N,P)$.

EXERCISE 2-16

Show that the equilibrium states for the model defined by Eqs. (2-15) are:

$(0,0)$, $(0,M)$, $(K,0)$, and $\left(\dfrac{K - bM}{1 - bg}, \dfrac{M - gK}{1 - bg}\right)$.

The equilibrium state $\left(\dfrac{K - bM}{1 - bg}, \dfrac{M - gK}{1 - bg}\right)$ determined in Exercise (2-16) will be of interest only when both of its coordinates are non-negative. There are various cases that need to be examined, and the computations become technical and somewhat tedious. Complete classification of the equilibrium states as stable or unstable will also require the examination of all such cases. It can be shown that (0,0) is always unstable (see Exercise 2-17) and that the other three equilibrium states could be stable or unstable depending on the values of the parameters. Figure 2-24 gives examples that demonstrate convergence to the states $\left(\dfrac{K - bM}{1 - bg}, \dfrac{M - gK}{1 - bg}\right)$, (0,$M$), and ($K$,0), respectively. Convergence to the state $\left(\dfrac{K - bM}{1 - bg}, \dfrac{M - gK}{1 - bg}\right)$ demonstrates coexistence of the populations, while convergence to either of the other states corresponds to one of the populations dying out. Notice that the initial conditions, the carrying capacities, and the values of the parameters $a = 0.03$ and $c = 0.04$ are the same for all fours panels in the figure. The difference in the long-term behavior of the models is due to the difference in the competition parameters b and g.

EXERCISE 2-17

Show that the equilibrium state (0,0) is always unstable, regardless of the values of the parameters.

EXERCISE 2-18

In the model defined by Eq. (2-15), for $K = 2000$, $M = 1000$, and $g = 0.51$, prove that the equilibrium state (K, 0) is asymptotically stable, regardless of the values of a, b, and c.

Before leaving this topic, we note there are many other models describing competition among species. Though the model we considered is quite simple, one could contemplate further refinements. The methods we used to develop and analyze this model were familiar and could also be used to develop models of symbiotic interactions between species.

VIII. APPENDIX: VALIDATION OF A MATHEMATICAL CLAIM

The epidemic models models considered in this chapter were developed assuming the groups are uniformly mixed and the amount of contact between two groups is proportional to the product of the number of

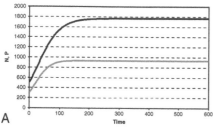

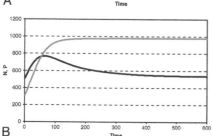

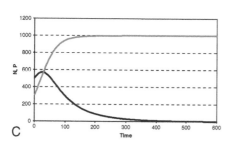

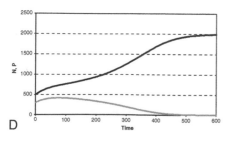

FIGURE 2-24.
Numerical solutions for *N* (black line) and *P* (gray line) of the competition model defined by Eq. (2-15) with initial conditions $N(0) = 500$ and $P(0) = 300$ and carrying capacities for the two populations of $K = 2000$ and $M = 1000$. The values for $a = 0.03$ and $c = 0.04$ are the same for all four panels. Panel A: $b = 0.25$, $g = 0.04$; panel B: $b = 1.5$; $g = 0.04$; panel C: $b = 2.7$; $g = 0.04$; panel D: $b = 2.7$; $g = 0.8$.

elements in each group. To complement the heuristic justification for this assumption presented earlier, we present its mathematical explanation. We begin by outlining the basic ideas; the first two come from probability, while the third uses calculus.

A. Idea 1

If we roll a standard 6-sided die, the probability that a particular number, say 4, comes up is 1/6. The probability that a 4 does not come up is $1 - 1/6$. If we roll 10 dice, the probability that none of the 10 dice is a 4 is $\left(1 - \dfrac{1}{6}\right)^{10}$.

The spread of an epidemic model is a good example of a problem where we need to measure the contact between groups. In what follows, when a susceptible and an infective interact we mean that the two contact one another and the disease is passed to the susceptible. We hypothesize that the probability that a particular susceptible and a particular infective interact in a unit of time is known, and we denote it by p. (Note that p is a number between 0 and 1. In most cases, it will be close to 0.) So the probability that two random members do *not* interact is $1 - p$.

Now keep the susceptible element fixed—call it s^*. Suppose we denote the number of infectives by I. Then the probability that s^* does *not* interact with any of the infectives is $(1 - p)^I$. In this description, we are assuming that both populations are uniformly mixed. Is this assumption reasonable?

B. Idea 2

Suppose that a basketball player makes 70% of her free throws over the course of a season. If she attempts 20 free throws in a game, we would expect her to make about:

$$20 \times 0.70 = 14$$

of these. This does not mean that she will make exactly 14 free throws in every game where she has 20 attempts, but it does mean that if she had many games in which she shoots 20 free throws, we would expect that the average number of successes would be close to 14.

This is an example of the following principle: If we have an experiment such that on each trial the probability of success is p, and we do k trials of the experiment, then the expected number of successes is kp.

Back to our example: Suppose we have two populations S and I, and we want to know the expected number of susceptibles that do *not* interact with an infective.

We assume:

1. The probability that a particular susceptible is infected by a particular infective is p (again implying the groups are uniformly mixed).

2. The number of susceptibles is denoted by S and the number of infectives by I.

With these assumptions, the expected number of susceptibles that do *not* interact with an infective is $S(1 - p)^I$.

C. Idea 3

There are some facts from calculus we now apply:

1. If x is close to 0, then $e^x \approx 1 + x$;

2. If p is close to 0, then $\ln(1 - p) \approx -p$;

3. If $x > 0$, then $x^Y = e^{\ln(x^Y)} = e^{Y \ln(x)}$.

We want to find an expression for $S(1 - p)^I$ that is easier to work with. From 3 above:

$$(1 - p)^I = e^{I \ln(1 - p)}.$$

Now, if p is close to 0, then $\ln(1 - p)$ is close to $-p$. Thus:

$$e^{I \ln(1 - p)} \approx e^{-Ip}$$

If, in addition to p being close to 0, Ip is also close to 0, then we would have:

$$e^{-Ip} \approx 1 - Ip$$

Therefore, under these assumptions,

$$S(1 - p)^I \approx S(1 - pI).$$

So, where does the differential equation come from? We explain this through a discrete model. To construct our difference equation, we let:

$$S_n = \text{the number of susceptibles at the } n\text{-th stage.}$$

$$I_n = \text{the number of infectives at the } n\text{-th stage.}$$

The disease spreads when a susceptible interacts with an infective. So S_{n+1} is the number of susceptibles in the n-th stage who did *not* interact with an infective. According to our work above:

$$S_{n+1} = S_n(1 - pI_n) = S_n - pS_nI_n,$$

leading to the difference equation:

$$S_{n+1} - S_n = -pS_nI_n.$$

This converts to $\dfrac{dS}{dt} = -pSI$ in the continuous model. The main idea of this conversion is outlined next.

For a very small number h, choose a value of t and pick the integer n for which nh is as close to t as possible (see Figure 2-25). Let $S_n = S(nh)$ and $I_n = I(nh)$. Then $S_{n+1} - S_n = S((n+1)h) - S(nh)$ and $S_nI_n = S(nh)I(nh)$.

Note that under the model assumptions, the probability of a susceptible being infected by an infective in an interval of time h is now approximated by ph. Thus:

$$S(nh + h) - S(nh) \approx hpS(nh)I(nh).$$

Because nh is very close to t, this is approximately:

$$S(t + h) - S(t) \approx -hpS(t)I(t).$$

So:

$$\frac{S(t + h) - S(t)}{h} \approx -pS(t)I(t).$$

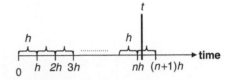

FIGURE 2-25.
Locating t in a subinterval. Fix a number $h > 0$ and consider the intervals $[0,h]$, $[h,2h]$, $[2h,3h]$, and so on. Next, choose a value $t > 0$, and find the integer n for which nh is closest to t.

Taking the limit as $h \to 0$ gives

$$\frac{dS}{dt} = \lim_{h\to 0}\frac{S(t + h) - S(t)}{h} = -pS(t)I(t).$$

We thus obtained that the rate of change from S to I is given by the differential equation $\dfrac{dS}{dt} = -pS(t)I(t)$ exactly as assumed in the SIS model from Eqs. (2-1) and (2-2) and the SIR model defined by Eqs. (2-4) (with p in place of α). As p is the probability that a particular susceptible is infected by a particular infective in a unit length of time, this also confirms the interpretation of the parameter α given in Section II.B.1.

REFERENCES

Beltrami, E. (1987). *Mathematics for dynamic modeling.* New York: Academic Press.

Crosby, A. W. (1989). *America's forgotten pandemic: The influenza of 1918.* Cambridge, UK: Cambridge University Press.

Gause, G. F. (1927). Experimental studies on the struggle for existence. I. Mixed population of two species of yeast. *Journal of Experimental Biology, 9,* 389–402.

Johnson, N. P. A. S., & Mueller, J. (2002). Updating the accounts: Global mortality of the 1918–1920 "Spanish" influenza pandemic. *Bulletin of the History of Medicine, 76,* 105–115.

Kermack, W. O., & McKendrick, A. G. (1932). Contributions to the mathematical theory of epidemics. *Proceedings of the Royal Society A: Mathematical, Physical & Engineering Sciences, 138,* 55–83.

Kermack, W. O., & McKendrick, A. G. (1927). Contributions to the mathematical theory of epidemics. *Proceedings of the Royal Society A: Mathematical, Physical & Engineering Sciences, 115,* 700–721.

Kermack, W. O., & McKendrick, A. G. (1933). Contributions to the mathematical theory of epidemics. *Proceedings of the Royal Society A: Mathematical, Physical & Engineering Sciences, 141,* 94–122.

Luckinbill, L. S. (1973). Coexistence in laboratory populations of *Paramecium aurelia* and its predator *Didinium nasutum. Ecology, 54,* 1320–1327.

Stenseth, N. C., Falck, W., Bjørnstad, O.N, & Krebs, C. J. (1997). Population regulation in snowshoe hare and Canadian lynx: Asymmetric food web configurations between hare and lynx. *Proceedings of the National Academy of Sciences of the United States of America, 94,* 5147–5152.

Taubes, C. H. (2001). *Modeling differential equations in biology.* Upper Saddle River, NJ: Prentice Hall.

The Communicable Disease Surveillance Centre and the Communicable Diseases (Scotland) Unit Report. (1978). Influenza in a boarding school. *British Medical Journal, March 4,* 587.

Utida, S. (1957). Cyclic fluctuations of population density intrinsic to the host-parasite system. *Ecology, 38,* 442–449.

FURTHER READING

Bernoulli, D. (1760). Essai d'une nouvelle analyse de la mortalité causée par la petite vérole, et des avantages de l'inoculation pour la prévenir. *Histoire de l'Acad. Roy. Sci. (Paris) avec Mém. des Math. et phys., Mém.,* 1–45.

Gause, G. F. (1964). *The struggle for existence.* New York: Macmillan (Hafner Press).(Original work published 1934).

Lotka, A. J. (1956). *Elements of physical biology.* New York: Dover Publications. (Original work published 1925).

Murray, J. D. (1993). *Mathematical biology* (2nd corrected ed.). New York: Springer-Verlag.

Odum, E. P. (1953). *Fundamentals of ecology.* Philadelphia: Saunders.

Snow, J. (1855). *On the mode of communication of cholera.* London: John Churchill.

Volterra, V. (1926). Fluctuations in the abundance of a species considered mathematically. *Nature, 118,* 558–560.

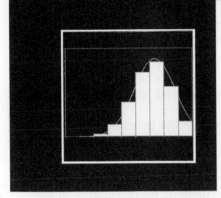

Chapter 3

MATHEMATICS IN GENETICS

Science never solves a problem without creating ten more.
George Bernard Shaw (1856–1950)

In this chapter, we continue examining how populations evolve, focusing on genetic changes. Two apparently contradictory characteristics are observable: a tendency for populations to preserve variability by maintaining the genetic status quo, versus the presence of continuing genetic adaptation and change. These characteristics represent the fundamental ideas of Mendelian genetics and Darwinian evolution that coexist in synthesis in contemporary biology. We shall consider mathematical models describing the maintenance of genetic variability in a population and analyze the change in the genetic constitution of a population subjected to natural selection.

I. INTRODUCTION

A. Early Experiments

The Earth is home to a tremendous variety of living organisms, but what is the origin of this diversity? By observing the similarities and differences among finches in the Galapagos Archipelago, Charles Darwin (1809–1882) came to question the prevailing idea that species are immutable, and, after years of work, came to the conclusion that species evolved over time. In 1859, he published his revolutionary book *On the Origin of Species by Means of Natural Selection*. Darwin, however, had no understanding of the molecular mechanisms that created these variations or of the cellular mechanisms of heredity which passed these variations onto the following generations. Unknown to Darwin, while he was writing his book, experiments were underway in a monastery garden that were to shed light on the rules of heredity.

Gregor Mendel (1822–1884), an Augustinian monk and schoolteacher, sought to understand the rules governing the inheritance of different characteristics by designing experiments using the garden pea, *Pisum sativum*. His careful selection of seven traits with alternate forms, his monitoring of the generations produced, and his counting of the different types of offspring allowed him to deduce the fundamental laws of genetics. Mendel took true-breeding varieties of peas and crossed those with well-defined alternative forms. He made multiple fertilizations of each cross, and his first observations were that the hybrids forming the F_1 or *first filial generation* all resembled one of their two parents. When he crossed round peas with wrinkled peas, for example, the F_1 hybrids were all round, so he called this round characteristic *dominant* and the wrinkled characteristic that disappeared *recessive*.

Mendel then bred a second generation from the hybrid generation. Interestingly, the parental characteristic that had disappeared in the F_1 generation reappeared in this new F_2 or *second filial* generation in a ratio of 3 dominant to 1 recessive. This was true for each of the seven traits Mendel examined. Figure 3-1 gives a schematic representation of the experiments. Mendel further observed that while all of the recessive F_2 individuals would breed true, only one third of the dominant individuals would breed true, while the remaining two thirds would give rise to dominant and recessive offspring in the previously observed 3:1 ratio.

From these observations, Mendel came to the conclusion that the alternative forms of each characteristic were passed along to the offspring unchanged. He concluded that the alternative forms, which he called "particulate factors" and we now call *alleles*, were separated or segregated in the formation of gametes, such that each gamete contained only one particulate factor (allele) for each character. This idea is called the *Principle of Segregation* or *Mendel's First Law*. The pair of alleles inherited by the offspring would determine its appearance. A pea plant with two recessive alleles would have the recessive appearance, whereas a pea plant with one or two dominant alleles would have the dominant appearance. Thus, Mendel differentiated between the appearance of the organism, called *phenotype*, and its genetic constitution, called *genotype*.

Mendel also made crosses between plants that differed in two traits, such as seed shape and plant height, called *dihybrid* crosses. He observed

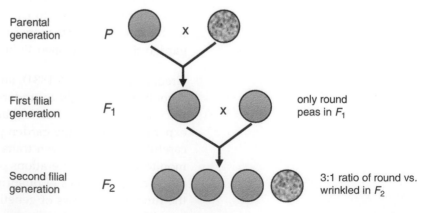

FIGURE 3-1.
Diagram illustrating Mendel's pea experiments and results.

that the alleles for the two different traits are distributed independently of each other in the formation of gametes. This idea is called the *Principle of Independent Assortment* or *Mendel's Second Law*. More information about Mendel's experiments can be found in traditional genetics texts, such as Russell (2006).

Mendel's work was an early application of quantitation to biology. His appreciation of the value of large sample sizes, which anticipated modern statistics, prompted him to obtain thousands of peas in his experiments and allowed him to formulate the explanation that alleles occurred in pairs and that each sperm or egg delivered one allele to the offspring. It is remarkable that he formed this model without any knowledge of the underlying physical and chemical processes. He never observed cell division and never saw a chromosome—the physical carriers of genetic information, where his particulate factors were to be found.

In 1865, Mendel presented a paper on his experiments and his interpretation of the results to the Natural Science Society in Brünn, and the following year published "Versuche über Pflanzen-hybriden" ("Experiments in Plant Hybridization") in the Society's journal. Unfortunately, the scientific community paid little attention to Mendel's work, and when he was named head of the monastery, his experimental career ended. Mendel died in 1884.

It was not until 1900 that Mendel's unparalleled achievements were independently resurrected by Hugo De Vries (1848–1935), Eric von Tschermak (1871–1962), and Carl Correns (1864–1933), and Mendel was recognized as the father of genetics. This new appreciation of Mendel's work was the result of advances in the science of *cytology* that made it possible to stain cell nuclei and observe chromosomes directly.
It also became possible to examine the behavior of chromosomes during *mitosis* (cell division) and *meiosis* (the reduction division that gives rise to gametes). In 1903, working independently, the German Theodor Boveri (1862–1915) and the American Walter Sutton (1877–1916) came to the conclusion that the behavior of chromosomes strongly resembled the behavior of genes. This realization led them to put forth the *Chromosome Theory of Inheritance*, which states that genes are found on chromosomes.

It soon became apparent that Mendel's work was insufficient to explain all inheritance patterns and that more complex genetic behavior was possible. For example, Thomas Hunt Morgan (1866–1945) used the white-eye color mutant of *Drosophila melanogaster* to show that some genes behave in a sex-linked manner. In 1909, Hermann Nilsson-Ehle (1873–1949) reported experiments in which multiple genes were involved in producing a single trait. We investigate the behavior of such polygenic traits later in the chapter.

Unfortunately, the laws of genetics have sometimes been purposefully misinterpreted and abused. What Mendel and other scientists had proved for the color of flower petals, the shape of peas, and the height of plants, pseudoscientists misapplied for political purposes to so-called traits such as intelligence and purity of race. In Nazi Germany, the project to purify the Aryan race assumed horrifying proportions with the attempted extermination of Jews—about one third of the world's Jews lost their lives in the concentration camps. In the 1950s in the United States, intelligence quotient tests were used by some to "prove" the intellectual superiority of the white race. In the Soviet Union during the Stalinist era, Mendel was declared a reactionary of no scientific value, and the problems of heredity were treated in accordance with the so-called progressive ideas of Trofim Denissovich Lysenko (1898–1976). Lysenko claimed that biological species, when raised under appropriate conditions, could change their hereditary profile and transform from one species into another—wheat plants could produce seeds of rye or barley, or corn could turn into wheat. Once again, history turned ugly, and those who disagreed with Lysenko's pseudoscientific discoveries were persecuted. It was only in 1964 (11 years after Stalin's death!) that Lysenko was declared a fraud, and Mendel's theory was reinstated in the Soviet Union.

II. CHROMOSOMES AND THE PHYSICAL BASIS OF HEREDITY

Chromosomes are the carriers of genetic information found in the cell nucleus. For most sexually reproducing organisms, chromosomes occur in pairs, one coming from the mother and one from the father. Each chromosome has areas that specify genes, and each member of a pair of chromosomes will specify the same genes in the same area. Figure 3-2 depicts this correspondence.

Even though each chromosome of the pair contains the same genes in the same areas, they may not necessarily contain the same version of each gene. Thus, in Figure 3-2, the maternal chromosome may have one form of the gene for eye color while the paternal chromosome may

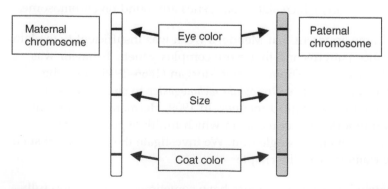

FIGURE 3-2.
Pair of chromosomes from an imaginary organism.

have another form. Different forms of the same gene, as mentioned above, are called alleles. For the present discussion, assume each gene has only two alleles. This is a reasonable assumption for many genes and is the basis of classical Mendelian genetics. An easy way to think about this case is to imagine one allele as representing a functional gene (the so-called normal gene) and the other allele as representing a nonfunctional mutant of the gene. As is shown in Figure 3-3, both chromosomes may have the same allele for the eye color gene (**A** and **A**), different alleles for the size gene (**B** and **b**), and the same allele for the coat color gene (**c** and **c**).

Some genes have a dominant allele and a recessive allele. A dominant allele is expressed if it is present in both copies or in one copy with a recessive allele on the opposite chromosome. A recessive allele is only expressed if it is present in both copies. Suppose in an imaginary creature, amber eyes are dominant to green eyes. The creature will be amber-eyed if it has either one or two dominant alleles (i.e., **AA** or **Aa**) and will only be green-eyed if it has two recessive alleles (**aa**). We distinguish between the genotype of the creature (**AA**, **Aa**, or **aa**) and its phenotype or appearance (amber-eyed or green-eyed). We call the genotypes *homozygous* if they have two of the same allele (**AA** or **aa**) and *heterozygous* if they have one of each (**Aa**). In Figure 3-4, we illustrate Mendel's result with the seed shape gene, found on the *Pisum* chromosome number 7. We use **A** to denote the dominant smooth-seed allele and **a** to denote the recessive wrinkled-seed allele.

The parents are both true-breeding and therefore homozygous. They can only produce a single type of gamete, as shown in Figure 3-4. The F_1 generation, resulting from the combination of egg and sperm in fertilization, is heterozygous (**Aa**) and dominant (smooth). The F_1 individuals can produce two different kinds of gametes, **A** and **a**, in approximately equal numbers (see Figure 3-5).

Keeping track of the gametes and ensuring that each one is paired appropriately can be done with a *Punnett square* (see Figure 3-6). In a

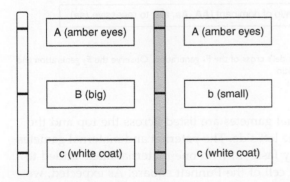

FIGURE 3-3.
Hypothetical allele composition for a pair of chromosomes of an imaginary organism.

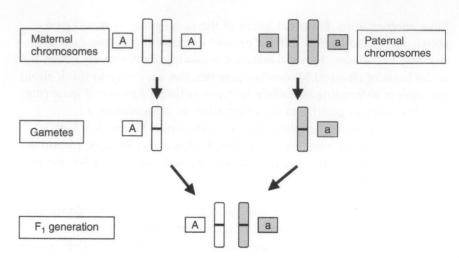

FIGURE 3-4.
Crossing a pure dominant parent and a pure recessive parent. Pea chromosomes in Mendel's first cross, where **AA** and **Aa** produce smooth seeds and **aa** produce wrinkled seeds. The F_1 are all **Aa** genotypically and are therefore smooth, which is the dominant phenotype.

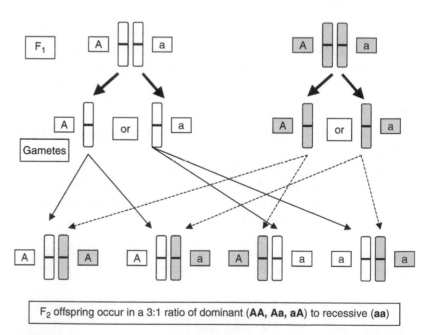

F_2 offspring occur in a 3:1 ratio of dominant (**AA, Aa, aA**) to recessive (**aa**)

FIGURE 3-5.
Diagram of the chromosomes in Mendel's cross of the F_1 generation. Observe the F_2 generation and the chromosomal basis of the 3:1 ratio.

Punnett square, the paternal gametes are listed across the top and the maternal gametes down the left side. The paternal and maternal gametes are combined in an orderly fashion, with one maternal allele added to one paternal allele in each cell of the Punnett square. As expected, we see 1/4 **AA**, 1/2 **Aa**, and 1/4 **aa** or 3/4 dominant and 1/4 recessive, giving the 3:1 ratio that Mendel observed.

We next examine the case of multiple genes, considering an example with two genes in detail. The ideas are easily generalized to three and more genes (see Exercise 2 below). Suppose we have two different genes on two different chromosomes, such as the seed shape gene on *Pisum* chromosome 7 and the plant height gene on *Pisum* chromosome 4. Let us further suppose we have two parents, one homozygous for dominant alleles of both genes, and the other recessive for both genes, as in Figure 3-7.

When reproduction occurs, one allele from each gene from each parent is randomly selected to contribute to the offspring. Because of the parents' genetic purity, the mother must contribute dominant alleles [**A, B**] and the father recessive alleles [**a, b**]. Therefore, this first filial generation (or F_1) will have the genotype **AaBb** and the dominant phenotype for both genes.

Now, suppose we mate two parents from the F_1 generation (**AaBb** × **AaBb**). They will each be able to produce four different types of gametes, because the first parent may deliver either an **A** or **a** allele for the first gene (two possibilities) and either a **B** or **b** allele for the second gene (two possibilities). So there are $2 \times 2 = 4$ possibilities from Parent 1. These possible gametes are **AB, Ab, aB,** and **ab**. Parent 2 has the same four possibilities, so there are a total of $4 \times 4 = 16$ possibilities for the offspring in the F_2 generation. Because in the F_1 there is a 1:1 ratio of recessive and dominant alleles, and because we assume random mating from the F_1, each of these possibilities is equally likely to occur in the F_2.

We want to know what the phenotypes of the F_2 generation are, and in what proportions they occur. We calculate this in two ways. First, we list the 16 possibilities using a Punnett square (see Figure 3-8).

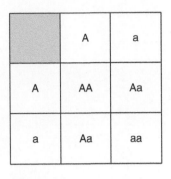

FIGURE 3-6.
Punnett square of Mendel's results diagrammed in Figure 3-5.

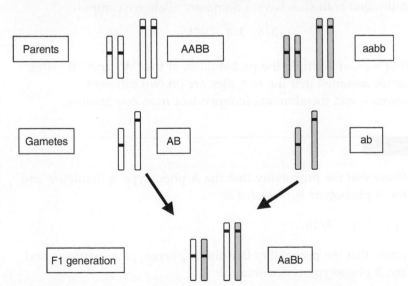

FIGURE 3-7.
Chromosomes of a cross of double homozygotes. **A** or **a**, seed shape; **B** or **b**, plant height.

	AB	Ab	aB	ab
AB	AABB	AABb	AaBB	AaBb
Ab	AABb	AAbb	AaBb	Aabb
aB	AaBB	AaBb	aaBB	aaBb
ab	AaBb	Aabb	aaBb	aabb

FIGURE 3-8.
Punnett square, showing the 16 possibilities for the F_2 offspring of two double heterozygotes (**AaBb**×**AaBb**).

The phenotype will have both dominant characteristics if there is at least one capital **A** and one capital **B**. The following nine combinations fulfill this condition: **AABB, AABb, AaBB, AAbB, aABB, AaBb, AabB, aAbB,** and **aABb**. The phenotype will have the dominant **A** characteristic and the recessive **B** characteristic if there is at least one capital **A** and two lower case **b**'s: **AAbb, Aabb,** and **aAbb**. The phenotype will have the recessive **A** characteristic and the dominant **B** characteristic if there are two lower case **a**'s and at least one capital **B**: **aaBB, aaBb,** and **aabB**. The phenotype will have both the recessive **A** characteristic and the recessive **B** characteristic if there are two lower case **a**'s and two lower case **b**'s: **aabb**. So the 16 possibilities give the phenotypes in the ratio 9:3:3:1.

The second way to see this result is less cumbersome but not as explicit. Consider how the phenotype with both dominant characteristics may be formed. The "**A**" site must have at least one dominant allele, which happens with probability 3/4, because there are four "**A**" possibilities: **AA, Aa, aA,** and **aa**. Three of these four have at least one **A**. Similarly, the "**B**" site has a dominant allele with probability 3/4. To find the probability that both sites have a dominant allele we compute:

$$3/4 \cdot 3/4 = 9/16.$$

Note that we can multiply the probabilities of the "**A**" and "**B**" sites, because we assumed that the two sites are on two different chromosomes and therefore are independent from one another.

EXERCISE 3-1

(a) Show that the probability that the **A** phenotype is dominant and the **B** phenotype is recessive is:

$$3/4 \cdot 1/4 = 3/16.$$

(b) Show that the probability that the **A** phenotype is recessive and the **B** phenotype is dominant is:

$$1/4 \cdot 3/4 = 3/16.$$

(c) Show that the probability that the phenotypes are recessive is:

$1/4 \cdot 1/4 = 1/16$.

(d) How does this imply the same proportion of the phenotypes as the Punnett square?

EXERCISE 3-2

Suppose we now have three sites on three different chromosomes occupied by **A** or **a**, **B** or **b**, and **C** or **c**, where upper case denotes the dominant allele. If we mate two triply heterozygous parents, **AaBbCc** × **AaBbCc**:

(a) What is the probability that each site has at least one dominant allele?

(b) What is the probability that the **A** and **B** sites are dominant and the **C** site is recessive?

(c) What is the probability that the **A** site is dominant and the **B** and **C** sites are recessive?

(d) What is the probability that all sites are recessive?

(e) List all of the possible phenotypes and the probability with which each occurs.

Next, we want to investigate how the genetic make-up evolves. We consider two mathematical models describing how the distribution of alleles at a particular site comes to equilibrium.

III. HARDY–WEINBERG LAW OF GENETIC EQUILIBRIUM

It may appear that, over time, the genotypes of each species should converge to a single, optimal genotype. The simplest explanation for why this is inaccurate is that our view of "optimal" is not nature's view. In fact, nature's view is that genetic diversity is valuable. Genetic diversity allows some members of a species to survive periods of catastrophic environmental change, and therefore promotes the perpetuation of the species. We begin with some examples that demonstrate how the lack of genetic variability can be disastrous.

In 1970, 15% of the U.S. corn (maize, *Zea mays*) crop was lost to infection by the fungus *Cochliobolus heterostrophus*, the southern corn leaf blight. This $1 billion loss occurred because a huge fraction of the U.S. corn crop was planted in a single type of corn, Texas male-sterile cytoplasm (or T-cytoplasm). This particular type was chosen because its seed was considerably cheaper to produce. A corn plant produces female flowers (containing ovules that will become the seeds) and male flowers (which make pollen). The female flowers, which become the "ears," form along the stem, and the male flowers, called tassels, form at the top. In order to produce the high-yielding, uniform plants demanded by modern mechanized agriculture, corn breeders remove the tassels from the desired female parent so its pollen does not compete with pollen from the desired male parent. Because male-sterile corn produces infertile pollen on its tassels, it is not necessary to detassel the female parent plants by hand, resulting in immense savings in labor costs. Unfortunately, the genetic similarity of these plants also made the crop more vulnerable to disease, meaning that if one plant was susceptible to it, they might all be—and they were.

The southern corn leaf blight infestation of 1970 was caused by a strain of *Cochliobolus heterostrophus* called "race T," which preferentially infected T-cytoplasm corn. The fungus found fertile hunting grounds in the nation's cornfields that year, and a spate of warm, moist weather allowed it to spread with shocking rapidity. Blight spores were carried by the wind and leapt from county to county and then from state to state. Farmers were left with fields of rotten, useless plants. The situation would have been even worse had the weather not turned cool and dry, stemming the spread of the disease. Additional details on this infestation can be found in Ullstrup (1972) and Hooker et al. (1970).

Another striking example of the danger of lack of diversity is the Irish potato famine of the mid-1840s. In this case, the pathogen was the water mold or oomycete *Phytophthora infestans*, the cause of late blight. The late blight infects both the leafy portions of the plant, causing reduced yields, and the potatoes themselves, causing them to rot in the ground or in the root cellar. By the mid-1840s, Ireland's mixed agricultural system marked by a variety of crops and livestock had been gradually replaced by a potato-based economy.

As with the southern corn leaf blight in 1970, a period of warm, wet weather in 1845 provided ideal conditions for the spread of this pathogen. Some potatoes survived the winter and were planted in 1846, but the blight returned, harbored in some of the seed potatoes or possibly in piles of discarded infected potatoes. As the germ theory of disease was unknown at the time, the importance of destroying infected plant materials was likewise unknown. Most of the crop failed, and the Irish population, who relied upon the potato for most of their caloric intake, went hungry. There were no stockpiles to relieve the famine because potatoes cannot be stored for more than a year. Hundreds of thousands of people starved, and many more emigrated.[1]

---

1. Detailed studies of the Irish Potato Famine can be found in Fraser (2003), Fry and Goodwin (1997), Goodwin (1997), Ristaino et al. (2001), and Garrett and Mundt (2000).

These examples underscore the importance of maintaining crop diversity to reduce the severity of infections and avoid massive crop failures. Nature preserves genetic diversity, and we humans eliminate it at our peril. We next create a mathematical model explaining why, under certain conditions, genetic diversity is preserved.

Suppose we focus on a particular site of a particular chromosome. We assume the gene at that locus has two possible alleles, **A** and **a**. How does the proportion of alleles in the gene pool change over time? We shall show that if certain idealized assumptions are met, then equilibrium in the proportion of allele combinations (**AA**, **Aa**, or **aa**) is reached after just one generation. We make the following assumptions:

1. The population is large (theoretically, infinite).

2. Mating is random.

3. All allele combinations have the same fitness (i.e., there is no natural selection occurring).

4. There is no net mutation.

5. There is no immigration or emigration.

Although this set of assumptions may appear restrictive, they provide a good approximation in many cases. We should not forget that whether these assumptions are appropriate depends very much on the particular genes being examined. For example, if one were looking at the genes for "tall, dark, and handsome," the random mating assumption might not apply. However, if one were considering a biochemical difference—one not readily apparent—mating would almost certainly be random with respect to that particular gene. In addition, mutation is always occurring, but it occurs at a small rate that may be safely ignored for many purposes. With these assumptions, we derive the result presented below, called the *Hardy–Weinberg Law of Genetic Equilibrium* because of its almost simultaneous publication in 1908 by British mathematician Godfrey H. Hardy (1877–1947) (Hardy [1908]) and German physician Wilhelm Weinberg (1862–1937) (for the English translation of Weinberg's paper, see Boyer [1963], pp. 4-15).

Theorem (Hardy–Weinberg Law of Genetic Equilibrium). Let assumptions 1–5 above be satisfied. Let P(**aa**), P(**Aa**), and P(**AA**) denote the proportions of the genotypes **aa**, **Aa**, and **AA**, respectively, in the parental generation. Assume these proportions have the values P(**aa**) $= x$, P(**Aa**) $= y$, and P(**AA**) $= 1 - x - y$, where $0 \leq x \leq 1$, $0 \leq y \leq 1$, and $x + y = 1$. Then:

1. The proportion of the **A** and **a** alleles in the parental generation is calculated, respectively, as

$$p = \frac{2(1 - x - y) + y}{2} \quad \text{and} \quad q = \frac{y + 2x}{2}. \tag{3-1}$$

2. For generation F_1, the proportion of the genotypes will be

$$P_1(\mathbf{aa}) = q^2, P_1(\mathbf{Aa}) = 2pq, \text{ and } P_1(\mathbf{AA}) = p^2. \qquad (3\text{-}2)$$

The proportion of the **A** and **a** alleles in the F_1, however, does not change and remains as it was in the parental generation.

3. The proportion of the genotypes in the F_1 from Eq. (3-2) will be maintained in all subsequent generations—that is, the system will be in genotypic equilibrium from generation F_1 on. The same is true for the proportions of the **A** and **a** alleles, which remain the same as in the parental generation.

Before giving a formal proof, we present a computation that incorporates all of the ideas of the abstract argument, but may be easier to follow.

Example 3-1
....

Imagine a "colony" of 50 people selected to begin a settlement. Suppose the distribution of genotypes is **AA** = 15, **Aa** = 30, and **aa** = 5. Thus, in this parental generation the proportions of the genotypes will be P(**AA**)= $15/50 = 0.3$, P(**Aa**) $= 30/50 = 0.6$, P(**aa**) $= 5/50 = 0.1$.

To determine the proportion of **A** and **a** alleles in the colony, notice that the total number of **A** and **a** alleles is 100. As each **AA** individual contributes two **A** alleles and each **Aa** individual contributes one **A** allele, there is a total of $15 \times 2 + 30 \times 1 = 60$ **A** alleles in the population. In the same way, there are 40 **a** alleles ($5 \times 2 + 30 \times 1 = 40$). Thus, the proportion of **A** alleles is calculated as $p = 60/100 = 0.6$, and the proportion of **a** alleles is $q = 40/100 = 0.4$.

We are now ready to proceed with establishing results 1 and 2 in the general case. Notice that even though we use symbols to denote the genotypic distribution, the ideas of the computations in the general case are identical. Again, imagine that we select a colony of N individuals that have genomes in the following proportions:

$$P(\mathbf{aa}) = x, \qquad P(\mathbf{Aa}) = y, \qquad P(\mathbf{AA}) = 1 - x - y.$$

Therefore, the number of individuals with genotype **Aa** is Ny, the number with genotype **aa** is Nx, and the number with genotype **AA** is $N(1 - x - y)$. We can now compute the proportions of alleles in this group to obtain the proportions of the **A** and **a** alleles in the parental generation:

$$p = \frac{2N(1 - x - y) + Ny}{2N} = \frac{2(1 - x - y) + y}{2}$$

$$q = \frac{Ny + 2Nx}{2N} = \frac{y + 2x}{2}. \qquad (3\text{-}3)$$

This establishes claim 1 of the theorem.

To prove part 2, we use a Punnett square approach to generalize from an individual mating to any random mating occurring in the population. This can be done by treating a randomly chosen diploid organism as being the result of a random pairing of two alleles, each drawn from the parental pool containing **A** and **a** alleles at proportions p and q, respectively. Because diploid organisms inherit one allele from each of their parents, and because we assumed random mating and equal fitness of the genotypes, this treatment is well justified. Thus, the probability of homozygous dominant P(**AA**) will be the probability of randomly drawing **A** twice from the allelic pool, and we obtain $P(\mathbf{AA}) = (p) \cdot (p) = p^2$. Notice that the assumption we made for an infinitely large population allows us to assume that the frequencies of the allele do not change from draw to draw, regardless of the fact that our selection is without replacement. In the same way, the probability for a heterozygous **Aa** organism will be the probability for drawing an **A** followed by an **a**, calculated as pq, and so on. The Punnett square, Table 3-1, helps us visualize this. In addition to the genotypes resulting from the cross (see Figure 3-6), we have now also included, in parentheses, the proportions of the **A** and **a** alleles and the proportions of the resulting genotypes.

From here, we see that the proportion of the genotypes in F_1 will be:

$$P(\mathbf{AA}) = p^2$$
$$P(\mathbf{aa}) = q^2 \tag{3-4}$$
$$P(\mathbf{Aa}) = pq + qp = 2pq.$$

We have calculated P(**Aa**) as the sum of the proportions of **Aa** and **aA**, both of which represent the same genotype. To finish the proof of claim 2, we need to show that the proportions of the **A** and **a** alleles in the F_1 remain the same as in the parental generation. The argument is the same as in our example and the proof of claim 1 above, but applied this time to the genotype distribution given by Eq. (3-4). If there are N_1 individuals in the F_1, of them $N_1 p^2$ will have the **AA** genotype; $N_1 q^2$ will have **aa**; and $2N_1 pq$ will have the **Aa** genotype. Thus, the proportion of the **A** allele in the F_1 will be:

$$\frac{2N_1 p^2 + 2N_1 pq}{2N_1} = \frac{2N_1 p(p+q)}{2N_1} = p(p+q) = p, \text{ since } p+q = 1.$$

	A (p)	**a** (q)
A (p)	AA $(p \cdot p = p^2)$	Aa (pq)
a (q)	aA (qp)	aa $(q \cdot q = q^2)$

TABLE 3-1.
Punnett square applied to genotype frequencies.

Similarly, the proportion of the allele in the F_1 will be:

$$\frac{2N_1 q^2 + 2N_1 pq}{2N_1} = \frac{2N_1 q(q+p)}{2N_1} = q(q+p) = q.$$

Thus, the allele proportions remain unchanged from the parental generation, and the proof of claim 2 is complete.

To prove claim 3, notice we obtained the genotypic distribution in the F_1 presented in Eq. (3-4) by using a Punnett square based on the proportions p and q of the **A** and **a** alleles in the previous generation. We now know, from claim 2, these proportions remain unchanged in the F_1 generation. The argument can now be repeated to show that the genotypic proportions in the F_2 will be $P_2(\mathbf{aa}) = q^2$, $P_2(\mathbf{Aa}) = 2pq$, and $P_2(\mathbf{AA}) = p^2$, implying, again, that the allelic proportions in the F_2 remain unchanged, and so on. Thus, beginning with the F_1, the system remains in genotypic equilibrium.

The Hardy–Weinberg Law of Genetic Equilibrium gives a mathematical explanation for a well-known biological fact—equally fit genotypes are generally preserved in nature and coexist in equilibrium. Although the equilibrium genotypic frequencies need not be exactly those of the original population, under the assumptions of the model, the equilibrium is reached in the first generation and is preserved for all later generations. In contrast, under the same assumptions, the allelic frequencies remain constant from the very beginning.

It is clear that changes in the genetic constitution of a population must occur under some conditions, or evolution would not occur. This gradual change is caused by the presence in the population of alleles with varying degrees of fitness. In this case, the genotype proportions change from generation to generation, and the dynamic behavior of the system is more complex. We next investigate the effect of natural selection in populations containing maladaptive alleles.

IV. THE EFFECT OF A MALADAPTIVE OR LETHAL GENE

The Hardy–Weinberg Law of Genetic Equilibrium assumes that each of the genotypes in the population is equally successful or equally fit, and therefore natural selection is not acting on any of the genotypes. However, there are many genetic diseases, such as Tay–Sachs, phenylketonuria, severe combined immune deficiency, and hemophilia, for which this is not the case. Cystic fibrosis (CF), for instance, is a genetic disease caused by a mutation in the gene for a chloride ion-transporting protein, the cystic fibrosis transmembrane regulator (CFTR). People who are homozygous for the recessive mutant CFTR allele will have CF, while those who carry only one CFTR allele will not.

The impaired CFTR function results in decreased secretion of fluids from gastrointestinal and respiratory epithelia. Without normal fluid secretion, the delivery of digestive enzymes from the pancreas is diminished, and the protective mucus in the lungs is rendered viscous and extremely difficult to clear from the airways. As a result, people with CF have frequent respiratory infections and impaired digestion, accompanied by chronic lung infection, infertility, and an early death. Obviously, this represents a significant selective disadvantage.

We now examine a model describing such situations and study the long-term behavior of allele proportions that, under these circumstances, are not in equilibrium and change over time. We care about such changes for several reasons, including their contribution to our understanding of the public health implications of genetic diseases. Information on gene frequencies can help guide our genetic testing and prenatal diagnosis efforts.

Suppose we have a gene with two alleles, **A** and **a**, and the **aa** combination is disadvantaged in that a certain predictable proportion of the **aa** individuals will die before they reproduce. We would like to know the effect of such a selective disadvantage for the **aa** genotype and examine the change in the proportion of the harmful **a** allele between generations. We would also like to know if there is an equilibrium state for the species' gene pool. The mathematical model we describe makes the following assumptions:

1. The population size is large.

2. Mating is random.

3. The recessive homozygous genotype is less fit to survive, and only a fraction, say α, (where $0 \leq \alpha < 1$), of the individuals with this genotype survives to reproduce.

4. There is no net mutation.

5. There is no immigration or emigration.

Notice the only difference between this set of assumptions and that for the Hardy–Weinberg Equilibrium is in assumption 3, reflecting differences in natural selection.

We begin with an example that illustrates the change in gene frequencies over time caused by selective disadvantage.[2]

_____

2. To simplify the calculations, in this example we assume that the initial population size is 100. Because this is not exactly a large number, as we explicitly require in assumption 1 for the model, the gene proportions and frequencies calculated in the example should be considered in terms of averages.

Example 3-2

Let us assume a population of 100 individuals of genotype **Aa**. The initial frequencies of the **A** allele, p_0, and the **a** allele, q_0, are both 0.5. This initial population reproduces, producing 100 offspring in the following proportions:

$$P(\mathbf{AA}) = p_0^2 = (0.5)^2 = 0.25.$$

$$P(\mathbf{Aa}) = 2p_0q_0 = 2 \times (0.5) \times (0.5) = 0.50.$$

$$P(\mathbf{aa}) = q_0^2 = (0.5)^2 = 0.25.$$

Because there are 100 offspring, we can calculate that $100 \times 0.25 = 25$ of them will have genotype **AA**; $100 \times 0.5 = 50$ will have genotype **Aa**; and $100 \times 0.25 = 25$ will have genotype **aa**. However, let us also assume the **aa** allele is associated with a selective disadvantage of 0.2, such that 20% of the **aa** offspring will not survive to reproduce. Thus, only 80% of the 25 **aa** will be able to pass their genes on to the next generation, and, therefore, the proportions of the **A** and **a** alleles will be different from the initial generation. Let us look at the genes that will be passed on:

The 25 **AA** individuals will contribute 2×25 or 50 **A** alleles.

The 50 **Aa** individuals will contribute 50 **A** alleles and 50 **a** alleles.

The $0.8 \times 25 = 20$ **aa** individuals will contribute 2×20 or 40 **a** alleles.

Therefore, there will be 95 individuals contributing their alleles to the next generation, and the total gene pool will be twice that number or 190 alleles. Of those, 100 are **A** alleles and 90 **a** alleles. If we denote the allele frequencies for **A** and **a** by p_1 and q_1, respectively, we can calculate that:

$$\boldsymbol{p}_1 = 100/190 = 0.5263 \quad \text{and} \quad \boldsymbol{q}_1 = 90/190 = 0.4737.$$

Notice the proportions of the **A** and **a** alleles have changed from $p_0 = 0.5$ and $q_0 = 0.5$ in the initial population to $p_1 = 0.5263$ and $q_1 = 0.4737$ in the population surviving until reproduction. To see how these proportions will change even further, suppose now that the 95 individuals reproduce, and 100 offspring result. What would be the allele frequencies in the next generation?

The **AA** individuals would be represented by Np_1^2, where N is the population size (in our case $N = 100$). This would be $100 \times (0.5263)^2 = 27.7$ or, rounding up, 28 people. The **Aa** individuals would be $N(2p_1q_1)$ or $100 \times 2 \times (0.5263) \times (0.4737) = 49.8$, or 50 people. The **aa** individuals would be Nq_1^2 or $100 \times (0.4737)^2 = 22.4$, or 22 people. We can see there are fewer **aa** individuals in this generation than in the previous one.

The 28 **AA** individuals will contribute 2×28 or 56 **A** alleles.

The 50 **Aa** individuals will contribute 50 **A** alleles and 50 **a** alleles.

Only 80% of the 22 **aa** individuals will reproduce. Therefore, $0.8 \times 22 = 17.6$, or 18 people, will pass along their **a** alleles, contributing 2×18 or 36 **a** alleles.

A total of $28 + 50 + 18 = 96$ individuals will reproduce, so the total gene pool will consist of 96×2 or 192 alleles.

p_2 (new frequency of the **A** allele) $= (56 + 50)/192 = 106/192 = 0.5521$.

q_2 (new frequency of the **a** allele) $= (50 + 36)/192 = 86/192 = 0.4479$.

After two generations, we see the **a** allele frequency has already fallen substantially because of the selective disadvantage of the **aa** genotype, showing the population is not in Hardy–Weinberg equilibrium. It also raises the following questions: Because the pool of the **a** allele will be reduced when the **aa** individuals die, it is intuitive that the **a** allele will diminish, but will it eventually die out? Does the answer depend on what fraction of those with paired **aa** alleles die?

To answer these questions, we derive a general formula describing the dependence of allele frequencies in any given generation upon those of the previous generation and enabling us to determine the long-term behavior of the population. To reinforce the informal notation used above, suppose the initial proportions of the alleles are:

$p_0 =$ proportion of **A** allele;

$q_0 = 1 - p_0 =$ proportion of **a** allele.

Then

$$P(\mathbf{AA}) = p_0^2;$$

$$P(\mathbf{Aa}) = 2p_0q_0, \quad \text{and}$$

$$P(\mathbf{aa}) = q_0^2.$$

Now suppose the **aa** combination is harmful so that only a fraction, say α (where $0 \leq \alpha < 1$), of the homozygous recessive genotype survives to reproduce. Thus, before reproduction, the genotype distribution in the population will change to:

$$P(\mathbf{AA}) = p_0^2;$$

$$P(\mathbf{Aa}) = 2p_0q_0, \text{and}$$

$$P(\mathbf{aa}) = \alpha q_0^2.$$

The decreased amount of **aa** genotype at the time of reproduction causes a decrease in the **a** allele frequencies in the next generation, calculated as follows: assume the total number of individuals in the original population was N (we assumed N to be very large), which implies there were Nq_0^2 homozygous recessive individuals to start with. Because only a fraction α of them survives, the number of those individuals has changed to $N\alpha q_0^2$ by the time of reproduction. Thus, the number of **A** alleles at that time will be $2p_0^2 N + 2p_0 q_0 N$, and the number of **a** alleles will be $2\alpha q_0^2 N + 2p_0 q_0 N$. Denoting the proportions of the **A** and **a** alleles in the first generation by p_1 and q_1, we now calculate that:

$$q_1 = \frac{2\alpha q_0^2 N + 2p_0 q_0 N}{2\alpha q_0^2 N + 2p_0 q_0 N + 2p_0^2 N + 2p_0 q_0 N}$$

$$= \frac{\alpha q_0^2 + p_0 q_0}{\alpha q_0^2 + 2p_0 q_0 + p_0^2} = q_0 \left(\frac{\alpha q_0 + p_0}{\alpha q_0^2 + 2p_0 q_0 + p_0^2} \right). \tag{3-5}$$

The proportion p_1 could be calculated similarly, but we assumed that **A** and **a** are the only alleles present in the population, so it is easier to state that $p_1 = 1 - q_1$. Similarly, if p_n and q_n denote the proportions of the **A** and **a** alleles in the nth generation, then because of weaker fitness of the **aa** genotype, these frequencies will change in the $(n + 1)$-st generation to:

$$q_{n+1} = q_n \frac{\alpha q_n + p_n}{\alpha q_n^2 + 2p_n q_n + p_n^2} \quad \text{and} \quad p_{n+1} = 1 - q_{n+1}. \tag{3-6}$$

Notice how different this situation is compared with the Hardy–Weinberg case discussed previously. The allelic frequencies of **A** and **a** now change from generation to generation, and, when we know the frequencies for any given generation, Eq. (3-6) allows us to compute their values for the following generation. Formulas such as Eq. (3-6) are called *recursive formulas*.

Notice that because:

$$\alpha q_n^2 + 2p_n q_n + p_n^2 = (\alpha q_n + p_n)q_n + p_n(q_n + p_n) = (\alpha q_n + p_n)q_n + p_n,$$

Eq. (3-6) could be rewritten as:

$$q_{n+1} = q_n \frac{\alpha q_n + p_n}{(\alpha q_n + p_n)q_n + p_n}. \tag{3-7}$$

Now, because $\alpha < 1, \alpha p_n < p_n$ and because $p_n = 1 - q_n$, we have $\alpha(1 - q_n) < p_n$, which is the same as $\alpha < \alpha q_n + p_n$. This implies that $\frac{\alpha q_n + p_n}{(\alpha q_n + p_n)q_n + p_n} < 1$ and, together with Eq. (3-7), proves that $q_{n+1} < q_n$. Thus, the allelic frequency of the harmful allele **a** does indeed decrease from generation to generation.

Our next goal is to establish what happens to the harmful allele **a** in the long run. Will it disappear from the gene pool or stabilize at a nonzero value? More importantly, are we sure the limit for the sequence $q_0, q_1, q_2, \ldots, q_n, \ldots$ exists?

The last question has an easy answer. Because the sequence $q_0, q_1, q_2, \ldots, q_n, \ldots$ is a decreasing sequence of numbers bounded below by 0, the limit $\lim_{n\to\infty} q_n$ must exist. Let $q = \lim_{n\to\infty} q_n$. Then $p = \lim_{n\to\infty} p_n = \lim_{n\to\infty}(1 - q_n) = 1 - \lim_{n\to\infty} q_n = 1 - q$. *Because we began with* $q_0 < 1$ *and proved the values of the* **a** *allele frequencies decrease from generation to generation, it is clear the limit value* q *will be such that* $q < 1$. *We now show that* q = 0.

Taking limits as $n \to \infty$ of both sides in Eq. (3-7), we obtain:

$$\lim_{n\to\infty} q_{n+1} = \lim_{n\to\infty}\left(q_n \frac{\alpha q_n + p_n}{(\alpha q_n + p_n)q_n + p_n}\right)$$

Because $q = \lim_{n\to\infty} q_n = \lim_{n\to\infty} q_{n+1}$ and $p = 1 - q$, this implies that:

$$q = q\frac{\alpha q + p}{\alpha q^2 + pq + p} = q\frac{\alpha q + (1-q)}{\alpha q^2 + (1-q)q + (1-q)} = q\frac{\alpha q - q + 1}{\alpha q^2 - q^2 + 1} = q\frac{(\alpha - 1)q + 1}{(\alpha - 1)q^2 + 1}.$$

Thus, the limit value q satisfies the following algebraic equation:

$$q = q\frac{(\alpha - 1)q + 1}{(\alpha - 1)q^2 + 1}. \qquad (3\text{-}8)$$

It is easy to see the value $q = 0$ satisfies Eq. (3-8). We next show that $q = 0$ is the only solution of Eq. (3-8) with $q < 1$. Assume Eq. (3-8) has another solution; that is, assume we can find a value $q \neq 0$ that satisfies Eq. (3-8). To solve for q, because we assumed $q \neq 0$, we can divide both sides of Eq. (3-8) by q to obtain:

$$1 = \frac{(\alpha - 1)q + 1}{(\alpha - 1)q^2 + 1}.$$

Then

$$(\alpha - 1)q^2 + 1 = (\alpha - 1)q + 1,$$

and so, because $\alpha \neq 1$,

$$q^2 = q.$$

This equation has two solutions: $q = 0$ and $q = 1$, establishing that $q = 1$ is the only nonzero solution of the algebraic Eq. (3-8). As we know that $q = 1$ cannot be the limit of the decreasing sequence of the allelic frequencies $q_n < 1$, this implies that Eq. (3-8) does not have a nonzero solution $q < 1$. Then $q = 0$ is the only solution of Eq. (3-8), and this implies $\lim_{n\to\infty} q_n = 0$.

We have therefore established the following fact: *Under the assumption that the recessive homozygous genotype* **aa** *is less fit than the homozygous dominant* **AA** *and the heterozygous* **Aa** *genotypes, the frequency of the harmful allele* **a** *will diminish from generation to generation, and allele* **a** *will be eventually eliminated from the gene pool.*

An interesting question here is how many generations it takes for the maladaptive allele to reach virtually negligible frequency levels in the gene pool. The exact answer depends on the survival level α—the lower the survival level, the faster the harmful allele is eliminated from the population. We examine this question in more detail in the laboratory manual project *Selection in Genetics: The Effect of a Maladaptive or Lethal Gene.*

Another question is why certain genetic diseases that confer a severe genetic disadvantage are still prevalent in the population, given that our results show that they should be decreasing and gradually disappear. For instance, CF is extremely common among the reproductively isolated population of the Old Order Amish, with frequencies as high as 1 in 500 live births. Why should an allele with such devastating effects be present at such a high frequency? The answer may lie in a phenomenon called heterozygote advantage. Some research suggests that individuals who have one copy of the CFTR allele are partially resistant to the devastating effects of cholera, typhoid fever, or other gastrointestinal infections. So individuals with two copies of the normal allele might be at a disadvantage whenever contracting bacterial diarrhea was a possibility.

An analogous example is sickle-cell anemia among people of African heritage. Sickle-cell anemia, a life-threatening disease, occurs at high frequencies in individuals whose ancestors come from those areas of the world where malaria is endemic. Sickle-cell anemia is caused by a single mutation in one of the genes for the oxygen-carrying red blood cell protein hemoglobin. Individuals inheriting two copies of the mutant allele will have sickle-cell disease, which causes red blood cells to change shape and clog capillaries, starving the tissues served by those capillaries of food and oxygen. On the other hand, individuals having two copies of the normal allele will have normal, circular, biconcave red blood cells, which carry lots of oxygen and provide a perfect environment for the malarial protozoan parasite. Malaria, like sickle-cell anemia, is life threatening. In areas where malarial infection is likely, heterozygotes with one normal and one mutant allele have a selective advantage in that the slightly reduced oxygen-carrying capacity of their red blood cells provides an inhospitable environment for the malarial parasite and thus gives them resistance against the disease. Among African Americans, however, the heterozygous state confers no advantage, because malaria is not common in the United States. For further details on the balance between some genetic and infectious human diseases, see Dean, Carrington, and O'Brien (2002).

The last two examples show once again that decisions regarding the use of a mathematical model should always be made with care. The above examples show there are situations when the use of the models developed in this section would not be prudent, as they may fail to capture the more complex genetic dynamics in a population.

V. MORE COMPLEX HEREDITARY PATTERNS

So far, we have been dealing with only the simplest Mendelian genetic systems, in which there are only two alleles at a particular locus, one of which is completely dominant to the other. Genetics, however, is not always that simple.

Sometimes, two alleles combine to give an entirely new phenotype, as with flower color in snapdragons. If one crosses a true-breeding red snapdragon with a true-breeding white snapdragon, the offspring will all have pink flowers. This phenomenon is called *incomplete dominance*. In the human ABO blood type system, we find another example. If a person who is homozygous for the type A allele ($I^A I^A$) marries a person who is homozygous for the type B allele ($I^B I^B$), their children will all be type AB, with a genotype $I^A I^B$. This situation is called *codominance*—the phenotypes (in this case, the particular sugars on the red blood cell membranes) of both parents are found in the child.

Some genetic loci have multiple alleles. We return to the human ABO blood type system for an example. In addition to the A, B, and AB blood types mentioned above, humans may also have blood type O, which has the genotype **ii**. This means that there are three alleles which may be found at the blood type locus—I^A, I^B, or **i**. The **i** allele is recessive to either of the I^A or I^B alleles. Thus, a person with type A blood may have genotype $I^A I^A$ or I^A**i**, and a person with type B blood may be $I^B I^B$ or I^B**i**.

Some traits are sex linked, in that they occur on one of the sex chromosomes of a species (such as the human X chromosome). Such traits are therefore present in a single copy in the sex with the unmatched "pair" of sex chromosomes (such as the human male, in which there are 22 pairs of chromosomes and an unmatched set, XY). When calculating the frequency of X-linked alleles in the human population, one needs to take into account the sexes of the individuals comprising the population.

EXERCISE 3-3

If p is the frequency of the I^A allele, q is the frequency of the I^B allele, and r is the frequency of the **i** allele, how would you state the frequencies of the blood types A, B, AB, and O, assuming Hardy–Weinberg conditions?

EXERCISE 3-4

Assume a gene **C** is found on the human X chromosome and that it occurs in only two alleles, **C** and **c**. Using p to represent the frequency of **C** and q to represent the frequency of **c**, how would you express the genotype frequencies in males and females?

The Hardy–Weinberg theorem and the selection observed with maladaptive genes might prompt us to believe the genetic make-up of a population will reach equilibrium, and that equilibrium state will be the single state most beneficial to the species. Obviously, this does not happen. Change is a fact of life. So what is wrong with our model? A good place to start is by reexamining the model's assumptions.

First of all, the Hardy–Weinberg assumptions and the model we developed to study the effect of a maladaptive gene are only approximations that do not fully reflect reality. Mating is usually not totally random, and one does not often have a population free from both immigration and emigration. Catastrophic events, such as disease or a meteor striking the earth, may wipe out entire species. Mutations occur, and some may propagate, becoming a non-negligible part of a population's gene pool—especially if the mutations confer some advantage. Of course, what may be an advantageous mutation for some members of a species may be a disadvantage for others, as amply illustrated by the sickle-cell anemia allele.

In a way, we are back to where we were modeling population growth, with initial attempts yielding limited insights, but failing to fully describe what actually happens. For better descriptions and more refined outcomes, our models need to be modified by adding new variables and/or parameters. Although carrying out such modifications falls beyond the scope of this text, a detailed discussion may be found in Falconer (1989). We shall now move forward to another application of mathematics to genetics.

VI. QUANTITATIVE TRAITS

A. Discontinuous Versus Continuous Traits

In each of our previous examples, genetic variations within the population were due to different genotypes involving one or two loci. These examples give rise to *discontinuous traits*, where just a few distinct phenotypes are observed: The seed coats of pea plants are either round or wrinkled; the seedpods are green or yellow; etc. Because of the relatively small number of phenotypes, they are easily separated from one another. Studying the phenotypes of the parents and offspring and examining the phenotypic ratios can readily establish the connection between the traits and the genes.

Many other traits, known as *continuous traits*, do not follow this pattern. These traits, such as human weight and height, exhibit a wide range of possible phenotypes. The color of human eyes, for example, varies from the lightest shades of green and blue to deep dark brown to nearly black. The branch of genetics that examines the inheritance of continuous traits is called *quantitative genetics*. It employs a variety of quantitative methods to study the genetic make-up of continuous traits and to determine to what extent variation in phenotype is caused by genotypic differences and to what extent it might be caused by differences in environmental factors. We take a closer look at these questions in the next chapter.

In this section, we present a mathematical model that explains continuous traits by assuming there are multiple genetic loci controlling the expression of a particular quantitative characteristic, with each individual gene contributing to the trait's magnitude. Each of the controlling genes can add to, or fail to add to, the magnitude of the specific characteristic. The model explains the so-called bell-shaped curve distribution of values exhibited by many continuous traits, such as height, weight, or intensity of flower petal color.

This hypothesis was formulated in 1909 by Hermann Nilsson-Ehle in relation to his study of wheat kernel colors. Similar to the experiments Mendel performed, Nilsson-Ehle started by crossing pure lines of white grain and red grain wheat. In the first generation, he only observed wheat with a grain color intermediate to that of the parents. An intercross of the first generation produced offspring with both white and red grain color, as expected. The puzzling fact in his experiments, however, was the presence of *different shades* of red among the red grain in the F_2, in addition to the color observed in the F_1—a departure from the Mendelian result. In one of the crosses, for example, a ratio of 15/16 red to 1/16 white was observed. There were about as many plants with grains as intensely red as in the grandparents as there were plants with white grains. The number of plants with pure red grains and white grains was not very large, and the largest number of plants in the F_2 had grains of the same color as the F_1. There were also two intermediate shades of red that were not present in either the P or F_1 generations. The different phenotypes appeared in a ratio of 1:4:6:4:1 from the dark red through the light red grains to white. In still other cases, Nilsson-Ehle found that 63/64 of the F_2 plants were red-kernelled and only 1/64 had white kernels. In these experiments, the range of intensity of the red was even wider and the variety of shades greater.

B. The Polygenic Hypothesis

To explain these experiments, Nilsson-Ehle made the conjecture, known as the *polygenic (or multiple-factor) hypothesis*, that the color of the kernel is controlled, not by one, but by *several different genes*. The genes are independent and contribute cumulatively to the red pigmentation of the wheat kernels. None of the genes is completely dominant over white, explaining the appearance of the F_1 as a blend of the characters of the two parents. Nilsson-Ehle conjectured that the intensity of the red pigmentation in the wheat kernel is controlled by *the number* of contributing alleles present at the loci controlling the production of red pigmentation. In general, he conjectured that quantitative traits are controlled by a number of genes, each of which could contribute a unit of height, weight, or other measurable characteristics. In the discussion that follows, we use grain color and the original work by Nilsson-Ehle as a reference, but the ideas adapt easily to other quantitative traits.

To determine whether the polygenic hypothesis holds promise, the assumptions just made should result in a model that could explain the following:

1. The increased variability of F_2 phenotypes, and

2. The F_2 phenotypic ratios arising from the experiments.

The effect of the polygenic hypothesis can be explained by the *binomial theorem* and the related *binomial distribution*, which we now review. Assuming that the parents are true-breeding lines, the F_1 generation has a genetic make-up formed of exactly 50% contributing and 50% noncontributing alleles. To determine the make-up of the F_2, the mathematical question we want to answer is: If we are filling N slots with one of two alleles **R** or **r** that are equally likely, what is the probability that there are k slots filled with **R**'s and $N - k$ slots filled with **r**'s? The answer is

$$\binom{N}{k} \cdot \left(\frac{1}{2}\right)^N, \text{ where } \binom{N}{k} = \frac{N!}{k!(N-k)!} \text{ and } k! = \begin{cases} 1 & \text{if } k = 0 \\ 1 \cdot 2 \cdots k & \text{if } k > 0. \end{cases}$$

$$(3\text{-}9)$$

The factor $\left(\frac{1}{2}\right)^N$ comes from the fact that we have N positions to fill independently with two alleles that are equally likely to be selected. This means that any string of length N composed of **R**'s and **r**'s has a probability of $\left(\frac{1}{2}\right)^N$. The $\binom{N}{k}$ term comes from the fact that there are $\binom{N}{k}$ different arrangements of k **R**'s and $N - k$ **r**'s.

The Binomial Theorem states that for a positive integer N,

$$(a + b)^n = \sum_{K=0}^{N} \binom{N}{k} a^k b^{N-k}.$$

If we set $a = b = \frac{1}{2}$, we see that

$$\sum_{k=0}^{N} \binom{N}{k} \left(\frac{1}{2}\right)^k \left(\frac{1}{2}\right)^{N-k} = \sum_{k=0}^{N} \binom{N}{k} \left(\frac{1}{2}\right)^N = \left(\frac{1}{2} + \frac{1}{2}\right)^N = 1,$$

so that we indeed have a probability density.

We now show how the polygenic hypothesis gives rise to a characteristic having several manifestations and why the "central manifestations" are most common. A helpful analogy is mixing red and white paint. If we are to choose two drops of red (R) or white (W) paint at random, there are three different colors arising from the following mixtures: WW, pure white; WR and RW, pink; and RR, pure red. If we were to choose four drops of paint, the possible colors are presented in Table 3-2. It seems apparent that as the number of drops increases, so does the number of possible colors. If there are m genes that contribute equally to the expression of a characteristic, then there will be $N = 2m$ alleles and $2m + 1$ different manifestations of the characteristic. According to Eq. (3-9), the probability that the manifestation corresponds to exactly k of a particular allele is $\binom{2m}{k} \cdot \left(\frac{1}{2}\right)^N$. Notice the 15:1 ratio between red- and white-kernelled corn in Table 3-2 and that the phenotypic ratios are 1:4:6:4:1, exactly as observed by Nilsson-Ehle. The model, therefore, describes the experimental results accurately.

EXERCISE 3-5

Compute $\binom{4}{k}$ for $k = 0, 1, 2, 3, 4$. Compare the results with column 3 of Table 3-2.

EXERCISE 3-6

Suppose a quantitative genetic trait is determined by six genes ($m = 6$), each of which has two alleles, **T** and **t**. Assuming each allele has an equal chance of appearing, calculate the proportions of the phenotypes in F_2 corresponding to:

(a) 12T (that is, the genotype comprised only of contributing alleles),

(b) 5T and 7t (the phenotype corresponding to genotypes comprised of 5 contributing and 7 noncontributing alleles),

Genotypes	Phenotype (grain color)	Number of Sequences	Proportion
WWWW	White	1	1/16
RWWW, WRWW, WWRW, WWWR	Light Pink	4	1/4
RRWW, RWRW, RWWR, WRRW, WRWR, WWRR	Pink	6	3/8
RRRW, RRWR, RWRR, WRRR	Dark Pink	4	1/4
RRRR	Red	1	1/16

TABLE 3-2.
Genotypes and phenotypes predicted by the polygenic hypothesis.

(c) **6T** and **6t** (the phenotype corresponding to genotypes comprised of 6 contributing and 6 noncontributing alleles),

(d) **2T** and **10t** (the phenotype corresponding to genotypes composed of 2 contributing and 10 noncontributing alleles).

If we examine the binomial distribution as shown in Figure 3-9, we see that the most common manifestation of the trait occurs when $k = m$ (that is, when the alleles are half of each type). This is the mathematical explanation for why most people seem to be near the average of a population trait.

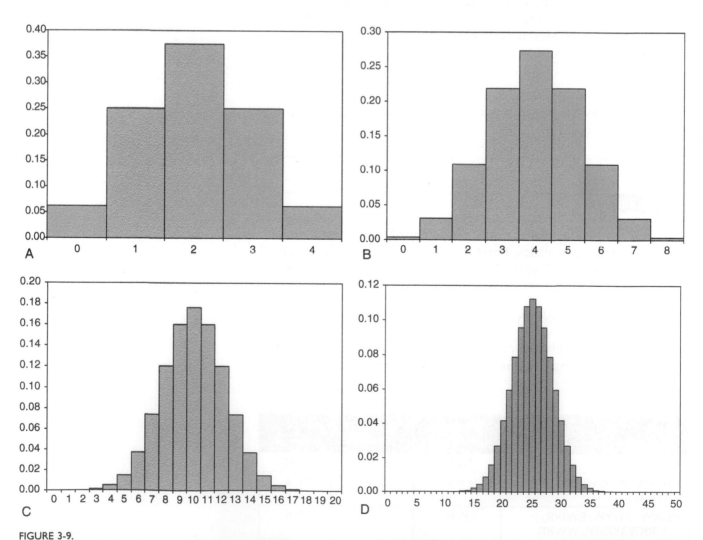

FIGURE 3-9.
The distribution of colors in F_2 for a different number of controlling genes m when the alleles **R** and **r** are equally likely. The number of different phenotypes is 5, 9, 21, and 51, respectively. Notice that (1) the distributions are symmetric; (2) the shade corresponding to exactly m contributing alleles (the shade of the F_1) is the most common; (3) the parental colors (corresponding to 0 and $2m$ contributing alleles) are the least common; (4) the closer the number of contributing alleles is to m, the more prevalent the color is; and (5) as m increases, the histograms involve a larger number of classes that correspond to increased variability of the characteristic in the F_2. Panel A: $m = 2$; panel B: $m = 4$; panel C: $m = 10$; panel D: $m = 25$.

In our discussion so far, we have only considered the case where each allele was equally likely. This isn't necessary. If we consider two alleles, one (say **R**) occurring with probability p and the other (**r**) with probability $1 - p$, the probability that exactly k of the $2m$ alleles will be **R** is

$$\binom{2m}{k} p^k (1 - p)^{2m-k}, \quad \text{where } k = 0, 1, 2, \ldots, 2m. \qquad (3\text{-}10)$$

Note that when $p = \dfrac{1}{2}$, then:

$$p^k (1 - p)^{2m-k} = \left(\frac{1}{2}\right)^k \left(1 - \frac{1}{2}\right)^{2m-k} = \left(\frac{1}{2}\right)^k \left(\frac{1}{2}\right)^{2m-k} = \frac{1}{2^{2m}}$$

and, thus, we obtain exactly the same results as in Eq. (3-9) for $N = 2m$.

Figure 3-10 presents the probability histograms for the binomial probabilities with varying values for p and N. In contrast with Figure 3-9, these histograms are not symmetric about the value $k = m$. The smooth bell-shaped curve in Figure 3-10 depicts the graph of a function known as a *normal* or *Gaussian curve*. The normal curve is a theoretical construction and presents a special case of the *Central Limit Theorem*. In this context, the central limit theorem guarantees that for sufficiently large values of m, the histogram of the binomial probabilities is well approximated by a certain Gaussian curve. In the symmetric case when $p = 1/2$, the approximation is very good, even for small values of n. When $p \neq 1/2$ the approximation gets better for larger values of m. A widely used rule of thumb advises that Gaussian approximation for $p \neq 1/2$ is appropriate when $2mp \geq 5$ and $2m(1 - p) \geq 5$.

The Gaussian approximation described above is useful, as it allows for a natural transition from a qualitative to a quantitative differentiation between the phenotypes. For example, if there are only three possible colors of wheat kernels (e.g., white, pink, and red), there is no problem referring to them as separate colors. With five different colors, the qualitative description of the phenotypes becomes more challenging. In our example above, we used "white," "light pink," "pink," "dark pink," and "red" to describe them. With seven, nine, or more different phenotypes, coming up with appropriate names for all possible colors becomes increasingly difficult. In such cases, it would be more convenient to describe the trait quantitatively as a deviation from the most commonly occurring characteristic.

The polygenic hypothesis, in combination with the central limit theorem, provides what could be called a mechanistic insight to the observation that certain quantitative traits exhibit approximately Gaussian distribution, namely, that if m, the number of genes controlling the trait is relatively large, the binomial histograms representing the actual distribution of the trait is well-approximated by a Gaussian curve. More details regarding this result will be presented in the next chapter. We note that the converse may not be true. If a trait is assumed to be

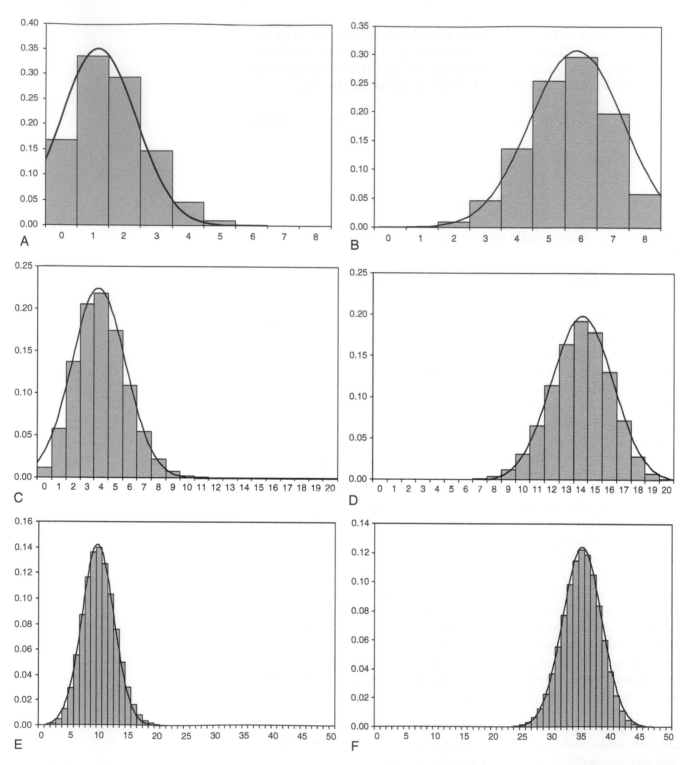

FIGURE 3-10.
The distribution of colors in F_2 for a different number of controlling genes m with probabilities p and $1 - p$ for the alleles **R** and **r**, respectively. While for $p = 1/2$ the histogram is symmetric, for $p \neq 1/2$ this is not the case. However, as the diagrams in Figure 3-10 exemplify, even for values $p \neq 1/2$, the distribution becomes quite symmetric when m is relatively large. Panel A: $m = 4$, $p = 0.2$; panel B: $m = 8$, $p = 0.7$; panel C: $m = 10$, $p = 0.2$; panel D: $m = 10$, $p = 0.7$; panel E: $m = 25$, $p = 0.2$; panel F: $m = 25$, $p = 0.7$.

quantitative, this does not necessarily imply that it is controlled by multiple loci. Instead, other factors, such as the environment, proper nutrition, or socioeconomic behaviors, may have a smoothing effect on the trait, introducing phenotypes that are not entirely determined by genetic inheritance. These observations raise some interesting and important questions.

D. GENES, ENVIRONMENTS, AND VARIATION IN A POPULATION

In the first part of this chapter, we examined the genetics of discontinuous traits. These are the kinds of traits described by Gregor Mendel as either dominant or recessive: the alternate forms of tall or short pea plants with green or yellow seeds which may be smooth or wrinkled. However, not all inheritance patterns are so straightforward. As discussed above, it has been experimentally determined that many traits, such as height, assume their values from a continuum of possibilities. Additionally, some of the variation in human height is not caused by genetic factors alone. Environmental factors, such as the availability and nutritive value of the food consumed by individuals during childhood, will certainly play a part in determining the height distribution of a population. Because of the continuous nature of quantitative traits such as height, it is much more difficult to determine the relative contributions of heredity and environment to the resulting variation observed in the population.

It is important to be able to assess the relative contributions of heredity and environment, because environmental factors have considerable impact on the development of many phenotypes. For example, genetic predisposition is known to increase the possibility for developing type II diabetes, but obesity is another critical factor. Thus, the development of the disease phenotype depends upon the interaction of genetic and environmental (including behavioral) factors. Knowing how much of this predisposition is genetic and how much is caused by environmental factors would allow individuals to make informed decisions about diet and exercise, and help public health officials to make appropriate recommendations to minimize chances of developing the disease.

Consider all of the women whose mothers had breast cancer. How can we quantify their risk of developing breast cancer? How would that quantification change if we were talking about women who pursue low-fat diets and healthy lifestyles? How would it change if the women in question were alcoholic chain-smokers? How much of their risk is caused by genetics, and how much is caused by environmental (specifically behavioral) factors? In the next chapter, we shall examine techniques that will help us quantify these kinds of questions.

REFERENCES

Darwin, C. (1964). *On the origin of species by means of natural selection, or the preservation of favoured races in the struggle for life.* Cambridge, MA: Harvard University Press. (Original work published 1859).

Falconer, D. S. (1989). *Introduction to quantitative genetics* (3rd ed). New York: John Wiley and Sons.

Fraser, E. D. G. (2003). Social vulnerability and ecological fragility: Building bridges between social and natural sciences using the Irish potato famine as a case study. *Conservation Ecology, 7,* 9. Retrieved September 21, 2006, from http://www.consecol.org/vol7/iss2/art9.

Fry, W. E., & Goodwin, S. B. (1997). Resurgence of the Irish potato famine fungus. *BioScience, 47,* 363–372.

Garrett, K. A., & Mundt, C. C. (2000). Host diversity can reduce potato late blight severity for focal and general patterns of primary inoculum. *Phytopathology, 90,* 1307–1312.

Goodwin, S. B. (1997). The population genetics of *Phytophthora. Phytopathology, 87,* 462–473.

Hooker, A. L., Smith, D. R., Lim, S. M., & Beckett, J. B. (1970). Reaction of corn seedlings with male-sterile cytoplasm to *Helminthosporium maydis. Plant Disease, 54,* 708–712.

Mendel, G. (1965). *Experiments in plant hybridisation.* Cambridge, MA: Harvard University Press.

Ristaino, J. B., Groves, C. T., & Parra, G. R. (2001). PCR amplification of the Irish potato famine pathogen from historic specimens. *Nature, 411,* 695–697.

Russell, P. J. (2006). *iGenetics: A molecular approach* (2nd ed.). San Francisco: Pearson Education/Benjamin Cummings.

Ullstrup, A. J. (1972). The impacts of the southern corn leaf blight epidemics of 1970–1971. *Annual Review of Phytopathology, 10,* 37–50.

FURTHER READING

Boyer, S. H. (1963). *Papers on human genetics.* Englewood Cliffs, NJ: Prentice-Hall.

Dean, M., Carrington, M., & O'Brien, S. J. (2002). Balanced polymorphism selected by genetic versus infectious human disease. *The Annual Review of Genomics and Human Genetics, 3,* 263–292.

Hardy, G. H. (1908). Mendelian proportions in a mixed population. *Science, 28,* 49–50.

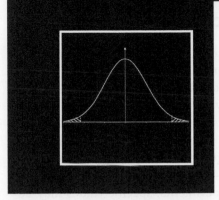

Chapter 4

QUANTITATIVE GENETICS AND STATISTICS

Probability Background

Relation of Probability Distributions to Statistical Testing

Statistical Testing

Science is the great antidote to the poison of enthusiasm and superstition.

Adam Smith (1723–1790)

A fundamental question, both philosophically and biologically, is what causes differences within a species? The issue is complicated, but commonly the causes are classified as genetic or environmental or as developmental "noise." Developmental noise is a random occurrence unrelated to genetic or environmental causes, such as the small physical differences between identical twins when they emerge from the womb. When considering genetic, environmental, or developmental factors, one would like to attribute a certain amount of variation to each factor, but this is usually not possible on an individual basis. Complicating the problem further, there will usually be some crossover between factors (called *covariance*). For example, a child whose parents are musicians may be genetically predisposed toward music (if such musical genes exist), but would also benefit from being reared in a musically nurturing environment and engaging in music-related behaviors. Similar confounds occur in analyzing intelligence levels.

In no sense have these questions been thoroughly answered in this book, but we shall pursue some worthwhile inquiries while presenting several statistical tools useful in many areas of biology.[1] These tools will enable us to make unbiased judgments on data-based hypotheses. We shall focus on three related topics:

1. How do we decide whether any observed differences are statistically significant?

2. Is the contribution of an underlying genetic factor significant relative to environmental factors?

3. What is the common mathematical thread linking all statistical tests comparing means, evaluating the contribution of various factors, or testing the linear dependence of an outcome on a set of predictive variables?

We begin by reviewing some terminology from probability.

---

1. For an in-depth analysis of this field, the reader is referred to Falconer (1989).

I. PROBABILITY BACKGROUND

Consider an experiment whose outcome depends on chance—such as flipping a coin or rolling a die, the number of chickens hatching on a farm each day, or how long after sunset a bat leaves a cave. If we repeat such experiments, the outcomes will vary randomly, and so we say the outcome is described by a *random variable*. Each time the experiment is performed, the random variable takes a specific value corresponding to that outcome. In most cases, this assignment is very natural. When rolling a die, the value rolled (1, 2, 3, 4, 5, or 6) will be the number assigned to the outcome. For some experiments, specific values may be more likely to occur than others. The chances of an expectant mother delivering twins are smaller than for delivering a single baby, but are higher than for septuplets. Each random variable has a *probability distribution* specifying how likely an outcome, or a set of outcomes, is to occur. The probability distribution of the random variable can be visualized as follows: If we make many (theoretically, infinitely many) trials and find the numbers the random variable assigns to the trials, the relative frequency histogram of these numbers will provide a good approximation of the random variable's distribution. It is common in probability to denote random variables by letters from the Greek alphabet, such as ξ and η.

In the previous chapter, we described how Nilsson-Ehle used this approach to obtain the distribution of phenotypes in the F_2 generation originating from a parental cross of white and red grain wheat and presented the results in a table (reproduced here as Table 4-1). The last column of Table 4-1 represents the probability distribution obtained from the approximate proportions 1:4:6:4:1 he observed between the phenotypes in the F_2. In this example, the random variable representing the grain color in the F_2 is *discrete*, as there is a finite number of values this random variable can take. The same will be true for the random variable giving the number of chickens hatching on a farm each day. The

Genotypes	Phenotype (Grain Color)	Number of Sequences	Probability
WWWW	White	1	1/16
RWWW, WRWW, WWRW, WWWR	Light pink	4	1/4
RRWW, RWRW, RWWR, WRRW, WRWR, WWRR	Pink	6	3/8
RRRW, RRWR, RWRR, WRRR	Dark pink	4	1/4
RRRR	Red	1	1/16

TABLE 4-1.
Genotypes and phenotypes predicted by the polygenic hypothesis.

binomial distribution introduced in the previous chapter provides another example of a discrete distribution.

In contrast, if a random variable represents the weight of a newborn baby, it can take on a continuous range of values. This is an example of a *continuous* random variable, and the tool that replaces the table of values with their corresponding probabilities is the *probability density function*. A function $f(x)$ is a probability distribution density function for a random variable ξ provided that:

1. $f(x) \geq 0$;

2. $\int_{-\infty}^{\infty} f(x)dx = 1$;

3. For any numbers $a < b$, the probability that ξ is between a and b is calculated as $P(a < \xi < b) = \int_a^b f(x)dx$.

The most common and important probability density function is the *normal* or *Gaussian* distribution. A special case is the *standard normal density function* defined by the expression

$$f(x) = \frac{1}{\sqrt{2\pi}} . e^{\frac{-x^2}{2}}. \tag{4-1}$$

The graph of this function is depicted in Figure 4-1(A).

The *mean* and *variance* of a random variable are two numerical characteristics associated with its distribution. The mean could be thought of as the average we would expect after doing many trials. The variance measures the spread of the data around the mean value. The mean is typically represented by μ and the variance by σ^2. The *standard deviation*, defined as the square root of the variance, is denoted by σ.

The following heuristic explanation outlines a fundamental principle: A normal distribution occurs when multiple independent random choices are made, each of them attempting to achieve a certain fixed average value, but each is vulnerable to errors that are symmetrical in both directions around the mean.

Several examples will clarify this point. Thousands of bats exit their cave about 2 hours after sunset. Because inside the cave there is no indication of when exactly sunset occurs, each bat relies on its biological clock to estimate the exit time. As we shall see in a later chapter, some biological clocks run fast and some slow, with larger errors less likely to occur than smaller ones. Experiments have shown that the number of bats exiting a cave per minute follows quite precisely a bell-shaped curve with a mean of about 2 hours after sunset. Similarly, if we place microscopic particles in a glass of water, they get hit by water molecules and travel various distances, depending on the

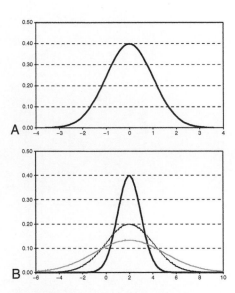

FIGURE 4-1.
Comparison of normal density functions with differing parameters. Panel A: Standard normal density; that is, normal density with mean $\mu = 0$ and standard deviation $\sigma = 1$; panel B: normal densities with $\mu = 2$ and $\sigma_1 = 3$ (gray line), $\sigma_2 = 2$ (black dotted line), and $\sigma_3 = 1$ (solid black line).

force of the impact. The average travel distance depends on the temperature (energy) of the water, while the distribution of the distances around this average is approximately normal. In other words, each bat's time estimate of 2 hours after sunset, or each particle's travel distance, is one outcome of an experiment described by a random variable with a normal density function. The general analytical expression of the normal (Gaussian) density is given by the function

$$f(x) = \frac{1}{\sqrt{2\pi \cdot \sigma}} e^{1 - 2\left(\frac{x - \mu}{\sigma}\right)^2}, \tag{4-2}$$

where x can take any value in the interval $(-\infty, \infty)$. When we say that a random variable ξ has a normal distribution with parameters μ and σ, this means the random quantity represented by ξ has a density function such as in Eq. (4-2). The values μ and σ define uniquely the shape of the curve and correspond to the mean and the standard deviation of the normal distribution. The mean value μ, $-\infty < \mu < \infty$, determines the position of the maximum for the bell-shaped graph of the normal density function from Eq. (4-1). The standard deviation σ, $\sigma > 0$, determines how sharp the peak is near the maximum. Figure 4-1(B) illustrates how the density graphs for the function in Eq. (4-2) change with the change of the parameters μ and σ. When $\mu = 0$ and $\sigma = 1$, the density function from Eq. (4-2) takes the form presented by Eq. (4-1). That is, for $\mu = 0$ and $\sigma = 1$, we obtain the *standard normal distribution*.

If the outcomes of two or more experiments do not influence one another, we refer to them as being *independent*. Similarly, we say that the random variables associated with these experiments are also independent. Consider three random variables describing the following quantities: (1) The time after sunset when a bat exits its cave located in Virginia; (2) the height of a corn plant in a cornfield in Mexico; and (3) the acidity of the soil on which the same plant grows. Because it appears unlikely that the characteristics of the cornfield in Mexico can affect the behavior of Virginia bats, the bat exit time can be considered independent from the corn height and also independent from the acidity of the cornfield. On the other hand, the acidity of the soil and the height of the corn plants cannot be considered independent because the acidity is expected to affect growth.

If we combine two random variables with the same distribution, the result will not necessarily have that same distribution. For example, if we flip a coin twice and let η_1 be the number of heads (H) on the first flip and η_2 be the number of heads on the second flip, then the probabilities that η_1 takes values 1 or 0, respectively, are $P(\eta_1 = 1) = 1/2$ and $P(\eta_1 = 0) = 1/2$. The random variable η_2 will have exactly the same distribution. But $\eta_1 + \eta_2$ is the number of heads on both flips. The possibilities are HH, HT, TH, and TT, so the probabilities that $\eta_1 + \eta_2$ takes values 2, 1, and 0, respectively, are $P(\eta_1 + \eta_2 = 2) = 1/4$, $P(\eta_1 + \eta_2 = 1) = 1/2$, and $P(\eta_1 + \eta_2 = 0) = 1/4$.

An important property of the normal distribution is that when we add two independent random variables ξ and η that are normally distributed, the sum $\xi + \eta$ also has a normal distribution with mean equal to the sum of the means and variance equal to the sum of the variances of ξ and η. Other operations performed on ξ and η, however, will no longer result in normal distribution. We next consider several other probability distributions that are derived from the normal distribution and used for various statistical tests. We shall introduce them briefly, without the cumbersome formulas for their densities. Figure 4-2 shows the graphs of these densities for various choices of parameters.

Chi-square (χ^2) distribution. This type of distribution arises when we consider squares of random variables with standard normal distribution. More specifically, if ξ is a random variable with a standard normal distribution, it may sometimes be necessary to consider $\eta = \xi^2$. Because ξ is a random variable, so is η, but the density function of η cannot be normal because the new random variable η takes only positive values. We say that $\eta = \xi^2$ has a χ^2 *distribution with one degree of freedom.* If we consider several independent random variables $\xi_1, \xi_2, \ldots, \xi_N$, all of which have standard normal distributions, then we say the sum of the squares of these random variables, namely $\eta = \xi_1^2 + \xi_2^2 + \ldots + \xi_N^2$, has a χ^2 *distribution with N degrees of freedom.* Figure 4-2-(A) presents the density function of a χ^2 distribution with $N = 7$ and $N = 21$ degrees of freedom, respectively.

t-(Student) distribution. This distribution arises when we need to consider specific ratios of random variables. In particular, if ξ is a random variable with a standard normal distribution and η is another random variable with a χ^2 distribution with N degrees of freedom, then their ratio $\zeta = \xi/\sqrt{\eta}$ will be a new random variable. This new random variable is said to have a *t-distribution* (or *Student distribution*) *with N degrees of freedom.* The graphs of the probability density for *t*-distribution with $N = 2$ and $N = 23$ degrees of freedom are shown in Figure 4-2(B).

F-(Fisher) distribution: This distribution also arises when ratios of random variables are considered, but this time it is the ratio of two random variables with χ^2 distributions. More specifically, if ξ is a random variable with a χ^2 distribution with M degrees of freedom and η is another random variable with a χ^2 distribution with N degrees of freedom, then their ratio $\zeta = \xi/\eta$ will be a new random variable said to have an *F-distribution* (or *Fisher distribution*) *with M, N degrees of freedom.*

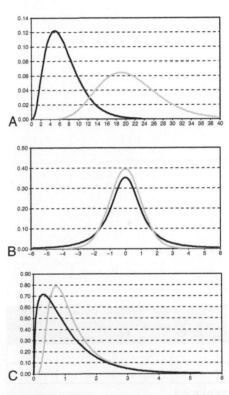

FIGURE 4-2.
χ^2, *t*- and *F*-distributions with varying parameters. Panel A: χ^2 distribution with 7 (black) and 21 (gray) degrees of freedom; panel B: *t*-distribution probability density with 2 (black) and 23 (gray) degrees of freedom; panel C: *F*-distribution with 3, 40 (black) and 23, 8 (gray) degrees of freedom.

Graphs of *F*-distribution densities for different choices of *M* and *N* are shown in Figure 4-2(C).

We now give an intuitive explanation of how these probability definitions relate to common problems of statistical testing. First, statisticians typically assume that their observations are collected by independent measurements of a random variable having a normal distribution. In this case, it is said the data have been derived from a population described by a normally distributed random variable. There are numerous statistical procedures that validate this assumption, and it is always important to check the normality of your sample before applying statistical tests, such as the *t*-test or *F*-test described below. If the normality assumption does not hold, the results from some statistical tests could be misleading.

II. RELATION OF PROBABILITY DISTRIBUTIONS TO STATISTICAL TESTING

Returning to the question of quantitative traits and the polygenic hypothesis, suppose that an agricultural company is attempting to alter a type of corn to produce a new variety (B) that will be superior to the original variety (A). First, we need to specify what we mean by superior. It could mean higher yield, greater resistance to drought or disease, less-intensive soil preparation or cultivation, or easier harvest. In this case, let us suppose we wish to improve the yield of the plants. Like height and weight in humans, we have reason to suspect corn yield is well described by a normal distribution; thus, we shall assume that the normality assumption holds.

Plant No.	A (Yield)	Plant No.	B (Yield)
1	2.1	1	2.4
2	2.6	2	2.5

TABLE 4-2.
The trial 1 data from an imaginary experiment involving two plants each of the varieties A and B.

The next step would be to design an experiment and collect data. This may sound simple, but several factors need to be considered. Would you take one of each type, plant them side-by-side under the specified conditions, and see which produces a higher yield? That may sound appealing (it is certainly simple), but one can never be sure that every plant of the same type will have exactly the same yield. In fact, it is very likely there will be some variance between plants of the same type, even under the same conditions.

Suppose, then, that we run trials with different numbers of plants and record the data in Tables 4-2 and 4-3.

Plant No.	A (Yield)	Plant No.	B (Yield)
1	2.2	1	2.4
2	2.8	2	2.8
3	1.9	3	3.1
4	3.2	4	2.6
5	2.6	5	2.5
6	2.1	6	2.8
7	2.7	7	3.2
8	2.4	8	3.4
9	2.5	9	2.9
10	2.0	10	2.7

TABLE 4-3.
The trial 2 data from an imaginary experiment involving 10 plants each of the varieties A and B.

Would you have more confidence making a conclusion based on the first or second trial? What we have done in both trials is take a *sample* from each of the plant varieties. What we are attempting to do with the sample is estimate the yields from the *population*. The fact we used more plants in the second trial would probably engender more confidence in its representation of the population. In other words, *sample size* is an important factor in determining our confidence in the outcome (the word "confidence" is an important part of the statistical vocabulary). There is a class of statistical methods called *power analysis* that allows an optimal sample size for achieving a desired confidence to be estimated prior to the experiment. While it is intuitively clear that a larger sample size would result in more powerful tests, power analysis is a convenient tool whenever data collection incurs expenses. In such a case, the problem is to find the minimal sample size to achieve the desired power of the test.

The mathematical formulation of the problem uses some specific language we outline next. The traditional way to describe a statistical problem begins like this: Let $x_1, x_2, \ldots, x_N$ be N independent observations of a normally distributed random variable ζ with unknown parameters μ and σ. This means we have collected some data by measuring a random quantity known to have a normal distribution, and the values from the measurements have been denoted by $x_1, x_2, \ldots, x_N$. These values are the *data points* forming our sample for the random variable ζ. In terms of our corn yield example, assume that the random variable ζ represents the population yield from variety A that we are attempting to estimate by sampling repeatedly. In the first trial, the sample size was $N = 2$, while in the second trial we considered a larger sample size of $N = 10$. The data points $x_1 = 2.1, x_2 = 2.6$ define the sample for ζ from the first trial, while $x_1 = 2.2, x_2 = 2.8, \ldots, x_{10} = 2.0$ define the sample from the second trial.

As already mentioned, the mean of a random variable could be thought of as the average after many trials. Thus, it is natural to expect that the average value of the data points, denoted by $\bar{x}$,

$$\bar{x} = \frac{x_1 + x_2 + \ldots + x_N}{N} = \frac{\sum_{i=1}^{N} x_i}{N} = \frac{1}{N} \sum_{i=1}^{N} x_i, \tag{4-3}$$

would be a good estimate of the mean value parameter μ of the normal distribution.[2] This is why the average value calculated in Eq. (4-3) is sometimes called the *empirical mean* or *sample mean* of the random variable ζ. It can be proven that this is the best estimate (in terms of statistical criteria), also called the *maximum likelihood* estimate. In these terms, the test average you earn in a class is a maximum likelihood estimate of your grade. Similarly, it can be shown that a maximum likelihood estimate of the variance σ^2 of a normal distribution is given by the formula

$$s^2 = s^2(N) = \frac{(x_1 - \bar{x})^2 + (x_2 - \bar{x})^2 + \ldots + (x_N - \bar{x})^2}{N-1} = \frac{1}{N-1} \sum_{i=1}^{N} (x_i - \bar{x})^2.$$

$$\tag{4-4}$$

---

2. As is customary in mathematics, we have used $\sum_{i=1}^{N} x_i$ to denote the sum $x_1 + x_2 + \ldots + x_N$.

The value s^2 is sometimes called the *empirical variance* or *sample variance*. The square root, s, is called the *empirical standard deviation* of the random variable ξ.

If we need to measure two different random variables ξ and η having a normal distribution, we record the data points for ξ and η as samples A and B. Sample A contains the data points measured for ξ and sample B contains the points measured for η. In our earlier example, samples A and B contained the yield data for corn varieties A and B, presented in Tables 4-2 and 4-3. To distinguish between the mathematical expressions using data from sample A from those using sample B, we use the name of the sample as part of the notation. For example, we use $\bar{x}_A$ to denote the average of the data points from A, and $s_B^2 = s_B^2(N)$ to denote the maximum likelihood estimate of the variance calculated by the formula in Eq. (4-4) with the data from sample B. In the latter case, the value of N will correspond to the number of data points in the sample B.

Recall that the sum of two independent, normally distributed variables also has a normal distribution, with a mean equal to the sum of the means and variance equal to the sum of the variances. It follows then, the empirical mean $\bar{x}$ and the difference $(x_i - \bar{x})$ will also have normal distributions. The same holds for the difference $\bar{x}_A - \bar{x}_B$, when we consider two samples. Particularly, $\bar{x}$ will have a mean equal to μ and a variance equal to σ^2/N. Further, because $(x_i - \bar{x})$ has normal distribution, the estimate of the variance s^2, which is a sum of the squares of N such quantities, will have an approximately χ^2 distribution with N degrees of freedom.

When these considerations are paired with the definitions we gave for χ^2 distribution, t-distribution, and F-distribution, in general, the following broad principles hold:

1. Any statistical test (such as the Z-test, as we shall see later) using the difference $\bar{x}_A - \bar{x}_B$, between the empirical means of two samples A and B as their test statistic requires the use of a normal distribution;

2. Any statistical test (such as the t-test, as we shall see later) using a variant of the ratio $\bar{x}/s$ (empirical mean/empirical standard deviation) as their test statistic, requires the use of a t-distribution (as it is approximately normal/$\sqrt{\text{Chi}-\text{square}}$); and

3. Any statistical test (such as the F-test, as we shall see later) using the ratio $s_A^2(M)/s_B^2(N)$ (empirical variance of one sample with M

readings/empirical variance of another sample with N readings) as their test statistic, requires the use of a F-distribution (as it is approximately Chi-square/Chi-square) with M, N degrees of freedom.

We will usually have a group of data for which we need to choose which statistical test to perform, based on the question at hand. Once this has been decided, we shall use standard statistical software, such as *MINITAB* or *SPSS*, to carry out the computations. When this is done, we should be able to interpret the output. We begin by outlining the fundamentals of formulating a hypothesis and performing statistical testing.

III. STATISTICAL TESTING

A. Testing a Hypothesis

In hypothesis testing, we make a claim and then gather data to evaluate this claim. The claim is called the *null hypothesis*. The negation of the null hypothesis is called the *alternative hypothesis*. The exact criteria used for the evaluation are based on knowledge of probability distributions coming from statistical theory. In this text, the claims will most often be about the value of a distribution parameter. The salient point is: Assuming that the null hypothesis is correct, we want to be able to calculate the probability that the sample we took could occur. For example, suppose the null hypothesis for the mean value μ of a probability distribution is that $\mu \leq 0$ and we take a sample of size 20 and find the sample mean $\bar{x}$ is 0.7. Suppose also statistical theory says that if the null hypothesis is correct, the probability distribution of the means of the samples of size 20 will be as shown in Figure 4-3(A). In Figure 4-3(B), we locate the particular value of our sample and shade the area under the probability density curve to the right of this value. The area of the shaded region represents the probability the sample mean we took would have been 0.7 or larger, if the null hypothesis is correct. This probability is the p-value corresponding to the empirical result of 0.7. A very small p-value would indicate the observed mean is very unlikely to have occurred among all samples of size 20 if the null hypothesis was, in fact, true, and therefore provides enough evidence for rejecting the null hypothesis. If the evidence from the data gathered provides enough confidence to reject the null hypothesis, this is also evidence in favor of the alternative hypothesis.

To illustrate these concepts with a specific example, we return to the two types of corn and ask whether the genetically engineered variety B produces higher average yields than the original variety A. The yields

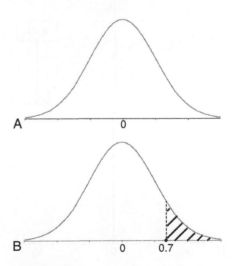

FIGURE 4-3.
Standard normal distribution. Panel A: Probability distribution for the means of samples of size 20; panel B: The area of the shaded region represents the probability that in taking a sample of size 20 we would get a sample mean of 0.7 or larger if the null hypothesis is correct.

from A and B can be considered random quantities and denoted by ξ and η, with the mean values of ξ and η denoted by μ_A and μ_B, respectively. In this case, rejecting a null hypothesis that $\mu_A \geq \mu_B$ will provide evidence supporting the alternative $\mu_A < \mu_B$, which would imply a higher yield from variety B. The null hypothesis is commonly denoted by H_0 and the alternative hypothesis by H. In general, hypothesis tests are classified as *one-tailed* or *two-tailed*, or, alternatively, as *one-sided* and *two-sided*. A one-tailed test is one in which the hypothesis is directional (i.e., uses either a "<" or a ">"). This corn problem is an example of a one-tailed test. In a two-tailed test, the hypothesis does not specify a direction and will use the "$\neq$" symbol.

An example of a two-tailed test with the same data would be "is there a difference between the yield of plants A and B?" In this case, the null hypothesis would be H_0: $\mu_A = \mu_B$ and H: $\mu_A \neq \mu_B$ would represent the two-tailed alternative.

It is important to note that the decision one makes in hypothesis testing is to "reject the null hypothesis" or "not reject the null hypothesis." One does not use the language "accept the alternative hypothesis," even though this might seem appropriate. The analogy often used to explain this philosophy is a criminal case in the United States court system, where the null hypothesis is "innocent" and a jury will vote to convict only if the evidence of guilt is compelling. A vote to acquit, therefore, does not imply that innocence was proven.

In hypothesis testing, there are two types of error one can make. A *type I error* is rejecting a true null hypothesis, and a *type II error* is failing to reject a false null hypothesis. A type I error is what we really want to avoid. In the court analogy, this would correspond to convicting an innocent person (a type II error would be to acquit a guilty person). It is intuitively clear that the lower we set the threshold for a type I error, the more certain we are that we will not reject a true null hypothesis. This, however, leads to the danger of increasing the type II error. In the example at the beginning of this section, the probability for type I error is the area of the shaded region in Figure 4-3(B). It is generally accepted that this *p*-value should be less than 0.05 for the null hypothesis to be rejected.

Back to our corn example, suppose we are testing H_0: $\mu_A \geq \mu_B$ versus the alternative H: $\mu_A < \mu_B$ or, equivalently, H_0: $\mu_A - \mu_B \geq 0$ versus the alternative H: $\mu_A - \mu_B < 0$. Suppose also that when testing our hypothesis, we want to limit the magnitude of type I error we may be making to 0.05. From the data gathered for the experiment, we compute $\bar{x}_B - \bar{x}_A$, which we denote by α. We then need to evaluate the probability of how likely it is the sampling distribution for $\bar{x}_B - \bar{x}_A$, determined assuming the correctness of the null hypothesis, would produce that value. Assume now that the sampling distribution for $\bar{x}_B - \bar{x}_A$ is represented by the density function depicted in Figure 4-4(A).

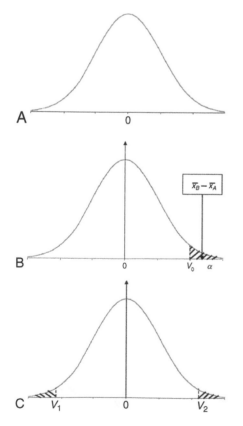

FIGURE 4-4.
Graphical representation of one-tailed and two-tailed *p*-values.

Because we are considering a one-sided hypothesis, we find a value V_0 such that the area under the curve to the right of V_0 is exactly 0.05 [see Figure 4-4(B)]. Next, we plot the value $\alpha = \bar{x}_B - \bar{x}_A$ on the horizontal axis. If the value of α falls to the right of V_0, this would mean that if we reject the null hypothesis, the chance we are wrong (that is, the chance we shall reject the null hypothesis when it is, in fact, true) would be smaller than 5%.

If we have a two-sided alternative hypothesis H: $\mu_B \neq \mu_A$, we need to find two numbers V_1 and V_2 [see Figure 4-4(C)] such that the area to the left of V_1 and to the right of V_2 is 0.025, and thus the total area under the curve outside the interval $[V_1, V_2]$ is 0.05. If the value $\alpha = \bar{x}_B - \bar{x}_A$ falls to the right of V_2 or to the left of V_1, then the chance we shall wrongly reject the null hypothesis will be less than 5%.

The p-value produced by statistical software such as *MINITAB* and *SPSS* is equal (in the case of one-sided hypotheses) to the area under the curve to the right of the value of our statistic $\alpha = \bar{x}_B - \bar{x}_A$. Therefore, if this p-value is less than 0.05, this means $\alpha = \bar{x}_B - \bar{x}_A$ is to the right of the value V_0, and therefore H_0 can be rejected with a chance of type I error less than 5%. In the case of two-sided hypotheses, the p-value represents the combined area under the curve outside of the interval $[-|\bar{x}_B - \bar{x}_A|, |\bar{x}_B - \bar{x}_A|]$. The important thing to remember is that, in all cases, the p-value is exactly the probability for type I error.

One question we have not addressed so far is how to determine the actual type of the sampling distributions under the assumption the null hypothesis is true. This choice is based on underlying assumptions for the populations, as well as on the type of parameters and claims referenced by the null hypothesis. As we shall see, the probability distributions we introduced above play a fundamental role in this process. The following cases will be quite common:

Case I. The null hypothesis deals with comparing mean values; and

Case II. The null hypothesis deals with comparing variances.

In the next two sections, we outline some basic statistical tests that allow for hypothesis testing of the above cases. They represent essential ideas that will be needed in later chapters to understand how to interpret the results from other statistical analyses. As a very broad rule, when the assumptions for normality are met and the sample size is large enough, hypotheses of the type outlined in case I would use a Z-test or t-test, while those outlined in case II would use an F-test. In the next section, we complete the corn example, which represents a special instance of case I. In Section C, we illustrate case II by examining testing for heritability.

B. Z-Test and Student's *t*-Test

The Z-test and Student's *t*-test[3] are used to compare the means of two
samples. Assume, as above, that we want to evaluate the claim
$H_0: \mu_A \geq \mu_B$ (or, equivalently, $\mu_A - \mu_B \geq 0$) versus the alternative
$H: \mu_A < \mu_B$ (or, equivalently $\mu_A - \mu_B < 0$). In this case, the sampling
distribution of $\bar{x}_B - \bar{x}_A$, will be approximately normal, and we would use
either a Z-test or a *t*-test. We use a *t*-test when the variance of either
group is unknown [in which case the unknown variance(s) are estimated
by the sample variance(s)], leading to the need to use a *t*-distribution as
the sampling distribution. In fact, in most situations, the group variances
are unknown, necessitating the use of the *t*-test. We should note,
however, that for large samples, the resulting *t*-distribution is closely
approximated by a standard normal distribution. It is clear from Figure 4-2
(B) that the density graphs of the *t*-distribution and the standard
normal distribution in Figure 4-1(A) are very similar. If the degrees of
freedom in the *t*-distribution exceed 30, the two distribution densities are
virtually indistinguishable.

To evaluate the probability of making a type I error, we compute a test
statistic as we did above. We first present the simpler procedure of using
a Z-test.

Z-test: Let's go back to our corn example and use the data from Trial 2 to
test the one-sided alternative hypothesis:

*H: Average yield of corn B is superior to the average yield of corn A (that is,
$\mu_A < \mu_B$),*

against the null hypothesis:

H_0: *Corn B has the same or inferior average yield to corn A (that is, $\mu_A \geq \mu_B$).*

To apply the Z-test, we need to have information about the population
variance. For purposes of the illustration, let's assume that we
know that the population variance of corn is equal to 0.16 (i.e., the
standard deviation is 0.4). If we use Eq. (4-3) with the data from
Trial 2, we find the empirical mean yield of corn A is $\bar{x}_A = 2.44$,
while the empirical mean of corn B is $\bar{x}_B = 2.84$. The variance of $\bar{x}_A$
and $\bar{x}_B$ is equal to the sum of variances of their components divided by
10 (the number of plants in each sample). If the null hypothesis is true,
the difference $\bar{x}_B - \bar{x}_A$ will have a normal distribution, with mean
$\mu = \mu_B - \mu_A \leq 0$ and variance 0.032 (equal to $(0.16 + 0.16)/10$). Thus,
the difference $\bar{x}_B - \bar{x}_A$ has a normal distribution with parameters $\mu \leq$
0 and standard deviation $\sigma = \sqrt{0.032} = 0.1789$. Following Figure 4-4(B),

---

3. The *t*-test was developed by W. S. Gossett (1876–1937), who worked in Dublin,
Ireland, at the Guinness Brewery and published under the pen name Student.

the probability of finding the result $\bar{x}_B - \bar{x}_A \geq 0.4$ from the data when $\bar{x}_B - \bar{x}_A$ is normal with parameters $\mu = 0$ and $\sigma = 0.1789$ is 0.0127. We find this probability using statistical tables for normal distribution or software that computes normal distribution. Thus, the *p*-value of our one-sided test is $p = 0.0127 < 0.05$ (see Figure 4-5), and we can reject the null hypothesis of equal yield from the two corn types (and, thus, also that the yield from type B is inferior to that from type A). When either of group variances are unknown, we use a *t*-test as exemplified next.

t-test: As before, we want to test the one-sided alternative

H: *Average yield of corn B is superior to the average yield of corn A* ($\mu_A < \mu_B$)

against the null hypothesis:

H_0: *Corn B has the same or inferior average yield to corn A* ($\mu_A \geq \mu_B$).

This time, however, we assume the variance in the yield of the corn is not known. This is, in fact, the more common practical situation. Now we must rely on empirical estimates of the variances of the two types of corn, using the formulas in Eq. (4-4) to obtain $s_A = 0.4033$ and $s_B = 0.3169$. The sample means and variances can also be computed using statistical software, such as *MINITAB* and *SPSS*. The output from *MINITAB* is given here.

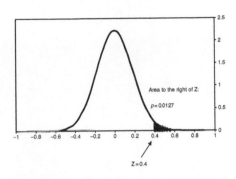

FIGURE 4-5.
Relating the *p*-value and the *Z*-score. Using a table, one associates a *Z*-score of the standard normal distribution with an area (i.e., a *p*-value). For example, $Z = 0.4$ gives a one-tailed *p*-value of 0.0127.

```
Descriptive Statistics: A, B

Variable   N    Mean    StDev   Minimum   Maximum
A         10   2.440    0.403    1.900     3.200
B         10   2.840    0.317    2.400     3.400
```

The sample distribution for testing the hypothesis H against the null hypothesis H_0 becomes a *t*-distribution necessitating use of the *t*-test. Because the computation of the *t*-statistics is quite complex and is routinely done by computer, we are not going to present the exact formula here. However, using *MINITAB* we find the *t*-statistics have a value of $t = 2.466$ with 17 degrees of freedom. We again follow the general paradigm depicted in Figure 4-4(B), this time using Student *t*-distribution.

Using appropriate software, we find that the *p*-value of this one-sided test is $p = 0.012 < 0.05$, meaning the probability of having an empirical result $\bar{x}_B - \bar{x}_A \geq 0.4$ assuming the null hypothesis holds is 0.012. Therefore, we can reject the null hypothesis. The *MINITAB* output from this test is presented here. Figure 4-6 presents a graphical illustration of the output.

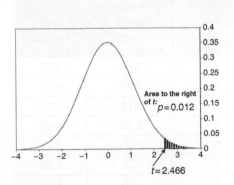

FIGURE 4-6.
p-value calculated from the *t*-value of the Student's *t*-distribution with 17 degrees of freedom.

```
Two-Sample t-Test and CI: B, A
Two-sample T for B vs A

        N      Mean     StDev      SE      Mean
B       10     2.840    0.317              0.10
A       10     2.440    0.403              0.13

Difference = mu (B) − mu (A)
Estimate for difference: 0.400000
95% lower bound for difference: 0.117824
T-Test of difference = 0 (vs >):
T-Value = 2.466 P-Value = 0.012; DF = 17
```

The Z-test and t-test help us answer the first of the three questions we posed at the beginning of this chapter; namely, how to find out whether any observed group differences are statistically significant. In our example, the Z-test and the t-test produce quite similar results because of the fact that the empirical variances used for the t-test (calculated above as 0.403 for type A corn and 0.317 for type B) are close to the population variance assumed in the Z-test.

C. F-Test

The second question we posed was how to decide whether the contribution of an underlying genetic factor is significant, relative to environmental factors. To answer this question, we have to compare the variance explained by the genotype to the entire variance observed in the phenotype and decide whether the genotype explains a significant portion of the entire variance. This brings us to one of the most important markers evaluated in genetic studies—the metric called *heritability*. Heritability is defined as the ratio of additive genetic variance (V_A) to the entire variance observed in the phenotype (V_P), and by tradition is denoted by $h^2 = V_A/V_P$. Numerous studies have been designed to evaluate the heritability of various traits. For example, the person's stature is a trait with relatively high heritability, $h^2 = 0.65$, while insulin resistance (a major factor in the development of type 2 diabetes) has heritability $h^2 = 0.31$ (see Bergman et al. [2003]).

In general, the methods used to estimate heritability include parent–offspring regressions and analysis of variance (ANOVA) comparing siblings to half-siblings, or identical to nonidentical twins. For the purposes of this chapter, the important property of heritability is that it is defined as the ratio of two variances. Therefore, we would expect that in statistical problems heritability would have an F-distribution.

We shall illustrate the evaluation of heritability of stature via linear regression. Consider the data in Table 4-4. To investigate the dependence of the child's stature on the average parental stature, we choose

Family No.	Average Parental Stature (X) [feet]	Child's Stature (Y) [feet]
1	5.60	5.70
2	5.90	6.10
3	6.10	6.20
4	5.30	5.60
5	5.70	5.45
6	6.20	5.90
7	6.40	6.10
8	5.50	5.80
9	5.20	5.40
10	6.10	6.20

TABLE 4-4.
Example of parent-child stature data.

parental stature to be the independent variable (X) and the grown child's stature to be the dependent variable (Y).

Plotting average parental stature versus child's stature gives the graph in Figure 4-7, suggesting a possible linear relationship $Y = aX + b$ between the variables X and Y. The line in the figure is the line that "best fits" the data set. This line is called the (least-squares) *regression line*. In Chapter 5, we shall examine a criterion for best fit and how the coefficients a and b for this line can be determined from the data.

Denote the vertical distances of the data points from the line $Y = aX + b$ by $r_1, r_2, \ldots, r_n$. These numbers, calculated as $r_i = |Y_i - (aX_i + b)|, i = 1, 2, \ldots, n$, give the variation in the Y variable from the straight line relationship (see Figure 4-7). The *sum of squared residuals (SSR) measure*, defined as:

$$SSR = r_1^2 + r_2^2 + \ldots + r_n^2 = \sum_{i=1}^{n} r_i^2, \qquad (4\text{-}4)$$

is the most frequently used measure to express the combined variance of the data from the regression line.

The regression line in the figure represents the mathematical model that explains the variance in the data caused by genetic factors. The value of SSR, on the other hand, represents the variance caused by other factors.

A second sum of squares, often called the *total sum of squares (TSS)*, can be used to assess the total variation among the observed Y values. It is calculated as the sum of the squared residuals around the mean $\overline{Y}$ of the Y values (see Figure 4-8):

$$TSS = (Y_1 - \overline{Y})^2 + (Y_2 - \overline{Y})^2 + \ldots + (Y_n - \overline{Y})^2 = \sum_{i=1}^{n} (Y_i - \overline{Y})^2.$$

It can be shown (and is somewhat obvious from the graphs) that for any set of points $SSR \leq TSS$, and the equality is only possible when the regression line is horizontal; that is, when the regression line is $Y = \overline{Y}$. The difference $TSS - SSR$ gives the variance explained by the model, in this case, the regression line.

The coefficients of the least squares regression line, together with the quantities SSR and TSS, can be obtained as part of the regression output from all standard statistical software. Here is the *MINITAB* output:

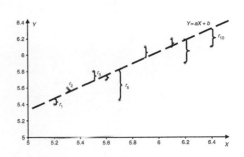

FIGURE 4-7.
Scatter plot of the parent–child data with a plot of the least-squares regression line and residuals.

```
Regression Analysis:Y versus X

The regression equation is
Y = 2.33+0.606 X

Predictor        Coef      SE Coef    T        P
Constant         2.3292    0.9092     2.56     0.034
X                0.6062    0.1564     3.88     0.005

S = 0.189009 R- Sq = 65.2% R- Sq (adj ) = 60.9%

Analysis of Variance

Source           DF      SS        MS        F       P
Regression       1       0.53646   0.53646   15.02   0.005
Residual Error   8       0.28579   0.03572
Total            9       0.82225
```

This output shows that for the dataset in Table 4-4, we find
SSR = 0.28579 and TSS = 0.82225. Thus, the sum of squares attributed to
the regression line is the difference TSS − SSR = 0.53646. This number
presents the portion of the variance in the data explained by the
genotype. Therefore, the heritability ratio $h^2 = V_A/V_P$, representing the
genetic variance divided by the entire variance of the phenotype,
translates, in our notation, to $h^2 = V_A/V_P = (\text{TSS} − \text{SSR})/\text{TSS}$. Thus, the
proportion of the variance in the child's stature data explained by
parental stature data is $h^2 = 0.53646/(0.82225) = 0.652$. In this case, the
heritability ratio h^2 is exactly the coefficient of determination, R^2, in the
MINITAB linear regression analysis output that gives the percentage of
variation of Y attributable to the approximate linear relationship
between X and Y.

The next question is whether this result is statistically significant.
More specifically, we want to see whether the portion of the variance
explained by the regression (genotype) is statistically significant. We
need to formulate a null hypothesis H_0 about the regression line and
define a statistic that allows us to decide whether we can reject it. The
hypothesis that we would like to reject is:

H_0: *There is no genotypic (inherited) component in the child's stature caused by
the parental stature.*

Mathematically, this would correspond to a horizontal regression line.
Therefore, the hypothesis H_0 can be stated in mathematical terms as:

H_0: *The slope of the regression line is equal to zero.*

If H_0 were true, the mean square error of the residuals would be equal
to the total mean square error of the residuals and the variance in the
data explained by the regression would be zero (see Figure 4-8).
Furthermore, recall that the mean square error of the residuals would

FIGURE 4-8.
Deviation of individuals' heights from the mean.
The data are plotted as in Figure 4-7, with
parental data on the x-axis and child's data plotted
on the y-axis.

have a χ^2 distribution with a parameter equal to the degrees of freedom. Therefore, the ratio (regression mean square error)/(residual mean square error), which is a quotient of two χ^2 distributions, would have an F-distribution. As the regression line has 1 degree of freedom, the χ^2 distribution corresponding to the regression mean square error has parameter 1. In our example, the value of the regression mean square error is 0.53646. The χ^2 distribution that corresponds to the residual mean square error has 8 degrees of freedom (because we have 10 data points and two parameters of the regression line), so the residual mean square error is $0.28579/8 - 0.03572$. Finally, this implies the quotient (regression mean square error)/(residual mean square error) has F-distribution with (1,8) degrees of freedom.

The value of the quotient (regression mean square error)/(residual mean square error) for our example is $F = 0.53646/0.03572 = 15.02$. Following the procedure illustrated with Figure 4-4(C), what is the probability of obtaining such a value if H_0 were true? The answer is found using software that computes the F-distribution or from F-distribution tables, taking into account the degrees of freedom we determined. The p-value, corresponding to the F value of 15.02 with degrees of freedom (1,8) is $p = 0.005$. As this is less than the standard confidence level of 0.05, we can conclude that the null hypothesis should be rejected. That is, we cannot assume zero heritability, showing the contribution of the genetic factor in this example is significant.

The F-test answers our second question—how to decide whether the contribution of an underlying genetic factor is significant, relative to environmental factors. In our example, we obtained an affirmative answer to this question under the assumption of a linear relationship between the factors.

We have now discussed two common types of hypothesis—whether there is difference between two means and whether there is a difference between two variances. In analyzing the question of a difference in the means, we shall usually analyze the data using a t-test, because the variances are usually unknown. When comparing variances, the appropriate statistical test is an F-test, because the underling sampling distribution is approximately an F-distribution. In any case, all calculations are carried out by statistical software. What is important is to know how to choose the appropriate statistical test and how to interpret the software output.

In closing, we can now answer the third question we asked at the beginning of the chapter; namely, what is the common mathematical thread that links all statistical tests comparing means, evaluating the contribution of various factors, or testing the linear dependence of an outcome on a set of predictive variables? The common mathematical background of all of the tests we considered is the underlying normal distribution of the data, the common paradigm of formulating a null hypothesis and an alternative, and the rejection of the null hypothesis if the test statistic has a low probability of belonging to the sampling distribution.

EXERCISE 4-1

Germination and growth of seeds are affected by many factors: Type of seed, type of soil, solar exposure, amount of water, and other environmental factors. Magnetism is a powerful force affecting many physical processes and could potentially affect the growth, development, and germination of seeds. Scientist U. J. Pittman concluded, "The roots of some plants (winter and spring wheat, and wild oats) normally align themselves in a North–South plane approximately parallel to the horizontal face of Earth's magnetic field" (Pittman [1970]). If the Earth's own magnetism affects the way plants grow, perhaps added magnetism will affect their rate and quality of growth as well. The experiment described below is designed to explore these questions.

Use the observations/data provided via the Internet resources for this chapter to formulate one or more hypotheses regarding the effect the presence of a magnetic field may have on the germination of the seeds and the growth of the seedlings. Use appropriate statistical techniques to corroborate or reject these hypotheses. Present a clear summary of your findings, including appropriate tables, plots, and charts. Based on your findings, what additional studies would you propose to investigate the stated hypotheses further?

EXPERIMENT

Materials

100 lentil seeds; one magnet (from an audio speaker, approximately 7 cm in diameter, 1 cm in height) to provide the magnetic field; two bowls (approximately 19 cm in diameter, 5 cm in height) to serve as flower pots, cotton as a medium for initial germination; water (1 cup every 2 days); and potting soil (2.5 cups per bowl).

Procedure

1. Lentil seeds were placed on a layer of damp cotton for 24 hours to help speed germination.

2. 50 seeds were then planted in each bowl, using the same type and amount of potting soil.

3. One of the bowls was placed on top of the magnet.

4. Bowls were placed 10 feet apart under otherwise similar environmental conditions, including light.

5. The seeds were watered for 2 weeks.

6. The number of seedlings was counted at end of the 2-week experimental period and their heights recorded.

Observations/Data

- After 24 hours in the cotton, most of the seeds had opened and had begun to sprout.

- After 2 weeks, 47 of 50 seeds sprouted in the bowl exposed to magnetism.

- After 2 weeks, 43 of 50 seeds sprouted in the bowl not exposed to magnetism (see Figure 4-9).

- The sprouts under the influence of magnetism appeared to be taller than the sprouts not under the influence of magnetism (by at least 2–3 cm, as a rough estimate). The actual data are recorded in the file *seedlab.xls*, which can be downloaded from *http://www.bio-math.sbc.edu/data.html*.

EXERCISE 4-2

When engaged in manual activities, most people favor one hand over the other. A smaller proportion of the human population is left-handed

FIGURE 4-9.
A photograph of the sprouts in the bowl exposed to magnetism (left) and the bowl not exposed to magnetism (right) taken 2 weeks after planting the lentil seeds.

than right-handed. Does the dominant hand differ in any way (other than the obvious ease of manipulation) from the nondominant hand?

One can imagine many different measurements that could be made to address these questions. One could measure for each hand the distance around the hand (as if measuring to determine glove size) or hand strength or hand span (the distance from the tip of the thumb to the tip of the little finger). A data set given below allows us to explore some of these questions.

MEASUREMENT OF HAND SPAN—PILOT STUDY

Population

Fifteen female college students in a class at Sweet Briar College.

Materials

A meter stick.

Procedure

1. To obtain the hand span, students were directed to open each hand as widely as possible.

2. Students placed the tip of the thumb on the zero mark of the meter stick and then measured to the tip of the little finger.

3. The measurement (in cm) was recorded on the data sheet.

4. Steps 1 through 3 were repeated for the other hand.

5. The dominant hand was indicated by checking the appropriate box on the data sheet.

Data

The data are recorded in the file *handspan.xls*, available from *http://www. biomath.sbc.edu/data.html*. The first sheet contains the data as collected; the second sheet has the left and right hand measurements in adjacent columns; and the third sheet has the nondominant and dominant hand measurements in adjacent columns.

Formulate one or more hypotheses regarding handedness and hand span, and use the data from the pilot study and the appropriate statistical techniques to corroborate or reject these hypotheses. Present a clear summary of your findings, including the appropriate tables, plots, and charts.

The above pilot study was conducted chiefly to test the mechanism of data collection and analysis. Based on your findings, propose additional studies for further investigating the stated hypotheses. Finally, propose additional hypotheses that could be tested using this study method. Consider factors such as age, occupation, or gender in your additional hypotheses, and suggest in as much detail as possible how you would conduct the test.

REFERENCES

Bergman, R. N., Zaccaro, D. J., Watanabe, R. M., Haffner, S. M., Saad, M. F., Norris, J. M., Wagenknecht, L. E., Hokanson, J. E., Rotter, J. I., & Rich, S. S. (2003). Minimal model-based insulin sensitivity has greater heritability and a different genetic basis than homeostasis model assessment or fasting insulin. *Diabetes, 52,* 2168–2174.

Falconer, D. S. (1989). *Introduction to quantitative genetics* (3rd ed.). New York: John Wiley & Sons.

Pittman, U. J. (1970). Magnetotropic responses in roots of wild oats. *Canadian Journal of Plant Science, 50,* 350.

INTERNET RESOURCES

http://www.biomath.sbc.edu/data.html
The Biomathematics Web site, containing data sets for lentil study and hand span study.

FURTHER READING

Darwin, E. (1796). *Zoonomia; or the laws of organic life* (vol 2). London: Johnson.

Devore, J., & Peck, R. (2005). *Statistics—The exploration and analysis of data* (5th ed.). Pacific Grove, CA: Brooks Cole-Thompson Learning.

Feller, W. (1968). *An introduction to probability theory and its applications* (3rd revised printing ed.). New York: John Wiley & Sons.

Hardy, G. H. (1908). Mendelian proportions in a mixed population. *Science, 28,* 49–50.

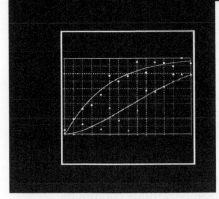

Chapter 5

RISK ANALYSIS OF BLOOD GLUCOSE DATA

Diabetes is a wonderful affection, being a melting down of the flesh and limbs. Life is short, disgusting and painful, death inevitable.

Aretaeus the Cappadocian, 2nd century A.D.

Diabetes is one of the most prevalent serious diseases in modern society. In America, the high level of obesity (a predisposing factor for diabetes) makes it likely that an increasing number of people will contract the disease.

Diabetes mellitus occurs when the body is unable to use glucose effectively. Insulin is a hormone that helps to get glucose from the blood into the cells, where it is used to provide energy for the body. There are three major types of diabetes: Type 1 diabetes mellitus (T1DM), type 2 diabetes mellitus (T2DM), and gestational diabetes. T1DM usually occurs in people under age 30. With T1DM, the beta cells of the pancreas have been destroyed or fail to function properly, and insulin is not produced. T2DM usually occurs in people over the age of 40. Those who are obese or have a family history of diabetes are at increased risk. With T2DM, insulin is produced by the pancreas, but the body is unable to use it properly, and progressively the body loses the ability to produce insulin. Gestational diabetes affects about 4% of all pregnant women—about 135,000 cases in the United States each year.

According to the American Diabetes Association, there are 21 million people in the United States (6.3% of the population) who have diabetes. Approximately 8% of these have T1DM, whereas more than 90% have T2DM. About one-third of the people with T2DM are *unaware* they have the disease. In addition, at least 20.1 million Americans are estimated to have *pre-diabetes*, a condition where a person's blood glucose (BG) levels are higher than normal but not high enough for a diagnosis of T2DM. From 1990 to 1998, there was a one-third increase in diabetes in U.S. adults.

The economic costs of diabetes are enormous. A recent study estimated that the direct medical and indirect expenditures attributable to diabetes in 2002 were $132 billion (American Diabetes Association [2003]). The study estimated that per capita medical expenditures totaled $13,243 for people with diabetes and $2560 for people without diabetes. Diabetes is the fifth-leading cause of death by disease in the United States. Diabetes also contributes to higher rates of morbidity—people with diabetes are at higher risk for retinal disease, which is the leading cause of adult blindness; renal disease, which represents half of all kidney failures; and neuropathy, which leads to more than 65,000 amputations annually. Cardiovascular disease is also two to four times more common in diabetics, and is also more morbid, more lethal, and benefits less from modern interventions, such as bypass surgery or stents.

In this chapter, we present a model, developed by some of the authors of this text, that represents a significant advance in predicting episodes of hypoglycemia or *low* BG—one of the most hazardous conditions resulting from the use of the hormone insulin to treat diabetes. Although the levels of mathematics and statistics used are elementary, the model, nonetheless, provides a solution to a problem previously pronounced virtually unsolvable.

Again, we begin by gathering data, then building a model based on the data, and finally testing the model with additional data. We shall use data collected by self-monitoring BG (SMBG) devices from people with diabetes. The major steps of model-building process are:

1. Rescale the data to obtain symmetric samples and thus ensure that certain well-known statistical techniques will be valid.

2. Define a "risk function" that measures the risk of dangerously low and high deviations of BG from clinically safe levels.

3. Test the risk function to determine whether it is indeed a superior tool for predicting future episodes of severe hypoglycemia.

I. HISTORICAL OVERVIEW

Aretaeus's grim description of diabetes given in the beginning of this chapter summarizes all that was known about this disease for nearly 1700 years, until physicians working in the late nineteenth century began to recognize the connection between the pancreas and diabetes. This connection was later narrowed to specific parts of the pancreas: the islets of Langerhans. The islets of Langerhans contain the beta cells that produce insulin—a hormone discovered in 1921 by the Canadian surgeon Frederick Banting (1891–1941) and his assistant, Charles Herbert Best (1899–1978).

The discovery of the action of insulin was one of the greatest achievements in medicine: diabetes, once an automatic death sentence, was no longer a fatal disease. The chronicles of medicine trace this discovery to Banting's 1920 visit to the University of Toronto, when he spoke to John J. R. Macleod (1876–1935), who was the head of the department of physiology and an expert in glucose metabolism and diabetes. Banting presented to Macleod an idea on how to find the cause of diabetes and also a treatment. Macleod initially rejected the idea, but Banting persisted, and Macleod conceded to give him laboratory space, experimental dogs, and a student assistant—Charles Best. Banting and Best began their experiments in May, 1921, and within a few months they had extracted a substance from the islets of Langerhans called "insulin," (the name comes from the Latin *insula*, which means *island*). When given to diabetic dogs, this substance lowered their high blood sugar levels. Repeated experiments confirmed the blood sugar lowering action of insulin, but the insulin preparation was not pure enough for human testing. Macleod added the biochemist James Bertram Collip (1892–1965) to the group to help with insulin purification. Within a few weeks, the substance was sufficiently purified and deemed safe for human application. The first patient who received insulin injections was a 14-year-old boy dying of diabetes. The injected insulin reduced his abnormally high blood sugar and alleviated other signs of the disease. The scientific community quickly recognized the significance of these findings, and, in 1923, the discovery of insulin action brought the Nobel Prize in Medicine to Banting and Macleod, which they shared with the other researchers on the team.

Although extremely important, the discovery of insulin did not solve all of the problems associated with diabetes. We now know that diabetes is a complex of disorders, characterized by the common element of high blood sugar, or hyperglycemia, that arise from and are determined in their progress by mechanisms acting at all levels of the biosystem—from molecular through hormonal to behavioral. The treatment of diabetes requires not only lowering extremely high blood sugar levels, but also avoiding low blood sugar (hypoglycemia) and optimizing blood sugar fluctuations within a certain target range.

II. CLINICAL BLOOD GLUCOSE OPTIMIZATION PROBLEM OF DIABETES

In a healthy person, the BG level is internally regulated through insulin released from the pancreas that counterbalances carbohydrate intake. Because patients with diabetes are unable to produce insulin (T1DM) or produce insufficient insulin combined with higher insulin resistance (T2DM), this internal self-regulation is disrupted. The standard daily control of T1DM involves multiple insulin injections or a continuous insulin infusion (insulin pump) that lowers BG. The daily control of T2DM also requires insulin or oral medications.

Large-scale research studies, including the 10-year Diabetes Control and Complications Trial (DCCT; 1993) and a similar European trial (Reichard and Phil 1994), have proved that intensive treatment with insulin and with oral medication is indeed the best strategy for optimal glycemic control. Such therapy has been proved effective in bringing BG to nearly normal levels and markedly reducing the chronic complications of both types of diabetes (except cardiovascular disease). However, the same studies have shown that external BG control is still not nearly as good as normal internal self-regulation. Too little insulin results in chronic high BG levels, causing complications in multiple body systems, whereas too much insulin triggers hypoglycemia. Without corrective action, hypoglycemia can rapidly progress to severe hypoglycemia, which may lead to brain abnormalities, cognitive dysfunction, accidents, coma, and even death. Severe hypoglycemia (SH) is defined as severe nutrient deprivation of the brain resulting in stupor, seizure, or unconsciousness that precludes self-treatment. A retrospective population survey from Norway investigating 246 deaths from 1981 to 1990 among diabetic patients younger than age 40 attributed 10% of these deaths to hypoglycemia (Bloomgarden [1999]). Hypoglycemia has been identified as the primary barrier to optimization of glycemic control in diabetics (Cryer et al. [1994]; Cryer [1999]).

In short, a number of important aspects of the pathogenesis of diabetes and its complications relate to *optimal control* of the insulin–carbohydrate balance. People with diabetes face the life-long *optimization problem* of maintaining strict glycemic control without increasing their risk of hypoglycemia. This optimization has to be based on data collection, data processing, and meaningful feedback readily available to individuals with T1DM and T2DM. The mathematical challenge is to create diabetes-specific mathematical methods and analytical procedures that continuously assess the biological and behavioral characteristics and precursors of hypoglycemia and hyperglycemia.

III. QUANTIFYING CHARACTERISTICS OF DIABETES

The range of BG in a living human is approximately 1.1 to 33.3 millimoles per liter (mmol/L), and the safe (target) range is considered to be 3.9 to 10 mmol/L, using Système International (SI) units. Low BG levels, with values below 3.9 mmol/L, correspond to a condition called hypoglycemia. High BG levels, with values above 10 mmol/L, correspond to a condition defined as hyperglycemia (see Figure 5-1).

The first thing to do with a patient with indications of diabetes is to determine his or her BG level. This is not as simple as it might seem, because the BG level fluctuates throughout the day. A healthy person's

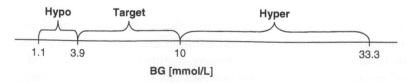

FIGURE 5-1.
BG ranges based on Système International (SI) millimole per liter units.

BG level will be generally high after eating and low when he or she wakes up in the morning (fasting BG). In diabetes, the timing and amount of insulin or oral medication additionally disturb these fluctuations. It has also been shown that the average of several readings made during the day do not give a good measure of the average BG level, except under very tightly controlled conditions, because of the significant irregularity of the BG fluctuations. For example, Figure 5-2 presents the BG fluctuations of a person with T1DM recorded over 17 days. Notice the extreme excursions from the target range in both hypoglycemia and hyperglycemia. (The target BG range for a person with diabetes was established by the DCCT in 1993: the normal range is 4.5–5.5 mmol/L, except during the two to three hours after a meal). This observation also confirms the imperfection of external insulin replacement discussed in the previous section.

There is a substantial difference in the clinical effects caused by excursions into the hyper- and the hypoglycemic ranges. Deviations of the BG into the hyperglycemic range are undesirable, but sharp peaks in the BG profiles are not immediately dangerous. In contrast, sharp nadirs could be extremely dangerous and potentially life threatening, because they indicate hazardously low levels of BG that may cause glucose (fuel) deprivation of the brain and seizures that preclude self-treatment. For example, such episodes could be particularly dangerous while operating a vehicle.

The classic marker of average glycemic status is HbA_{1c}, introduced 22 years ago by Aaby Svendsen (Svendsen et al. [1982]). While HbA_{1c} has been linked to long-term complications in both T1DM and T2DM, in 1993, it was also confirmed that the mean BG level for the previous four to six weeks can be determined by a test of glycosylated hemoglobin,

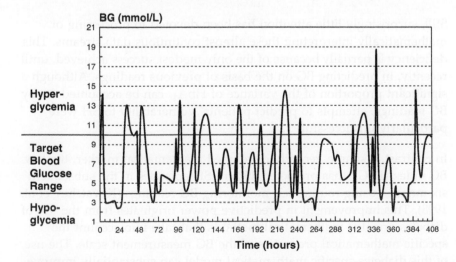

FIGURE 5-2.
BG fluctuations of a patient with T1DM. Note the excursions both above and below the target BG range.

and especially its component HbA_{1c} (Santiago [1993]). Since then, HbA_{1c} has been confirmed as the gold-standard assay for people with T1DM and T2DM. The guidelines specify that HbA_{1c} of 7% corresponds to a mean BG of 8.3 mmol/L (150 mg/dl), an HbA_{1c} of 9% corresponds to a mean BG of 11.7 mmol/L (210 mg/dl), and a 1% increase in HbA_{1c} corresponds to an increase in mean BG of 1.7 mmol/L (30 mg/dl).

HbA_{1c}, however, only captures average glycemia. The radical fluctuations in BG levels represented in Figure 5-2—especially those that take the BG levels into the hypoglycemic range—are very hazardous, but are not recognized in the HbA_{1c} test. In fact, the DCCT concluded in 1997 that only about 8% of future severe hypoglycemia episodes can be predicted from known variables, including HbA_{1c}. Predictions improved to 18% with a more recent model using history of SH, hypoglycemia awareness, and autonomic score (see DCCT Research Group [1997]; Gold et al. [1997]). Given that intensive therapy increases the risk for hypoglycemia, strict control of T1DM implies that BG levels should be closely monitored for large deviations at both the low and the high end of the BG scale. It also follows that the risk for hypoglycemia needs to be monitored by means other than HbA_{1c}.

The rapid development of home BG monitoring devices provides a means for monitoring BG fluctuations, and, in particular, monitoring for hypoglycemia. Contemporary memory meters can store several hundred BG readings and can calculate various statistics, including the mean of these BG readings. Increasingly, research is focused on developing devices for continuous, or nearly continuous, non-invasive self-monitoring of BG. Two new journals, *Diabetes Technology & Therapeutics* and *Diabetes Science & Technology*, were launched in 1999 and 2006, respectively, to report technological advances, including information processing.

Still, surprisingly little attention has been devoted to processing or mathematically interpreting these almost continuous data streams. This deficiency is partially because of the only modest success achieved, until recently, in predicting BG on the basis of previous readings. Although a significant proportion of the variance of HbA_{1c} can be accounted for by BG readings, attempts to predict patients' vulnerability to SH were particularly unsuccessful.

In contrast, a simple, recently developed mathematical marker, the low BG index (LBGI), has predicted 40% of SH episodes in the subsequent six months, using routine self-monitoring BG readings (Kovatchev et al. 1998). This improvement in predictive power originates from the use of diabetes-oriented mathematical methods that take into account the specific mathematical properties of the BG measurement scale. The use of this diabetes-specific mathematical model can substantially improve the forecasting of hypoglycemia and the overall quality of the monitoring to control diabetes.

IV. SELF-MONITORING OF BLOOD GLUCOSE

Diabetes nowadays is a disease controlled at home. Contemporary home BG meters (see Figure 5-3) provide a convenient means for frequent and accurate BG determination through SMBG.

SMBG devices store large amounts of data—hundreds of BG readings with the date and time for each—and can compute some summary statistics, such as estimates of the mean BG over the previous two weeks. The meters are usually accompanied by software with expanded capabilities for data analysis, review, and graphical representation. Given a set of SMBG readings downloaded from a subject's meter, various SMBG characteristics are routinely computed: mean, standard deviation, minimum, maximum, and range of BG, as well as percentage of SMBG readings below or above certain BG levels. However, there is a missing link between the data collected by the BG meters on one side, and the evaluation of HbA_{1c} and the risk for hypoglycemia on the other. Currently, there are no reliable methods for evaluating HbA_{1c} and recognizing imminent hypoglycemia based on SMBG readings. One can speculate that one reason for the missing link is that these advanced home monitoring devices, as well as the clinical methods for assessment and data collection, are infrequently supported by diabetes-specific, mathematically sophisticated quantitative procedures.

 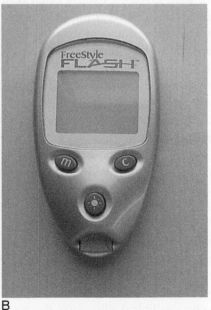

A   B

FIGURE 5-3.
Self-monitoring devices. Panel A: Lifescan OneTouch UtraSmart™; panel B: Abbott Diabetes Care FreeStyle FLASH™.

A. SMBG and Average BG Levels

As already mentioned, the value of HbA_{1c} accurately reflects the average BG over the preceding five to six weeks. A natural, but incorrect, conclusion would be that the mean BG derived from SMBG readings would also accurately reflect the real BG values. In reality, the mean SMBG can deviate substantially from the real mean BG value since a patient could measure at fixed times of day when his or her BG is at its extreme high or low. In this case, the average of the SMBG readings will be an overestimate or underestimate of the real mean BG.

B. SMBG and Prediction of Hypoglycemia

Little attention was paid to the evaluation of the risk for hypoglycemia from SMBG before the first announcement in 1994 that SH could be predicted from SMBG (Cox et al. [1994]). The reason is that, theoretically, routine SMBG three to four times a day would rarely capture rapidly developing events, such as descent into hypoglycemia. This is one incentive behind the development of systems for continuous BG monitoring. However, we found that specific analysis of SMBG data can capture *trends* towards increased risk for hypoglycemia and can identify *periods* of increased risk for hypoglycemia.

In this chapter, we present the mathematics behind these methods for analysis of SMBG data. The mathematical foundation of our techniques is based on the following general biomathematical concept: The struggle for tight glycemic control often results in great BG fluctuations over time. This process is influenced by many external factors, including the timing and amount of insulin injected, food eaten, physical activity, etc. In other words, fluctuations of the BG level over time are the measurable result of the action of a complex dynamic system, influenced by many internal and external factors. Observed over short periods of time, this system is nearly deterministic, and its fluctuations can be predicted by knowing the state of its components and their interaction. Over longer periods of time, the system has nearly random behavior that includes extreme transitions, such as SH episodes. Consequently, different analytical strategies would quantify long-term characteristics of diabetes, such as HbA_{1c}, long-term risk for SH, and patient behavior, and short-term characteristics such as imminent moderate or severe hypoglycemia. Following this concept, this chapter offers a system of quantitative methods simultaneously evaluating three important components of glycemic control: HbA_{1c} and long-term and short-term risk for hypoglycemia. In order to be clinically useful, these methods utilize readily available SMBG data and relatively simple algorithms. In order to prove clinical relevance, the results are correlated with established measures of glycemic control, such as HbA_{1c}, and risk for upcoming hypoglycemia.

V. SYMMETRIZATION OF THE BLOOD GLUCOSE MEASUREMENT SCALE

BG fluctuations are often the object of statistical description and various data analyses in research and clinical practice. Most statistical techniques, however, require assumptions about the shape of the underlying distribution of the data being analyzed. For example, the routine statistical practice of reporting in the format "mean value ± standard deviation" assumes a symmetric distribution of the data readings. This is not the case with BG readings. For example, Figure 5-4 presents a typical BG data distribution for a subject with T1DM (186 readings downloaded from his memory meter). The distribution is substantially skewed, and the superimposed bell curve (normal density) describes the data poorly.

This problem is not new, and it arises often in statistics. There are well-developed techniques that provide transformations converting nonsymmetric samples to approximately symmetric ones (Box and Cox [1964]). The statistical analyses are performed with the symmetric data, and then an inverse transformation is used to translate the results so that they correspond to the original sample. It is important to be aware that such transformations are *sample-dependent* (i.e., different samples will be symmetrized by different transformations). Therefore, this approach will be impractical for implementing in a SMBG device, because the transformation should be known in advance.

An alternative approach that eliminates sample dependency is to *change the scale* of the BG readings so that in the new scale the BG sample is symmetric. In Figure 5-4, notice the following:

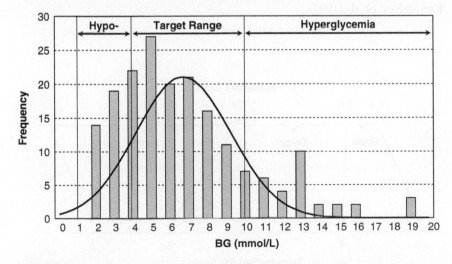

FIGURE 5-4.
The distribution of BG levels. The bar graph represents the BG readings of a person with T1DM. A normal distribution (the line graph) does not fit the data well (© 1997 American Diabetes Association. From *Diabetes Care*, 20, 1655–1658. Reprinted with permission from The American Diabetes Association.)

1. The range of scores in the hypoglycemic range is much smaller than the range of hyperglycemic scores; and

2. The target range is not in the center of the data range.

We want to convert this scale to a scale as shown in Figure 5-5. The idea is to expand the hypoglycemic range, squeeze the hyperglycemic range, and position the target BG range symmetrically about zero. More specifically, we want the transformed scale to satisfy the following conditions:

1. The directions of the original and transformed scales are the same;

2. The target range is centered at 0; and

3. The entire BG range is centered at 0.

First, we must find such a transformation, and, second, establish the transformation's validity by testing it with BG reading samples from a sufficiently large number of people.

We seek a transformation in the form:

$$f(BG, \alpha, \beta) = [(\ln(BG))^{\alpha} - \beta], \alpha, \beta > 0. \qquad (5\text{-}1)$$

Why choose this particular analytic expression? Such a question is not always easy to answer. Developing a good mathematical model may sometimes border on artistic creativity, and it may not always be possible to retrace every single step of the process. In the case of model (5-1), we began with a widely accepted skewness correction formula, and then modified it to fit our needs. The specific details can be found in Kovatchev et al. (1997).

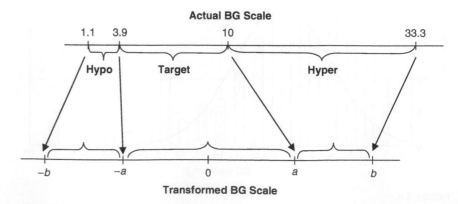

FIGURE 5-5.
Conversion of the BG scale.

We determine specific values for α, β from the conditions 1 through 3. The symmetric conditions require:

$$f(33.3, \alpha, \beta) = -f(1.1, \alpha, \beta) \text{ and } f(10, \alpha, \beta) = -f(3.9, \alpha, \beta)$$

These give:

$$(\ln(33.3))^\alpha - \beta = -[(\ln(1.1))^\alpha - \beta] \qquad (5\text{-}2)$$

$$(\ln(10))^\alpha - \beta = -[(\ln(3.9))^\alpha - \beta] \qquad (5\text{-}3)$$

Subtracting Eq. (5-3) from Eq. (5-2) gives:

$$(\ln(33.3))^\alpha - (\ln(10))^\alpha = -(\ln(1.1))^\alpha + (\ln(3.9))^\alpha.$$

This equation can be (approximately) solved for α by a computer, giving $\alpha = 1.0329$. Substituting this value into Eq. (5-2) and solving (approximately) for β gives $\beta = 1.8707$. With these values of α and β, we have now centered the whole BG range about 0 and the target BG range about 0.

One further embellishment will prove useful: to calibrate the new scale and make the total BG range from $-\sqrt{10}$ to $\sqrt{10}$. We do this for the following reasons. First, if our data satisfy some hypothesis that we must verify, then 99.8% of the readings should be between $-\sqrt{10}$ to $\sqrt{10}$. Second, this will enable us to calibrate the risk function that we shall define shortly to be a function with values from 0 to 100%. Thus, we seek a value γ for which:

$$\gamma[(\ln(33.3))^{1.0329} - 1.8707] = \sqrt{10}.$$

From this condition, we find that $\gamma = 1.774$. We thus obtain the following transformation:

$$f(BG) = 1.774 \cdot [(\ln(BG))^{1.0329} - 1.8707]. \qquad (5\text{-}4)$$

Figure 5-6 shows how the function (5-4) transforms the BG scale.

EXERCISE 5-1

In our measurement of BG levels we used the SI system. In the United States, the most popular scale is milligram per deciliter (mg/dl). The scales are related by $18 \text{ mg/dl} = 1 \text{ mmol/L}$[1].

---

1. The weight of a molecule is the sum of the weights of the atoms of which it is made. A mole is the quantity of a substance whose weight in grams is equal to the molecular weight of the substance. Thus, 1 mole of glucose weighs 180 g. Accordingly, 1 mmol glucose = 180 mg. Thus, 1 mmol of glucose per liter equals 18 mg glucose per deciliter.

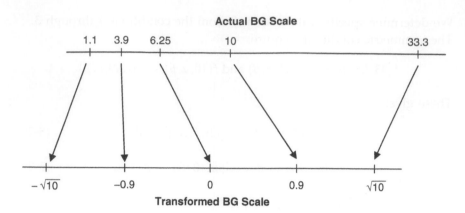

FIGURE 5-6.
Transformation of the BG scale that uses the function $f(BG)$ from Eq. (5-4).

(a) Find the target range and the whole range of BG values in the U.S. system.

(b) Find the values of α and β that will transform the target and whole ranges into intervals centered at 0 for the U.S. system.

(c) Find the value of γ that will scale the whole BG range to be between $-\sqrt{10}$ to $\sqrt{10}$.

It is now time to test the transformation to determine its validity. We again consider Figure 5-4, which presents the distribution of 186 BG readings from the SMBG device of a subject with T1DM. As evident from the graph, the distribution is substantially skewed. In fact, when we calculate the basic statistics for the data, we find that the mean is 6.7 mmol/L and the standard deviation is 3.6. In applying statistical tests, it is standard practice to assume that 95% of the data lies within two standard deviations of the mean. For this data, $\bar{x} = 6.7$, SD $= 3.6$, so $\bar{x} \pm 2(\text{SD}) = 6.7 \pm 7.2$ gives the range -0.5 to 13.6 mmol/L. Now about 2.5% (or four readings from the total 186 readings of this data) should lie below -0.5 mmol/L, which cannot happen.

We expect this skewed distribution will appear nearly normal in the transformed BG measurement scale, and Figure 5-7 confirms this. It presents the histogram of the same data over the transformed symmetrized scale. Notice how symmetric the data now appear. If we find the mean $\bar{x}$ and the SD of the data in this scale, we find $\bar{x} = -0.13$ mmol/L and SD $= 1.02$. So $\bar{x} \pm 2(\text{SD}) = -0.13 \pm 2.4$, which gives the range -2.17 to 1.91 mmol/L. Now four readings fall below -2.17 and three above 1.91. *This is nearly an exact fit with a normal distribution.*

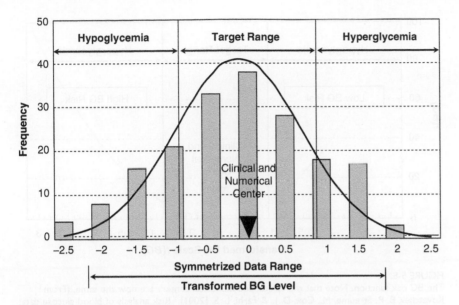

FIGURE 5-7.
Distribution of BG levels in the transformed scale. Note the symmetry of the bar graph and how well the normal distribution fits the transformed data. (© 1997 American Diabetes Association. From *Diabetes Care, 20,* 1655–1658. Reprinted with permission from The American Diabetes Association. Used by permission of Taylor & Francis, Ltd. [http://www.informaworld.com]).

We now know our transformation works for one subject. Are we done? Of course not! We must test other subjects. We ran SMBG data sets for 205 people with diabetes and looked at their individual BG distributions with the transformed scale. All of the histograms in the transformed scale resulted in symmetric distributions, and the normality hypotheses for only two out of the 205 were rejected at a *p*-level of 0.005 (note that with more than 200 tests this *p*-level practically guarantees this should happen).

In summary, we solved the following problem: the typical distribution of SMBG readings of a person with T1DM is substantially skewed; that is, the numerical center of the data is substantially separated from its clinical center. Thus, clinical conclusions based on numerical methods will be less accurate for the constricted hypoglycemic range. The solution was to introduce a data transformation that symmetrizes the BG scale around a single numerical/clinical center of 6.25 mmol/L and converts a typical distribution of BG readings into a normal distribution. This approach establishes a mathematical foundation for risk analysis of BG data through introduction of the BG risk function.

VI. THE BLOOD GLUCOSE RISK FUNCTION

Now we want to create a risk function that will assign a risk value to each BG level from 1.1 to 33.3 mmol/L. Figure 5-8 presents a quadratic risk function superimposed over the transformed BG

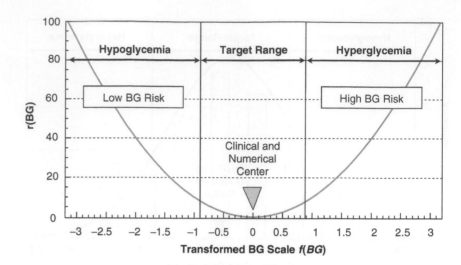

FIGURE 5-8.
The BG risk function. Note that the clinical and numerical centers are now the same. (From Kovatchev, B. P., Straume, M., Cox, D. J., & Farhi, L. S. [2001]. Risk analysis of blood glucose data: A quantitative approach to optimizing the control of insulin dependent diabetes. *Journal of Theoretical Medicine, 3*, 1–10. Used by permission of Taylor & Francis, Ltd. [http://www.informaworld.com]).

scale. The equation of the BG risk function in Figure 5-8 is:
$$r(BG) = 10[f(BG)]^2.$$

Now we can better understand the calibration condition imposed on the transformation $f(BG)$ earlier. Because $f(BG)$ ranges from $-\sqrt{10}$ to $\sqrt{10}$, the risk function $r(BG)$ ranges from 0 to 100. Its minimum value of 0 is achieved at $f(BG) = 0$ or, in the original scale, BG = 6.25 mmol/L. The maximum is reached at the extreme ends of the BG scale, $f(BG) = \sqrt{10}$, i.e., BG = 1.1 mmol/L in the original scale (extreme hypoglycemia) and $f(BG) = \sqrt{10}$, i.e., BG = 33.3 mmol/L in the original scale (extreme hyperglycemia). Thus, $r(BG)$ can be interpreted as a measure of the risk associated with a certain BG level. The left branch of this parabola identifies the risk of hypoglycemia, whereas the right identifies the risk of hyperglycemia. Notice, again, that because in this scale the hypo- and hyperglycemic ranges of the BG scale are symmetric about 0, the symmetric risk function in Figure 5-8 would be equally sensitive to hypoglycemic and to hyperglycemic readings.

For comparison, Figure 5-9 presents $r(BG)$ in the original BG scale. As you may have expected, the risk function in this scale increases much more rapidly in the hypoglycemic range and thus is not equally sensitive to hypoglycemic and to hyperglycemic readings.

EXERCISE 5-2

Compute the risk function for the following BG readings: 1.8 mmol/L, 3.9 mmol/L, 6.25 mmol/L, and 24.6 mmol/L.

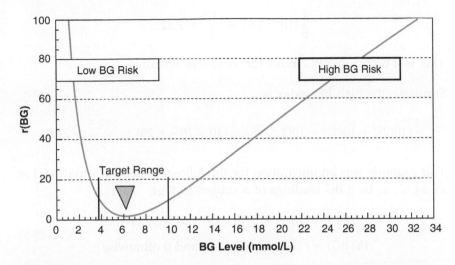

FIGURE 5-9.
The BG risk function in the original scale. Note the asymmetrical nature of the function. (From Kovatchev, B. P., Straume, M., Cox, D. J., & Farhi, L. S. [2001]. Risk analysis of blood glucose data: A quantitative approach to optimizing the control of insulin dependent diabetes. *Journal of Theoretical Medicine, 3,* 1–10. Used by permission of Taylor & Francis, Ltd. [http://www.informaworld.com]).

Based on the BG risk function, we are now ready to develop two new SMBG characteristics:

- The LBGI—a measure of the frequency and extent of low BG readings, and

- The high BG index (HBGI)—a measure of the frequency and extent of high BG readings.

VII. THE LOW AND HIGH BLOOD GLUCOSE RISK INDICES

We want to assess the risk caused by low readings and high readings separately. To do this, we separate the low scores [those for which $f(BG) < 0$] from the high scores [those for which $f(BG) > 0$]. We begin with an example and then show the general formula.

Example 5-1

Suppose the $f(BG)$ readings for a patient are -1, 2, 0.4, 1, and 2.5. Then the low readings are -1 and -0.4, and the risk function values are:

$$10[f(BG)]^2 = 10(-1)^2 = 10 \quad \text{and} \quad 10[f(BG)]^2 = 10(-0.4)^2 = 1.6.$$

We now sum these values and divide by 5 (the total number of readings). This gives:

$$\frac{1}{5}(10 + 1.6) = \frac{11.6}{5} = 2.32.$$

This will be the low BG index for these five readings.

Similarly, the high BG index is:

$$\frac{1}{5}\left(10(2)^2 + 10(1)^2 + 10(2.5)^2\right) = \frac{1}{5}(40 + 10 + 6.25) = \frac{112.5}{5} = 22.5.$$

This example was an illustration for the following general situation. Let $x_1, x_2, \ldots x_n$ be n BG readings of a subject and let:

$$rl\,(BG) = r\,(BG) \text{ if } f(BG) < 0 \text{ and } 0 \text{ otherwise;}$$

$$rh\,(BG) = r\,(BG) \text{ if } f(BG) < 0 \text{ and } 0 \text{ otherwise.}$$ (5-5)

The LBGI and the HBGI are then defined as:

$$LBGI = \frac{1}{n}\sum_{i=1}^{n} rl\,(x_i)$$

$$HBGI = \frac{1}{n}\sum_{i=1}^{n} rh\,(x_i).$$

The LBGI is based on the left branch of the BG risk function, whereas the HBGI is based on the right branch of the BG risk function (see Figure 5-8).

EXERCISE 5-3

Give factors that will cause the LBGI and HBGI to increase.

VIII. MODEL VALIDATION STRATEGIES

In the preceding sections, we developed a mathematical model of a quantitative risk measure that, we claim, holds promise in assessing the clinical risk for BG deviations from the safe target range. As with any new mathematical model, however, the burden of proof for its validity and usefulness lies with its creators, and in this section we address the validation question.

To assess the performance of our model, we need to test it on data. The model uses assumptions on SMBG readings; thus, the data we need must be of this form. Several questions arise.

First, how are these data obtained? This question is logistic in nature—when humans are used as test subjects, there are strict government regulations that must be followed to ensure their protection and their well-being.

Second, after we have the data and have computed the HBGI and LBGI, what will the results tell us? How do we know whether our risk indices provide a good fit to the clinical reality? This question is quite deep. It encapsulates the essential difference between the risk index models developed in this chapter and the models for tracking changes in quantitative variables (such as population sizes and concentrations) considered in Chapters 1 and 2. The ultimate test for a good fit was made by estimating the deviation of the model values from the observed data. Fundamentally, the difference stems from the fact the model variables in Chapters 1 and 2 represented physical quantities measurable (directly or indirectly) by means of standardized procedures. In this chapter, however, we have stressed that the mean BG level over a four- to six-week period cannot be estimated accurately from SMBG data, but could be determined by a HbA_{1c} test that is the standard for measuring the average BG.

But how do we physically measure the risk for hypo- and hyperglycemia? We have already emphasized that there are no established standards. How can a model be validated in the absence of a norm?

IX. VALIDATION OF THE BLOOD GLUCOSE RISK INDICES

In the absence of a quantitative standard, new measures could be initially validated by testing their ability to reflect verifiable medical distinctions. For example, because of the physiological differences between T1DM and T2DM, patients with T1DM are known to be at a much higher risk of experiencing both hyper- and hypoglycemic episodes, and their BG profiles are generally marked with frequent excursions into the hyper- and hypoglycemic BG zones. In contrast, the BG fluctuations of patients with T2DM have smaller amplitudes and span a narrower range about the target zone, because large BG fluctuations are caused by the instability of the insulin–glucose dynamics. A critical factor for such instability is the insulin sensitivity of the body, measured as the amount of glucose metabolized per unit of insulin. Heuristically, it is natural to expect that insulin will produce a stronger effect with higher insulin sensitivity (or lower insulin resistance), causing larger BG fluctuations. Because T2DM is a disease of increased insulin resistance, the BG fluctuations in T2DM are less extreme (refer to Bergman et al. [1979] and Bergman [2003]).

Figure 5-10 graphically illustrates this difference. The data in Figure 5-10(A) are from a 20-year-old man with T1DM from the age of 5. The data in Figure 5-10(B) are from a 70-year-old man who had had

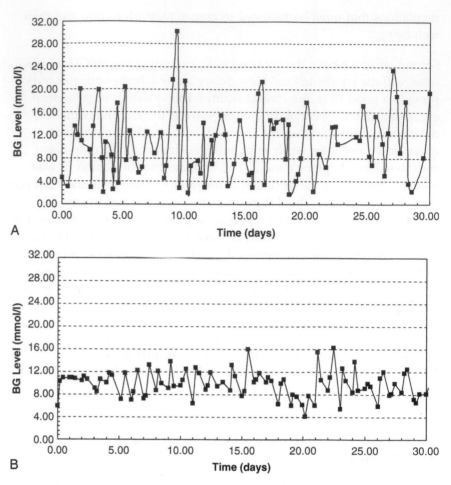

FIGURE 5-10.
BG fluctuations in T1DM vs. T2DM. Panel A shows the greater variation of BG levels in T1DM and panel B shows the BG levels in a patient with T2DM. (From Kovatchev, B. P., Cox, D. J., Gonder-Frederick, L. A., & Clarke, W. L. [2002]. Methods for quantifying self-monitoring blood glucose profiles exemplified by an examination of blood glucose patterns in patients with type 1 and 2 diabetes. *Diabetes Technology and Therapeutics, 4*, 295–303. Used by permission.)

T2DM for 14 years. Each figure presents 30 days of SMBG data, representing 90 readings collected, an average of three times per day.

As is evident from the graphs, the BG levels for both subjects fluctuate about comparable average values. This was confirmed by their practically identical HbA_{1c} values (9.1% for patient A versus 9.2% for patient B), reflecting the average BG control throughout the study. This comparison confirms once again that average BG levels cannot be used to quantify BG fluctuations. Therefore, one possible validation of the BG risk indices would be to demonstrate their capability of quantitatively describing the increased risk for hypo- and hyperglycemia associated with such fluctuations.

Indeed, after computing the LBGI and HBGI for both patients, we found the LBGI and HBGI of patient A (T1DM) were both substantially

Variable	T1DM	T2DM
Age (years)	20	70
Duration of diabetes (years)	15	14
Average HbA$_{1c}$ during study	9.1%	9.2%
Low BG Index	4.79	0.07
High BG Index	14.6	8.4

TABLE 5-1.
Quantitative comparison of subjects with T1DM versus T2DM.

higher, exactly as we anticipated! The specific values are given in Table 5-1.

EXERCISE 5-4

Do you think the results in Table 5-1 provide sufficient validation for the BG risk indices? Why or why not? Consider ways for performing a more credible validation.

The two data sets used in Figure 5-10 and Table 5-1 only serve as *illustrations* of the capabilities of the LBGI and HBGI; they do not provide a sound validation of these measures. Because we considered only two subjects, it is likely some (or all) of the observed phenomena are caused by chance. A credible validation should include a large number of T1DM and T2DM subjects to minimize the element of chance because of unavoidable differences in the patients' BG control. We present the results of these analyses next.

1. Group Comparisons

We use data collected over three months for 600 subjects (277 with T1DM and 323 with T2DM), all of whom used insulin to manage their diabetes. HbA$_{1c}$ was tested twice during the course of the study, at 1.5 and 3 months. The participants collected three to five BG measurements per day for the entire period.

The following variables were recorded for each participant:

- HbA$_{1c}$ at 1.5 months

- HbA$_{1c}$ at 3 months

- Minimal BG value for the 3-month period

- Maximal BG value for the 3-month period

The following statistics were computed for each participant at the end of the study:

- Mean value and standard deviation of all BG readings

- Maximal BG range (calculated as the difference Maximal − Minimal value for each subject)

- LBGI

- HBGI

Our goal is to compare the T1DM group with the T2DM group with respect to these variables and basic statistics. Put in a different way, instead of comparing a specific person with T1DM with a specific person with T2DM as we did before, we now want to compare the "average person with T1DM" with the "average person with T2DM" and assess the similarities and differences. As before, we would like to verify that LBGI and HBGI for the T1DM group are substantially (and statistically significantly) higher than those for the T2DM group.

In Chapter 4, we examined the problem of group comparisons in terms of their average values and the specifics of performing a t-test. We shall apply the same statistical procedure here. We use t-test to reject (at a specified confidence level p) the null hypothesis for equality of the group averages. The smaller the p-value we choose, the smaller the chance we will reject the null hypothesis when it is, in fact, true.

Table 5-2 presents the mean values for the T1DM and T2DM groups and the results of performing a t-test. The last column contains both the

Variable	T1DM − Mean (SD)	T2DM − Mean (SD)	Significance
A: Glycemic control averages			
HbA$_{1c}$ at 1.5 months	9.6 (1.2)	9.7 (1.2)	$t = 0.7, p = 0.48$
HbA$_{1c}$ at 3 months	9.2 (1.2)	9.3 (1.1)	$t = 1.6, p = 0.11$
Average BG	10.2 (1.9)	10.4 (2.2)	$t = 1.5, p = 0.13$
BG Standard deviation	4.8 (0.9)	3.3 (1.0)	$t = 18.3, p < 0.0001$
B: Blood glucose range			
Minimal BG (mmol/L)	2.2 (0.7)	3.4 (1.2)	$t = 15.0, p < 0.0001$
Maximal BG (mmol/L)	24.9 (3.8)	21.2 (4.3)	$t = 11.0, p < 0.0001$
BG Range (mmol/L)	22.7 (3.9)	17.8 (4.5)	$t = 14.1, p < 0.0001$
C: Risk characteristics			
Low BG Index	2.7 (2.0)	0.8 (1.1)	$t = 14.5, p < 0.0001$
High BG Index	13.1 (5.8)	12.0 (7.1)	$t = 2.0, p = 0.05$

TABLE 5-2.
Group comparisons of T1DM versus T2DM subjects.

value of the t-statistics and the p-value corresponding to this value in the t-distribution. Recall that, in general, a p-value of 0.05 or smaller is considered to reflect statistically significant differences between the group averages, although any smaller p-value diminishes the possibility of incorrectly rejecting the null hypothesis.

More specifically, Table 5-2(A) presents comparisons of T1DM versus T2DM on glycemic control averages, such as mean BG derived from SMBG readings and HbA$_{1c}$. As expected, no significant difference between the groups is observed for these variables. However, the standard deviation is markedly and significantly higher for the T1DM group. This is not surprising, given that patients with T1DM more frequently experience substantial deviations from the safe target BG range and the average BG levels.

Table 5-2(B) presents BG characteristics and demonstrates that T1DM subjects have both significantly lower *and* significantly higher BG readings than T2DM subjects.

Finally, Table 5-2(C) presents risk characteristics of the SMBG data in terms of the LBGI and the HBGI. Table 5-2(C) demonstrates that T1DM subjects had significantly increased risk for severe hypoglycemia and increased risk for hyperglycemia.

Note that the significance level for the HBGI is exactly 0.05. Given the multiple parallel t-tests made in this study, such a significance level cannot indicate a rejection of the null hypothesis (i.e., cannot signify the HBGI in T1DM is greater than the HBGI in T2DM). The reason is that when multiple parallel comparisons are performed on the same data, simply by chance one of these comparisons may turn out to be significant, because (if we try many times) a low-probability event may actually happen. This fact was mathematically formulated by the Italian mathematician Carlo Emilio Bonferroni (1892–1960), who introduced an inequality, stating that the probability of a sum of events is less than the sum of the probabilities of these events. Based on this inequality, statisticians introduce *Bonferroni corrections* for the significance level of multiple parallel tests, dividing the significance level by the number of tests. In our case, we have nine parallel tests, so it will be prudent to only reject null hypotheses that meet a significance level of 0.005.

Based on Table 5-2, we now have statistical results that provide initial validation of the low and high BG index as markers reflecting a medical reality—the differences between the BG patterns in T1DM and T2DM. However, these statistical analyses do not provide immediate evidence that these measures would be useful in assessing the *risk* for hypo- and hyperglycemia.

EXERCISE 5-5

Table 5-2(A) demonstrates that the standard deviation of SMBG data is significantly higher for T1DM. Give reasons why the standard deviation cannot be effectively used as a risk measure for hyper- and hypoglycemia separately?

2. Validation of the Low Blood Glucose Index as a Predictor of Upcoming Severe Hypoglycemia

Recall that in 1997, the DCCT's consensus statement concluded that only about 8% of future SH episodes could be predicted from SMBG data. In contrast, we now present the results of a validation trial for the LBGI that predicted about 40% of future SH episodes within the following 6 months.

This LBGI validation trial was performed with a data set containing about 13,000 SMBG readings from 96 adults having T1DM for at least two years and using insulin for BG control. Participants measured their BG three to five times per day for 1 month. Upon completion of the data collection, the LBGI of each participant was computed. For the next 6 months, patients recorded the date and time of all SH episodes on diary sheets they mailed in monthly. Patients were instructed to telephone the investigators whenever an SH episode occurred to schedule a structured interview designed to verify that an episode of SH had, in fact, taken place.

Regression analysis was applied to determine the significance of the LBGI as a predictor of SH episodes.[2] This analysis showed the LBGI was the most significant predictive variable for SH, predicting 40% of the variance of the SH episodes in the subsequent 6 months. In addition, among patients who reported at least one SH episode during the 12 months before the study, this rose to 43%.

The LBGI also provided a means for classifying the subjects for their risk for SH in the subsequent 6 months: subjects with a LBGI below 2.5 experienced, on average, 0.6 SH episodes; subjects at moderate risk (LBGI of 2.5–5) experienced 1.5 SH episodes; and subjects at high risk (LBGI above 5) experienced 5.8 SH episodes (see Figure 5-11). Given these results and a review of the available literature, we conclude the LBGI is the best predictor to date of SH using SMBG data.

Finally, an exploratory analysis confirmed the LBGI was capable of differentiating patients at near-term risk for SH from patients at risk for future, but not imminent, SH. We compared patients who reported SH

---

2. The details of this standard statistical procedure are too technical to present here, but can be found in Draper and Smith (1998).

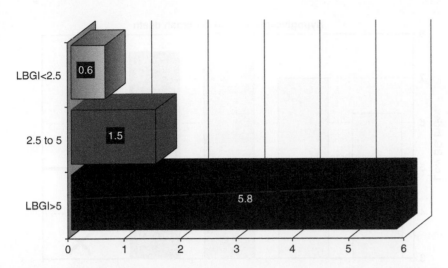

FIGURE 5-11.
Correlation of the LBGI with the number of severe hypoglycemic episodes in the subsequent six months. The individuals with the lowest LBGI suffered the fewest episodes of SH.

within three months of the LBGI calculation with those who reported SH during months 4 through 6, but not within the first 3 months. The LBGI of the first subgroup was significantly higher—6.4 versus 3.5 ($p = 0.005$). At the same time, the two subgroups reported similar numbers of SH episodes within the last year—9.6 and 7.6 ($p = 0.5$), and did not differ in HbA_{1c}, age, or duration of diabetes. This suggests that high-risk status indicates the likelihood of an SH event sooner rather than later.

As a result of this and other similar studies, it is now known that about 130 SMBG readings, spread over 4 to 5 weeks, are sufficient to permit the calculation of an accurate LBGI. This index has been shown to be reliable, internally and across studies, and to be a significant predictor of future SH. It can be used separately or in combination with other variables to create more complex models with higher levels of sophistication. In addition, the computation of a patient's LBGI is quite uncomplicated, which encourages its clinical and research use. We anticipate that the calculation of a person's LBGI could be incorporated directly into future SMBG devices. A patient's LBGI exceeding a critical value could then be used to provide an immediate alert for an increased risk for SH.

Our last comment raises a serious ethical question. Let's assume that in a clinical trial a researcher records a dangerously high LBGI value. What should the researcher do? During the studies described above, two fatal automobile accidents occurred where the driver's loss of control was attributed to hypoglycemia. The values of the LBGI for these two people (shown in Figure 5-12) indicate a continuous increase of the LBGI during the three months preceding the accidents. Should researchers in this situation contact the participant to convey a concern or issue a warning?

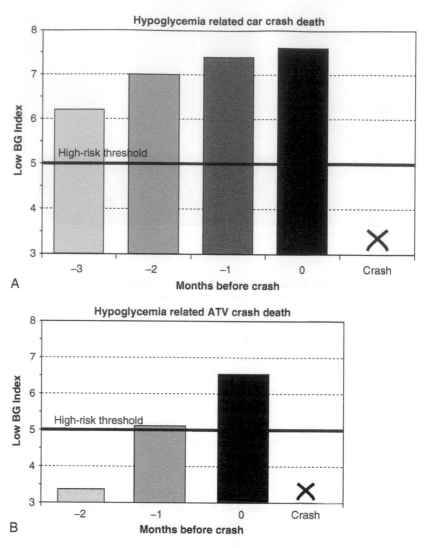

FIGURE 5-12.
The LBGI of two people with T1DM who died in an automobile crash and an ATV accident later attributed to hypoglycemia. It is seen that the risk increased in the months prior to their accidents.

Should the researchers get in touch with the patient's physician to suggest a change in medical management? Or should they simply allow events to take their course without intervening?

There are strict government guidelines addressing these issues. Before any new study with human subjects begins, a detailed protocol needs to be worked out to address every aspect of the study, from recruiting methods, to data collection procedures, confidentiality issues, information-sharing, and documenting and reporting of adverse events. Academic institutions where human subjects research is performed are required to have human investigation committees (or institutional review boards) that examine the protocols and rule on the suitability of the proposed procedures, taking into account both the expected scientific impact of the study and the safety of the participants. Once the protocols are approved, government guidelines strictly prohibit any deviation from the outlined procedures. In particular, unless the need for information-sharing between researchers and participants is documented as essential for achieving the main objective of the study, and made explicit in the protocol, no results of any experimental measurements can be disclosed to the participants.

4. Validation of the High Blood Glucose Index

High BG levels not lasting very long are not nearly as dangerous as similarly low BG levels that may trigger SH. The risk for hyperglycemia only becomes significant when chronically high BG levels are maintained over time. Therefore, because we hypothesized that HBGI accounts for the trends towards hyperglycemia observed in patients' SMBG records, one possible validation of this measure is to evaluate its correlation with HbA_{1c}—the standard for assessing high average BG levels.

We analyzed SMBG and HbA_{1c} data provided by Amylin Pharmaceuticals (San Diego, CA), from 600 subjects with type 1 ($N = 277$) and type 2 ($N = 323$) diabetes. The subjects collected more than 300,000 SMBG readings and had 4180 HbA_{1c} assays taken over six months. The overall correlation between the HBGI and HbA_{1c} was 0.73, $p < 0.0001$, demonstrating a strong linear relationship between these two variables. Further, we identified five categories for the HBGI (below 7, 7–12, 12–15, 15–20, above 20) and computed 95% confidence intervals for HbA_{1c} corresponding to these categories, establishing almost one-to-one correspondence between HBGI and HbA_{1c}. (The statistical details can be found in Kovatchev et al. [2000].)

In summary, we are now in a position to conclude that LBGI and HBGI are valuable quantitative characteristics that could be used for assessment and maintenance of glycemic control. Because these parameters are directly quantifiable from routine SMBG data, they can provide accurate information for analysis and assessments of the effects of changes in therapeutic regimens. In addition, the LBGI and HBGI could be used as building blocks for more sophisticated mathematical models.

We present one such model next.

X. MORE COMPLEX MODELS

The 40% success rate for predicting SH from SMBG data using the LBGI provided a substantial improvement over the 8% rate achieved by the DCCT. In this section, we build a more complex model that utilizes the LBGI, some basic probability laws, and curve-fitting to achieve even greater success in predicting future SH episodes.

In our study, we determined the LBGI for 600 people with type 1 ($N = 277$) and type 2 ($N = 323$) diabetes (the same data used above). Each person had collected three to five measurements per day for 4 to 6 weeks. As in the study described in the previous section, patients also recorded the date and time of all SH episodes during the following 6 months.

I. Developing and Testing a More Sophisticated Model for Predicting Severe Hypoglycemia

In developing the model, we first divided the 600 LBGI values into 15 groups of 40 subjects each, with consecutive risk ranges. The resulting intervals had the following cutoffs: 0.1, 0.5, 0.8, 1.1, 1.4, 1.7, 2.1, 2.5, 2.9, 3.5, 4.2, 5.0, 6.0, and 7.8. Thus, the lowest-risk range is LBGI < 0.1, and the highest-risk range is LBGI > 7.8. We numbered the classes 0, 1, 2, ..., 14. For example, class 1 corresponds to the range $0.1 \leq$ LBGI < 0.5; class 2 corresponds to the range $0.5 \leq$ LBGI < 0.8; and so on.

Next, we determined the probability that a person from each class had an episode of SH in the following month. Suppose 14 of the 40 subjects in class 5 reported *at least one* episode of SH. Then $14/40 = 0.35$ of the people from class 5 had an episode or, said another way, the (empirical) probability of a person in class 5 having at least one episode of SH is 0.35. With this, we have the point (5,0.35); the first coordinate is the class, and the second coordinate is the probability for at least one episode of SH in that class. We do this for each of the 15 classes and obtain the points shown by black triangles in Figure 5-13.

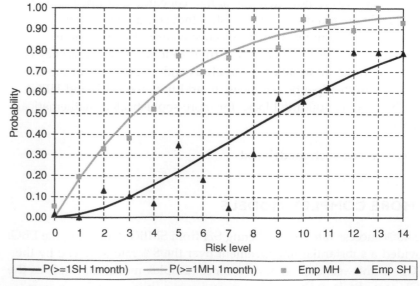

FIGURE 5-13.
Probability for at least one SH episode within 1 month.

We now follow a sequence of steps familiar from previous chapters:

1. Construct a curve (a theoretical model) that describes the change in the data points.

2. Determine values for the model parameters that provide the best fit with the data points.

3. Test the model with additional subjects.

The form of the curve we will use is guided by our experience with engineering problems where one wants to predict future failures of devices that depend slightly on whatever previous failures have occurred. The formula for the curve is called a *Weibull distribution*. Because it is known there is dependence between SH episodes (see, for example, Cryer [1993]), we hypothesized that such a curve would provide a good fit for the probability in question.

The analytic form of the Weibull curve is:

$$F(x) = 1 - \exp(-ax^b), \tag{5-5}$$

where x is the class 0, 1,..., 14. For us, $F(x)$ is the probability that a subject in class x will have at least one episode of SH within 1 month. For those familiar with the theory of probability distributions, note that the value of the parameter b controls the level of dependence between the SH events. In the special case $b = 1$, we obtain the well-known exponential distribution.

As in Chapters 1 and 2, we next used a computer program to determine the values of the model parameters a and b that provide the best fit for $F(x)$ to the data points. For our data, the best fit gave the following probability that a person in class x will have one or more episodes of SH within a month from the end of the data collection phase:

$$F(x) = 1 - \exp(-\exp(-4.1947)\, x^{1.7472}). \tag{5-6}$$

The solid line in Figure 5-13, which appears to provide a good fit to the data, gives the graph of this function. This visual estimate was confirmed by a statistical analysis of the goodness-of-fit of the model given by the so-called coefficient of determination (D^2). This statistic has a meaning similar to that of R^2 in linear regression (R^2 is not applicable to nonlinear models). The coefficient of determination computed for our model was 87.3, meaning that 87.3% of the variation of the probability for at least one episode of SH is explained by our mathematical model. The probability for moderate hypoglycemia (defined by similar symptoms as for SH but that do not preclude self-treatment) was approximated even better by a Weibull model, with $D^2 = 95.9\%$. The theoretical and observed probabilities are presented by the gray line and data triangles, respectively, in Figure 5-13.

The model was also tested with various time intervals (e.g., 3 months or 6 months of prediction) and various criteria (e.g., probability of more than two or more than three episodes) and gave similar results. We concluded that the risks for subsequent SH are described well by a Weibull distribution with scale parameter a in the range of -5 to -3 and a shape parameter b in the range of 1.5 to 2.2, depending on the time period and probability under consideration. Similarly, the risk for subsequent moderate hypoglycemia was described well by a Weibull distribution with scale parameter a in the range of -1.4 to -2 and a shape parameter b in the range of 1.0 to 1.3. These parameters have a direct physiological meaning and reflect clinical reality: the scale parameter reflects the frequency of events [e.g., more frequent events will result in a larger parameter a (and moderate hypoglycemia is more frequent than SH)], and the shape parameter b reflects the degree of dependence between sequential events (e.g., a larger parameter means a more dependent event). For example, it is known that past SH is a major factor for future SH, but this is not true for moderate hypoglycemia.

2. Risk Categories for Future Significant Hypoglycemia

Consistent with the validation data, when the risk classes identified by the probability model were aggregated into four clinically justified risk categories for future significant hypoglycemia—minimal risk (LBGI $\leq$ 1.1); low risk (1.1 < LBGI $\leq$ 2.5); moderate risk (2.5 < LBGI $\leq$ 5), and high risk (LBGI > 5)—it became evident the number of all prospectively observed hypoglycemic episodes increased significantly as the risk category increased. Table 5-3 presents the number of symptomatic SH episodes in each of the risk categories.

We believe that this four-risk category version of our model is much more convenient for patients and health care providers to interpret. This categorization allows for distinguishing between subjects who have practically no chance for significant hypoglycemia from subjects at progressively increasing risk.

We conclude that routine SMBG data with a frequency of three to five readings per day contain valuable information for the metabolic control of people with T1DM and T2DM. Our models suggest such

		Minimal Risk (LBGI $\leq$ 1.1)	Low Risk (1.1 < LBGI $\leq$ 2.5)	Moderate Risk (2.5 < LBGI $\leq$ 5)	High Risk (LBGI > 5)
Number of severe hypoglycemic episodes	T1DM	0	0.35	0.68	4.24
	T2DM	0.09	0.18	1.42	1.85

TABLE 5-3.
Number of prospectively observed significant hypoglycemic episodes per person per month, by risk category and by type of diabetes.

data have very high predictive power for SH. This power, combined with the minimal computing power the models require (calculating the LBGI is no more difficult than computing average BG values), indicates that hand-held SMBG devices should be entirely capable of carrying out the computations. Therefore, implementation of these models in future-generation SMBG devices is entirely possible. It is our hope that the use of some of the models studied in this chapter will, in the near future, become as widespread and as routine in improving BG control as the use of the SMBG devices themselves is now.

REFERENCES

American Diabetes Association. (2003). Economic costs of diabetes in the U.S. in 2002. *Diabetes Care, 26,* 917–932.

Bergman, R. N. (2003). The minimal model of glucose regulation: a biography. *Advances in Experimental Medicine & Biology, 537,* 1–19.

Bergman, R. N., Ider, Y. Z., Bowden, C. R., & Cobelli, C. (1979). Quantitative estimation of insulin sensitivity. *American Journal of Physiology, 236,* E667–E677.

Bloomgarden, Z. T. (1998). International Diabetes Federation meeting, 1997 and Metropolitan Diabetes Society of New York meeting. *Diabetes Care, 21,* 658–665.

Box, G. E. P., & Cox, D. R. (1964). An analysis of transformations (with discussion). *Journal of the Royal Statistical Society, Series B (Methodological), 26,* 211–252.

Cryer, P. E. (1993). Hypoglycemia begets hypoglycemia. *Diabetes, 42,* 1169–1693.

Cryer, P. E. (1999). Hypoglycemia is the limiting factor in the management of diabetes. *Diabetes/Metabolism Research and Reviews, 15,* 42–46.

Cryer, P. E., Fisher, J. N., & Shamoon, H. (1994). Hypoglycemia. *Diabetes Care, 17,* 734–755.

Gold, A. E., Frier, B. M., MacLeod, K. M., & Deary, I. J. (1997). A structural equation model for predictors of severe hypoglycaemia in patients with insulin-dependent diabetes mellitus. *Diabetes Medicine, 14,* 309–315.

Kovatchev, B. P., Cox, D. J., Gonder-Frederick, L. A., & Clarke, W. L. (1997). Symmetrization of the blood glucose measurement scale and its applications. *Diabetes Care, 20,* 1655–1658.

Kovatchev, B. P., Cox, D. J., Gonder-Frederick, L. A., Young-Hyman, D., Schlundt, D., & Clarke, W. L. (1998). Assessment of risk for severe hypoglycemia among adults with IDDM: Validation of the Low Blood Glucose Index. *Diabetes Care, 21,* 1870–1875.

Kovatchev, B. P., Cox, D. J., Straume, M., & Farhy, L. S. (2000). Association of self-monitoring blood glucose profiles with glycosylated hemoglobin. In Johnson, M., & Brand, L. (Eds.), *Methods in Enzymology* (vol. 321, pp. 410–417). New York: Academic Press.

Santiago, J. V. (1993). Lessons from the Diabetes Control and Complications Trial. *Diabetes, 42,* 1549–1554.

The Diabetes Control and Complications Trial Research Group (DCCT). (1997). Hypoglycemia in the Diabetes Control and Complications Trial. *Diabetes, 46,* 271–286.

FURTHER READING

Aaby Svendsen, P., Lauritzen, T., Soegard, U., & Nerup, J. (1982). Glycosylated haemoglobin and steady-state mean blood glucose concentration in type 1 (insulin-dependent) diabetes. *Diabetologia, 23,* 403–405.

Draper, N. R., & Smith, H. (1998). *Applied regression analysis.* (3rd ed., includes disk). New York: John Wiley & Sons.

Reichard, P., & Phil, M. (1994). Mortality and treatment side-effects during long-term intensified conventional insulin treatment in the Stockholm Diabetes Intervention study. *Diabetes, 43,* 313–317.

The Diabetes Control and Complications Trial Research Group (DCCT) (1993). The effect of intensive treatment of diabetes on the development and progression of long-term complications of insulin-dependent diabetes mellitus. *New England Journal of Medicine, 329,* 978–986.

U.K. Prospective Diabetes Study Group (UKPDS). (1998). Intensive blood-glucose control with sulphonylureas or insulin compared with conventional treatment and risk of complications in patients with type 2 diabetes. *Lancet, 352,* 837–853.

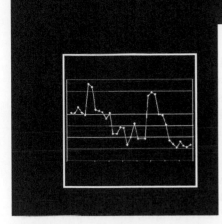

Chapter 6

PREDICTING SEPTICEMIA IN NEONATES

I had a kind of blind faith. I believed in the collaboration between the firm will of my one-pound-twelve-ounce daughter and the expertise of modern medicine.

Wendy Wasserstein. *Complications.*
The New Yorker, *Feb. 21, 2000, pp. 87–109*

Between five and six of every 100 births in the United States are premature, requiring specialized medical assistance or emergency intervention. Medical advances over the last 2 decades now allow infants born up to 18 weeks prematurely and with birth weights (BWs) as low as 500 g to survive. Although the survival rates for premature babies weighing less than 750 g have increased tremendously, to about 50%, every low birth weight (LBW) premature baby faces a high risk of serious complications, diseases, and long-term developmental problems.

In this chapter, we employ mathematical techniques and models to answer the following question: Can heart rate (HR) be used to predict potentially deadly bacterial infections in premature babies before other symptoms manifest themselves? To answer this question, we shall study symmetric and asymmetric distributions, statistical measures of data variability and symmetry, and measures of irregularity in time series.

I. PREMATURE BIRTHS, LOW BIRTH WEIGHTS, AND HEALTH RISKS

Gestation is the period between the conception and birth of a baby during which the fetus is growing within the mother's uterus. Justifiably considered one of the most awe-inspiring of natural phenomena, the period of gestation has been studied for centuries. *Gestational age* (GA) is the time the fetus has been *in utero* (inside the uterus) and is usually measured in weeks. A normal pregnancy lasts about 40 weeks, and during this period a human being arises from a single fertilized egg. At birth, every infant is classified as *premature* (<37 weeks gestation), *full-term* (37–42 weeks gestation), or *post-term* (>42 weeks gestation). In the medical literature, infants are also referred to as *neonates*. Physicians specializing in neonatal care are called *neonatologists*.

Following birth, assessment of an infant's weight, head circumference, and vital signs are used to determine a *developmental GA.* This developmental age may differ from the calendar GA and is based on the level of the infant's physical development and muscle tone. For example, an infant born with a GA of 35 weeks may at birth be assessed to have a developmental GA of 37 weeks. BW is perhaps the most important developmental characteristic. Neonates that are "small for GA" or "birth restricted" may be full-term but underweight.

In the medical literature, babies born weighing less than 2500 g (5 lbs, 8 oz) are referred to as LBW infants. Babies weighing less than 1500 g (3 lbs, 5 oz) are classed as very LBW (VLBW) infants, and it is this VLBW group for which the risk of health problems and long-term complications is highest.

Premature infants may face a number of medical complications because of their LBW and underdeveloped organ systems. For example, underdevelopment of the lungs and the digestive and nervous systems presents major health hazards, such as respiratory complications, difficulty coordinating sucking and swallowing, food intolerance, susceptibility of bleeding into the brain, and episodes of breathing cessation (apnea). Infection is another significant threat to premature infants, because their immature immune system is less able to fight germs that can cause serious illness. Some infections can be life threatening, even with early initiation of the most advanced treatments.

For premature babies, hospitals have special neonatal intensive care units (NICUs). Virtually all NICUs have neonatologists and neonatology nurses on duty around the clock, as well as ready access to respiratory therapists, pharmacists, occupational therapists, dietitians, and anesthesiologists, among other highly specialized medical staff. These experts provide a full range of neonatal intensive care, from treatment of infection to complex neurologic surgery.

Approximately 40,000 VLBW infants are born in the United States annually. Although advances in newborn medical care have greatly reduced the deaths and disabilities associated with LBW, a small percentage of survivors are still left with problems, such as mental retardation, cerebral palsy, and impairments in lung function, sight, and hearing. Infections such as sepsis (see below) continue to be a major cause of morbidity and mortality (Tortora and Grabowski [2003]; Gray et al. [1995]).

II. SEPSIS: MEDICAL OVERVIEW, CLINICAL DIAGNOSIS, AND DIAGNOSTIC CHALLENGES

Sepsis is a life-threatening illness caused by an overwhelming infection of the bloodstream by toxin-producing bacteria. According to the American Academy of Pediatrics (AAP), the incidence of proven sepsis is approximately 2 in 1000 live births. Alarmingly, 7% to 13% of neonates show symptoms or signs of infection and are evaluated for sepsis. Infection is a major cause of fatality during the first month of life, contributing to 13% to 15% of all neonatal deaths. The mortality rate in neonatal sepsis may be as high as 50% for infants not receiving prompt treatment.

Neonatal sepsis is categorized as *early* or *late onset*. Eighty-five percent of newborns with early-onset sepsis have symptoms within their first 24 hours of life, with the remaining 15% of cases occurring within 5 to 6 days after birth. Early onset sepsis is connected with an infant's acquisition of microorganisms from the mother, perhaps at delivery by passage through a colonized birth canal. The microorganisms most commonly associated with early onset infection include group B *Streptococcus* (GBS), *Escherichia coli, Haemophilus influenzae,* and *Listeria monocytogenes.*

Late-onset sepsis syndrome occurs between 7 and 90 days after birth and is acquired from the caregiving environment. Organisms implicated in causing late-onset sepsis include coagulase-negative staphylococci, *Staphylococcus aureus, E. coli, Klebsiella, Pseudomonas, Enterobacter, Candida,* GBS, *Serratia, Acinetobacter,* and anaerobes. The onset of sepsis is most rapid and severe in premature infants because their immune system is generally underdeveloped.

The clinical syndrome of sepsis is brought about by the infant's response to the insults of the bacterial infection and has been named the *systemic inflammatory response syndrome* (SIRS) by the ACCP/SCCM Consensus Conference Committee (1992). Neonatal sepsis occurs in as many as 25% of VLBW infants. Neonates who develop late-onset sepsis have a mortality rate of 17% (more than twice the 7% mortality rate of noninfected infants), as well as increased morbidity, according to findings of the National Institute of Child Health & Human Development Neonatal Research Network (see Bone et al. [1997] and Stoll et al. [1996]).

Figure 6-1 shows the monitoring equipment typically utilized in the care of premature infants. Despite the extreme efforts made by NICUs to eliminate every possibility of bacterial contamination, risk factors for late-onset sepsis are ever present. Each interventional device is a potential source of infection and increases the risk of infectious illness for the newborns. For example, the following medical interventions have been independently associated with sepsis: intubation, umbilical catheters, prolonged mechanical ventilation, nutrition via venous catheters, respiratory distress syndrome, intraventricular hemorrhage, and nasogastric and tracheal cannulae. As a result, sepsis is common in neonates, and infected infants spend significantly more days on the ventilator and an average of 25 more days in the hospital.

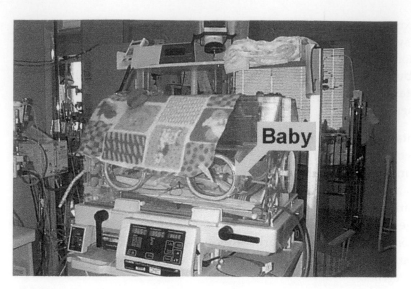

FIGURE 6-1.
Equipment monitoring the vital signs of a premature baby. (Photo courtesy of Drs. J. Randall Moorman and M. Pamela Griffin, University of Virginia.)

Sepsis may predispose newborns to other hazardous medical conditions that may be more likely fatal for premature infants. For example, the risk for death or meningitis from sepsis is much higher in infants with LBW than in full-term neonates. In addition, *necrotizing enterocolitis* (NEC), a serious gastrointestinal disease, affects up to 4,000 infants in the United States yearly, and an estimated 10% to 50% of infants who develop NEC die (Kellogg et al. [1997]). Infants who develop NEC may require intubation and increased respiratory support. Survivors are often left with strictures and short-bowel syndrome. Thus, premature infants require careful vigilance so that sepsis or SIRS can be detected early and therefore treated more effectively.

Unfortunately, the early clinical signs of sepsis and NEC in newborns are neither uniform nor specific, being associated with characteristics of the causative bacteria and the body's response to the infection. This makes the early diagnosis of neonatal sepsis extremely difficult and results in many unnecessary procedures, including blood cultures, short courses of antibiotics administered to infants without bacterial infection, and interruptions in neonatal nutrition. Many newborns undergo diagnostic studies and treatment with antibiotics before the diagnosis has been confirmed, and this is clinically justified because of the rapid development of the infection and its potentially lethal consequences.

Drawing blood and evaluating the culture in a laboratory is the gold standard for establishing the diagnosis of sepsis caused by systemic bacterial infection. However, there are also concerns regarding its reliability, especially if single, small-volume samples are submitted, as is often the practice with critically ill newborns. For example, Schelonka et al. (1996) estimated that as many as 60% of culture results may be falsely negative if only 0.5 ml of blood is obtained from infants with low–colony-count sepsis. And drawing more blood from an infant who weighs less than 1500 g may be quite tricky—after all, her entire blood volume is less than 100 ml.

The situation is further complicated because not all infants with clinical signs of sepsis have positive blood cultures. In general, the AAP, the American Academy of Obstetrics and Gynecology (AAOG), and the Centers for Disease Control and Prevention (CDC) all recommend sepsis screening and/or treatment for various risk factors related to GBS diseases. The diagnostic challenge is manifested even more clearly by the fact the CDC allows for a diagnosis of neonatal "clinical sepsis" with either a negative blood culture or no blood culture at all (see Cabal et al. [1980] for more details). Following these recommendations, many neonates now are subjected to routine treatment, even when blood culture results are negative. Because the mortality rate for untreated sepsis can be as high as 50%, the hazard of untreated sepsis is too great to wait for confirmation by positive cultures. Clinical neonatologists caring for VLBW infants recognize sepsis and NEC as potentially catastrophic illnesses and do not hesitate to begin antibiotic treatment empirically at the first appearance of symptoms. In general, most clinicians initiate treatment while awaiting culture results from the laboratory. Gerdes and Polin (1987) estimated that 10 to 20 infants are treated for sepsis for every infant with a positive blood culture.

In summary, the high risk of sepsis or NEC in VLBW infants, the high mortality rate, and the lack of reliable clinical means for early diagnosis of these diseases all underscore the need for new diagnostic tools that would provide sufficient warning to ensure successful treatment. Identifying one or more biomedical variables that would provide comparatively early danger signals for these babies is therefore of utmost importance.

An optimal surveillance strategy should: (1) Be based on non-invasive monitoring methods; (2) utilize continuous monitoring of the newborns; and (3) provide dynamic estimates of the infant's risk for developing sepsis or SIRS. Because as a rule all infants in the NICU have certain vital characteristics that are continuously monitored, such as temperature, HR, and blood pressure, a method for assessing the risk of sepsis and SIRS based on these characteristics would be highly desirable and easy to use.

We now focus on a proposed solution using HR characteristics. This novel method, proposed by University of Virginia researchers, can successfully predict sepsis and SIRS 12 to 24 hours before the clinical diagnosis is made (Griffin et al. [2003]). Our next section serves as an introduction to understanding the terminology used in describing HR data and patterns.

III. HEART RATE AND HEART RATE VARIABILITY

In searching for a surveillance strategy for early warning of sepsis or NEC, it is important to keep in mind the clinical understanding of these conditions. As mentioned previously, the current hypothesis is that the clinical syndrome of sepsis, SIRS, is brought about by the infant host's response to such insults as bacterial infection. The major host response at the molecular level is the release of cytokines—small circulating peptides that serve as mediators of the inflammatory response. They play the role of invisible messengers or ligands that bind to specific targets (receptors) on the cell surface. Various studies suggest that SIRS may be caused by an imbalance between cytokines' proinflammatory and anti-inflammatory effects. In sepsis, circulating cytokines are important in triggering and maintaining the inflammatory response, and their quantities are correlated with the severity of illness. For example, elevated levels of circulating cytokines have been found up to 2 days before the diagnosis of clinical sepsis—see Kuster et al. (1998). Of particular interest to our investigation are the widespread effects of cytokines on signal transduction processes and, specifically, their potential to interfere with the normal control of HR by the sympathetic and parasympathetic nervous systems.

Signal transduction at the cellular level refers to the mechanism by which signals are transmitted from the outside of the cell to the inside. One of the mechanisms of signal transduction may involve small ion movement, either into or out of the cell. These ion movements result in changes in the electrical potential of the cells that propagate the signal along the cell. Other mechanisms of signal transduction may involve the coupling of ligand–receptor interactions to cascades of intracellular events that alter enzyme activities and protein conformations. Signal transduction leads to alterations in cellular activity and to changes in the program of genes expressed within the responding cells. For example, it has been shown by Oddis and Finkel (1997) that specific cytokines (acting as ligands and binding receptors on the cell surface) may increase HR. Interestingly, the same cytokines have also been shown to reduce HR responses to certain drugs.

Figure 6-2 presents an electrocardiogram (EKG), which is a record of the electrical activity in the heart muscle during the heartbeat cycle. The markers at the peaks pinpoint the time of occurrence of sequential heartbeats. The characteristic that quantifies HR is the interbeat interval (RRI). The RRI is the elapsed time (usually measured in milliseconds) from one heartbeat to the next (e.g., the distance between two sequential markers [so a larger RRI corresponds to a slower HR]).

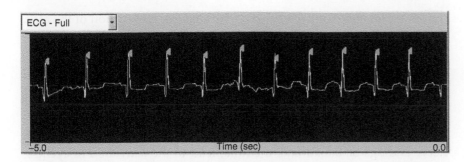

FIGURE 6-2.
An EKG recording. An EKG records the electrical activity of the heart muscle as it contracts. The triangular markers indicate the peak of each beat. (Courtesy of Dr. J. Randall Moorman, University of Virginia.)

Figure 6-2 presents 5 seconds of EKG data. It is important to note that even in healthy individuals the duration of sequential RRIs is not constant (and, as we see below, is normally even more variable than in some diseases). This natural HR variability (HRV) arises from the interplay between the two arms of the autonomic nervous system—the system responsible for those necessary bodily functions not under voluntary control. The autonomic nervous system controls HR through the sympathetic pathway, which increases HR, and the parasympathetic pathway, which slows it.

In general, some variability in the HR is a healthy sign, because internal fluctuations give the organism greater freedom to adapt to external challenges. In contrast, in newborn infants, as in adults, HRV is substantially reduced during periods of severe illness (see Burnard, [1959]; Rudolph et al. [1965]; Cabal et al. [1980]; Griffin et al. [1994]).

There are various theories concerning the mechanisms of reduced HRV aimed at explaining and quantifying the differences in the mathematical characteristics of RRIs for normal and low HRV. We focus on the theory, developed at the University of Virginia by Drs. J. Randall Moorman and M. Pamela Griffin, which explains the mechanism of observed HRV abnormalities using signal transduction cascades (see Nelson et al. [1998]). We begin with the physiological mechanisms causing and regulating heartbeat.

The heart is a muscular pump consisting of four chambers (two atria and two ventricles) that contract in a coordinated fashion to push blood through the circulatory system. Blood returning from the body enters the right atrium, passes through to the right ventricle, and is then pumped to the lungs. Blood returning from the lungs enters the left atrium, passes through to the left ventricle, and is then pumped to the body. The heartbeat (the rhythmic contraction of the heart muscle) is controlled by the sinoatrial node (or pacemaker), which causes contraction of the atria. The contraction of the main pumping chambers, the ventricles, is controlled by the atrioventricular node.

Recall that autonomic nervous system exerts control over HR through the sympathetic and parasympathetic pathways. The sinoatrial node cell membrane has beta-adrenergic (epinephrine- and norepinephrine-binding) receptors which, on binding ligands released from sympathetic nerve endings or the adrenal medulla, lead to the activation of cyclic adenosine monophosphate (cAMP)-dependent protein kinase. The active cAMP-dependent protein kinase phosphorylates cardiac ion channels and results in cell depolarization, an action potential, and a heartbeat. Consequently, HR rises after sympathetic stimulation. The sinoatrial node cell membrane also contains muscarinic acetylcholine receptors. When bound with acetylcholine from parasympathetic nerve endings, they inhibit the process described above and HR falls. Thus, for as long as the complex steps of intracellular signal transduction can be successfully completed, the sinoatrial node can be viewed as an amplifier of the input signals of the autonomic nervous system and HR as the output signal. As the amounts of sympathetic and parasympathetic activity vary, HR varies as well.

Now consider a severe illness like sepsis. In such an unfavorable metabolic milieu, the optimal conditions for signal transduction are unlikely; HRV becomes abnormal because sinus node cells, like all other cells, are unable to respond normally to sympathetic and parasympathetic inputs. In general, reduced HRV is a sign of system isolation and disruption of the control mechanisms.

IV. QUANTIFYING HEART RATE AND HEART RATE VARIABILITY

It is a cinematic cliché to use the sound of a thumping heart in suspenseful scenes—the beats becoming very slow and regular when the character gets into trouble. Following the scientific literature (and the movies), we could expect that before sepsis and SIRS the HR would become more regular and would sometimes slow down. Longer RRIs represent HR decelerations, while shorter RRIs represent HR accelerations. Formulated scientifically, this means we should see reduced baseline HRV and transient decelerations. To investigate whether these expectations are correct, we need two components: (1) Lots of HR data for healthy hearts and for hearts before sepsis; and (2) methods quantifying HRV and distinguishing normal from abnormal HR series, paying particular attention to HR decelerations.

We begin by visually inspecting the HR time series of a baby who developed sepsis. The baby was born 5 months prematurely, at

24 weeks gestation, and weighed 720 g at birth. The baby had evidence of lung disease but was clinically stable when the HR observations began. The three plots in Figure 6-3 illustrate normal HRV and the abnormal changes occurring before sepsis.

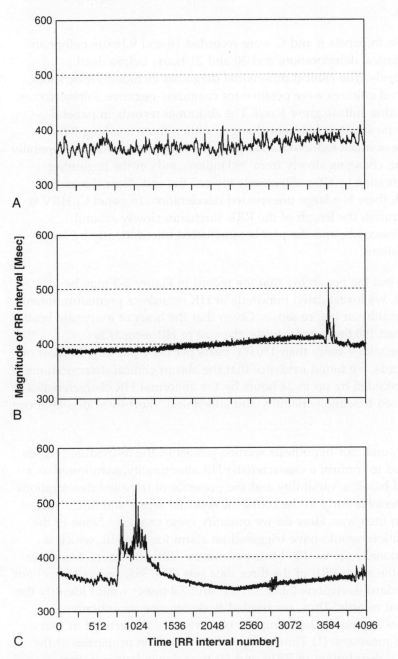

FIGURE 6-3.
Plot of RRI time series. The three panels are from the same infant while clinically stable (panel A) and from 18 (panel B) and 9 (panel C) hours before an acute clinical deterioration leading to death. (From Kovatchev, B. P., Farhy, L. S., Hanging, C., Griffin, M. P., Lake, D. E., Moorman, J. R. [2003]. Sample asymmetry analysis of heart rate characteristics with application to neonatal sepsis and systemic inflammatory response syndrome. *Pediatric Research, 54*, 892–898. Used by permission.)

Each panel in Figure 6-3 shows 4096 consecutive RRIs, approximately 20 to 30 minutes of heart beat data. The data in panel A, representing normal HRV, were recorded well before any signs of sepsis occurred. As expected, the length of the RRIs oscillated around 360 to 370 milliseconds, with the persistent small changes showing the accelerations and decelerations characteristic of normal, healthy variability in heart rhythm.

The data in panels B and C were recorded 18 and 9 hours before an acute clinical deterioration, and 30 and 21 hours before death, respectively. The clinical differential diagnosis included septic shock, and blood cultures were positive for coagulase-negative *Staphylococcus*, and a urine culture grew *E coli*. The abnormal records in panel B display no small or frequent changes; rather, there is a long-lived baseline of much reduced variability—the length of the RRIs is generally constant, changing slowly from 390 milliseconds in the beginning of observation to 410 milliseconds toward the end. Near the end of panel B, there is a large unexpected deceleration. In panel C, HRV is also reduced, the length of the RRIs fluctuates slowly around 350 milliseconds, and the plot is punctuated intermittently by large decelerations.

To rule out the possibility that the record in Figure 6-3 may be atypical, we investigated hundreds of HR records of premature infants when healthy or before sepsis. Given that the heart of a neonate beats more than 100 times per minute, the size of HR records is overwhelming—more than 150,000 beats per baby per day. In most of the records, we found evidence that the abrupt clinical deteriorations were preceded by up to 24 hours by the abnormal HR characteristics of reduced baseline variability and subclinical, short-lived decelerations in HR.

At this point, our hypothesis seemed plausible: the individual records appeared to confirm a characteristic HR abnormality, manifested as a reduced baseline variability and the presence of transient decelerations was detectable early in the course of neonatal sepsis and SIRS. The question then was: How do we quantify these changes? None of the decelerations would have triggered an alarm for low HR, which is conventionally set for 100 beats per minute (RRI 600 msec). Moreover, neither the mean RRIs of the three data sets (369, 398, and 362 msec) nor the standard deviations (SDs; 3.7, 5.0, and 7.4 msec) would identify the abnormal records. Thus, we needed to design new quantitative measures accurately depicting our observations. We arrived at two types of measures: (1) Time-independent, based on properties of the statistical distribution of RRIs; and (2) time-dependent measures of temporal regularity. We describe these measures in the next sections.

V. TIME-INDEPENDENT MEASURES: INTERBEAT INTERVAL DISTRIBUTION, STANDARD DEVIATION, AND SKEWNESS

As discussed above, the length of sequential RRIs varies from beat to beat, and the degree of its variability is related to health or illness. A time-independent approach to investigating the variability of a sample of RRIs focuses on their statistical distribution (this approach is time-independent because changing the order of RRIs within the sample will not change its statistical distribution). In this setting, the length of a RRI is assumed to be a random quantity with a certain unknown distribution. A histogram of a certain sample of RRIs depicts one realization of this distribution. The shape and the symmetry of this histogram provide information about the properties of the random RRIs. Changes in the histogram may, therefore, signify changes in health.

We want to develop a measure of how at-risk a baby is for the imminent onset of sepsis based on HRV. If we examine the panels in Figure 6-3, some things stand out. In panel A, where the baby is clinically stable, there is a pattern of "regular irregularity." Compared with panels B and C, there is a substantial amount of variation throughout the period in the RRIs, but the variation seems to be under control. In panel B, when the baby's health has begun to deteriorate, there is very little variation during most of the period, with a few large decelerations. In panel C, as health continues to worsen, these changes become more pronounced.

The fact that panels B and C depict RRI data with little variation suggests that the corresponding histograms will have a relatively low SD. The isolated decelerations, on the other hand, suggest a certain asymmetry in the histogram of the RR data. To better understand why, we briefly review the relation between the following basic descriptors of a histogram—the mean, the median, and the skewness.

The two most common ways to measure the center of a body of quantitative data are the mean and the median. The mean is what people often refer to as the average, and the median is the 50th percentile. In some ways, the median can be more descriptive of the data's center because it is not affected by extreme values (as the mean can be). A body of data is symmetric if the histogram can be divided into two halves that are mirror images of one another. If the data are symmetric, the mean and the median coincide.[1] In reality, it would be unusual to find a completely symmetric body of data, but many phenomena give data that are approximately symmetric.

Some data that are not symmetric can be described as skewed to the left or right. To get an intuitive idea of skewness, suppose we begin with

---

1. It is possible, however, for the mean and the median to coincide when the data are not symmetric.

a symmetric histogram as in Figure 6-4(A). Note that the mean and the median coincide. Now suppose that we take a data point that is to the right of the center and move it farther to the right [Figure 6-4(B)]. The mean moves to the right, but the median says the same. The data in Figure 6-4(B) are skewed to the right.

This example illustrates the following ideas:

1. The mean is sensitive to extreme values, but the median is less so.

2. If a body of data has a long tail to the right but not to the left, the data is skewed to the right.

3. If the data is skewed to the right, then the mean will lie to the right of the median.

One way to quantify the balance between the left and right tails of a histogram is to use its skewness. The skewness γ of a distribution is defined as the quotient of the third moment μ_3 about the mean $E(X)$ and the third power of the SD σ and is given by the expression $\gamma = \dfrac{\mu_3}{\sigma^3} = \dfrac{E\{(X - E(X))^3\}}{[E\{(X - E(X))^2\}]^{3/2}}$. The skewness for symmetric distributions is zero; it is positive if the distribution develops a longer tail to the right; and it is negative if the distribution develops a longer tail to the left. Accordingly, distributions skewed to the right are called *positively skewed* and distributions skewed to the left are called *negatively skewed*.

In Figure 6-5(A), we have a symmetric distribution. The degree of variability is quantified by the SD – reduced variability corresponds to

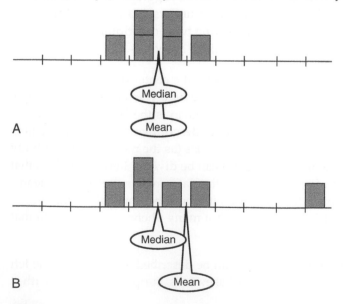

A

B

FIGURE 6-4.
A schematic representation of skewness as illustrated by the relative position of the mean versus the median. In panel A, the mean and median coincide, whereas in panel B the mean is to the right of the median.

a smaller SD. If a sample of RRIs has predominantly decelerations, and few accelerations, their histogram would have a longer right tail and almost no bars to the left of its mode (thus, its shape will be similar to that in panel B). Conversely, if the sample of RRIs includes mostly accelerations, but no decelerations, the histogram would be similar in shape to that in panel C—with a long left tail and almost no bars to the right of its mode.

However, the SD and the skewness of the distribution of RRIs have certain limitations. In particular, SD and skewness are computed with respect to the mean of the distribution, which is quite vulnerable to large deviations and does not accurately represent the center of a skewed distribution. Figure 6-6 exemplifies this point by presenting a sample of RRIs that includes transient decelerations. These decelerations result in an asymmetric distribution, with a longer right tail. As a result, the mean and the median of that sample do not coincide—the mean is influenced by a few decelerations and is substantially shifted to the right, poorly describing the center of the RRI distribution. This and other limitations make SD and skewness insufficiently accurate for the life-saving task of predicting sepsis in prematurely born infants.

In order to overcome this limitation, we need to compute measures that quantify deviations in the length of RRIs from their median. In addition, such measures (unlike SD or skewness) will need to allow for a *separate quantification* of HR accelerations (deviations less than the central RRI, forming the left-hand portion of the histogram) and decelerations (deviations greater than the central RRI, forming the right). The next

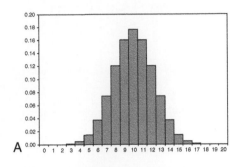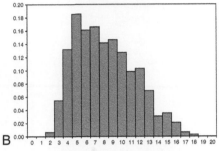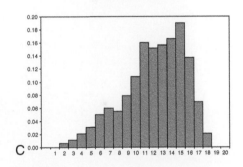

FIGURE 6-5.
Typical histogram of a symmetric (panel A), skewed to the right (panel B), and skewed to the left (panel C) statistical distributions.

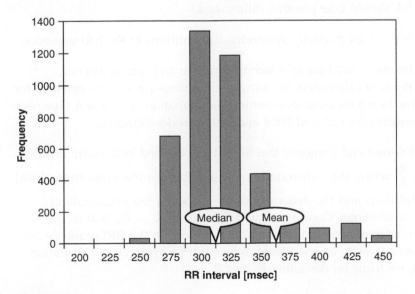

FIGURE 6-6.
Sample of RRIs that includes transient decelerations. Note the skewness of the distribution.

section introduces the sample asymmetry (SA), a measure possessing these desired properties.[2]

EXERCISE 6-1

Describe why the mean value does not represent a good measure for assessing the variability of the RRI series.

EXERCISE 6-2

Give examples of two data sets having equal SDs, one corresponding to a RRI series with transient decelerations and the other to one with transient accelerations.

VI. TIME-INDEPENDENT MEASURES: SAMPLE ASYMMETRY OF A RANDOM VARIABLE

We design the new SA measure to comply with the following conditions:

1. SA should grow when there are more decelerations in the RRI sequence;

2. SA should decrease as there are more accelerations in the RRI sequence;

3. SA should take positive values; and

4. SA = 1 for perfectly symmetric distributions of the RRI sequence.

Conditions 3 and 4 are of a technical nature and are, in essence, conditions of calibration. In contrast, conditions 1 and 2 are essential for constructing a measure overcoming the limitations of SD and skewness as measures for reduced HRV and transient decelerations.

Conditions 1 and 2 suggest that SA may be defined in the form of a ratio $SA = \dfrac{R_2}{R_1}$ where the numerator R_2 is a measure for the magnitude of RRI decelerations and the denominator R_1 measures the magnitude of RRI accelerations. Condition 4 will then mean $R_1 = R_2$, that is, the magnitudes of accelerations and decelerations in the RRI sequence are the same, exactly as one would expect from a symmetric distribution. Thus, we focus on designing R_1 and R_2.

---

2. This measure was first introduced by Kovatchev et al. (2003).

Let m be the median for the lengths of a sequence of observed RRIs and x be the length of a single RRI. For this interval, the quantity $(x - m)^2$ could be used as a measure describing the deviation of the RRI length from the median length m. In the language we employed in Chapter 5, we can view this quantity to be the risk assigned to the RRI of length x because of its deviation from the median value m.

The problem we faced in Chapter 5 was quite similar. The question there was to separately assess the risk for hypoglycemia because of low blood glucose readings and the risk for hyperglycemia because of high blood glucose readings. Fundamentally, our problem here is identical—we need to assess the magnitude of the risk for RRI decelerations as measured by positive deviations from m, and the risk for RRI accelerations, as measured by negative deviations from m. In that sense, R_1 corresponds to the Low BG Index, and R_2 corresponds to the High BG Index from Chapter 5. Thus, following the same approach, for any RRI of length x, we define the quantities:

$$rd(x) = \begin{cases} (x - m)^2 & for \quad x > m \\ 0 & for \quad x \leq m \end{cases}$$

and   (6-1)

$$ra(x) = \begin{cases} (x - m)^2 & for \quad x < m \\ 0 & for \quad x \geq m. \end{cases}$$

The function $rd(x)$ describes the degree of deviation to the right from the median value (risk for deceleration) while the function $ra(x)$ describes the degree of deviation to the left from the median (risk for acceleration).

EXERCISE 6-3

Compare the definitions in Eq. (6-1) with the definitions for rl and rh given by Eq. (5-5) in Chapter 5. What are the similarities and differences?

Consider now a sequence of RRIs of lengths $x_1, x_2, \ldots x_n$ and let m denote the median of the data. Using the weighing functions in Eq. (6-1), define two quantities representing the sum of the weighted deviations to the left and to the right of the median μ as follows:

$$R_1 = \frac{1}{n} \sum_{i=1}^{n} ra(x_i) \quad and \quad R_2 = \frac{1}{n} \sum_{i=1}^{n} rd(x_i). \qquad (6-2)$$

EXERCISE 6-4

Compare the quantities in Eq. (6-2) with the low blood glucose index and high blood glucose index measures defined in Chapter 5. What are the differences? What are the similarities?

EXERCISE 6-5

Find R_1 and R_2 for the following time series: 666.7, 666.7, 659.3, 666.7, 689.7, 652.2, 666.7, 681.8, 666.7, and 645.2.

Definition. The SA of the data set $x_1, x_2, \ldots, x_n$ is defined by the ratio:

$$SA = R_2 \Big/ R_1,\tag{6-3}$$

where R_1 and R_2 are the quantities defined by Eq. (6-2).

In Figure 6-7, we present two examples of data sets—one approximately symmetric (panel A) and one positively skewed (panel B). The graph of the weighing function $(x - m)^2$ is given by a solid black line. As anticipated, the SA for the skewed distribution is higher compared with the SA for the approximately symmetric distribution (3.5 vs 1.1). Notice that for the data set with an approximately symmetric histogram, the SA is close to 1. We leave it as an exercise to verify that the SA measure, as defined by the Eq. (6-3), satisfies conditions 1 through 4 at the beginning of this section.

EXERCISE 6-6

Verify that the measure SA defined by Eq. (6-3) satisfies conditions 1 through 4 listed at the beginning of this section.

Figure 6-8 contains the histograms of the distributions of RRIs depicted in Figure 6-3. The histograms present the distribution of the RRI lengths with regard to their deviation from the median value m. Note that we used a logarithmic vertical scale. As expected, the distribution of RRIs in the period of health is approximately symmetric [Figure 6-8(A)], while the HR abnormalities observed before sepsis cause marked asymmetry of the histograms in panels B and C. The positive (right) skewness of the histograms is caused by a reduction in the number of shorter-than-median RRIs. The skewness of the distribution increased from 0.82 in panel A to 1.92 in panel B and to 1.55 in panel C. This change, however, is not consistent with the worsening medical

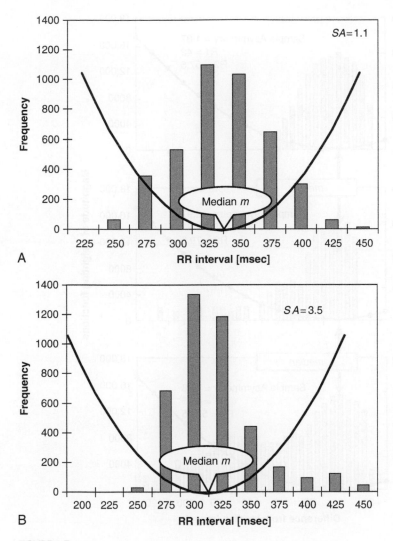

FIGURE 6-7.
SA for an approximately symmetric versus nonsymmetric distribution of RRIs.

condition of the baby, as described earlier. A better description of the observed phenomenon is offered by considering the SA of the three data sets.

In Figure 6-8, the value of the SA for the data in panel A is 1.37, showing approximately symmetric distribution of the RRIs in a condition of health. In panel B, the value of SA increases to 2.97, and the value of the left weighting R_1 decreases to 27, pointing to certain abnormalities (reduced accelerations and increased decelerations) beginning to appear in this infant's HR 18 hours before any clinical signs of sepsis. Nine hours before the onset of sepsis (panel C), the value of the SA has increased dramatically to 11.8, mainly because of a large number of substantial heart decelerations. This is confirmed by the 10-fold increase in the value of R_2 over its healthy baseline.

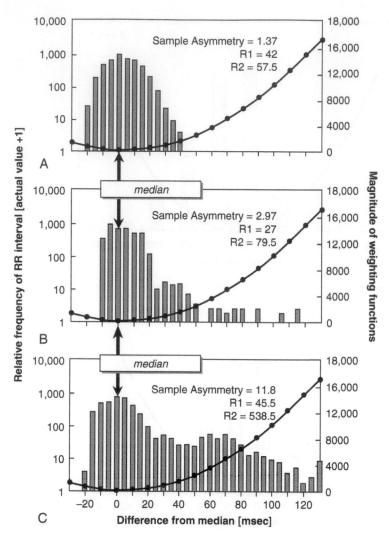

FIGURE 6-8.
Plot of RRI distribution. The three panels are from the same infant while clinically stable (panel A) and from 18 (panel B) and 9 (panel C) hours before an acute clinical deterioration leading to death. (From Kovatchev, B. P., Farhy, L. S., Hanging, C., Griffin, M. P., Lake, D. E., Moorman, J. R. [2003]. Sample asymmetry analysis of heart rate characteristics with application to neonatal sepsis and systemic inflammatory response syndrome. *Pediatric Research, 54*, 892–898. Used by permission.)

The data depicted in Figure 6-8 suggest that the SA may hold promise as a marker quantifying HRV related to upcoming sepsis and SIRS episodes. In the last example, the SA was markedly higher 18 hours before the emergence of clear clinical symptoms. It could be hypothesized that if treatment had been initiated at this time, there would have been a possibility of saving this infant's life.

The SA has performed well in one example. Does this mean that it will perform the same way for another individual data set, for a group of data sets, or for all data sets? The appropriate way to answer such questions is through statistical analyses, and we discuss these and other validation questions next.

VII. DATA VALIDATING THE UTILITY OF SAMPLE ASYMMETRY ANALYSIS OF HEART RATE VARIABILITY

1. Experimental design and methods

In this section, we describe a study involving 158 infants admitted to the NICU at the University of Virginia that focuses on the following question: Could SA be considered a measure for the risk of approaching episodes of sepsis and SIRS? In this section, we follow a fundamental statistical approach involving an experiment designed to compare two groups of infants—those with and without episodes of sepsis and SIRS. For each group, HRV is collected throughout the study. For the group of infants developing sepsis, the SA is calculated for every day of the five days preceding the medical diagnosis. We seek to corroborate our hypothesis that SA levels increase significantly 24 hours before sepsis.

The participants in the study were infants with high risk factors for acquiring late-onset sepsis, including LBW, prematurity, the need for central venous access, and NICU stays longer than 2 weeks. Fifty of these infants had a total of 75 episodes of sepsis or SIRS during the study, defined to be present when a physician suspected sepsis or SIRS, obtained a blood culture, and administered antibiotic therapy for seven or more days. These 50 infants formed our *experimental group*.

A *control group* of 50 healthy infants (i.e., infants who did not develop sepsis and SIRS during the study) was selected from the remaining 108 consecutive admissions to the NICU to precisely match the experimental group by birth weight and GA.[3] Table 6-1 includes the statistical comparisons between the experimental and the control groups. A group comparison *t*-test (Table 6-1, last column) indicates that there is no evidence for rejecting the hypotheses for equality of the mean BW and GA of the two groups.

---

3. To do this, we had to tackle the following problem: The entire group of healthy infants had, on average, both a higher BW and GA, which was to be expected because LBW is a risk factor for sepsis. Thus, our random selection of control infants used a nonuniform random generator with a higher probability of selecting LBW infants and lower probability of selecting higher BW infants. Specifically, the probability of selecting a control infant ranged from 1 for VLBW (BW ≤ 656 g) to 0.16 for BW > 1451 g. These probability values were determined by the relative distribution of BW in the experimental versus control group. Once experimental versus control infants were matched by BW, the matching by GA occurred naturally (Table 6-1).

	Experimental Group: Episodes of Sepsis and SIRS	Control Group: No Episodes	Group Comparison (*t*-test)
Infants	50	50	—
Mean BW in grams (SD)	1227 (847)	1228 (760)	$t = 0.01$, $p > 0.99$
Mean GA in days (SD)	198 (34)	196 (26)	$t = 0.37$, $p = 0.72$

TABLE 6-1.
Study population.
Abbreviations: BW, birth weight; GA, gestational age; SD, standard deviation.

2. Results

Table 6-2(A) shows results for SA values for experimental-group infants in 24-hour blocks beginning 5 days before sepsis. Column 2 of this table contains the average SA values for each of the 24-hour blocks for the experimental group. Columns 3 and 4 of the table present

	Mean Sample Asymmetry (SD)	$\ln(R_1)$ (accelerations)	$\ln(R_2)$ (decelerations)
A: Changes Before Sepsis for the Experimental Group			
5 Days before sepsis	3.28 (1.59)	4.40 (0.94)	4.41 (0.77)
4 Days before sepsis	3.59 (1.63)	4.36 (0.92)	4.45 (0.76)
3 Days before sepsis	3.50 (1.94)	4.27 (0.92)	4.39 (0.80)
2 Days before sepsis	3.66 (2.12)	4.19 (0.85)	4.37 (0.71)
1 Day prior to sepsis	4.20 (2.32)	4.10 (0.87)	4.28 (0.65)
Contrast 1: 5 days versus 1 day before sepsis	$F = 5.5$, $p = 0.02$	$F = 5.8$, $p = 0.02$	$F = 1.9$, $p = 0.2$
B: Values in Health			
Infants Posttreatment	3.28 (1.34)	4.60 (0.87)	4.65 (0.74)
Contrast 2: 5 days before sepsis versus posttreatment	$F = 0.01$, $p > 0.99$	$F = 2.1$, $p = 0.2$	$F = 4.2$, $p = 0.05$
Control Infants	2.90 (0.93)	4.75 (0.67)	4.67 (0.62)
Sepsis versus control— ANOVA with covariates BW and GA	$F = 2.40$, $p = 0.13$	$F = 0.35$, $p = 0.6$	$F = 0.01$, $p > 0.99$

TABLE 6-2.
Sample asymmetry of heart rate and asymmetries caused by accelerations and decelerations before neonatal sepsis and SIRS, and in good health.

separately the mean values of contributions caused by accelerations (measured by the logarithm of R_1) and decelerations (measured by the logarithm of R_2).

It is evident that SA began increasing approximately 3 to 4 days before sepsis, with the steepest increase in the last 24 hours. To statistically compare the mean values of the baseline values of SA, $\ln(R_1)$, and $\ln(R_2)$ 5 days before sepsis and those for the last 24 hours, a statistical test called repeated measures analysis of variance (ANOVA) was applied. This analysis is used when repeated measurements are made on the same subject as we do for the SA values. Recall that the t-test is used for testing the equality of the means for two groups. In effect, repeated measures ANOVA extends the two-sample t-test for equality of two independent population means to the more general setting of comparing the means of groups formed by repeated measurements of the SA. The data in Table 6-2(A) show that the difference between the baseline group (5 days before sepsis) and the group containing the SA values from the last 24 hours before sepsis was significant: $F = 5.5$, $p = 0.02$.

Within the experimental group, analysis of the contribution of accelerations (measured as the natural logarithm of R_1) and of decelerations (natural logarithm of R_2) shows the surprising finding that there was a significant fall in R_1 ($F = 5.8, p = 0.02$) before sepsis, but no significant change in R_2 [the last two columns in Table 6-2(A)]. This is readily interpreted to mean that a decrease in the extent and duration of HR accelerations is more marked than an increase in decelerations, which are more easily identified by eye.

As presented in Table 6-2(B), posttreatment SA returned to exactly presepsis levels. A comparison of the two periods of health, 5 days before sepsis versus posttreatment, showed no difference ($F = 0.01$, $p > 0.99$). A similar return toward baseline was observed for both R_1 (accelerations) and R_2 (decelerations), although the second index, R_2, remained somewhat elevated. As expected, a comparison of experimental-group infants' data when healthy to control-group infants revealed no significant difference in SA ($F = 2.4, p = 0.13$) and no significance of the covariates BW and GA (p-levels of 0.2 and 0.7, respectively). Similarly, no significant differences in R_1 and R_2 were observed between healthy and experimental infants in health [Table 6-2(B)].

In closing, these analyses show that SA becomes elevated in the 24 hours before sepsis and then, after a successful treatment, returns to its baseline. Thus, SA may be a valid tool in the detection of sepsis in prematurely born infants before the occurrence of any clinical symptoms.

VIII. VALIDATING THE PROPERTIES OF SAMPLE ASYMMETRY THROUGH COMPUTER SIMULATION

In addition to testing with real data, a commonly accepted strategy for validating a new statistical measure is to simulate data with certain properties and then investigate whether the new measure depicts these properties accurately. In the case of SA, we need to confirm its ability to distinguish, in a graded fashion, HR time series with varying degrees of transient decelerations. So, we constructed simulations consisting of a baseline signal with the same frequency components as clinically observed neonatal HR data, and then added to this signal scaled versions of clinically observed deceleration. In this way, we were able to simulate a wide range of HRV abnormalities. Figure 6-9 shows an example of simulated data with baseline SD 7 milliseconds and four decelerations of magnitude 100 milliseconds, corresponding to moderately reduced HRV with subclinical decelerations from 150 beats per minute (RRI 400 msec) to 120 beats per minute (RRI of 500 msec). This is a realistic simulation—inspection of Figure 6-3(B) suggests a deceleration might easily be 20 times larger than the SD of a stable, low-variability neonatal HR.

The results are plotted in Figure 6-10. SA is plotted as a function of the height of the deceleration measured in SD of the baseline signal. We simulated an increasing number of decelerations, from 0 to 8. Figure 6-10 shows that both larger and more frequent decelerations lead to increasing values of SA, with values exceeding 10 in many conditions. For the example in Figure 6-9, with four decelerations

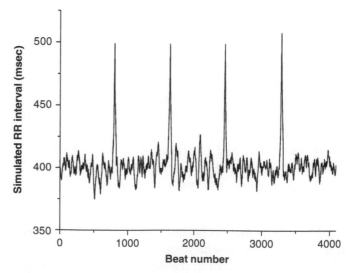

FIGURE 6-9.
Simulated heart rate data with baseline variability with SD = 7 milliseconds and 4 decelerations. (From Kovatchev, B. P., Farhy, L. S., Hanging, C., Griffin, M. P., Lake, D. E., Moorman, J. R. [2003]. Sample asymmetry analysis of heart rate characteristics with application to neonatal sepsis and systemic inflammatory response syndrome. *Pediatric Research, 54,* 892–898. Used by permission.)

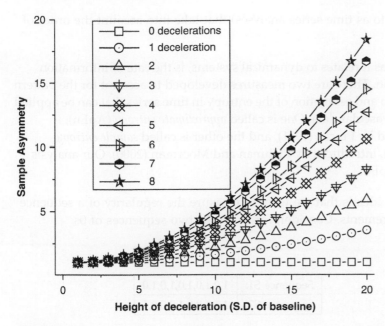

FIGURE 6-10.
Increase in SA as a function of number (different curves) and magnitude of HR deceleration. (From Kovatchev, B. P., Farhy, L. S., Hanging, C., Griffin, M. P., Lake, D. E., Moorman, J. R. [2003]. Sample asymmetry analysis of heart rate characteristics with application to neonatal sepsis and systemic inflammatory response syndrome. *Pediatric Research, 54,* 892–898. Used by permission.)

of approximate height 14 SD, SA is 5.7. Thus, SA is sensitive to gradually increasing frequency and magnitude of transient decelerations, which makes it a promising measure for abnormality in HR data samples.

IX. TIME-DEPENDENT MEASURES: SAMPLE ENTROPY

It should be noted that the timing component in our discussion so far appeared only in Figure 6-3, where consecutive samples of 4096 RRIs were sequentially recorded. The measures subsequently introduced— SD, skewness, and SA—are based solely on the statistical distribution of RRIs. Thus, none of these measures takes into account the order of RRIs within each sample and the potential contribution of the exact *timing sequence* of the RRIs is ignored. Recall, however, Figure 6-3, which suggests that a change in the time pattern of the RRIs may be indicative of an upcoming sepsis episode. The statistical approach may be insufficient for thoroughly addressing the problem. To emphasize this point even further, if we take a sample of 4096 RRIs and calculate the samples' SD, skewness, and SA, they will be exactly the same as the SD, skewness, and SA of any reshuffling of the sample. Clearly, however, the evolution of the RRI sequences with time is a factor that should not be ignored, but to study this factor we shall need to employ methods (referred to as time series analyses) that take into account the order of the RRIs.

Entropy, as it relates to dynamical systems, is the rate of information production. There are two measures developed to account for the pattern of the data and estimation of the entropy in time series that can be applied to cardiovascular data. One is called *approximate entropy* (ApEn), introduced by Pincus (1991), and the other is called *sample entropy* (SampEn), introduced by Richman and Moorman (2000). Our analysis uses SampEn.

The basic idea is that we want to measure the regularity of a sequence of measurements. Consider, for example, two sequences of 0s and 1s:

| Sequence S1: | 1,0,1,0,1,0,1,0,1,0 |
| Sequence S2: | 1,1,1,0,0,1,0,0,0,1 |

Both sequences have five 0s and five 1s, but sequence S1 seems to have a pattern, and sequence S2 seems random. Sample entropy assigns a non-negative number to each sequence. The more regular a sequence is, the lower the sample entropy will be. Thus, SampEn for sequence S1 will be lower than SampEn for sequence S2.

Intuitively, in a regular sequence, if we know two or three adjoining terms, then we have a good idea of what the next term will be. In sequence S1, there are nine adjoining two-term subsequences (only eight with a following term). Each subsequence is either 1,0 or 0,1, and if it is 1,0, the following term is always 1; if it is 0,1, the following term is always 0.

EXERCISE 6-7

Determine all different types of two-term subsequences for the sequence S2.

In sequence S1, the 1,0 subsequence appears five times. Two identical subsequences are called *matches*. When considering RRI data, it would be highly improbable to ever have an exact match, even in subsequences of length 2, so instead we say there is a match when the subsequences agree within a certain tolerance. The customary way to assign this tolerance depends on the SD of the sample. Usually, the tolerance t is chosen to be the product, $t = r \cdot SD$, where r is a number between 0.1 and 0.25 and SD is the standard deviation of the sample.

EXERCISE 6-8

Compute the tolerance t for sequence S1 with $r = 0.2$. What is the
tolerance for sequence S2 if $r = 0.2$?

Now suppose we have determined a tolerance value t. We say
the subsequences (a,b) and (a',b') are matches if $|a - a'| < t$
and $|b - b'| < t$ (i.e., if the distance between each component differs by
less than the tolerance). In a regular sequence, if subsequences (a,b) and
(a',b') are matches, it is likely the sequences (a,b,c) and (a',b',c') will also
be matches. Conversely, if the time series is irregular, having matches of
length 2 is not necessarily a precursor to matches of length 3. The
following definition of SampEn generalizes this observation.

For a fixed integer m, SampEn is defined as the negative natural
logarithm of the ratio A/B, where $A =$ number of matches of length
$m + 1$, and $B =$ number of matches of length m (see Figure 6-11). For
those familiar with calculating conditional probabilities, the quantity
A/B is precisely the conditional probability that two sequences within a
tolerance r for m points remain within r of each other at the next point.
Because the computation of SampEn depends on the order of the data
points in the RRI time series, SampEn is a true time-dependent measure,
capturing temporal complexity in sequences of data. ApEn is defined
similarly, but it counts self-matches as well (see Pincus [1991] for
details).

Notice that parameters m and r are critical in determining the value of
SampEn. It is, in fact, more appropriate to use the notation SampEn =
SampEn(m, r) to reflect the dependence. However, no guidelines exist for
optimizing their values, and this is a shortcoming in the current theory.

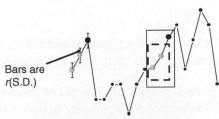

Bars are
r(S.D.)

$A =$ number of matches of length $m+1$
$B =$ number of matches of length m

ApEn » –ln $(1+A)/(1+B)$
SampEn = –ln A/B

For regular, repeating data, A/B nears 1 and entropy
nears 0.

FIGURE 6-11.
Defining sample entropy of RRIs. (Courtesy of Dr. J. Randall Moorman, University of Virginia.)

The various existing rules of thumb generally lead to the use of values of m of 1 or 2 for data records of length N ranging from 100 to 5000 data points and values of r between 0.1 and 0.25 to determine the tolerance. In general, the accuracy and confidence of the entropy estimate improve as the numbers of matches of length m and $m + 1$ increase. By this token, the number of matches can be increased by choosing a small m (short templates) and a larger r (wide tolerance). There are penalties, however, for criteria that are too relaxed: as r increases, the probability of matches tends toward 1, and SampEn tends to 0 for all processes, thereby reducing the ability to distinguish any salient features in the data set; and as m decreases, underlying physical processes that are not optimally apparent at smaller values of m may be obscured.

This being said, in most current applications the parameter values of choice are $m = 2$ and $t = 0.2*\text{SD}$, which means we are counting templates with a length of 2 to calculate B and templates with a length of 3 to calculate A, and the tolerance for matches is set to 0.2 times the SD of the process. In most cases, all readings in the observed sample are first divided by the SD of that sample, so the SD of the sample becomes exactly $\text{SD} = 1$, in which case $t = r = 0.2$. This preprocessing of the data eliminates the influence of the variance of the sample on the irregularity (or complexity) of the process, thus leaving SampEn to pick up only characteristics strictly related to the sequential timing of the observations and generally independent from the distribution of the observations. More details, including the strict definition of SampEn, can be found in Lake et al. (2002).

Although the computation of SampEn for long time series certainly requires appropriate software (see Internet Resources at the end of the chapter), one simple numerical example using the short sequences considered earlier should clarify the template counting algorithm.

Example 6-1

For $m = 2$ and $r = 0.2$, calculate the SampEn and SD for the sequences S1: 1,0,1,0,1,0,1,0,1,0 and S2: 1,1,1,0,0,1,0,0,0,1, and compare the results.

We begin with the periodic series S1. The SD of this sample is 0.527. Thus, for $r = 0.2$, the tolerance for similarity between two templates would be $t = (0.2)(0.527) = 0.1054$. With $m = 2$, all subsequences of length 2 (beginning at up to N-m) in the series are 10,01,10,01,10,01,10,01. Given a similarity tolerance of $t = 0.1054$, two subsequences would be matches only if they are identical. Thus, the total number of template matches of length $m = 2$ is B = 3+3+3+3+3+3+3+3 = 24 (each template 10 or 01 has exactly three matches, excluding self-matches). All subsequences of length $m + 1 = 3$ in the above series are 101,010,101,010,101,010,101,010. Thus, the total number of template matches of length $m = 3$ is A $= 3+3+3+3+3+3+3+3 = 24$ (each template 101 or 010 has three matches). SampEn is then computed as SampEn $= -\ln(A/B) = -\ln(24/24) = 0$, signifying complete order.

Now consider series S2, also having five 1s and five 0s, but in a more random order: 1,1,1,0,0,1,0,0,0,1. The SD of this series is again 0.527 because SD does not depend on the order of observations. Thus, the similarity tolerance would be again $t = 0.1054$. All subsequences of length 2 in the series are 11,11,10,00,01,10,00. Thus, the total number of template matches of length $m = 2$ is B$=1+1+1+2+0+1+2+2=10$. All subsequences of length $m + 1=3$ are 111,110,100,001,010,100,000,001, and the total number of template matches of length $m = 3$ is A $= 0+0+1+1+0+1+0+1 = 4$. Then, SampEn $= -\ln(4/10) = 0.9163$, signifying a certain degree of irregularity.

Thus, two time series that otherwise have identical distributions of 0s and 1s (and therefore identical distribution-based characteristics) are distinguished solely on the basis of the order of their observations. In the first case, the series is periodic (i.e., there is apparent order), leading to 0 SampEn, while in the second there is no apparent order, and the SampEn is substantially larger.

With regard to HRV, Griffin and Moorman (2001) and Lake et al. (2002) have shown that SampEn is lower for RRI records with reduced variability and transient decelerations. Thus, SampEn is another potential predictor of sepsis and SIRS in prematurely born infants. Because a lower value of SampEn indicates reduced complexity and more self-similarity in the RRI time series, lower SampEn before sepsis would capture decreased system complexity, disruption in the signal transduction pathways controlling the HR, and increased system isolation.

SampEn has been validated in clinical studies recording HR in premature infants. Figure 6-12 presents a plot of the changes in HR SampEn of an infant during the 9 days before sepsis. There is an apparent transient reduction in SampEn 4 to 5 days before sepsis, indicating temporarily increasing HR regularity, and larger and more sustained reduction in SampEn during the 24 hours before the clinical manifestation of sepsis. As presented here, SampEn, combined with the other measures of HRV discussed in this chapter, is a powerful tool for predicting potentially deadly clinical situations.

EXERCISE 6-9

For $m = 2$ and $r = 0.2$, calculate SampEn for the sequences 0,0,1,0,0,1,0,0,1 and 1,0,0,0,1,0,1,0,0. What can you say about their SDs?

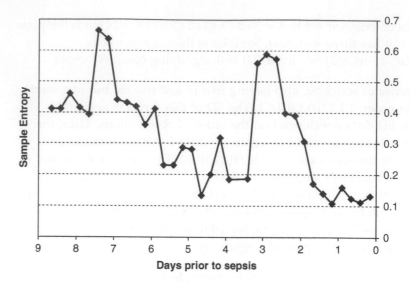

FIGURE 6-12.
Changes in HR sample entropy before sepsis.

X. COMBINING VARIOUS MEASURES OF HEART RATE VARIABILITY ABNORMALITY

Abnormalities in HRV characteristic of illness can be identified by comparing the HRV parameters we have discussed (SD, skewness, SA, and SampEn) using more advanced mathematical models, such as logistic regression models, neural networks, multiple variable analysis, nearest neighbor analysis, or other predictive mathematical instruments. It is most important to track the data of each at-risk infant and compare sequential HR time series. Experiments show that changes observed in sequential 4096-beat time series tracked over time provide a good indication and reliable warning of upcoming episodes of sepsis. In order to conduct such experiments, the HR of each at-risk baby is recorded continuously, using bedside monitors.

A large clinical trial involving 316 neonates in the University of Virginia NICU and 317 neonates in the Wake Forest University NICU tested and validated the utility of the HRV characteristics presented in this chapter. In this trial, clinical data were prospectively collected, and RRIs were continuously recorded in all infants in these two NICUs who stayed for more than 7 days. Episodes of sepsis and sepsis-like illness were defined as acute clinical deteriorations that prompted physicians to obtain blood cultures and start antibiotics. During the trial, 273 such episodes were observed in 194 of the infants, a staggering number, which demonstrates the high risk for sepsis in these premature babies.

A predictive statistical model was developed using data from the derivation cohort in the University of Virginia NICU, and then tested on the validation cohort at Wake Forest University. An HR characteristics (HRC) index was defined as the output of a regression model combining the RRI characteristics described in this chapter to predict sepsis and sepsislike illness. The HRC index is a unitless number between 0 and 1 proportional to the risk of an acute clinical deterioration in the next 24 hours. The HRC index: (1) Showed highly significant association with impending sepsis and sepsislike illness ($p < 0.001$); and (2) added significantly to the demographic information of BW, GA, and days of postnatal age in predicting sepsis and sepsislike illness ($p < 0.001$). The conclusion from this clinical trial was that continuous HRC monitoring is a generally valid and potentially useful noninvasive tool in the early diagnosis of neonatal sepsis and sepsislike illness. Additional details regarding the clinical trial and the exact definition of the HRC index can be found in Griffin et al. (2003).

EXERCISE 6-10

For an RRI series with low SampEn, what additional tests could be used to determine whether there is an increased proportion of decelerations (of interest in predicting sepsis) or accelerations? Could you suggest two such tests?

REFERENCES

Bone, R. C., Grodzin, C. J., & Balk, R. A. (1997). Sepsis: A new hypothesis for pathogenesis of the disease process. *Chest, 112,* 235–243.

Burnard, E. D. (1959). Changes in heart size in the dyspnoeic newborn infant. *British Medical Journal, 1,* 1495–1500.

Cabal, L. A., Siassi, B., Zanini, B., Hodgman, J. E., & Hon, E. E. (1980). Factors affecting heart rate variability in preterm infants. *Pediatrics, 65,* 50–56.

Gerdes, J. S., & Polin, R. A. (1987). Sepsis screen in neonates with evaluation of plasma fibronectin. *The Pediatric Infectious Disease Journal, 6,* 443–446.

Gray, J. E., Richardson, D. K., McCormick, M. C., & Goldmann, D. A. (1995). Coagulase-negative staphylococcal bacteremia among very low birth weight infants: Relation to admission illness severity, resource use, and outcome. *Pediatrics, 95,* 225-230.

Griffin, M. P., & Moorman, J. R. (2001). Toward the early diagnosis of neonatal sepsis and sepsis-like illness using novel heart rate analysis. *Pediatrics, 107,* 97–104.

Griffin, M. P., Scollan, D. F., & Moorman, J. R. (1994). The dynamic range of neonatal heart rate variability. *Journal of Cardiovascular Electrophysiology, 5,* 112–124.

Griffin, M. P., O'Shea, T. M., Bissonette, E. A., Harrell, F. E., Jr, Lake, D. E., & Moorman, J. R. (2003). Abnormal heart rate characteristics preceding neonatal sepsis and sepsis-like illness. *Pediatric Research, 53,* 920–926.

Kellogg, J. A., Ferrentino, F. L., Goodstein, M. H., Liss, J., Shapiro, S. L., & Bankert, D. A. (1997). Frequency of low level bacteremia in infants from birth to two months of age. *The Pediatric Infectious Disease Journal, 16,* 381–385.

Kovatchev, B. P., Farhy, L. S., Hanging, C., Griffin, M. P., Lake, D. E., & Moorman, J. R. (2003). Sample asymmetry analysis of heart rate characteristics with application to neonatal sepsis and systemic inflammatory response syndrome. *Pediatric Research, 54,* 892–898.

Kuster, H., Weiss, M., Willeitner, A. E., Detlefsen, S., Jeremias, I., Zbojan, J., Geiger, R., Lipowsky, G., & Simbruner, G. (1998). Interleukin-1 receptor antagonist and interleukin-6 for early diagnosis of neonatal sepsis 2 days before clinical manifestation. *Lancet, 352,* 1271–1277.

Lake, D. E., Richman, J. S., Griffin, M. P., & Moorman, J. R. (2002). Sample entropy analysis of neonatal heart rate variability. *American Journal of Physiology, 283,* R789–R797.

Members of the ACCP/SCCM Consensus Conference Committee. (1992). American College of Chest Physicians/Society of Critical Care Medicine Consensus Conference: Definitions for sepsis and organ failure and guidelines for the use of innovative therapies in sepsis. *Critical Care Medicine, 20,* 864–874.

Nelson, J. C., Rizwan-uddin, J. C., Griffin, M. P., & Moorman, J. R. (1998). Probing the order of neonatal heart rate variability. *Pediatric Research, 43,* 823–831.

Oddis, C. V., & Finkel, M. S. (1997). Cytokines and nitric oxide synthase inhibitor as mediators of adrenergic refractoriness in cardiac myocytes. *European Journal of Pharmacology, 320,* 167–174.

Pincus, S. M. (1991). Approximate entropy as a measure of system complexity. *Proceedings of the National Academy of Sciences of the United States of America, 88,* 2297–2301.

Richman, J. S., & Moorman, J. R. (2000). Physiological time-series analysis using approximate entropy and sample entropy. *American Journal of Physiology, 278,* H2039–H2049.

Rudolph, A. J., Vallbona, C., & Desmond, M. M. (1965). Cardiodynamic studies in the newborn, III: Heart rate patterns in infants with idiopathic respiratory distress syndrome. *Pediatrics, 36,* 551–559.

Schelonka, R. L., Chai, M. K., Yoder, B. A., Hensley, D., Brockett, R. M., & Ascher, D. P. (1996). Volume of blood required to detect common neonatal pathogens. *Journal of Pediatrics, 129,* 275–278.

Stoll, B. J., Gordon, T., Korones, S. B., Shankaran, S., Tyson, J. E., & Bauer, C. R. (1996). Late-onset sepsis in very low birth weight neonates: A report from the National Institute of Child Health and Human Development Neonatal Research Network. *Journal of Pediatrics, 129,* 63–71.

Tortora, G. J., & Grabowski, S. R. (2003). *Principles of anatomy and physiology.* ((10th ed.)). New York: John Wiley & Sons.

INTERNET RESOURCES

http://physionet.incor.usp.br/physiotools/sampen/
Link to a page containing open-source MATLAB and C code, written by Lake, D. K., Moorman, J. R., and Hanqing, C., for computing SampEn

FURTHER READING

Akselrod, S., Gordon, D., Ubel, F. A., Shannon, D. C., Barger, A. C., & Cohen, R. J. (1981). Power spectrum analysis of heart rate fluctuation: A quantitative probe of beat-to-beat cardiovascular control. *Science, 213,* 220–222.

Goldberger, A. L., & West, B. J. (1987). Chaos in physiology: Health or disease? In Degn, H., Holden, A. V., & Olsen, L. F. (Eds.), *Chaos in biological systems* (pp. 1–4). New York: Plenum Press.

Goldberger, A. L., Rigney, D. R., & West, B. J. (1990). Chaos and fractals in human physiology. *Scientific American, 262,* 42–46.

Goldberger, A. L., Bhargava, V., West, B. J., & Mandell, A. J. (1985). On a mechanism of cardiac electrical stability: The fractal hypothesis. *Biophysics Journal, 48,* 525–528.

Malik, M., & Camm, A. J. (1993). Heart rate variability: From facts to fancies. *Journal of the American College of Cardiology, 22,* 566–568.

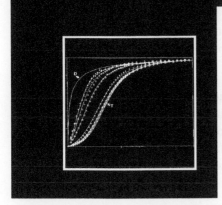

Chapter 7

COOPERATIVE BINDING: HOW YOUR BLOOD TRANSPORTS OXYGEN

We'd like to know the molecular basis of the oxygenation process, not only as a way to understand how hemoglobin functions, but also as a model to find out how other important proteins and enzymes regulate metabolic processes.
Chein Ho, Director, Pittsburgh NMR Center for Biomedical Research

Hemoglobin is a protein constituting more than 95% of the solid mass of red blood cells. Its function is to carry oxygen (O_2) from the lungs to peripheral tissues and return carbon dioxide (CO_2) to the lungs. Consequently, it is one of the most important and most studied of all proteins. Myoglobin (sometimes called muscular hemoglobin) is a structurally analogous protein whose function is to bind and store O_2 in muscle tissue and provide O_2 to the working muscle.

Hemoglobin is made up of four subunits or *monomers* called globins. In adult hemoglobin, two of these are alpha globin proteins and two are beta globin proteins, expressed symbolically as $\alpha_2\beta_2$. Both alpha and beta globins are members of gene families that resulted from gene duplication, mutation, and selection of an ancestral globin gene. This duplication, mutation, and selection have allowed functional differentiation. For example, consider epsilon (ε) and gamma (γ), the beta globin family members found in embryonic and fetal hemoglobins. Early in embryonic development, hemoglobin consists of $\alpha_2\varepsilon_2$, and is followed by $\alpha_2\gamma_2$ or fetal hemoglobin. These hemoglobins have higher affinities for O_2 than adult hemoglobin $\alpha_2\beta_2$, permitting more efficient uptake of O_2 from the placenta.

Human hemoglobin tetramers have four sites where a molecule of O_2 may bind. A salient characteristic of hemoglobin is that this binding is cooperative. That is, after the first molecule of O_2 is bound, a second is more likely to bind. This affinity increases as more O_2 molecules bind, such that the affinity for the fourth O_2 molecule is approximately 300 times that of the first.

Hemoglobin becomes saturated with O_2 in the lungs, where the partial pressure of the O_2 is high. As hemoglobin circulates through the body, the level of O_2 decreases while the level of CO_2 increases. Hemoglobin's affinity for O_2 decreases in the presence of CO_2. In this environment, a change in hemoglobin structure occurs, facilitating O_2 release. CO_2 will then

211

bind to the hemoglobin. When the hemoglobin returns to the lungs, where CO_2 levels are lower and CO_2 is released, its affinity for O_2 increases. Thus, hemoglobin again binds O_2, and the cycle repeats.

This chapter focuses on creating mathematical models describing hemoglobin–oxygen binding. We begin by describing the structures and functions of myoglobin and hemoglobin to help conceptualize the models. Similar models could be used to study drugs binding to receptors. In the next chapter, we present numerical methods for appropriately testing the models.

I. INTRODUCTION

To produce the large amounts of adenosine triphosphate (ATP) required for cells to maintain organization and perform useful work, the cells of higher organisms must have access to sufficient O_2 to support aerobic respiration. Anaerobic respiration (fermentation) only yields two ATP per glucose, which is simply not sufficient to support the complicated organization of a higher eukaryote. Aerobic respiration allows the cells to convert the energy in each glucose molecule into the equivalent of 36 molecules of ATP, most of which is generated by the electron transport chain. Simple unicellular organisms, and those multicellular organisms one or two cell layers thick, may obtain sufficient O_2 via diffusion from the environment. More complex animals require a circulatory system, with specific molecules to bind and transport O_2. In vertebrates, this molecule is hemoglobin.

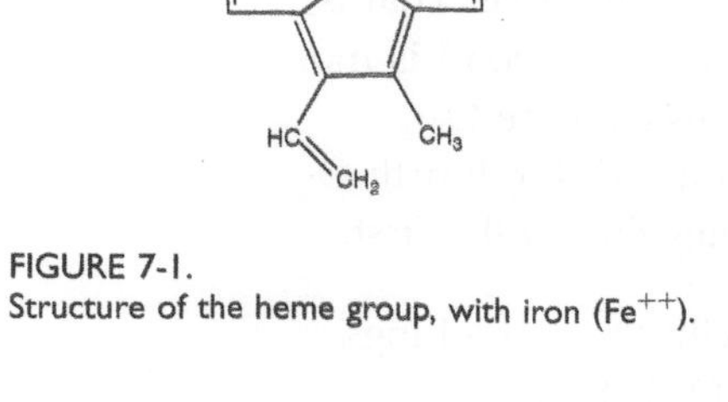

Deciphering the secrets of hemoglobin has challenged generations of organic chemists, biophysicists, biochemists, and physiologists. Until the late 1880s, it was unclear whether hemoglobin was a low–molecular weight compound or a macromolecule. Emil Fischer (1852–1919; Nobel laureate, 1902) was the first to establish that hemoglobin and all other proteins are biopolymers, called *polypeptides*, built of 20 different alpha-amino acids.[1] Another major obstacle in deciphering the structure of hemoglobin was overcome by Hans Fischer (1881–1945; Nobel laureate 1930), who established the presence of an iron-containing complex called *heme* in the macromolecule (Figure 7-1). Both myoglobin and hemoglobin contain heme.

Heme consists of a planar, ring-like structure with resonating double bonds, bound to an iron atom. The iron is bound by four nitrogen atoms at the center of the ring. Because the coordination number of iron is 6,

FIGURE 7-1.
Structure of the heme group, with iron (Fe^{++}).

---

1. For those with some chemistry background: The amino acids are linked with one another by means of peptide bonds: $-C(=O)NH-$. The general form of the amino acids is $R-CH(NH_2)COOH$, where different amino acids are represented by different radicals R. Peptides consisting of more than 50 amino acids are classified as proteins.

two vacant positions remain. In myoglobin, a globin protein molecule occupies one of the two positions. The remaining position can bind an O_2 molecule, so myoglobin could be thought of as a monomeric oxygen-binding heme protein. Hemoglobin, on the other hand, is a tetramer: it contains four protein subunits, $\alpha_2\beta_2$, each similar to myoglobin, each with its own heme. Because each hemoglobin subunit can bind an O_2 molecule, the tetramer can bind up to four O_2 molecules at a time. Figure 7-2 shows the three-dimensional hemoglobin structure determined by John Kendrew (1917–1997) and Max Perutz (1914–2002). In 1962, they shared a Nobel Prize in Chemistry for this work.

The ferrous ion in the deoxygenated hemoglobin is situated slightly to one side of the plane of the other heme atoms. When an O_2 molecule is attached, a new geometry arises in which the ferrous ion becomes coplanar with the rest of the heme. This causes a change increasing O_2 affinity and making consecutive oxygenation easier. This effect is called *cooperativity*. Figuratively speaking, as the process repeats, the whole macromolecule pulsates like a heart, the ferrous ion wobbling to and fro across the heme plane. The cooperative binding of O_2 results from molecular interactions at many levels: the movement of the iron in the heme, structural changes within the individual subunits, and a reorientation of the subunits within the tetrameric hemoglobin. Hemoglobin, therefore, is not just a simple mix of four independent oxygen-binding subunits but, rather, a "molecular machine" with its structure directly related to its O_2 transport function.

As hemoglobin–oxygen binding is just one of many binding reactions that occur in living organisms, we now examine some concepts fundamental to this larger class of molecular interactions.

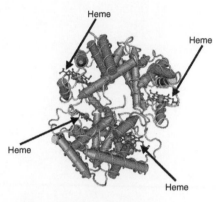

FIGURE 7-2.
Structure of the complete hemoglobin protein. The polypeptide chains of the four subunits are given in the form of coiled ribbons. The small spheres linked together are the atoms within the heme. The whole hemoglobin complex is globular (i.e., the polypeptide chains are coiled to an almost spherical configuration).

II. BINDING REACTIONS

Binding reactions are the most common types of molecular interactions taking place within living organisms. A binding reaction occurs when two or more molecules prefer each other's company more than the company of other molecules within a solution. Put simply, if we could look with a very high-powered microscope, we would find the bound molecules had a higher probability of being together than would be predicted by a random distribution of all the molecules within the solution. Binding reactions are common in almost all fields of chemistry, biochemistry, medicine, physiology, and physics. For example, in pharmacology the *simple mechanism of drug action* is for a drug to bind reversibly to a specific receptor to form a *drug-receptor complex,* initiating an effect that is usually proportional to the concentration of the complex. Symbolically, this process could be expressed as:

$$[\text{Drug}] + [\text{Receptor}] \leftrightarrow [\text{Drug-Receptor}] \rightarrow \text{Effect}, \qquad (7\text{-}1)$$

where square brackets [ ] are used to denote concentration. Thus, [Drug] implies the concentration, or more rigorously, the activity of the drug that is free in solution (i.e., unbound, or not bound to the receptor). Here a reversible reaction is one in which the drug–receptor complex can easily dissociate into the unbound forms [shown by the double-headed arrow in Eq. (7-1)]. The drug and receptor can also easily reform the drug–receptor complex. The pharmacological effect is assumed to be proportional to the concentration of the drug–receptor complex. This is the simplest of all possible drug–receptor interaction mechanisms.

The binding of the drug and receptor follows the *law of mass action*, which states that the concentration of the drug–receptor complex is proportional to the product of the molar concentrations of the unbound drug and unoccupied receptors. The law of mass action was experimentally derived and reported by Waage and Guldberg in 1864.

Once a conceptual model is rigorously stated, it can be translated into a mathematical model. Here, the concentration of the drug–receptor complex can be formulated in terms of the unbound (free) concentrations of the drug and receptor as:

$$[\text{Drug–Receptor}] = K_a[\text{Drug}]\,[\text{Receptor}]. \tag{7-2}$$

The proportionality constant K_a is called *association equilibrium constant*. These reactions are also commonly written in terms of a *dissociation constant*, K_d, where:

$$K_a = 1/K_d.$$

We shall derive Eq. (7-2) from the differential equations governing the reaction. Assume that the reaction has the following general form:

$$A + B \leftrightarrow C.$$

Then, the rates of change in the concentrations of A, B, and C will be given by the following differential equations:

$$\frac{d[C]}{dt} = k_1[A][B] - k_2[C]$$

$$\frac{d[A]}{dt} = k_2[C] - k_1[A][B] \tag{7-3}$$

$$\frac{d[B]}{dt} = k_2[C] - k_1[A][B].$$

Here, k_1 and k_2 are the respective reaction rate constants for the reactions $A + B \to C$ and $C \to A + B$. These equations are derived using considerations similar to those discussed in Chapters 1 and 2 for population sizes. For instance, in the first of Eq. (7-3), we see that the rate of increase of [C] is proportional to the product of [A] and [B], while the rate of decrease of [C] is proportional to [C.] The second and third equations follow from conservation of quantities arguments:

Because [A] + [C] and [B] + [C] are constants, we obtain:
$\frac{d}{dt}([A]+[C]) = 0$, and $\frac{d}{dt}([B]+[C]) = 0$, which shows that
$\frac{d}{dt}[A] = -\frac{d}{dt}[C]$ and $\frac{d}{dt}[B] = -\frac{d}{dt}[C]$.

In equilibrium, the rates of change of the concentrations are zero; that is, $\frac{d[C]}{dt} = 0, \frac{d[A]}{dt} = 0$, and $\frac{d[B]}{dt} = 0$. From Eq. (7-3), we obtain $k_1[A][B] - k_2[C] = 0$, which yields the law of mass action
$[C] = \frac{k_1}{k_2}[A][B] = K_a[A][B]$ with $K_a = \frac{k_1}{k_2}$.

When the law of mass action is applied to a balanced biochemical equation of the type

$$mA + nB \leftrightarrow C,$$

where m and n are the stoichiometric coefficients, the mathematical formulation of the law of mass action is:

$$[C] = K_a[A]^m[B]^n. \qquad (7-4)$$

We shall use this formulation in creating some of the mathematical models that follow.

EXERCISE 7-1

(a) Write the differential equations representing the rates of change for the reaction $mA + nB \leftrightarrow C$, and use them to obtain the law of mass action in the form of Eq. (7-4).

(b) Explain why the stoichiometric coefficients m and n appear as exponents in the formulation (7-4) of the law of mass action.

Binding reactions of the type in Eq. (7-1) are saturable because of the fixed number of receptors available. As the unbound drug concentration increases, the mass action equilibrium between the unoccupied receptors [Receptor] and the receptors with the drug bound [Drug–Receptor] will shift towards the latter. This follows directly from Eq. (7-2), which implies that the ratio:

$$\frac{[Drug–Receptor]}{[Drug][Receptor]} = K_a$$

is constant at constant temperature. The higher the drug concentration, the higher the concentration of drug-bound receptors should be for the quotient to remain unchanged. As the drug concentration increases, a point will be reached where virtually all of the receptors will be saturated with the drug. When this occurs, further increases in drug concentration will produce no additional effect.

To understand the action of a drug, it is important to understand how the *fractional saturation* of the receptor is related to the concentration of the unbound drug. The receptors' fractional saturation, $\overline{Y}$, is the concentration of the drug–receptor complex divided by the total receptor concentration:

$$\overline{Y} = \frac{[\text{Drug–Receptor}]}{[\text{Receptor}] + [\text{Drug–Receptor}]} = \frac{K_a[\text{Drug}][\text{Receptor}]}{[\text{Receptor}] + K_a[\text{Drug}][\text{Receptor}]}$$

$$= \frac{K_a[\text{Drug}]}{1 + K_a[\text{Drug}]} = \frac{[\text{Drug}]/K_d}{1 + [\text{Drug}]/K_d}. \tag{7-5}$$

In the special case of hemoglobin–oxygen binding, human hemoglobin can at most bind four O_2 molecules, and the coordinative bond between the O_2 molecule and the central ferrous ion in the heme is very weak. The number of oxygenated sites depends strongly upon the level of O_2 present (i.e., upon the partial O_2 pressure). In the lungs, where the partial pressure of the O_2 is relatively high, the prevailing form is the oxygenated hemoglobin that causes arterial blood's pure red color. Traveling through the capillaries of the tissues, where the O_2 partial pressure is low,[2] the oxygenated hemoglobin is subjected to considerable deoxygenation. This is the basic mechanism of the so-called hemoglobin shuttle: hemoglobin loads O_2 in the lungs, transports it to the tissues, where it is released, and the cycle repeats. In muscles, the O_2 is taken up by myoglobin, where it is then available to rapidly provide the large amounts of O_2 required by active muscles.

The molecular reaction mechanisms involved in O_2 binding to hemoglobin and myoglobin are the most thoroughly studied in biochemistry, and are the test cases for every mathematical model of binding reactions and cooperativity. If we replace the drug with O_2 and the receptor with myoglobin in Eq. (7-5), the mathematical model applies to the binding of O_2 by myoglobin. The fractional saturation of myoglobin is described by:

$$\overline{Y} = \frac{K_a[O_2]}{1 + K_a[O_2]} = \frac{[O_2]/K_d}{1 + [O_2]/K_d}. \tag{7-6}$$

---

2. The O_2 in the tissues is used to oxidize glucose to carbon dioxide and water. As a result, the partial pressure of the O_2 decreases, and the partial pressure of the CO_2 increases.

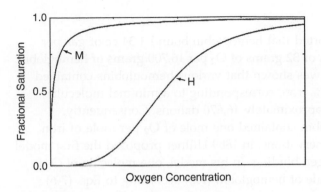

FIGURE 7-3.
Examples of the noncooperative myoglobin fractional saturation as a function of oxygen concentration (M) and the positive cooperative hemoglobin fractional saturation as a function of oxygen concentration (H).

Figure 7-3 depicts typical fractional saturation as a function of O_2 concentration for both myoglobin (M) and hemoglobin (H). Curve M represents the fractional saturation of myoglobin and curve H a typical sigmoid-shaped hemoglobin–oxygen fractional saturation curve. The difference in shapes should not be surprising, as hemoglobin's complex tetrameric structure allows cooperative interactions that the simple monomeric myoglobin cannot have.

Our next section presents a brief overview of the history of the mathematical modeling of hemoglobin–oxygen binding. We should note that the mathematical modeling of hemoglobin–oxygen binding developed concurrently with the decoding of the chemical and three-dimensional structure of the hemoglobin molecule. In many ways, these parallel efforts have been complementary, with advancements made in one area accelerating progress in the other.

III. MATHEMATICAL MODELS OF HEMOGLOBIN–OXYGEN BINDING

As George Santayana noted, "Progress, far from consisting in change, depends on retentiveness ... Those who cannot remember the past are condemned to repeat it" (Santayana [1905]). This is especially true for scientific research. It is nearly impossible to understand the conceptual models of hemoglobin–oxygen binding and cooperativity without understanding the context within which they were developed. The hemoglobin literature dates back almost 200 years, and almost every experimental technique used in biophysical chemistry was developed using hemoglobin. Many of the models of macromolecular interactions were developed in attempts to describe the functional properties of the binding of O_2 by hemoglobin. The early literature on hemoglobin was examined in a wonderful review by Edsall (1972).

A. Hüfner's Model

In 1884, Hüfner reported that hemoglobin bound 1.34 cc of gas per gram of hemoglobin, or 32 grams of O_2 per 16,700 grams of hemoglobin. A few years later, it was shown that various hemoglobins contained approximately 0.335% iron, corresponding to a minimal molecular weight per iron of approximately 16,670 daltons. Consequently, oxygenated hemoglobin contained one mole of O_2 per mole of iron. Based upon these observations, in 1889 Hüfner proposed the first model for hemoglobin–oxygen binding. In his model, one molecule of O_2 is bound to one molecule of hemoglobin (Hb) according to Eqs. (7-6) and (7-7). This predicts a nonsigmoid binding curve like myoglobin (i.e., M in Figure 7-3):

$$Hb + O_2 \leftrightarrow HbO_2. \tag{7-7}$$

However, Hüfner made a serious mistake. He believed the simple mathematical model must be correct and reasoned that only a single data point near half-saturation on the oxygen-binding curve was needed to evaluate the binding affinity. He collected one experimental data point and determined the binding affinity based upon that single point, without testing its validity. Therefore, Hüfner failed to observe the quintessential sigmoid-shaped binding curves (i.e., H in Figure 7-3) characteristic of the cooperative hemoglobin–oxygen binding system. In Hüfner's defense, his work was done 40 years before human hemoglobin was known to be a unique molecule.

B. Bohr's Approach

Christian Bohr (1855–1911) had a different approach to the study of hemoglobin–oxygen binding. Bohr was an experimentalist. As Edsall wrote, "Bohr's motto was that every experiment had a value, nothing which was obtained as the result of a test in the laboratory was set aside on the grounds of its inherent unlikelihood, of its failure to fit general principles or theories" (Edsall [1972]). Bohr's curves, obtained point by point from experimental measurements of oxygen pressure relative to oxygenated and deoxygenated amounts of hemoglobin, were reported in Bohr et al. (1904). They differed fundamentally from Hüfner's curve, having the characteristic sigmoid shape of a cooperative interaction (see the H curve in Figure 7-3). Although the sigmoid nature of the oxygen-binding curves for hemoglobin was an extremely important observation, it was not what made Bohr famous. In the same paper, Bohr also documented the effect CO_2 binding has on O_2 binding. He showed that the increased amount of CO_2 bound to the hemoglobin in the tissues lowers the affinity of the hemoglobin for the O_2 it is carrying and consequently aids the transfer of O_2 to tissues. This is now known as the *Bohr effect*.

Bohr and his collaborators tried to measure whether the binding of O_2 alters the binding of CO_2. They were unable to measure this reciprocal effect, however, because of the inadequacy of available experimental techniques, and thus concluded incorrectly that O_2 did not affect CO_2 binding. Again, remember these experiments were carried out more than 100 years ago.

Bohr's steadfast belief in the experimental approach probably prevented him from seeing what now seems an obvious inference; namely, that the reciprocal effects of the binding of O_2 and CO_2 by blood *must* exist. Clearly, if CO_2 affects the binding of O_2, then O_2 must affect the binding of CO_2. It appears Bohr trusted his data too much, while Hüfner trusted his theoretical model too much. Evidently, both experiments and theory are required, and neither should be favored to the exclusion of the other.

C. The Hill Equation

The apparent inconsistency between Hüfner's theory of noncooperative binding, Eqs. (7-6) and (7-7), and Bohr's experimentally observed cooperative O_2 binding data was difficult to resolve. Both seemed to be correct. Then in 1910, A. V. Hill (1886–1977; Nobel laureate, 1929) developed a conceptual model that appeared to reconcile experiment and theory. Hill realized Hüfner's theory would be correct if hemoglobin contained only a single O_2 binding site, but the sigmoid-shaped O_2 binding data of Bohr contradicted this theory, and, therefore, hemoglobin must have more than one binding site. Hill, however, did not know how many binding sites there were—his research predated the concept of unique multi-subunit high–molecular weight protein molecules by two decades. At this time, proteins were thought to be heterogeneous aggregates. Hill postulated that aggregates with n monomers would bind n molecules of O_2 according to a reaction like:

$$Hb_n + nO_2 \leftrightarrow Hb_n(O_2)_n, \tag{7-8}$$

where Hb_n is used to denote the assumption made by Hill that hemoglobin is constructed of n monomers. Hill did not know what the actual value of n was and wanted to determine the value experimentally.

Hill's scheme assumes that n molecules of O_2 are bound simultaneously in a single step. This is equivalent to assuming no intermediate states ever exist where the number of O_2 molecules bound is greater than zero and less than n. Given this assumption and the assumption that only one aggregate (i.e., a single value of n) exists, Hill formulated an equation for the fractional saturation as:

$$\overline{Y} = \frac{k[O_2]^n}{1 + k[O_2]^n}. \tag{7-9}$$

This *Hill equation* is still in use today. The n is the *Hill coefficient* that is sometimes used as a measure of cooperativity. For $n > 1$, Eqs. (7-8) and (7-9) predict a sigmoid-shaped binding curve like the H curve in

Figure 7-3. It was soon observed, however, that the value of n required to describe the actual experimental observations was approximately 2.5, when the theoretical assumption implied that n should be an integer. Hill rationalized the value of 2.5 as the statistical average of a mixture of different-sized aggregates (i.e., a mixture of monomers, dimers, trimers, tetramers, and so on).

In 1913, Hill noted that the limiting slope of the Hill equation is zero when the O_2 concentration approaches zero and $n > 1$. He showed that the following mathematical fact follows from Eq. (7-9):

$$\lim_{[O_2] \to 0} \frac{d\overline{Y}}{d[O_2]} = 0, \ n > 1, \ k > 0 \qquad (7\text{-}10)$$

(see Exercise 7-4 below). In 1913, accurate experimental measurement of this limiting slope was impossible, and the limited experimental data that existed indicated the limiting slope might be zero. Hill noted that if the limiting slope was actually not equal to zero, then models of the form of Eq. (7-9) were incorrect and should not be used. It is now known the limiting slope of the hemoglobin–oxygen binding curves is, in fact, not equal to zero. While this experimental observation demonstrated the Hill Equation should not be used for the study of binding phenomena, the Hill Equation is still commonly used because it gives a reasonable approximation of binding behavior.

EXERCISE 7-2

Use Eq. (7-8) and the law of mass action to derive Hill's model for the fractional saturation given by Eq. (7-9).

EXERCISE 7-3

Plot the fractional saturation $\overline{Y}$ from Eq. (7-9) as a function of the O_2 concentration using different values for the number of binding sites $n > 1$ and different values of the association constant k. Then answer the following questions:

(a) If the value of k is kept fixed, what is the effect on the fractional saturation $\overline{Y}$ when the number of binding sites n increases? Is this to be expected in the context of the problem? Explain why or why not.

(b) If the value of n is kept fixed, what is the effect on the fractional saturation $\overline{Y}$ when the association constant k increases? Is this to be expected in the context of the problem? Explain why or why not.

EXERCISE 7-4

Derive the limit in Eq. (7-10) from the Hill equation (7-9) by verifying the following steps:

(a) Show that for $n > 1$,

$$\frac{d\overline{Y}}{d[O_2]} = \frac{n\,k\,[O_2]^{n-1}}{(1+k[O_2]^n)^2}.$$

(b) Use part (a) to show that for $n > 1$

$$\lim_{[O_2]\to 0}\frac{d\overline{Y}}{d[O_2]} = \lim_{[O_2]\to 0} n\,k\,[O_2]^{n-1} = 0.$$

D. The Adair Equations

Shortly after World War I, osmotic pressure measurements by G. S. Adair and sedimentation equilibrium measurements by T. Svedberg demonstrated that human hemoglobin is a distinct molecule with a molecular weight of approximately 67,000 daltons, and not simply a mixture of aggregates. It was subsequently established that hemoglobin contains four polypeptide chains and four oxygen-binding sites, and that hemoglobin binds O_2 in a cooperative fashion, as described above. Thus, it became evident that Hill's equation, with no intermediates, does not explain the experimental data. Consequently, Adair formulated an equation for the fractional saturation of O_2 assuming that hemoglobin contained four oxygen-binding sites while allowing for all of the intermediate oxygenation stages. Adair's reaction scheme with intermediates is as follows:

$$\begin{aligned}
Hb_4 + O_2 &\leftrightarrow Hb_4O_2 \\
Hb_4 + 2O_2 &\leftrightarrow Hb_4(O_2)_2 \\
Hb_4 + 3O_2 &\leftrightarrow Hb_4(O_2)_3 \\
Hb_4 + 4O_2 &\leftrightarrow Hb_4(O_2)_4.
\end{aligned} \qquad (7\text{-}11)$$

The fractional saturation equation has four equilibrium-binding constants, one for each reaction defined in Eq. (7-11). Because of the effect of cooperativity, the values of these binding constants are different. According to the law of mass action, these equilibrium constants are defined as:

$$K_{4i} = \frac{[Hb_4(O_2)_i]}{[Hb_4][O_2]^i} \qquad i = 1, 2, 3, 4, \qquad (7\text{-}12)$$

and are commonly referred to as *product Adair-binding constants*. The subscript 4 refers to tetrameric hemoglobin, discussed in the next section. The Adair's four-site fractional saturation function in terms of the product Adair-binding constants is:

$$\overline{Y}_4 = \frac{1}{4} \frac{K_{41}[O_2] + 2K_{42}[O_2]^2 + 3K_{43}[O_2]^3 + 4K_{44}[O_2]^4}{1 + K_{41}[O_2] + K_{42}[O_2]^2 + K_{43}[O_2]^3 + K_{44}[O_2]^4}. \tag{7-13}$$

Heuristically, this equation reflects the concept that the fractional saturation $\overline{Y}$ of an entity with four binding sites must take into account the occupation of each of the sites. We present the derivation of Eq. (7-13) later in the chapter.

Eq. (7-11) is often written in the form:

$$\begin{aligned} Hb_4 + O_2 &\leftrightarrow Hb_4O_2 \\ Hb_4O_2 + O_2 &\leftrightarrow Hb_4(O_2)_2 \\ Hb_4(O_2)_2 + O_2 &\leftrightarrow Hb_4(O_2)_3 \\ Hb_4(O_2)_3 + O_2 &\leftrightarrow Hb_4(O_2)_4. \end{aligned} \tag{7-14}$$

The equilibrium constants, called *stepwise Adair-binding constants* (lower-case k's), are:

$$k_{4i} = \frac{[Hb_4(O_2)_i]}{[Hb_4(O_2)_{i-1}][O_2]} \qquad i = 1, 2, 3, 4. \tag{7-15}$$

The four-site fractional saturation function in terms of the stepwise Adair-binding constants is:

$$\overline{Y}_4 = \frac{1}{4} \frac{k_{41}[O_2] + 2k_{41}k_{42}[O_2]^2 + 3k_{41}k_{42}k_{43}[O_2]^3 + 4k_{41}k_{42}k_{43}k_{44}[O_2]^4}{1 + k_{41}[O_2] + k_{41}k_{42}[O_2]^2 + k_{41}k_{42}k_{43}[O_2]^3 + k_{41}k_{42}k_{43}k_{44}[O_2]^4}. \tag{7-16}$$

The stepwise binding constants are thus related to the product Adair-binding constants as:

$$K_{4i} = k_{41}k_{42} \ldots k_{4i}, \qquad i = 1, 2, 3, 4. \tag{7-17}$$

For example, if $i = 3$, then $K_{43} = k_{41}k_{42}k_{43}$; if $i = 2$, then $K_{42} = k_{41}k_{42}$, and so on.

Adair-binding equations continue in common use today. We emphasize that the only assumption needed for their derivation is that tetrameric human hemoglobin contains four oxygen-binding sites.

E. Coupling Between Subunit Assembly and Oxygen Binding

In the 1970s, Gary K. Ackers and coworkers first theoretically predicted and then experimentally observed another interesting property of O_2 binding by hemoglobin: That it is dependent upon the concentration of

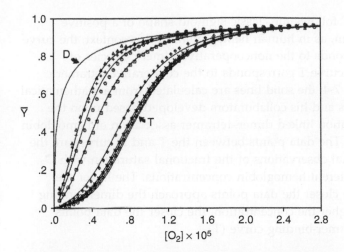

FIGURE 7-4.
Oxygenation curves determined at a series of hemoglobin concentrations. Each symbol represents experimental data from an oxygenation curve measured at a given hemoglobin concentration. The solid lines are calculated curves based on the oxygenation linked dimer–tetramer association scheme. The hemoglobin monomer concentration ranges from 4×10^{-8} M for the data on the left, to 4×10^{-4} M for the data on the right.
(Reprinted with permission from Mills, F.C., Johnson, M.L., and Ackers, G.K. [1976]. Oxygenation-linked subunit interactions in human hemoglobin: Experimental studies on the concentration dependence of oxygenation curves. *Biochemistry, 15*, 5350–5362. © 1976 The American Chemical Society.)

hemoglobin (see Figure 7-4). This observation cannot be explained by either the Hill or the Adair equations, but only by a mechanism whereby the hemoglobin is involved in a reaction dependent upon the hemoglobin concentration, because hemoglobin is not *always* a tetramer, even at the high concentrations found within red blood cells. The tetramer structure of hemoglobin, determined by Perutz using radiograph crystallography, describes only hemoglobin molecules in crystal form. In solution (e.g., in the blood), its degree of subunit assembly varies based on the hemoglobin concentration. But if under such conditions hemoglobin is not a tetramer, then what is it?

It is now known that hemoglobin may also exist as dimers, each consisting of one alpha chain and one beta chain. The alpha and beta chains have different amino acid sequences, but similar three-dimensional structures. Each contains a heme group, and each can bind to O_2. The binding of O_2 to dimers is not cooperative, whereas the binding to tetramers is. As the binding affinity of tetramers is different from that of dimers, the oxygenation curves depend on the ratio between dimers and tetramers. This ratio, in turn, depends on the hemoglobin concentration and the oxygen concentration.

Figure 7-4 presents experimental and calculated fractional-binding oxygenation curves, determined at a series of human hemoglobin concentrations, and contains examples of both noncooperative and positive cooperative binding curves. Curve D is an example of the nonsigmoid shape of a noncooperative curve, as in myoglobin and

Eq. (7-6). Curve T follows the classic sigmoid shape of a positive cooperative system, as in human hemoglobin. In this context, the curve labeled D corresponds to the noncooperative dimeric species of hemoglobin, and curve T corresponds to the cooperative tetrameric species. In Figure 7-4, the solid lines are calculated from a mathematical model that Ackers and his collaborators developed based upon the observed oxygenation linked dimer–tetramer association of hemoglobin (next paragraph). The data points between the T and D curves are the actual experimental observations of the fractional saturation with O_2 determined at different hemoglobin concentrations. The lower the concentration, the closer the data points approach the dimer-binding curve (D). The higher the concentration, the closer the data points approach the tetramer-binding curve (T).

We shall now present a conceptual description of the Ackers model. Hemoglobin tetramers are formed from two identical $\alpha\beta$ dimers. These dimers undergo a reversible association equilibrium, $2\alpha\beta \leftrightarrow \alpha_2\beta_2$. The $\alpha + \beta \leftrightarrow \alpha\beta$ association reaction has such a large association constant that it is, in effect, complete and virtually irreversible. In addition, the major structural transformation associated with O_2 binding occurs at the interface between the two $\alpha\beta$ dimers. The linkage between the dimer to tetramer subunit assembly and the O_2 binding properties of dimeric and tetrameric hemoglobin are shown in Figure 7-5. Each vertical arrow depicts an oxygenation step, and each horizontal arrow depicts a hemoglobin dimer–tetramer association step. For example, the horizontal arrow at the top represents the reaction that forms an unoxygenated $\alpha_2\beta_2$ hemoglobin tetramer from two unoxygenated $\alpha\beta$ dimers. The vertical arrow in the lower right depicts the reaction that combines a triply-oxygenated hemoglobin tetramer ($\alpha_2\beta_2(O_2)_3$) with O_2 to form a quad-oxygenated hemoglobin tetramer ($\alpha_2\beta_2(O_2)_4$).

Ackers and coworkers derived a mathematical model, Eqs. (7-18) and (7-19) below, for the fractional saturation of hemoglobin undergoing the oxygenation linked $2\alpha\beta \leftrightarrow \alpha_2\beta_2$ subunit assembly scheme shown in Figure 7-5. These equations represent a weighted average of a four–binding-site Adair equation [Eq. (7-16)], describing the oxygen-binding properties of the tetrameric hemoglobin ($\alpha_2\beta_2$), and an analogous two–binding-site Adair equation describing the oxygen-binding properties of the dimeric hemoglobin ($\alpha\beta$). The relative weights for these two Adair-binding equations are determined by the hemoglobin dimer to tetramer association reactions and are themselves a function of both the O_2 and hemoglobin concentrations (see Figure 7-6).

The model yields the following result for the fractional saturation:

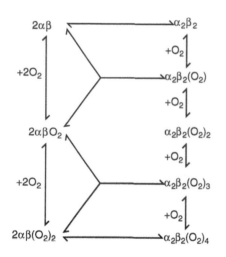

FIGURE 7-5.
The oxygenation linked $2\alpha\beta \leftrightarrow \alpha_2\beta_2$ subunit assembly scheme.
(Reprinted with permission from Mills, F.C., Johnson, M.L., and Ackers, G.K. [1976]. Oxygenation-linked subunit interactions in human hemoglobin: Experimental studies on the concentration dependence of oxygenation curves. *Biochemistry, 15*, 5350–5362. © 1976 The American Chemical Society.)

$$\overline{Y} = \frac{\Xi_2' + \Xi_4'\left(\sqrt{(\Xi_2)^2 + 4^0K_2\,\Xi_4[Hg]} - \Xi_2\right)/(4\,\Xi_4)}{\Xi_2 + \sqrt{(\Xi_2)^2 + 4^0K_2\,\Xi_4[Hg]}}, \qquad (7\text{-}18)$$

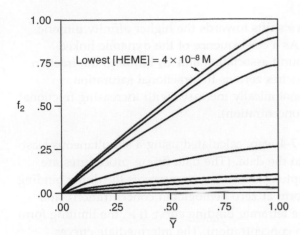

FIGURE 7-6.
The fraction of dimeric hemoglobin present in solution as a function of the fractional saturation of the oxygenation linked $2\alpha\beta \leftrightarrow \alpha_2\beta_2$ subunit assembly reaction. The upper curve corresponds to the lowest hemoglobin concentration of 4×10^{-8} M hemoglobin monomers while the lower curve is for 4×10^{-4} M hemoglobin monomers.
(Reprinted with permission from Mills, F.C., Johnson, M.L., and Ackers, G.K. [1976]. Oxygenation-linked subunit interactions in human hemoglobin: Experimental studies on the concentration dependence of oxygenation curves. *Biochemistry, 15*, 5350–5362. © 1976 The American Chemical Society.)

where [Hg] is the total molar concentration of hemoglobin (monomers), [O$_2$] is the molar concentration of unbound O$_2$, the constant $^0K_2$ is the equilibrium constant for dimer–tetramer assembly $2\alpha\beta \leftrightarrow \alpha_2\beta_2$, and:

$$\Xi_2 = 1 + K_{21}[O_2] + K_{22}[O_2]^2$$

$$\Xi_2{}' = K_{21}[O_2] + 2\ K_{22}[O_2]^2$$

$$\Xi_4 = 1 + K_{41}[O_2] + K_{42}[O_2]^2 + K_{43}[O_2]^3 + K_{44}[O_2]^4 \qquad (7\text{-}19)$$

$$\Xi_4{}' = K_{41}[O_2] + 2\ K_{42}[O_2]^2 + 3\ K_{43}[O_2]^3 + 4\ K_{44}[O_2]^4.$$

The quantities Ξ_2 and Ξ_4 in Eqs. (7-18) and (7-19) are sometimes referred to as *binding polynomials*. The quantities $\Xi_2{}'$ and $\Xi_4{}'$ in Eqs. (7-18) and (7-19) are related to Ξ_2 and Ξ_4, as we shall explore later. Although deriving Eq. (7-18) is beyond the scope of this textbook, we shall use binding polynomials to derive Adair's equation [Eq. (7-16)].

As mentioned, Eq. (7-18) could be thought of as the appropriately weighted average of the Adair-binding equations for the dimeric hemoglobin species (the $2i$ subscripts) and the tetrameric hemoglobin species (the $4i$ subscripts). The only assumption required for the derivation of Eqs. (7-18) and (7-19) is the reaction scheme depicted in Figure 7-5. Thus, Eqs. (7-18) and (7-19) apply to normal human hemoglobin, which exists in solution as $\alpha\beta$ dimers that undergo a reversible association reaction to form $(\alpha\beta)_2$ tetramers. Human hemoglobin dimers bind O$_2$ with higher affinities than tetramers. As the concentration of hemoglobin decreases, the fraction of dimers present increases (see Figure 7-6), and the hemoglobin concentration-dependent fractional saturation curve shifts towards the higher affinity dimeric species (see Figure 7-4). As a consequence of the dynamic linked equilibrium between subunit association and O_2 binding, the fraction of dimeric species approaches zero as the fractional saturation approaches zero and monotonically increases with increasing fractional saturation (i.e., oxygen concentration).

The solid lines in Figure 7-4 were calculated using a simultaneous least-squares fit of Eq. (7-18) to the data. (The data-fitting procedures are presented in the next chapter.) The D curve represents the dimer binding curve (i.e., the limiting form at zero hemoglobin concentration), and the T curve represents the tetramer binding curve (i.e., the limiting form at an infinite hemoglobin concentration). The intermediate curves correspond to the binding curve at intermediate hemoglobin concentrations as predicted by Eqs. (7-18) and (7-19).

The observation that O_2 binding by hemoglobin is dependent upon hemoglobin concentration might appear to be a serious complication requiring a much more complex model for data analysis. However, it also provides another means to probe the structure and function of hemoglobin. The properties of hemoglobin can now be studied as a function of O_2 concentration, or as a function of hemoglobin concentration, or as a simultaneous function of both. Ackers and coworkers studied the simultaneous function of both O_2 and hemoglobin concentrations, giving us a better understanding of how hemoglobin transports O_2 in our bodies.

IV. DERIVING FRACTIONAL SATURATION FUNCTIONS WITH BINDING POLYNOMIALS

There are multiple ways to approach the derivation of the fractional saturation equations. The easiest and the most generally applicable approach is based on binding polynomials. *A binding polynomial, Ξ, is simply a mathematical description of the sum of the concentrations (i.e., probabilities) of each of the hemoglobin species in the solution.* Given the mathematical form of the binding polynomial and a little calculus, it is easy to derive any desired fractional saturation function.

The binding polynomial approach for modeling cooperativity and oxygen-binding problems is based on Eqs. 1 through 67 from Hill's book (1960). The following equation relates the mean number of O_2 molecules bound by a macromolecule, $\overline{N}$, to the natural logarithm of the binding polynomial, $\ln \Xi$, and the O_2 concentration, $[O_2]$:

$$\overline{N} = [O_2]\frac{\partial \ln \Xi}{\partial [O_2]} = \frac{\partial \ln \Xi}{\partial \ln[O_2]}. \qquad (7\text{-}20)$$

Although Eq. (7-20) may not appear intuitive, its derivation is elementary. To illustrate its importance, we shall use it to justify

Adair's four-site fractional saturation model from Eq. (7-13).
The rationalization of Eq. (7-20) is presented in Section V.

Observe that the mean number of bound O_2 molecules $\overline{N}$ is the
fractional saturation, $\overline{Y}$, times the number of binding sites (i.e., four for
hemoglobin tetramers and two for hemoglobin dimers). Combining this
observation with Eq. (7-20) and a mathematical statement of the total
concentrations of all the species present in solution (i.e., the binding
polynomial Ξ), an equation for the fractional saturation of O_2 is easily
derived. In this example, the fractional saturation function for $\alpha_2\beta_2$
tetramers [e.g., the T curve in Figure 7-5 and Eqs. (7-13) and (7-16)] is:

$$\overline{Y}_4 = \frac{\overline{N}}{4} = \frac{1}{4}[O_2]\frac{\partial \ln\Xi_4}{\partial[O_2]} = \frac{1}{4}\frac{[O_2]}{\Xi_4}\frac{\partial \Xi_4}{\partial[O_2]}. \tag{7-21}$$

Using the expression $\Xi_4 = 1 + K_{41}[O_2] + K_{42}[O_2]^2 + K_{43}[O_2]^3 + K_{44}[O_2]^4$
from Eq. (7-19)] and calculating the appropriate derivatives, Eq. (7-21)
now yields Adair's four-site fractional saturation function:

$$\overline{Y}_4 = \frac{1}{4}\frac{K_{41}[O_2] + 2K_{42}[O_2]^2 + 3K_{43}[O_2]^3 + 4K_{44}[O_2]^4}{1 + K_{41}[O_2] + K_{42}[O_2]^2 + K_{43}[O_2]^3 + K_{44}[O_2]^4}. \tag{7-22}$$

The analogous fractional saturation for $\alpha\beta$ dimers (e.g., the D curve in
Figure 7-4) is:

$$\overline{Y}_2 = \frac{\overline{N}}{2} = \frac{1}{2}[O_2]\frac{\partial \ln\Xi_2}{\partial[O_2]} = \frac{1}{2}\frac{[O_2]}{\Xi_2}\frac{\partial\Xi_2}{\partial[O_2]} \tag{7-23}$$

$$\overline{Y}_2 = \frac{1}{2}\frac{K_{21}[O_2] + 2K_{22}[O_2]^2}{1 + K_{21}[O_2] + K_{22}[O_2]^2}. \tag{7-24}$$

Equation (7-24) is analogous to the form of Eq. (7-22), except that it
applies to dimeric hemoglobin with two binding sites instead of
tetrameric hemoglobin with four binding sites.

EXERCISE 7-5

Use the expressions for Ξ_4 and Ξ_2 given by Eq. (7-19):

$$\Xi_4 = 1 + K_{41}[O_2] + K_{42}[O_2]^2 + K_{43}[O_2]^3 + K_{44}[O_2]^4$$

$$\Xi_2 = 1 + K_{21}[O_2] + K_{22}[O_2]^2$$

to:

(a) Supply the calculations deriving Eq. (7-22) from Eq. (7-21); and

(b) Supply the calculations deriving Eq. (7-24) from Eq. (7-23).

How are the binding polynomials Ξ_2 and Ξ_4 obtained? They are simply the sum of the concentrations of all the binding species present in solution. For example, if dimeric hemoglobin contained two identical, nondistinguishable oxygen-binding sites, then in solution only three possible oxygenation states would exist; with zero, one, or two O_2 molecules bound. Thus, its binding polynomial could be written as:

$$\Xi_2 = [\alpha\beta] + [\alpha\beta O_2] + [\alpha\beta(O_2)_2]. \tag{7-25}$$

Using the law of mass action to express the concentration of the oxygenated species in terms of the concentrations of the unoxygenated dimeric hemoglobin and the unbound O_2 concentration, we obtain:

$$\Xi_2 = [\alpha\beta] + K_{21}[\alpha\beta][O_2] + K_{22}[\alpha\beta][O_2]^2. \tag{7-26}$$

In this case, the definition of K_{21} is the average (i.e., macroscopic) Adair-binding constant for the first O_2 being bound to either of the O_2 binding sites. As we shall see below, this is not the binding constant to either the α subunit or the β subunit, but rather the sum of the two.

An interesting property of this application of binding polynomials is that the units of hemoglobin and O_2 concentration are arbitrary. This is a consequence of the natural logarithms contained in Eqs. (7-21) and (7-23). We take the concentration of the reference state to be the unoxygenated dimeric hemoglobin concentration. Because the units of hemoglobin concentration are arbitrary, we simply express these units as a fraction of the unoxygenated hemoglobin concentration. In these units, Eq. (7-25) is transformed into Eq. (7-27), since in these units $[\alpha\beta] = 1$.

$$\Xi_2 = 1 + [\alpha\beta O_2] + [\alpha\beta(O_2)_2]$$
$$\Xi_2 = 1 + K_{21}[O_2] + K_{22}[O_2]^2. \tag{7-27}$$

Notice this is exactly the quantity Ξ_2 presented in Eq. (7-19).

EXERCISE 7-6

Use binding polynomials to derive Eq. (7-6).

EXERCISE 7-7

Assuming, as we did for the dimeric hemoglobin in Eq. (7-25), that the binding polynomial is defined as the sum of the concentrations of all the binding species present in solution, show that for the tetrameric hemoglobin, $\Xi_4 = 1 + K_{41}[O_2] + K_{42}[O_2]^2 + K_{43}[O_2]^3 + K_{44}[O_2]^4$.

In order to understand the mechanistic relationships involved in the hemoglobin subunit coupling, it is essential to understand the exact meaning of the average macroscopic Adair-binding constant for the first O_2 being bound to either of the oxygen-binding sites. It is not the binding constant to either the α subunit or the β subunit. To understand this last statement, consider a dimeric hemoglobin that contains two nonidentical, distinguishable oxygen-binding sites, such as the α subunit and the β subunit. In solution, four possible oxygenation states exist. The states containing none and two O_2 molecules are the same as the previous example. The other two states include the one where the O_2 is bound to the α subunit and the one where the O_2 is bound to the β subunit. If we assume the intrinsic binding affinities of the individual subunits are K_α and K_β, we can write the four-term binding polynomial as:

$$\Xi_2 = 1 + K_\alpha[O_2] + K_\beta[O_2] + K_\alpha K_\beta[O_2]^2. \tag{7-28}$$

By applying Eq. (7-20) to Eq. (7-28) we obtain the fractional saturation function for dimeric hemoglobin in terms of the Adair-binding constants of the individual α and β subunit binding constants:

$$\overline{Y}_2 = \frac{1}{2}\ \overline{N} = \frac{1}{2}\ \frac{(K_\alpha + K_\beta)[O_2] + 2K_\alpha K_\beta[O_2]^2}{1 + (K_\alpha + K_\beta)[O_2] + K_\alpha K_\beta[O_2]^2}. \tag{7-29}$$

By comparing the forms of Eqs. (7-24) and (7-29), it is apparent that the average macroscopic Adair-binding constant for the first O_2 being bound to either of the oxygen-binding sites is equal to the sum of the intrinsic binding affinities of the individual subunits

$$K_{21} = K_\alpha + K_\beta. \tag{7-30}$$

Before leaving the subject of utilizing binding polynomials to derive fractional saturation functions, we need to consider one additional case: When a protein has two identical binding sites and there is a cooperative interaction between them. In this situation, the binding of the first ligand (e.g., O_2) will enhance or inhibit the binding of the second ligand, even though the binding sites are identical in the unbound species. An example of this might be where the two binding sites are physically close to each other and the ligands are highly charged. The binding of the second ligand would be somewhat inhibited by the charge of the first ligand (i.e., it would be a negatively cooperative system). There would again be four possible ways to put the two identical ligands onto the molecule: Unbound, a ligand on the first site, a ligand on the second site, and a ligand on both sites. The binding polynomial for this system is given by:

$$\Xi = 1 + K_i[X] + K_i[X] + K_c(K_i[X])^2, \tag{7-31}$$

where K_i is the intrinsic affinity for either site and K_c is the cooperativity constant. The ligand concentration is expressed as $[X]$ because the mathematical form also applies to ligands other than O_2. The second and third terms on the right of Eq. (7-31) are identical and represent one ligand, X, on each of the two binding sites. The last term contains the $(K_i [X])^2$ term describing the binding of two ligands to two identical sites and the K_c cooperativity term. If K_c is less than 1, the system will exhibit negative cooperativity; if K_c is greater than 1, the system will exhibit positive cooperativity; and if K_c is equal to 1, there is no cooperativity. Eq. (7-32) presents the fractional saturation function for this system:

$$\overline{Y} = \frac{1}{2}\overline{N} = \frac{1}{2}\frac{2K_i[X] + 2K_c(K_i[X])^2}{1 + 2K_i[X] + K_c(K_i[X])^2}. \tag{7-32}$$

The 2 in the $2K_i[X]$ terms is included because the first ligand can go onto either of the two identical sites. This means that the average macroscopic Adair-binding constant for the first ligand being bound to either of the identical binding sites is equal to twice the intrinsic binding affinities of the individual sites.

V. APPENDIX: JUSTIFYING EQUATION (7-20)

We now use some basic probabilistic arguments to justify Eq. (7-20) for the case of hemoglobin tetramers. Recall that we linked Eq. (7-20) with the total number of hemoglobin species present in the solution. For Hb_4, it follows from Eq. (7-11) that the following oxygenated states of hemoglobin will be present: HbO_2, $Hb(O_2)_2$, $Hb(O_2)_3$, and $Hb(O_2)_4$. In addition, there will also be nonoxygenated Hb_4. If $p(i)$ denotes the concentration of $Hb(O_2)_i$, $i = 0,1,2,3,4$, we will have:

$$p(i) = \frac{[Hb_4(O_2)_i]}{[Hb_4] + [Hb_4(O_2)] + [Hb_4(O_2)_2] + [Hb_4(O_2)_3] + [Hb_4(O_2)_4]}.$$

Using Eq. (7-12), this can be written as

$$p(i) = \frac{K_{4i}[Hb_4][(O_2)]^i}{[Hb_4] + K_{41}[Hb_4][(O_2)] + K_{42}[Hb_4][(O_2)]^2 + K_{43}[Hb_4][(O_2)]^3 + K_{44}[Hb_4][(O_2)]^4}.$$

Simplifying yields

$$p(i) = \frac{K_{4i}[(O_2)]^i}{1 + K_{41}[(O_2)] + K_{42}[(O_2)]^2 + K_{43}[(O_2)]^3 + K\,44[(O_2)]^4}. \tag{7-33}$$

Using Eq. (7-33) gives the probabilities for i, $i = 0, 1, 2, 3,$ or 4, O_2 molecules to be bound to a hemoglobin macromolecule; the average number $\overline{N}$ of O_2 molecules bound by a macromolecule will be given by

$$\overline{N} = \sum_{i=0}^{4} i\,p(i) = \frac{1}{1 + K_{41}[(O_2)] + K_{42}[(O_2)]^2 + K_{43}[(O_2)]^3 + K_{44}[(O_2)]^4} \sum_{i=0}^{4} iK_{4i}[O_2]^i.$$

$$\tag{7-34}$$

From Eq. (7-19) recall that:

$$\Xi_4 = 1 + K_{41}[O_2] + K_{42}[O_2]^2 + K_{43}[O_2]^3 + K_{44}[O_2]^4.$$

We can now rewrite Eq. (7-34) as:

$$\overline{N} = \sum_{i=0}^{4} i p(i) = \frac{1}{\Xi_4} \sum_{i=0}^{4} i K_{4i}[O_2]^i = \frac{1}{\Xi_4}[O_2] \sum_{i=1}^{4} i K_{4i}[O_2]^{i-1}$$

$$= \frac{1}{\Xi_4}[O_2] \frac{\partial \Xi_4}{\partial[O_2]} = [O_2] \frac{\partial \ln \Xi_4}{\partial[O_2]}.$$

(7-35)

The chain of calculations in Eq. (7-35) now establishes Eq. (7-20) for tetramers.

REFERENCES

Bohr, C., Hasselbach, K. A., & Krogh, A. (1904). Uber einen in biologischer beziehung wichtigen einfluss, den die kohlensaürespannung des blutes auf dessen sauerstoffbindung übt. *Skandinavisches Archiv für Physiologie, 16,* 401–411.

Edsall, J. T. (1972). Blood and hemoglobin: The evolution of knowledge of functional adaptation in a biochemical system. Part I: The adaptation of chemical structure to function in hemoglobin. *Journal of the History of Biology, 5,* 205–257.

Santayana, G. (1905). *The life of reason.* New York: Charles Scribner's Sons.

FURTHER READING

Adair, G. S. (1925). The osmotic pressure of haemoglobin in the absence of salts. Proceedings of the Royal Society of London. Series A, Containing Papers of a Mathematical and Physical Character, *109,* 292–300.

Barcroft, J. (1914). *The respiratory function of the blood* (1st ed.). Cambridge, UK: Cambridge University Press.

Hill, A. V. (1910). The possible effects of aggregation of the molecules of haemoglobin on its dissociation curves. *Journal of Physiology, 40,* iv–vii.

Hill, A. V. (1913). The combinations of haemoglobin with oxygen and with carbon monoxide. I. *The Biochemistry Journal, 7,* 471–480.

Hill, T. L. (1960). *An introduction to statistical thermodynamics.* Reading, MA: Addison-Wesley Publishing.

Hüfner, G., & Hoppe-Seyler, S. (1884). *Z. Physiol. Chemie, 8,* 338–365.

Johnson, M. L. (1995). Statistical thermodynamic modeling of hemoglobin cooperativity. *Advances in Biophysical Chemistry, 5,* 179–231.

Mills, F. C., Johnson, M. L., & Ackers, G. K. (1976). Oxygenation-linked subunit interactions in human hemoglobin: Experimental studies on the concentration dependence of oxygenation curves. *Biochemistry, 15,* 5350–5362.

Modell, B., Khan, M., & Darlison, M. (2000). Survival in beta-thalassaemia major in the UK: Data from the UK Thalassaemia Register. *Lancet, 355,* 2051–2052.

Svedberg, T., & Nichols, J. B. (1927). The application of the oil turbine type of ultracentrifuge to the study of the stability region of carbon monoxide-hemoglobin. *Journal of the American Chemistry Society, 49,* 2920–2934.

Chapter 8

LIGAND BINDING, DATA FITTING, AND LEAST-SQUARES ESTIMATES OF MODEL PARAMETERS

We cannot solve the problems we have created with the same thinking that created them.

Albert Einstein (1879–1955)

In the preceding chapters, we have discussed the need for using mathematical models; outlined some of the important questions involved in constructing models; and analyzed a variety of models related to population ecology, epidemiology, genetics, endocrinology, and neonatology. Whatever the application, all of these models contained a set of numerical quantities referred to as the *model parameters*. For example, our first population growth model assumed a constant net per capita growth rate r, which was a parameter for the model. The new infections rate α and the recovery rate β we used to construct the SIS and SIR models are parameters; the association constant K_a used in the hemoglobin oxygenation models is likewise a parameter.

We have emphasized that every model is built upon certain assumptions and involves a certain number of parameters. The specific values of the model parameters may be unknown initially or be group-dependent. For example, it should be expected that Mexico's net per capita growth rate is different from Sweden's, because population growth is driven by socioeconomic, cultural, environmental, and other factors that differ substantially. Also, not all assumptions in a dynamic model may be valid for all time ranges; with the unlimited population growth model, we modified some assumptions and improved the model.

Model validation is a critical part of the modeling process. In general, validation requires gathering sufficient data through carefully designed experiments and then applying statistical techniques to determine how well the model describes the data. As we have seen, model predictions generally differ from experimental measurements, but model parameters can be estimated from the data to provide the best fit between actual and predicted values. For the population growth models, we estimated model parameters using averaging techniques aimed at obtaining the best visual fit.

This approach had several major limitations. First, it did not provide any information on whether the calculated parameter estimates could be improved, because no

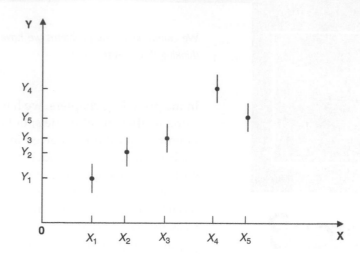

FIGURE 8-1.
Data points considered as intervals of confidence for each measurement.

specific mathematical criteria for optimization were formulated. Second, our approach made the unrealistic assumption that the experimental measurements were 100% accurate. Even the most carefully designed and conducted experiments contain inaccuracies caused by equipment sensitivity, experimental conditions, and similar factors. In this regard, it may be more appropriate to consider each measurement not as an actual data point but, rather, as an interval within which the true value may be found with a certain probability (Figure 8-1). The length of each interval is determined by estimating the errors in data collection. Generally, repeated measurements will not produce the same exact values, because of inherent measurement errors. In such scenarios, we take the mean of all of our readings and make it our data point. This mean is generally known with a certain precision that, intuitively, would increase with additional measurements. To account for this effect, the standard error of the mean[1] (SEM) is computed as the standard deviation of all readings at a data point divided by the square root of the number of readings. We shall not go into the details of why such a formula was chosen. However, it defines one common method for determining data point intervals. In Figure 8-1, the vertical lines are centered on the observed values and represent the ±1 SEM of experimental uncertainties for the particular data point. Finally, uncertainties in data measurements could affect the parameter estimates. The estimates obtained for the values of the parameters therefore should not be viewed as absolute and fixed values but, instead, as statistical estimates themselves.

---

1. In this context, SEM is also called standard error of the measurement.

In this chapter, we present one of the most popular mathematical criteria for estimating model parameters from data: the *least-squares criterion*. We begin with a mathematical introduction, describing a measure used to determine the *optimal fit* of the model. We consider linear models first, and then extend the definitions to general nonlinear models, using models of ligand binding and hemoglobin–oxygen binding as examples.

I. DATA-FITTING TERMS, DEFINITIONS, AND EXAMPLES

Once a data set has been acquired and a hypothesis-driven mathematical model developed, the next step is to fit the model to the data and obtain values for the parameters that provide the best description of the data.

Consider a hypothetical situation in which a linear model of the form $Y = aX + b = G(a, b; X)$ has been determined to provide a description of a biological phenomenon and the experimental data have been collected. The variable X is said to be the *independent variable*, while Y is the *response* or *dependent variable*. One of the objectives of the fitting process is to determine the values of a and b that will fit the data points best. If all the data points lie on a straight line, the slope a and the vertical intercept b of this line will provide the best fit. In practice, however, for any set of more than two data points, it is unrealistic to expect them all to lie on a straight line. Even if the linear model $Y = aX + b$ describes the dependence between the variables X and Y accurately, there will be some discrepancies (if nothing else, rounding errors are always present). Figure 8-2 illustrates this situation. The vertical deviations from the line $Y = aX + b$ are denoted by $r_1, r_2, \ldots r_n$

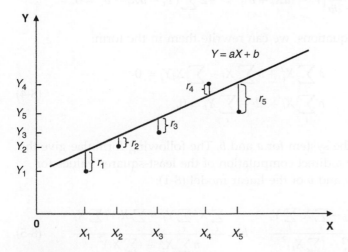

FIGURE 8-2.
Vertical residuals for the linear model $Y = aX + b$ and five data points.

and are calculated as $r_i = Y_i - G(a, b; X_i) = Y_i - (aX_i + b)$, $i = 1, 2, \ldots, n$. It is desirable to determine values for a and b that minimize the combined deviation between model and data. The *sum-squared residuals (SSR) measure* is the one most frequently used to express this deviation:

$$SSR(a, b) = r_1^2 + r_2^2 + r_3^2 + \ldots + r_n^2 = \sum_{i=1}^{n} r_i^2 = \sum_{i} [Y_i - G(a, b; X_i)]^2. \quad (8\text{-}1)$$

In an attempt to simplify the notation, the initial and final value of the summation index i are often ignored, as in the last expression of Eq. (8-1), the understanding being that the sum is taken over the whole range of available data points.

Notice that the measure is a function of model parameters a and b. The *least-squares data-fitting criterion* can now be stated: Using the experimental data, determine the parameter values minimizing the least-squares measure SSR(a,b). To do this, we use a basic idea of calculus: At a minimum value of a function, the derivative (if there is only one variable) or all the partial derivatives (if there is more than one variable) will be zero. The function in question is the SSR, and the variables are the model parameters.

For the linear a model

$$Y = aX + b \quad (8\text{-}2)$$

taking the partial derivatives for the function from Eq. (8-1) and setting them equal to zero leads to the following equations:

$$\frac{\partial(SSR)}{\partial a} = \sum_{i} \frac{\partial}{\partial a} [Y_i - (aX_i + b)]^2 = -2 \sum_{i} (Y_i - aX_i - b)X_i = 0$$

$$\frac{\partial(SSR)}{\partial b} = \sum_{i} \frac{\partial}{\partial b} [Y_i - (aX_i + b)]^2 = -2 \sum_{i} (Y_i - aX_i - b) = 0. \quad (8\text{-}3)$$

To solve these equations, we can rewrite them in the form:

$$a \sum_{i} X_i^2 + b \sum_{i} X_i - \sum_{i} X_i Y_i = 0$$

$$a \sum_{i} X_i + nb - \sum_{i} Y_i = 0, \quad (8\text{-}4)$$

and then solve the system for a and b. The following formulae give the result and allow a direct computation of the least-squares values for the parameters a and b of the linear model (8-1):

$$a = \frac{n \sum_{i} X_i Y_i - \sum_{i} X_i \sum_{i} Y_i}{n \sum_{i} X_i^2 - (\sum_{i} X_i)^2}, \quad b = \frac{\sum_{i} X_i^2 \sum_{i} Y_i - \sum_{i} X_i Y_i \sum_{i} X_i}{n \sum_{i} X_i^2 - (\sum_{i} X_i)^2}. \quad (8\text{-}5)$$

EXERCISE 8-1

Derive Eq. (8-5) for the solution (a,b) of the system of linear Eq. (8-4).

EXERCISE 8-2

Derive a formula for the least-squares value of the parameter a in the linear model $Y = G(a; X) = aX$.

EXERCISE 8-3

Consider the data in Table 8-1, collected to fit the model $Y = aX + b$. Determine the least-square estimates for parameters a and b.

X	0.25	0.5	0.75	1.0
Y	1.3	2.7	3.3	5.1

TABLE 8-1.
Data for exercise 8-3.

Thus far, we have considered the specific linear model $Y = aX + b = G(a, b; X)$. The principle definitions of the least-squares measure remain the same, however, for any model given by a function y of the form $Y = G$ (*parameters*; X). Consider, for example, Eq. (8-6), a rewritten form of Eq. (7-24) from Chapter 7:

$$Y = \frac{1}{2} \frac{K_{21}X + 2K_{22}X^2}{1 + K_{21}X + K_{22}X^2} = G(K_{21}, K_{22}; X). \qquad (8\text{-}6)$$

In this case, the measured quantity is the fractional saturation Y, the dependent variable. The experimentally manipulated quantity, the oxygen concentration X, is the independent variable. The model G defined by Eq. (8-6) has two parameters, K_{21} and K_{22}, to be adjusted by the data-fitting procedure.

Using the model from Eq. (8-6), we can write

$$Y_i \approx G(K_{21}, K_{22}; X_i) = \frac{1}{2} \frac{K_{21}X_i + 2K_{22}X_i^2}{1 + K_{21}X_i + K_{22}X_i^2} \qquad (8\text{-}7)$$

where the best possible fit between the model and the data is determined by minimizing the sum of squared residuals

$$SSR(K_{21}, K_{22}) = \sum_i [Y_i - G(K_{21}, K_{22}; X_i)]^2. \qquad (8\text{-}8)$$

Equation (8-6) is assumed to hold only approximately because it ignores the measurement errors always present in the experimental values of Y_i. By finding the values for K_{21} and K_{22} that minimize the SSR in Eq. (8-8), we find the best possible approximation for the entire set of data points. Notice that in this case the function SSR is not a linear function of K_{21} and K_{22}.

As our next example shows, this non-linearity complicates the computations significantly.

Example 8-1

Recall the population model $P(t) = P_0 e^{rt} = 5.3\, e^{rt}$, defined by Eq. (1-4) in Chapter 1. In this model, the independent variable is the time t (decades), and the dependent variable is the U.S. population P (millions). The net per capita rate of U.S. population growth r is the only parameter. In our current notation, the model can be rewritten as

$$P = G(r; t) = 5.3 e^{rt}. \tag{8-9}$$

Table 8-2 reproduces the U.S. population data from Chapter 1. We want to find the least-squares estimate for the parameter r.

Each column of Table 8-2 represents an experimental point of the form (t_i, P_i). The least-squares measure for this model can be written as:

$$SSR = SSR(r) = \sum_i [P_i - G(r; t_i)]^2 = \sum_i [P_i - 5.3 e^{rt_i}]^2.$$

To find the least-squares estimate for r, we need to solve the equation $\dfrac{\partial(SSR)}{\partial r} = 0$; that is:

$$\frac{\partial(SSR)}{\partial r} = 2\sum_i [P_i - 5.3 e^{rt_i}](-5.3)e^{rt_i}t_i = -10.6 \sum_i t_i[P_i - 5.3t_i e^{rt_i}]e^{rt_i} = 0, \tag{8-10}$$

or

$$\sum_i t_i[P_i - 5.3 e^{rt_i}]e^{rt_i} = 0.$$

Using the data points (t_i, P_i) from the table above, we obtain the equation:

$$[7.2 - 5.3e^r]e^r + 2[9.6 - 5.3e^{2r}]e^{2r} + 3[12.9 - 5.3e^{3r}]e^{3r}$$
$$+ 4[17.1 - 5.3e^{4r}]e^{4r} + 5[23.2 - 5.3e^{5r}]e^{5r} + 6[31.4 - 5.3e^{6r}]e^{6r} = 0. \tag{8-11}$$

Notice that this equation is not linear with respect to r. For general nonlinear models, there are no closed-form expressions to determine the solution, as there are for linear models. However, computers can be used to numerically calculate the solution of the equation.

Various methods for calculating the roots of nonlinear equations have been developed and are typically studied in courses on numerical analysis. One of the most popular methods is *Newton's method*, which provides an iterative technique for finding the roots of an algebraic

Time i (decades)	U.S. Population $P_i = P(i)$ (millions)
0	5.3
1	7.2
2	9.6
3	12.9
4	17.1
5	23.2
6	31.4

TABLE 8-2.
Population of the United States from 1800 to 1860.

equation of the form $f(r) = 0$. The idea is as follows: To begin the iterative process, we make an initial guess at a root. If our initial guess is not a root, the function and the initial guess will provide a second guess that is usually closer to the root than our first guess. The process is then repeated, producing increasingly accurate approximations to the root. We describe the details in the following sections.

Applying the least-squares fitting criterion for models with more than one parameter results in a system of nonlinear equations for the parameters, and the generalized method used for calculating the solutions is known as the Gauss–Newton approach. The details of this method will also be given later.

Example 8-2
·····················

Consider the model $P(t) = ce^{rt}$ again, but assume that this time we do not know the initial population value c at time $t = 0$ (1800) and would like to estimate this parameter from the data in Table 8-3.

The model now will have two parameters and can be written as:

$$P(t) = G(c, r; t) = ce^{rt}, \tag{8-12}$$

and the corresponding SSR is:

$$SSR = SSR(c, r) = \sum_i [P_i - G(c, r; t_i)]^2 = \sum_i [P_i - ce^{rt_i}]^2.$$

To find the least-squares estimate for the parameters c and r, we need to solve the system of equations $\dfrac{\partial(SSR)}{\partial c} = 0$ and $\dfrac{\partial(SSR)}{\partial r} = 0$. We calculate

$$\frac{\partial(SSR)}{\partial c} = -2 \sum_i [P_i - ce^{rt_i}]e^{rt_i} \tag{8-13}$$

$$\frac{\partial(SSR)}{\partial r} = 2 \sum_i [P_i - 5ce^{rt_i}](-c)e^{rt_i}t_i = -2c \sum_i t_i[P_i - ct_ie^{rt_i}]e^{rt_i},$$

and we need to solve this system for c and r:

$$\sum_i [P_i - ce^{rt_i}]e^{rt_i} = 0$$

$$\sum_i t_i[P_i - ce^{rt_i}]e^{rt_i} = 0.$$

Using the data points (t_i, P_i) from Table 8-3, we obtain the equations:

$$[7.2 - ce^r]e^r + [9.6 - ce^{2r}]e^{2r} + [12.9 - ce^{3r}]e^{3r}$$
$$+ [17.1 - ce^{4r}]e^{4r} + [23.2 - ce^{5r}]e^{5r} + [31.4 - ce^{6r}]e^{6r} = 0 \tag{8-14}$$

Time i (decades)	U.S. Population $P_i = P(i)$ (millions)
1	7.2
2	9.6
3	12.9
4	17.1
5	23.2
6	31.4

TABLE 8-3.
Population of the United States from 1810 to 1860.

$$[7.2 - ce^r]e^r + 2[9.6 - ce^{2r}]e^{2r} + 3[12.9 - ce^{3r}]e^{3r}$$

$$+ 4[17.1 - ce^{4r}]e^{4r} + 5[23.2 - ce^{5r}]e^{5r} + 6[31.4 - ce^{6r}]e^{6r} = 0.$$

Unlike the system of equations for the linear model from Eq. (8-4), this system of equations is nonlinear, and there is no exact formula for the solution such as given by Eq. (8-5). However, an approximation of the solution (c^*, r^*) can be obtained through computational approaches such as the Gauss–Newton algorithm.

Historically, the numerical challenges of solving nonlinear equations or systems of equations, such as Eqs. (8-11) and (8-14), have been overcome by transforming the experimental data to conform to a linear model. For example, if we take natural logarithms of both sides of Eq. (8-12), we obtain

$$\ln(P) = \ln(ae^{rt}) = \ln(c) + \ln(e^{rt}) = \ln(c) + rt.$$

This is a linear model of the form $Y = aX + b$ with $a = r$, $b = \ln(c)$, $X = t$, and $Y = \ln(P)$, for which the minimum of the least-squares sum of the residuals is easily found. Thus, in this case, transforming the data to the form $(X_i, Y_i) = (t_i, \ln(P_i))$ eliminates the technical difficulties arising from the need to solve the nonlinear Eq. (8-14) for c and r. This successful data transformation is caused by the specific exponential form of the model in Eq. (8-12). For general models $Y = G$ (*parameters*; X), finding a linearizing transformation may be difficult or impossible. In addition, such transformations often lead to circumstances in which the statistical validity of the least-squares procedure is violated. Thus, linearizing transformations should generally be avoided because they often lead to incorrect results, as our next example illustrates.

II. A LIGAND-BINDING EXAMPLE

Consider the data shown in Figure 8-3. It could represent the effect of a drug as a function of the drug concentration or an enzyme kinetic response as a function of ligand concentration. These two examples belong to a general category of biomedical investigations known as *ligand-binding experiments*. The mathematics and numerical analysis of these experiments are essentially identical and are discussed here.

Recall that vertical bars around the data point reflect the possibilities for errors in measurement. In Figure 8-3, the vertical lines are centered on the observed values and represent positive and negative deviations equal in magnitude to the standard error of measurement SEM. The estimated precision of each data point, SEM_i, can be different, allowing the data points to be known to variable precision. Different data points

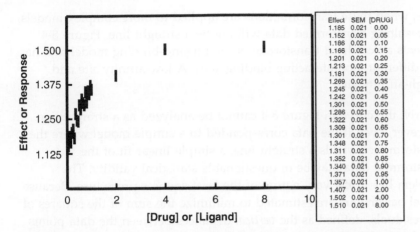

Effect	SEM	[DRUG]
1.185	0.021	0.00
1.152	0.021	0.05
1.186	0.021	0.10
1.166	0.021	0.15
1.201	0.021	0.20
1.213	0.021	0.25
1.181	0.021	0.30
1.269	0.021	0.35
1.245	0.021	0.40
1.242	0.021	0.45
1.301	0.021	0.50
1.286	0.021	0.55
1.322	0.021	0.60
1.309	0.021	0.65
1.301	0.021	0.70
1.348	0.021	0.75
1.311	0.021	0.80
1.352	0.021	0.85
1.340	0.021	0.90
1.371	0.021	0.95
1.357	0.021	1.00
1.407	0.021	2.00
1.502	0.021	4.00
1.510	0.021	8.00

FIGURE 8-3.
A typical example of a ligand-binding experimental measurement. The y-axis (arbitrary units) corresponds to the measurements of a drug effect or some other response that can be assumed to be proportional to the amount of the drug or other ligand that is bound. The x-axis is the concentration of the unbound drug or other ligand. The vertical lines are centered on the observed values and represent the ± 1 standard error of the measurement (SEM) of experimental uncertainties for the particular data points.

could have different measurement errors. Thus, it is important to consider a data point as a triplet (Y_i, SEM_i, X_i) consisting of the dependent variable, the precision of the dependent variable, and the independent variable.

Historically, two general approaches have been applied to the analysis of ligand-binding data. The earliest was to perform a transformation of the data such that the transformed data were reasonably described by a straight line. The resulting nearly linear data could then be analyzed by fitting a straight line to it and deducing the desired properties from the slope and intercept of that line. Now that high-speed computers are ubiquitous, the more common and more statistically valid approach is to fit the nonlinear equations to the original experimental data without transformation. We illustrate the shortcomings of the first approach next, and then examine the second approach in detail.

It is important to realize that transformation methods only apply to the simplest models, such as Eq. (8-12), where there was a unique linearizing transformation. In most other situations, this will not apply. Recall, for example, Eq. (7-5) from Chapter 7:

$$\overline{Y} = \frac{[\text{Drug}]/K_d}{1 + [\text{Drug}]/K_d}. \tag{8-15}$$

If the data can adequately be described by the simple mechanism of drug action represented by Eq. (8-15), then the data can be linearized several ways, such as a double reciprocal plot ($1/\overline{Y}$ vs. $1/[\text{Drug}]$) or a Scatchard plot ($\overline{Y}/[\text{Drug}]$ vs. $\overline{Y}$). In addition, if the model contains more than two parameters, it cannot be transformed into a two-parameter line.

Thus, when these transformations are applied to more complex models, the resulting transformed data will not be a straight line. Figure 8-4 presents a Scatchard transformation of a ligand-binding model that has two different, noninteracting binding sites: A low-affinity site and a high-affinity site.

Clearly, the curve in Figure 8-4 cannot be analyzed as a straight line. However, even if the data corresponded to a simple model where the transformation yields a straight line, a simple linear fit of the transformed data may be of questionable statistical validity. The problem lies in the very nature of the least-squares procedure. Because model parameters are estimated to minimize the sum of the squares of the residuals, defined as the *vertical* distances between the data points and the model, this assumes all of the uncertainties in the data can be attributed to the y-axis. With the transformed data, however, this is not always the case. For example, in Figure 8-4, the uncertainties at the left side of the graph are in the y-axis, while on the right the uncertainties are mostly in the x-axis.

So why were these linearizing transformations developed? Better methods have been available for a long time but require a lot of computer power. At the time the linearizing methods were developed, computers were not available, and calculations had to be performed by hand. The data transformations required in the past are no longer needed or desired. Biology is not linear, and our methods of analysis should not be linear either.

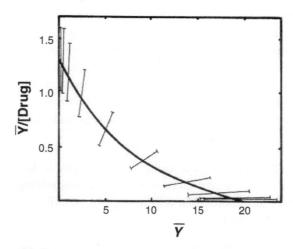

FIGURE 8-4.
A typical Scatchard plot of data containing both low- and high-affinity binding sites. The precision of the individual data points is represented by lines (error bars) that radiate from the origin. These error bars were generated by assuming that all of the uncertainties within the data are in the measured fractional saturation. When expressed as fractional saturation versus free concentration, all of these error bars are vertical. However, with the Scatchard transformation, both axes contain errors because the Scatchard transformation includes the fractional saturation on both the y-axis and the x-axis.
(Adapted from Johnson, M. L. and Frasier, S. G. [1985]. Nonlinear least-squares analysis. *Methods in Enzymology, 117*, 301–342, with permission from Elsevier.)

We now outline the computational methods used to determine the least-squares values of the parameters from the data.

III. A PRIMER FOR SOLVING NONLINEAR EQUATIONS

One way to use computers to solve nonlinear equations is by an iterative process that assumes a function can be expanded as a *Taylor series*. In one variable, this means that if we know the value of the function $f(x)$ at a point $x = x_0$ and we want to find the value of the function at another point x that is close to x_0, we can use the expression

$$f(x) = f(x_0) + f'(x_0)(x - x_0) + \frac{f''(x_0)}{2!}(x - x_0)^2 + \frac{f'''(x_0)}{3!}(x - x_0)^3 + \cdots.$$

What makes our technique work in most cases is that if our guess x_0 is close to the value x we seek, then $x - x_0$ is small, and the sum of higher order terms

$$\frac{f''(x_0)}{2!}(x - x_0)^2 + \frac{f'''(x_0)}{3!}(x - x_0)^3 + \cdots$$

will be negligible compared to $f(x) = f(x_0) + f'(x_0)(x - x_0)$. Thus:

$$f(x) \approx f(x_0) + f'(x_0)(x - x_0). \tag{8-16}$$

In two or more variables, the idea is similar. If we know the value of $f(x,y)$ at a point $(x,y) = (x_0, y_0)$ and we want to find the value of the function at another point (x,y) that is close to (x_0, y_0), we can use the following expression:

$$f(x,y) = f(x_0, y_0) + \frac{\partial f(x_0, y_0)}{\partial x}(x - x_0) + \frac{\partial f(x_0, y_0)}{\partial y}(y - y_0)$$

$$+ \frac{1}{2!}\frac{\partial^2 f(x_0, y_0)}{\partial x^2}(x - x_0)^2 + \frac{1}{2!}\frac{\partial^2 f(x_0, y_0)}{\partial x \partial y}(x - x_0)(y - y_0)$$

$$+ \frac{1}{2!}\frac{\partial^2 f(x_0, y_0)}{\partial y^2}(y - y_0)^2 + \cdots.$$

Like before, if $x - x_0$ and $y - y_0$ are both small, then the expression
$f(x_0, y_0) + \dfrac{\partial f(x_0, y_0)}{\partial x}(x - x_0) + \dfrac{\partial f(x_0, y_0)}{\partial y}(y - y_0)$ provides
a good approximation for $f(x,y)$, and we write

$$f(x,y) \approx f(x_0, y_0) + \frac{\partial f(x_0, y_0)}{\partial x}(x - x_0) + \frac{\partial f(x_0, y_0)}{\partial y}(y - y_0). \tag{8-17}$$

A. Newton's Method for One Variable

Suppose we have a function $f(x)$ and want to find a point x^* where $f(x^*) = 0$. We make an initial guess $x = x_0$, and then find the point where the line l_1 tangent to the graph of $f(x)$ at $(x_0, f(x_0))$ crosses the x-axis, denoting this point x_1 (see Figure 8-5). The equation of the tangent line l_1 is

$$y - f(x_0) = f'(x_0)(x - x_0)$$

or

(8-18)

$$y = f(x_0) + f'(x_0)(x - x_0).$$

Note that Eq. (8-18) is the truncated form of the Taylor series (8-16). Now, because the tangent line l_1 crosses the x-axis at $x = x_1$, we have $y = 0$ at this point, and Eq. (8-18) becomes

$$0 = f(x_0) + f'(x_0)(x_1 - x_0).$$

Solving for x_1, we obtain the following equation for the "better guess":

$$x_1 = x_0 - \frac{f(x_0)}{f'(x_0)}.$$

(8-19)

If the value x_1 is not the root, we consider the line l_2 tangent to the graph of $f(x)$ at $(x_1, f(x_1))$. The point x_2 at which the line l_2 crosses the x-axis is calculated from

$$x_2 = x_1 - \frac{f(x_1)}{f'(x_1)},$$

(8-20)

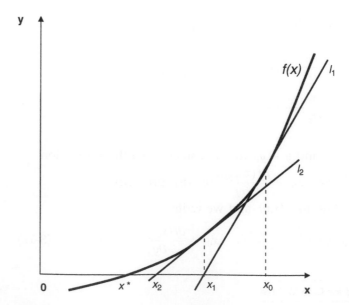

FIGURE 8-5.
Successive iterations for obtaining improved guesses for the point x^* where $f(x^*) = 0$ by using Newton's method.

and so on. In general, for any guess x_n we make, the "improved guess" can be calculated from the formula

$$x_{n+1} = x_n - \frac{f(x_n)}{f'(x_n)}. \qquad (8\text{-}21)$$

The process terminates when two successive iterations of the same value are produced.[2]

Example 8-3
·····················

Use Newton's method to solve $x + e^x = 5$.

SOLUTION:

We denote $f(x) = 5 - x - e^x$ and then want to solve the equation $f(x) = 0$. Since $f(0) = 4$ and $f(2) = 5 - 2 - e^2 < 0$, a root must lie between $x = 0$ and $x = 2$. We make an initial guess $x_0 = 1$. Since $f'(x) = -1 - e^x$, using Eq. (19), we calculate

$$x_1 = x_0 - \frac{5 - x_0 - e^{x_0}}{-1 - e^{x_0}} = 1 - \frac{5 - 1 - e}{-1 - e} = 1.344707.$$

With this value for x_1 and Eq. (8-20), we calculate

$$x_2 = x_1 - \frac{5 - x_1 - e^{x_1}}{-1 - e^{x_1}} = 1.307128,$$

and so on. Applying Eq. (8-21) in this case gives

$$x_{n+1} = x_n - \frac{5 - x_n - e^{x_n}}{-1 - e^{x_{n1}}}.$$

The process terminates when $x_{n+1} = x_n$. Table 8-4 presents the values of the consecutive iterations. Thus, we have found (to five decimal places) that $x + e^x = 5$ when $x = 1.306558$.

Choosing a good initial guess can be critical to the success of Newton's method and is usually based on the experimental data and the model. Recall the model $P = G(r; t) = 5.3\, e^{rt}$ and that the least-squares estimate for the parameter r is the solution of Eq. (8-11). In Chapter 1, we described a way of estimating the value of the parameter r from the data and found that the best value should be close to $r^* = 0.3$ (see Table 1-3 in Chapter 1 and the preceding text). We can now use this approximation as our initial guess r_0 for Newton's method to determine the least-squares value of r^*. The results of the iterative process described by

Iteration i	Guess x_i
0	1
1	1.344707
2	1.307128
3	1.306558
4	1.306558

TABLE 8-4.
Values of iterations for example 8-3.

----

2. In practice, the process is terminated when two successive iterations become closer than a small *tolerance value* (e.g., 0.00000001) chosen in advance.

A:	Iteration i	Guess x_i
	0	0.3
	1	0.29604731216119
	2	0.29591462767617
	3	0.29591448299412
	4	0.29591448299395
	5	0.29591448299395
	6	0.29591448299395

B:	Iteration i	Guess x_i
	0	0.5
	1	0.42811073165450
	2	0.36676718238807
	3	0.32253692486329
	4	0.30074060321933
	5	0.29609881338742
	6	0.29591476150979
	7	0.29591448299458
	8	0.29591448299395
	9	0.29591448299395

TABLE 8-5.
Role of the initial guess on outcome of Newton's method.

Eq. (8-21) are presented in Table 8-5(A). Thus, the least-squares value for r is $r^* = 0.29591448299395$. Our initial guess of $x_0 = 0.3$ was rather close to the root. If we start with a less accurate guess, more iterations may be required before finding the root, as illustrated in Table 8-5(B) with the guess $x_0 = 0.5$. In addition, because nonlinear equations may have more than one solution, if the initial guess is chosen at random, the method may converge to a false root, finding a minimum for the SSR that results in a value for r that is meaningless in the context of the problem.

In principle, Newton's method for one variable can be generalized to two or more variables and used to solve systems of nonlinear equations, such as the system defined by Eq. (8-14). However, for several reasons, such methods are not computationally optimal for determining the least-squares parameter estimates. First, minimizing the SSR function requires that its partial derivatives with respect to the model parameters be calculated and set to zero. This process may lead to complicated systems of equations, where the lengths of the algebraic expressions grow with the number of experimental data points. Second, to use Newton's method to solve these equations would require yet another differentiation. That is, the cumbersome expressions defining the systems of equations for the parameters will need to be differentiated again, and their derivatives used to calculate the iterations approximating the solutions. Instead, improved versions of the procedures, such as the following, are usually applied.

B. The Gauss–Newton Method for One Variable

We consider a one-parameter model of the general form $Y = G(r;X)$, experimental data $(X_1, Y_1), (X_2, Y_2), \ldots, (X_n, Y_n)$ and the SSR, which in this case is defined by

$$SSR(r) = \sum_{i=1} [Y_i - G(r; X_i)]^2. \tag{8-22}$$

As with the Newton's method, the process is iterative. We begin by making a guess $r = r_0$ of the parameter's value. Next, we use the Taylor series approximation for $G(r; X)$, as in Eq. (8-16). The variable of interest is r, so we obtain

$$G(r; X_i) \approx G(r_0; X_i) + \frac{dG(r_0; X_i)}{dr}(r - r_0). \tag{8-23}$$

Because we assume the model $Y = G(r;X)$ is correct, we seek the value for the parameter r for which:

$$Y_i = G(r, X_i) + \text{experimental uncertainties.}$$

Ignoring the experimental uncertainties, we write:

$$Y_i \approx G(r, X_i). \tag{8-24}$$

Combining Eqs. (8-23) and (8-24) yields:

$$Y_i \approx G(r_0; X_i) + \frac{dG(r_0; X_i)}{dr}(r - r_0). \qquad (8\text{-}25)$$

We use this approximation to begin an iterative procedure and produce a better guess, $r = r_1$, for the parameter. As in Newton's method, we expect using this better guess in place of r_0 will produce an even better guess, and so on. The process terminates when two consecutive iterations return the same value, which is the "answer." Generalizing, we can write:

$$G(answer; X_i) \approx G(guess; X_i) + \frac{dG(guess; X_i)}{d(guess)}(answer - guess).$$

We now have one equation of the form (8-25) for each data point. For illustration, assume we have only three data points: $(X_1, Y_1), (X_2, Y_2), (X_3, Y_3)$. The formula would yield three equations for r:

$$Y_1 = G(r_0; X_1) + \frac{dG(r_0; X_1)}{dr}(r - r_0)$$

$$Y_2 = G(r_0; X_2) + \frac{dG(r_0; X_2)}{dr}(r - r_0)$$

$$Y_3 = G(r_0; X_3) + \frac{dG(r_0; X_3)}{dr}(r - r_0),$$

which can be rewritten as:

$$\frac{dG(r_0; X_1)}{dr}(r - r_0) = Y_1 - G(r_0; X_1)$$

$$\frac{dG(r_0; X_2)}{dr}(r - r_0) = Y_2 - G(r_0; X_2) \qquad (8\text{-}26)$$

$$\frac{dG(r_0; X_3)}{dr}(r - r_0) = Y_3 - G(r_0; X_3).$$

We can use matrix notation to rewrite Eq. (8-26) more compactly. If we denote

$$P = \begin{bmatrix} \dfrac{dG(r_0; X_1)}{dr} \\[2mm] \dfrac{dG(r_0; X_2)}{dr} \\[2mm] \dfrac{dG(r_0; X_3)}{dr} \end{bmatrix}, Y^* = \begin{bmatrix} Y_1 - G(r_0; X_1) \\ Y_2 - G(r_0; X_2) \\ Y_3 - G(r_0; X_3) \end{bmatrix}, \text{ and } \varepsilon = r - r_0, \qquad (8\text{-}27)$$

Eq. (8-26) can be reduced to the single matrix equation:

$$P\varepsilon = Y^*. \qquad (8\text{-}28)$$

In general, when the data set contains n data points, the matrices P and Y^* will be:

$$P = \begin{bmatrix} \dfrac{dG(r_0; X_1)}{dr} \\[2mm] \dfrac{dG(r_0; X_2)}{dr} \\[2mm] \vdots \\[2mm] \dfrac{dG(r_0; X_n)}{dr} \end{bmatrix} \quad \text{and} \quad Y^* = \begin{bmatrix} Y_1 - G(r_0; X_1) \\ Y_2 - G(r_0; X_2) \\ \vdots \\ Y_n - G(r_0; X_n) \end{bmatrix}. \qquad (8\text{-}29)$$

To solve Eq. (8-28) for ε, we multiply both sides by the transposed matrix P^T:

$$P^T P \varepsilon = P^T Y^*. \qquad (8\text{-}30)$$

Now, $P^T P$ is a square matrix, and, if it is invertible, we can solve Eq. (8-30) and obtain

$$\varepsilon = (P^T P)^{-1}(P^T Y^*). \qquad (8\text{-}31)$$

Because $\varepsilon = r - r_0$, the next guess is calculated from $r = \varepsilon + r_0$. We call this improved guess r_1, use it in place of r_0 in Eq. (8-26), and then iterate. Schematically, the process can be represented as $\varepsilon + guess \Rightarrow$ *better guess*.

The process terminates when the calculated value for *better guess* is the same as *guess*; that is, when $\varepsilon = 0$. We have then found the *answer* for the parameter r.

The Gauss–Newton method just described is not based upon minimizing the SSR defined in Eq. (8-22), so how is it a least-squares procedure? The answer is found in Eq. (8-31). When $\varepsilon = 0$, we have $(P^T P)^{-1}(P^T Y^*) = 0$. Because $(P^T P)^{-1}$ cannot be zero, as the matrix $(P^T P)$ was inverted, it must be that $P^T Y^* = 0$. For the matrices P^T and Y^* defined in Eq. (8-29), we then have for $r = answer$, the product

$$P^T Y^* = \sum_i [Y_i - G(r; X_i)]\frac{dG(r; X_i)}{dr} = 0.$$

On the other hand, differentiating Eq. (8-22) gives $\dfrac{dSSR(r)}{dr} =$ $-2\sum_i [Y_i - G(r; X_i)]\dfrac{dG(r; X_i)}{dr}$. Thus, when r is such that $P^T Y^* = 0$, we will also have $\dfrac{dSSR(r)}{dr} = 0$. This shows that we have found the least-squares estimate for the parameter r.

Example 8-4

We use the Gauss–Newton algorithm to estimate parameter r in the model $G(r;t) = 5.3\,e^{rt}$ from the U.S. population data in Table 8-3.

The matrices are

$$P = \begin{bmatrix} t_0 5.3 e^{rt_0} \\ t_1 5.3 e^{rt_1} \\ t_2 5.3 e^{rt_2} \\ t_3 5.3 e^{rt_3} \\ t_4 5.3 e^{rt_4} \\ t_5 5.3 e^{rt_5} \\ t_6 5.3 e^{rt_6} \end{bmatrix} - \begin{bmatrix} 0 \\ 1(5.3)e^{r} \\ 2(5.3)e^{2r} \\ 3(5.3)e^{3r} \\ 4(5.3)e^{4r} \\ 5(5.3)e^{5r} \\ 6(5.3)e^{6r} \end{bmatrix} \text{ and } Y^* = \begin{bmatrix} P_0 - 5.3 e^{rt_0} \\ P_1 - 5.3 e^{rt_1} \\ P_2 - 5.3 e^{rt_2} \\ P_3 - 5.3 e^{rt_3} \\ P_4 - 5.3 e^{rt_4} \\ P_5 - 5.3 e^{rt_5} \\ P_6 - 5.3 e^{rt_6} \end{bmatrix} = \begin{bmatrix} 0 \\ 7.2 - (5.3)e^{r} \\ 9.6 - (5.3)e^{2r} \\ 12.9 - (5.3)e^{3r} \\ 17.1 - (5.3)e^{4r} \\ 23.2 - (5.3)e^{5r} \\ 31.4 - (5.3)e^{6r} \end{bmatrix}.$$

$$(8\text{-}32)$$

We begin with an initial guess of $r_0 = 0.3$ and compute $\varepsilon = (P^T P)^{-1}(P^T Y^*) = -0.0040379399$. The improved guess, r_1, is then calculated to be

$$r_1 = r_0 + \varepsilon = 0.3 - 0.0040379399 = 0.295962601.$$

Using this value for r in Eq. (8-31), we compute $\varepsilon = (P^T P)^{-1}(P^T Y^*) = -0.000048084$. The value for the next guess will now be

$$r_2 = r_1 + \varepsilon = 0.295962601 - 0.000048084 = 0.295914517.$$

The process continues until the desired accuracy is achieved. Assume we only want to calculate an answer accurate to at least three decimal places. The value $\varepsilon = 0.000048084$ calculated above is then $\varepsilon = 0.000$, and, therefore, r_2 is the answer.[3]

C. The Gauss–Newton Method for Two and More Variables

For models involving two or more parameters, the idea behind the Gauss–Newton method is the same, but the matrices P and Y^* need to be changed appropriately. We outline the process for two parameters and then discuss how it generalizes for an arbitrary number of parameters.

Consider the model $Y = G(parameters; X) = G(r,c;X)$, where the goal is to find the least-squares values of the parameters c and r estimated from the data $(X_1, Y_1), (X_2, Y_2), \ldots, (X_n, Y_n)$.

---

3. Compare this value with the least-squares value for r we obtained for the same model and data set earlier (Table 8-5) using Newton's method.

In this case,

$$SSR(r,c) = \sum_{i=1} [Y_i - G(r,c;X_i)]^2. \tag{8-33}$$

We begin with initial guesses $r = r_0$ and $c = c_0$ and the Taylor approximation from Eq. (8-17) for the values $r = r_0$ and $c = c_0$:

$$G(r,c;X_i) \approx G(r_0,c_0;X_i) + \frac{\partial G(r_0,c_0;X_i)}{\partial r}(r - r_0) + \frac{\partial G(r_0,c_0;X_i)}{\partial c}(c - c_0). \tag{8-34}$$

Because for the least-squares values of r and c, we want

$$Y_i = G(r,c;X_i) + \text{experimental uncertainties,}$$

the equations used to find those values are

$$Y_i = G(r_0,c_0;X_i) + \frac{\partial G(r_0,c_0;X_i)}{\partial r}(r - r_0) + \frac{\partial G(r_0,c_0;X_i)}{\partial c}(c - c_0).$$

As in the one-parameter case, we have one equation of this form for every experimental data point (X_i, Y_i) and can express this set of equations more conveniently in matrix notation as $P\varepsilon = Y^*$, where

$$P = \begin{bmatrix} \dfrac{\partial G(r_0,c_0;X_1)}{\partial r} & \dfrac{\partial G(r_0,c_0;X_1)}{\partial c} \\[2ex] \dfrac{\partial G(r_0,c_0;X_2)}{\partial r} & \dfrac{\partial G(r_0,c_0;X_2)}{\partial c} \\[2ex] \vdots & \vdots \\[2ex] \dfrac{\partial G(r_0,c_0;X_n)}{\partial r} & \dfrac{\partial G(r_0,c_0;X_n)}{\partial c} \end{bmatrix}, \quad Y^* = \begin{bmatrix} Y_1 - G(r_0,c_0;X_1) \\ Y_2 - G(r_0,c_0;X_2) \\ \vdots \\ Y_n - G(r_0,c_0;X_n) \end{bmatrix}, \text{ and}$$

$$\varepsilon = \begin{bmatrix} r - r_0 \\ c - c_0 \end{bmatrix}.$$

The solution ε can be obtained as in Eq. (8-31), with the iterative process continuing until $\varepsilon + \text{guess} \Rightarrow \text{better guess}$ returns $\varepsilon = 0$.

EXERCISE 8-4

Demonstrate that the values for parameters r and c obtained from the Gauss–Newton method are the least-squares estimates for r and c by showing these values provide a minimum for the SSR from Eq. (8-33).

Hint: Show that the values obtained for r and c are such that

$$\frac{\partial SSR(r,c)}{\partial r} = 0 \quad \text{and} \quad \frac{\partial SSR(r,c)}{\partial c} = 0.$$

Example 8-5

We now use the Gauss–Newton algorithm to estimate parameters r and c in the model $G(r, c; t) = ce^{rt}$ from the U.S. population data in Table 8-3. Now $\dfrac{\partial G}{\partial r} = cte^{rt}$ and $\dfrac{\partial G}{\partial c} = e^{rt}$, and the matrices P and Y^* become

$$P = \begin{bmatrix} ct_1 e^{rt_1} & e^{rt_1} \\ \vdots & \vdots \\ ct_6 e^{rt_6} & e^{rt_6} \end{bmatrix} = \begin{bmatrix} ce^{rt} & e^r \\ 2ce^{2r} & e^{2r} \\ 3ce^{3r} & e^{3r} \\ 4ce^{4r} & e^{4r} \\ 5ce^{5r} & e^{5r} \\ 6ce^{6r} & e^{6r} \end{bmatrix} \text{ and } Y^* = \begin{bmatrix} P_1 - ce^{rt_1} \\ \vdots \\ P_6 - ce^{rt_{61}} \end{bmatrix} = \begin{bmatrix} 7.2 - ce^r \\ 9.6 - ce^{2r} \\ 12.6 - ce^{3r} \\ 17.1 - ce^{4r} \\ 23.2 - ce^{5r} \\ 31.4 - ce^{6r} \end{bmatrix}.$$

Choosing initial guesses of $c_0 = 5$ and $r_0 = 0.3$ in the matrices above gives

$$\varepsilon = (P^T P)^{-1}(P^T Y^*) = \begin{bmatrix} -0.0009 \\ 0.2053 \end{bmatrix},$$

where now ε is the vector $\varepsilon = \begin{bmatrix} r - r_0 \\ c - c_0 \end{bmatrix}$.

Thus, the next guesses for the parameters are

$$r_1 = r_0 - 0.0009 = 0.2991 \text{ and } c_1 = c_0 + 0.2053 = 5.2053.$$

Substituting these values for r and c gives

$$\varepsilon = (P^T P)^{-1}(P^T Y^*) = \begin{bmatrix} 0.0000157 \\ 0.00016 \end{bmatrix}.$$

With three digits of accuracy, we can terminate the process at this step and use the values r_1 and c_1 as the least-squares estimates for the parameters based on the data in Table 8-3.

When models involve more than two parameters, the notation becomes quite cumbersome. Describing the steps of the computational process becomes a bit easier if we think of a set of *guesses*, one for each parameter, being produced at each step in the search for the set of *answers* that minimize the SSR. Eq. (8-34), for instance, can be written as

$$\begin{aligned} G(answers; X_i) \\ \approx G(guesses; X_i) + \frac{\partial G(guesses; X_i)}{\partial guess_1}(answer_1 - guess_1) \\ + \frac{\partial G(guesses; X_i)}{\partial guess_2}(answer_2 - guess_2). \end{aligned}$$

Here, $answer_1$ and $guess_1$ refer to the answer and guess for the first parameter (r), and $answer_2$ and $guess_2$ refer to the answer and guess for the second parameter (c). Using Σ-notation, we write

$$G(answers; X_i) \approx G(guesses; X_i) + \sum_j \frac{\partial G(guesses; X_i)}{\partial guess_j}(answer_j - guess_j),$$

(8-35)

and the equations used for the computation now become

$$\sum_j \frac{\partial G(guesses; X_i)}{\partial guess_j}(answer_j - guess_j) = Y_i - G(guesses; X_i).$$ (8-36)

As before, there are as many such equations as there are experimental data points. The index j takes as many values as the number of model parameters. Equation (8-7), for example, has two parameters, and, if it were being fit to 100 data points, then Eq. (8-36) would actually be 100 equations (one for each data point) in two unknowns (K_{21} and K_{22}).

EXERCISE 8-5

Identify matrices P, Y^*, and ε such that Eq. (8-36) can be written in matrix form as $P\varepsilon = Y^*$.

IV. WEIGHTED LEAST-SQUARES CRITERION AND THE GAUSS–NEWTON METHODS FOR WEIGHTED LEAST SQUARES

Thus far, we have not tried to account for different measurement errors in the data points. The most common approach to situations where the experimental measurements are known with different degrees of accuracy is to apply a *weighted least-squares parameter estimation criterion*. Under this criterion, model parameters are calculated so they minimize the following *weighted sum of squared residuals* (WSSR):

$$WSSR = \sum_i \left(\frac{Y_i - G(parameters; X_i)}{SEM_i} \right)^2 = \sum_i [W_i(Y_i - G(parameters; X_i))]^2 = \sum_i r_i^2,$$ (8-37)

where the weights, W_i, are reciprocal to the measurement errors. Thus, the larger the measurement error for a particular data point, the smaller the weight W_i, and consequently the smaller this point's contribution to the WSSR. The *residuals*, r_i, are defined to be weighted differences between the data points and the fitted curve.

Various choices are possible for W_i, and what values are used depends upon the design of the experiment. If each point is measured only once, and there is no known distribution of the errors, it is reasonable to assume all weights are equal to one, which makes the measure WSSR from Eq. (8-37) equivalent to the measure SSR from Eq. (8-1). Thus, WSSR in Eq. (8-37) is a generalization of SSR introduced in Eq. (8-1). Similarly, the two formulas are equivalent if the measurement errors of all data points have the same distribution, with a certain known standard deviation. When each data point is measured only once and there are different known errors in different measurements, it is reasonable to assign each data point a weight that is reciprocal to its measurement error. Thus, more uncertain data points contribute less to the least-squares estimates. Finally, if each data point is a result of several (>15) measurements, then we can calculate the standard error of measurement SEM as an empirical estimate of the measurement errors, and the weights W_i can be computed as $W_i = 1/SEM_i$ (see, for example, Johnson and Frasier [1985]).

The weighted least-squares estimates for the model parameters are those that minimize the function WSSR from Eq. (8-37). Different values of the model parameters correspond to different values of the WSSR. For example, Figure 8-6 is a two-dimensional contour map of the WSSR as a function of the two-parameters of the fitting equation for dimeric hemoglobin given in Eq. (8-7). Each of the contours represents a constant value for the WSSR, with contours nearer the center denoting lower WSSR values. The objective of the fitting procedure is to find the optimal parameter values corresponding to the dot in the center, which represents the lowest WSSR value. In the case of normally distributed errors, these values are also called *maximum likelihood solutions*.

As before, the minimization procedure for determining those solutions is based on series expansions and is not much different from the Gauss–Newton methods described earlier for data points with equal weights. Specifically, the Gauss–Newton method is based on the Taylor expansion:

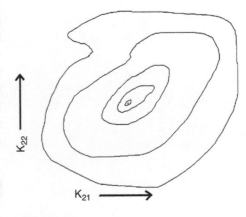

FIGURE 8-6.
Contour map of WSSR. The contours represent constant values of WSSR, with the values getting smaller as we approach the center contour.

$$G(answers, X_i) = G(guesses, X_i) + \sum_j \frac{\partial G(guesses, X_i)}{\partial guess_j} (answer_j - guess_j) + \ldots,$$

$$(8\text{-}38)$$

using the same notation as in Eq. (8-37). Again, Eq. (8-38) consists of one equation for each data point, hence the subscript i. The objective of the least-squares fitting is to determine the *answers* for which

$$Y_i = G(answers, X_i) + \text{experimental uncertainties.} \qquad (8\text{-}39)$$

By neglecting the experimental uncertainties in Eq. (8-39), ignoring the higher derivatives terms (i.e., the . . .) in Eq. (8-38), dividing Eq. (8-38) by SEM_i, and combining the results, we obtain

$$\sum_j \left[\frac{1}{SEM_i} \frac{\partial G(guesses, X_i)}{\partial guess_j} (answer_j - guess_j) \right] \approx \frac{Y_i - G(guesses, X_i)}{SEM_i}.$$

(8-40)

Equation (8-40) expresses the desired optimal answers in terms of the data points (Y_i, SEM_i, X_i), the fitting equation (G), and the initial estimates of the answers (*guesses*). The SEM_i is now included to allow each data point to have a different level of experimental uncertainty and thus a different statistical weight.

It is important to note that when we neglect the higher order derivative terms, represented by . . . in Eq. (8-38), Eq. (8-40) is only approximately correct, and thus the iteration of Eq. (8-40) must be performed many times so that it can converge to the optimal parameter values. Linear least-squares is a special case where all of the higher-order derivatives are exactly zero, so the solution to Eq. (8-40) is exact, and only a single cycle of the algorithm is required. In this context, the term *linear* refers to the form of Eq. (8-38), not the form of the fitting equation $Y = G$ (*parameters*; X). For example, fitting data to a straight line (i.e., $Y = a + bX$) is a linear fit, as is fitting to all of the higher degree polynomials. In these cases, the problem of finding the minimum WSSR leads to the problem of solving a system of equations that is linear with respect to all of the unknowns. In the same way, fitting a single Fourier wave (i.e., $Y = a\sin(2\pi X/d) + b\cos(2\pi X/d) + c$) is also a linear fit if d is a constant while only a, b, and c are being estimated. If, however, d is also being estimated, then the second- and higher-order derivatives are not all zero, and the fitting process becomes nonlinear.

EXERCISE 8-6

(a) Show that the model $Y = G(a, b, c, d, e, f; X) = aX^5 + bX^4 + cX^3 + dX^2 + eX + f$ is a linear model as far as least-squares fit for the parameters a,b,c,d,e,f is concerned.

(b) Generalize for a polynomial model of arbitrary (but fixed) degree with coefficients that are being estimated from the data.

EXERCISE 8-7

Verify that if d is fixed (and thus not a parameter to be estimated from the data), the model $Y = a\sin(2\pi X/d) + b\cos(2\pi X/d) + c$ is a linear model.

EXERCISE 8-8

(a) Show that the set of equations from Eq. (8-40) can be written in matrix form as

$$P\varepsilon = Y^*, \tag{8-41}$$

where the matrices P, Y^*, and ε are :

$$P = \begin{pmatrix} \dfrac{1}{SEM_1}\dfrac{\partial G(guesses, X_1)}{\partial\, guess_1} & \dfrac{1}{SEM_1}\dfrac{\partial G(guesses, X_1)}{\partial\, guess_2} & \cdots \\[2mm] \dfrac{1}{SEM_2}\dfrac{\partial G(guesses, X_2)}{\partial\, guess_1} & \dfrac{1}{SEM_2}\dfrac{\partial G(guesses, X_2)}{\partial\, guess_2} & \cdots \\[2mm] \cdots & \cdots & \cdots \end{pmatrix} \tag{8-42}$$

$$\varepsilon = \begin{pmatrix} answer_1 - guess_1 \\ answer_2 - guess_2 \\ \cdots \end{pmatrix} \quad and \quad Y^* = \begin{pmatrix} \dfrac{Y_1 - G(guesses, X_1)}{SEM_1} \\[2mm] \dfrac{Y_2 - G(guesses, X_2)}{SEM_2} \\[1mm] \cdots \end{pmatrix}. \tag{8-43}$$

The number of rows for the matrices P and Y^* are determined by the number of data points. The number of parameters determines the number of columns for P. The same is true for the number of rows for the matrix ε.

(b) Show that the representation in Eq. (8-41) leads to

$$(P^T P)\varepsilon = (P^T Y^*), \tag{8-44}$$

$$\varepsilon = (P^T P)^{-1}(P^T Y^*). \tag{8-45}$$

The iterative scheme

$$\varepsilon + guess \Rightarrow better\ guess, \tag{8-46}$$

terminates when ε is found.

EXERCISE 8-9

Show that when it converges, the Gauss–Newton algorithm from Eq. (8-40) produces parameter estimates that minimize the function WSSR defined in Eq. (8-37).

Hint: Notice that

$$P^T Y^* = \begin{pmatrix} \sum_i \frac{1}{SEM_i^2} \frac{\partial G(guesses, X_i)}{\partial guess_1} [Y_i - G(guesses, X_i)] \\ \sum_i \frac{1}{SEM_i^2} \frac{\partial G(guesses, X_i)}{\partial guess_2} [Y_i - G(guesses, X_i)] \\ \vdots \end{pmatrix}, \qquad (8\text{-}47)$$

and that the individual elements of the $P^T Y^*$ are proportional to the derivatives of the WSSR with respect to each of the parameters being estimated.

We need to stress that the Gauss–Newton approach is not guaranteed to converge. If the higher-order terms (... in Eq. [8-38]) do not converge sufficiently rapidly to zero, then this algorithm might actually diverge, because if higher-order terms cannot be ignored, then their omission in Eq. (8-40) might cause irreparable error. The Gauss–Newton approach will converge rapidly in most cases, and, when it does not, there are many adaptations, such as the Marquardt–Levenberg and damped Gauss–Newton algorithms, which specifically correct for the failure to converge (see Johnson and Frasier [1985]). The damped Gauss–Newton algorithm simply checks that the new value of the WSSR, Eq. (8-37), is lower for the *guesses* $+ \varepsilon$ than it was for the previous *guesses*. If it is not lower, then ε was too big—so it is divided by 2, and this new value of ε is used in Eq. (8-46). This process of dividing ε by 2 if the WSSR has increased is repeated until it decreases.

There are many weighted nonlinear least-squares algorithms in addition to the Gauss–Newton. Some converge faster, and some require more computer memory; but when correctly implemented, they all provide equivalent results.

V. OBJECTIVES OF THE DATA-FITTING PROCEDURES

We have explained why the data-fitting procedure provides parameter values affording the best description of a data set and have described some computational methods for finding the best least-squares fit. To obtain a complete analysis of the experimental data, any data-fitting procedure will have multiple objectives, which include estimating:

1. Optimal model parameters with respect to the desired criteria;

2. Cross-correlation of the estimated model parameters;

3. Precision of the model parameters;

4. Goodness-of-fit; and

5. Uniqueness of the parameters.

Having discussed the first objective in considerable detail with regard to the weighted least-squares criterion, we move to another optimization method—maximum likelihood—and the conditions under which the two are equivalent. We also outline objectives 2 through 5 with regard to their goals and features essential to the data-fitting analyses.

A. Conditions for Maximum Likelihood

The parameter values determined through least-squares minimization of the WSSR are estimates, based on the data, for the true parameter values. It is not unusual, however, for the parameter estimates to be derived from a different criterion that maximizes the likelihood of the parameter values. That is, the values sought by this criterion are those that have the highest probability of being correct based on the data. If the data are known to satisfy the following set of relatively broad conditions, the least-squares values for the parameters are also those that guarantee maximum likelihood:

1. The X_i values do not contain any measurement errors;

2. The Y_i values contain measurement errors that follow a bell-shaped, or Gaussian, distribution with a mean of zero;

3. The fitting function, G, is correct; and

4. The measurement error for each data point is independent of other measurement errors.

Determining whether the measurement errors satisfy those conditions is not a trivial task. Figure 8-3, for example, presents a typical situation illustrating how the least-squares methods can be applied to experimental data. Let's assume we would like to perform a least-squares fit of the data to the ligand-binding fitting equations derived in Chapter 7. The following general problem then becomes apparent: Actual experimental data are rarely formulated in exactly the correct form for the algorithms and fitting function to be applied. For example, in Figure 8-3, the dependent variable (i.e., the y-axis) is not a fractional saturation, but is in arbitrary units determined by the experimental protocol (in this case, a drug response). Consequently, either the data or the fitting equation must be transformed to match the other. The decision of what to transform and how to transform it should be determined by the nature of the experimental uncertainties in the data. The idea here is either to not alter the noise distribution within the data or, in the case where the existing experimental error distribution is not Gaussian, to perform the transform so as to make the noise distribution more Gaussian.

Here, we shall assume the distribution of experimental uncertainties meets the basic assumptions of the least-squares fitting procedure.

Assume we attempt to extrapolate the values presented in Figure 8-3 to zero and to an infinite concentration of the drug. The experimentally measured value at zero in Figure 8-3 cannot simply be used as the zero extrapolation because it contains experimental measurement error. These two limits could then be used in a linear transform of the data, such that the values range from 0 to 1, as in Eq. (8-48). The data would then be in a form that could potentially be fit to one (or more) of the fractional saturation functions above:

$$\text{Fractional data} = \frac{\text{Original data-Zero limit}}{\text{Infinite limit-Zero limit}}. \tag{8-48}$$

This approach is not optimal, because the extrapolated values of the data at zero and at infinite drug concentrations both contain uncertainties. They have not been determined to infinite precision and, as a consequence, will introduce an unknown systematic uncertainty into the transformed fractional data. A better approach is to perform the inverse transform of the fitting equation, as in Eq. (8-49), and then fit the original untransformed data to the transformed fitting equation. Incorporating these limits in the fitting equations will introduce two additional fitting parameters into the fitting process:

$$\text{Transformed function} = (\text{Infinite Limit-Zero Limit})\overline{Y} + \text{Zero Limit}. \tag{8-49}$$

For example, when Eq. (8-15) is modified by Eq. (8-49) and least-squares fit to the data in Figure 8-3, there are three parameters estimated simultaneously: the zero limit, the infinite limit, and the K_d. The resulting values of the zero and infinite drug concentration limits are 1.140 and 1.578, respectively, which are clearly not equal to the first and last data point values. The reasons for this are that the data values contain experimental uncertainties and the last data point is not at an infinite concentration. The estimated value of the dissociation constant, K_d, is 1.073.

B. Cross-Correlation of the Estimated Parameters

Usually, there will appear to be a correlation between estimated parameters. For example, when Eq. (8-7) is fit to a data set, the two estimated parameters, K_{21} and K_{22}, will appear to be correlated (i.e., the estimated value of K_{21} is linearly dependent upon the value of K_{22} and vice versa). This correlation is not caused by the molecular mechanism of hemoglobin action but is, rather, a consequence of fitting a complex equation to a small number of data points spanning a limited range of oxygen concentrations. It is important to be aware of the magnitude of these parameter correlations because they are associated with the difficulties encountered by any data-fitting procedure.

The cross-correlation coefficient for the i-th and k-th parameter can be evaluated from the elements of the inverse of the $P^T P$ matrix that was already evaluated by the Gauss–Newton least-squares parameter estimation procedure, namely,

$$\text{Cross Correlation}_{ik} = \frac{(P^T P)_{ik}^{-1}}{\sqrt{(P^T P)_{ii}^{-1}(P^T P)_{kk}^{-1}}} \qquad i \neq k. \qquad (8\text{-}50)$$

These cross-correlation coefficients have a range of ± 1, with zero being optimal. As the cross-correlation approaches $+1$ or -1, the fitting procedure becomes increasingly more difficult and the results more questionable, because the $P^T P$ matrix is becoming nearly singular and cannot easily be inverted for use in Eq. (8-45). For practical purposes, if the magnitudes of the cross-correlation coefficients are less than ± 0.97, the least-squares procedure can usually function adequately. However, ± 0.97 should not be considered an absolute threshold with everything acceptable below ± 0.97 and everything unacceptable outside this range. All fitting procedures get progressively worse as the magnitude of the cross-correlations increase toward 1.

C. Precision of the Model Parameters

Finding estimates of the precision of the estimated parameters is also of paramount importance because this allows investigators to test the significance of their results. For example, consider an experiment and subsequent analysis that determines the molecular weight of hemoglobin to be 67,000 daltons. In reality, this information tells us nothing new about hemoglobin, because virtually all proteins have a molecular weight of 67,000 $\pm$ 50,000 daltons. If all we know is that the molecular weight of hemoglobin is approximately 67,000 daltons, then all we can say about hemoglobin is that it appears to be a typical protein. However, if we know the molecular weight of hemoglobin is 67,000 $\pm$ 1,000 daltons, we have a lot more useful information. For example, because we also know that hemoglobin contains one iron atom per 16,700 $\pm$ 500 daltons, we can easily conclude that the hemoglobin molecule contains four irons and thus four oxygen-binding sites. Conversely, if our estimate of the molecular weight of hemoglobin is 67,000 $\pm$ 50,000 daltons, then we would have to conclude that the hemoglobin molecule contains 4 $\pm$ 3 irons and thus 4 $\pm$ 3 oxygen-binding sites.

The most common but least accurate approach is to use the *asymptotic standard errors*, which assume the fitting equation is linear and are calculated as follows:

$$\text{Asymptotic Standard Error}_i = \sqrt{\frac{WSSR}{N}(P^T P)_{ii}^{-1}}. \qquad (8\text{-}51)$$

The asymptotic standard error for the i-th estimated parameter is related to the weighted sum of squared residual of the fit, WSSR, the number of data points, N, and the ii-th element of the inverse of the $P^T P$ matrix that was already evaluated by the Gauss–Newton procedure and used in Eq. (8-45). It is commonly used because it requires almost no additional computer time to evaluate. There are, however, three assumptions required to utilize these asymptotic standard errors as realistic estimates of the precision of the estimated parameters: (1) The fitting equation must be linear; (2) a large number of data points are required; and (3) the parameter correlation should be near zero. The consequence of these required assumptions is that the asymptotic standard errors usually significantly underestimate the actual precision of the estimated parameters. This means the significance of the results will be overestimated, and conclusions not justified by the data will be made. More sophisticated methods, beyond the scope of this text, such as the *support plane* method (see Johnson and Frasier [1985]) and the *bootstrap approach* (see Efron and Tibshirani [1993] for the details), can be used for better precision.

D. Goodness-of-Fit

Parameter estimation procedures, such as weighted least-squares, can find an optimal fit of almost any equation to almost any data set. This does not mean, however, that the fitted curve accurately describes the experimental data. For example, the hemoglobin–oxygen binding data shown in Figure 7-4 of Chapter 7 could be least-squares fit to a straight line, but it would not provide a realistic description of the data points. Likewise, the slope and intercept of this optimal straight line would provide no information about the molecular mechanism of hemoglobin function. Goodness-of-fit tests provide rigorous statistical criteria to decide if the fitted equation actually provides a good description of the experimental data. Furthermore, if the form of the fitting equation is based upon mechanistic hypotheses about what is being measured, then goodness-of-fit tests also provide rigorous statistical criteria to test the mechanistic hypotheses. Because of this, the choice of the fitting equation should always be dictated by the mechanistic hypotheses under study.

Most goodness-of-fit criteria are based on the distribution of the residuals—the weighted differences r_i between the data points and the fitted curve in Eq. (8-37). If the justifying assumptions for the least-squares approach from Section A above are satisfied, then the residuals should follow a Gaussian distribution; if they do not, then one or more of these assumptions is not valid. Assumptions 1, 2, and 4 are within the control of the experimental protocol and thus can be verified independently. If, in a carefully controlled and performed experiment,

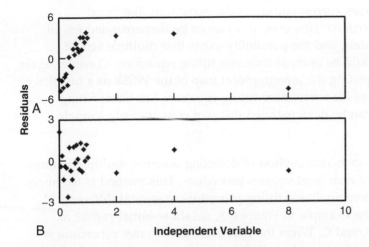

FIGURE 8-7.
The residuals for a straight line (panel A), or Eq. (8-15) (panel B), fit to the data in Figure 8-3.
The better fit gives more randomly distributed residuals.

this verification passes and the residuals do not follow a Gaussian distribution, then it is likely that the fitting equation is incorrect.

One of the most useful tests is to simply plot the residuals as a function of the independent and dependent variables. A visual inspection can usually provide an indication of a problem with the analysis. For example, Figure 8-7 presents the residuals for two different fits of the data shown in Figure 8-3. Panel B is a fit of Eq. (8-15), and it appears the residuals are random. By comparison, panel A corresponds to the fit of a straight line, and it is clearly not random. Thus, by inspection we can conclude this data cannot be described by a straight line.

There are many quantitative goodness-of-fit tests, such as the runs test, autocorrelation, and the Kolmogorov–Smirnov test. The runs test has proven to be very useful. A run is one or more residuals in a row with the same sign. Panel A of Figure 8-7 contains five runs: the first seven residuals are all negative; the 8th residual is positive; the 9th and 10th residuals are negative; residuals 11 to 23 are all positive; and the 24th is negative. Panel B contains 16 runs. This test is statistically based, and, if the residuals follow a Gaussian distribution, the expected number of runs and the variance of the expected number of runs can be computed. A Z score and probability are then calculated, determining the likelihood of the observed number of runs. More details for this and the other quantitative methods mentioned above can be found in Straume and Johnson (1992).

E. Uniqueness of the Least-Squares Values of the Parameters

For some types of fitting equations (i.e., linear models), it can be algebraically demonstrated that only a single set of unique model parameters exists corresponding to the maximum likelihood optimization criteria. However, this cannot be demonstrated for all nonlinear models, and the possibility exists that multiple sets of parameters could be optimal for some fitting equations. Consider again Figure 8-6 depicting the topographical map of the WSSR as a function of two parameters for a nonlinear fitting equation. For linear fitting equations, it can be demonstrated this plot contains only a single minimum.

Figure 8-8 presents one method of detecting whether multiple minima exist for a nonlinear least-squares procedure. This method is to simply start the iterative nonlinear fitting procedure at several different locations. In the example in Figure 8-8, we show initial *guesses* at locations A, B, and C. When the iterative least-squares procedure is started at either location A or B, the algorithm converges to the same minimum, but if the procedure is started at location C, the algorithm converges to a different minimum in the topographical map. Also note that a minimum exists that was not found when starting at these positions.

The potential for multiple minima always exists when fitting to nonlinear equations but, unfortunately, no computational method exists that will guarantee locating all of these minima. It is, however, common for some of the multiple minima to have parameter values that are physically unrealistic. For example, a negative molecular weight has no physical meaning. If multiple, physically meaningful minima are found, they must all be described in your report of the results.

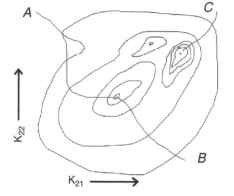

FIGURE 8-8.
This figure presents a topographical contour map of the variance-of-fit as a function of two parameters, K_{21} and K_{22}. In this example, three minimal points exist. Note that this is a nonlinear fitting equation and thus multiple minima can exist. For linear models, only a single minimum will exist.

VI. APPENDIX: BASIC MATRIX ARITHMETIC

In this chapter, we expressed a system of equations as a matrix equation and used matrix algebra to solve the system of equations. This is a convenient and common technique because hand-held calculators and computers are equipped to do matrix computations. In this appendix, we outline some basic matrix arithmetic that the reader needs to be familiar with in order to follow the matrix computations presented in the chapter.

A matrix is a rectangular array of numbers. An $m \times n$ matrix is one that has m rows and n columns. For example, the matrix $\begin{pmatrix} 1 & -1 & 0 \\ 2 & 4 & 5 \end{pmatrix}$ is a 2×3 matrix. Matrices are equal when they have the same dimensions and each corresponding entry is equal. The following arithmetic operations are fundamental to matrix arithmetic: addition, multiplication by a number, and multiplication of a matrix by a matrix (matrix multiplication). For our purposes, matrix multiplication is the most important operation.

We multiply a row matrix $(a_1, a_2, \ldots, a_n)$ by a column matrix $\begin{pmatrix} \varepsilon_1 \\ \varepsilon_2 \\ \vdots \\ \varepsilon_n \end{pmatrix}$ according to:

$$(a_1, a_2, \ldots, a_n) \begin{pmatrix} \varepsilon_1 \\ \varepsilon_2 \\ \vdots \\ \varepsilon_n \end{pmatrix} = a_1 \varepsilon_1 + a_2 \varepsilon_2 + \ldots a_n \varepsilon_n.$$

Note there are the same number of entries in both matrices. Also, the "row matrix" must be on the left, and the "column matrix" must be on the right. In the context of our discussions, we would like to think of the a_i's as numbers and the ε_i's as unknowns.

Now suppose we have two linear equations where the unknowns are $\varepsilon_1, \varepsilon_2, \ldots, \varepsilon_n$:

$$\begin{aligned} a_{11}\varepsilon_1 + a_{12}\varepsilon_2 + \ldots + a_{1n}\varepsilon_n &= b_1 \\ a_{21}\varepsilon_1 + a_{22}\varepsilon_2 + \ldots + a_{2n}\varepsilon_n &= b_2 \end{aligned}. \tag{8-60}$$

We define:

$$\begin{pmatrix} a_{11} & a_{12} & \ldots & a_{1n} \\ a_{21} & a_{22} & \ldots & a_{2n} \end{pmatrix} \begin{pmatrix} \varepsilon_1 \\ \varepsilon_2 \\ \vdots \\ \varepsilon_n \end{pmatrix} = \begin{pmatrix} a_{11}\varepsilon_1 + a_{12}\varepsilon_2 + \ldots + a_{1n}\varepsilon_n \\ a_{21}\varepsilon_1 + a_{22}\varepsilon_2 + \ldots + a_{2n}\varepsilon_n \end{pmatrix},$$

so we could write the system of Eq. (8-38) as the matrix equation.

$$\begin{pmatrix} a_{11} & a_{12} & \ldots & a_{1n} \\ a_{21} & a_{22} & \ldots & a_{2n} \end{pmatrix} \begin{pmatrix} \varepsilon_1 \\ \varepsilon_2 \\ \vdots \\ \varepsilon_n \end{pmatrix} = \begin{pmatrix} b_1 \\ b_2 \end{pmatrix}.$$

This is often written in the more compact form $A\varepsilon = b$, where A, b, and ε are the following matrices:

$$A = \begin{pmatrix} a_{11} & a_{12} & \ldots & a_{1n} \\ a_{21} & a_{22} & \ldots & a_{2n} \end{pmatrix}, \varepsilon = \begin{pmatrix} \varepsilon_1 \\ \varepsilon_2 \\ \vdots \\ \varepsilon_n \end{pmatrix}, b = \begin{pmatrix} b_1 \\ b_2 \end{pmatrix}.$$

Likewise, we could write the system of m linear equations in n unknowns $\varepsilon_1, \varepsilon_2, \dots, \varepsilon_n$:

$$
\begin{aligned}
a_{11}\varepsilon_1 + a_{12}\varepsilon_2 + \dots + a_{1n}\varepsilon_n &= b_1 \\
a_{21}\varepsilon_1 + a_{22}\varepsilon_2 + \dots + a_{2n}\varepsilon_n &= b_2 \\
&\;\;\vdots \\
a_{n1}\varepsilon_1 + a_{n2}\varepsilon_2 + \dots + a_{nn}\varepsilon_n &= b_n
\end{aligned}
\tag{8-61}
$$

as the matrix equation:

$$
\begin{pmatrix}
a_{11} & a_{12} & \cdots & a_{1n} \\
a_{21} & a_{22} & \cdots & a_{2n} \\
& & \vdots & \\
a_{m1} & a_{m2} & \cdots & a_{mn}
\end{pmatrix}
\begin{pmatrix}
\varepsilon_1 \\
\varepsilon_2 \\
\vdots \\
\varepsilon_n
\end{pmatrix}
=
\begin{pmatrix}
b_1 \\
b_2 \\
\vdots \\
b_m
\end{pmatrix}.
$$

This is often written in the more compact form:

$$
A\varepsilon = b.
\tag{8-62}
$$

We say that Eq. (8-62) is the matrix form of the system of Eq. (8-61).

If one examines what we have done, a requirement for the dimensions in the matrices appears, namely, an $m \times n$ matrix multiplying an $n \times 1$ matrix gives an $m \times 1$ matrix.

Everything we have done is a special (but very important) case of the following rules governing multiplication of matrices:

(i) If A is an $m \times n$ matrix and B is an $n \times k$ matrix, then AB is an $m \times k$ matrix; and

(ii) The entry in the i-th row and j-th column of the matrix AB is:

$$
(a_{i1}, a_{i2}, \dots, a_{in})
\begin{pmatrix}
b_{1j} \\
b_{2j} \\
\vdots \\
b_{nj}
\end{pmatrix}
= a_{i1}b_{1j} + a_{i2}b_{2j} + a_{in}b_{nj}.
$$

How does this help us solve a system of linear equations? Actually, we need to do one more thing before we can accomplish this. Assume that in the matrix Eq. (8-62), we know the entries of A and b and want to find the entries of ε. If there were a matrix A^{-1} for which $A^{-1}A\varepsilon = \varepsilon$, then multiplying both sides of Eq. (8-62) by A^{-1} would give the solution

$$
\varepsilon = A^{-1}A\varepsilon = A^{-1}b.
$$

If A is a square matrix (i.e., has the same number of rows as columns), then sometimes there is such a matrix A^{-1}, called the inverse of A. Thus, we somehow need to create a square matrix in Eq. (8-62) in order to solve for ε. We now describe how to do this.

Associated with each matrix A is its transposed matrix A^T. The matrix A^T is obtained by forming the matrix whose rows are the columns of A.

Thus if, for example, $A = \begin{pmatrix} 1 & 4 \\ 2 & 5 \\ 3 & 6 \end{pmatrix}$, then $A^T = \begin{pmatrix} 1 & 2 & 3 \\ 4 & 5 & 6 \end{pmatrix}$. Notice, if A is an $m \times n$ matrix, then A^T is an $n \times m$ matrix, so A^TA ($n \times m$ multiplied by an $m \times n$) is an $n \times n$ matrix, and AA^T ($m \times n$ multiplied by an $n \times m$ matrix) is an $m \times m$ matrix. Thus, either product is a square matrix.

There is a possibility that $(A^TA)^{-1}$ exists (again, computers will routinely check this), and, if so, we can solve in the following way. First multiply both sides of the equation by A^T from the left to get

$$A^TA\varepsilon = A^Tb. \tag{8-63}$$

Next, find the inverse $(A^TA)^{-1}$ and multiply both sides of Eq. (8-63) by this matrix from the left to obtain the vector of the unknowns ε:

$$(A^TA)^{-1}(A^TA)\varepsilon = (A^TA)^{-1}A^Tb$$

$$\varepsilon = (A^TA)^{-1}A^Tb. \tag{8-64}$$

Therefore, if the inverse matrix $(A^TA)^{-1}$ exists, the solution ε of Eq. (8-62) is given by Eq. (8-64).

REFERENCES

Efron, B., & Tibshirani, R. J. (1993). *An introduction to the bootstrap.* New York: Chapman and Hall.

Johnson, M. L., & Frasier, S. G. (1985). Nonlinear least-squares analysis. In Hirs, C. H. W., & Timasheff, S. N. (eds.), *Methods in Enzymology* (vol. 117, pp. 301–342). New York: Academic Press.

Straume, M., & Johnson, M. L. (1992). Analysis of residuals: Criteria for determining goodness-of-fit. In Brand, L., & Johnson, M. L. (eds.), *Methods in Enzymology* (vol. 210, pp. 87–105). New York: Academic Press.

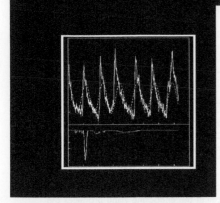

Chapter 9

ENDOCRINOLOGY AND HORMONE PULSATILITY

Introduction

Experimental Design, Data Collection, and Errors of Measurement

Classical Methods for Analyzing Hormone Concentration Time Series

Deconvolution Methods

I am turned into a sort of machine for observing facts and grinding out conclusions.

Charles Darwin (1809–1882)

In single-celled organisms, all life functions, such as metabolism, response to stimuli, and reproduction, are of necessity performed by the cell itself. In multicellular organisms, groups of cells become specialized to perform particular functions. The proper functioning of multicellular organisms, therefore, requires efficient mechanisms for cell-to-cell communication for controlling and coordinating the actions of disparate and often distant cell types. In mammals, communication functions are performed by the nervous and endocrine systems.

The endocrine system controls important physiological processes, including growth, metabolism, reproduction, and development, by means of secreted chemical agents called *hormones* that are distributed throughout the body by the bloodstream. An endocrine communication pathway is diagrammed in Figure 9-1. Endocrine communication is composed of three parts: (1) Endocrine glands containing secretory cells; (2) the hormones they secrete; and (3) the cells that are the targets of the secreted hormones. Although each hormone comes in contact with multiple cell types after its secretion, it only influences those targeted cells with appropriate receptors for that hormone.

Hormone secretion patterns are determined by the frequency of secretion events, the amount secreted, and the length of time the secretion event lasts. They encode messages for the target cells that control vital physiological processes, and an alteration of a secretion pattern may impede one or more of these processes. Understanding hormone secretion and developing the capability to recognize both normal and pathological patterns of hormone production is of utmost importance for establishing medical diagnoses, initiating treatment, and assessing the effects of treatment.

We begin with a brief introduction to the mechanisms of the human endocrine system. The reader is encouraged to refer to a textbook of human physiology (such as Guyton and Hall [2005]) or endocrinology (such as Williams et al. [2002]) for additional information. We then discuss

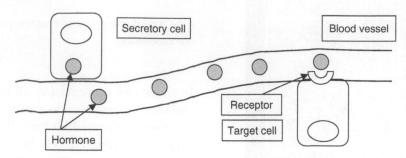

FIGURE 9-1.
Simple model of endocrine function. The secretory cell in the endocrine gland produces the hormone that is carried by the bloodstream. The hormone binds to a receptor on or in the target cell. Hormone binding causes a change in the behavior of the target cell.

designing experiments, collecting data, and analyzing hormone secretion patterns.

So far, our focus has been primarily on model development and data analyses, largely ignoring the data-collection process. For two reasons, however, data collection is not straightforward when determining hormone secretion patterns. First, it is generally not possible to collect data directly from the endocrine glands, where the hormones are secreted, for reasons discussed below. Instead, information about hormone secretion is inferred from data representing the hormone concentration in the blood. Second, even these data do not directly give an accurate picture of the secretion patterns, because once the hormone is secreted and has entered the bloodstream, its physiological elimination from the blood (because of binding, excretion, and/or biotransformation) begins immediately. Figuratively speaking, such data only provide a glimpse through a "dirty window," as the hormone secretion patterns are distorted because of the ongoing elimination processes. We need to "clean" the window by removing the effect of hormone elimination in order to be able to "see" the actual secretion amplitudes and frequencies.

We shall present a number of mathematical methods, both classical and novel, aimed at quantifying various aspects of this problem. The methods are divided into two groups. The first covers statistical approaches for analysis of hormone concentrations as observed in the blood. The second employs deconvolution methods to deduce hormone secretion patterns from the hormone concentration in the blood. Although some of the classical methods, such as the Fourier methods outlined in this chapter, are now known to be of limited use for analyzing general hormone data, we have included them because they are still routinely used in the literature to analyze specific aspects of hormone pulsatility.

Several of the mathematical methods we describe require data in the form of a *time series* (i.e., measurements of hormone concentration equally spaced in time). For example, data may be collected every 10 minutes or every hour, because equally spaced data points are perceived as best for capturing the dynamic nature of the secretion processes. This procedure, however, may be too restrictive at times, and it is also common for data points to be taken at times not separated by intervals of equal length. In describing mathematical methods, we shall be careful to separate those requiring time series data from those that do not. The latter class of methods is certainly more general.

Throughout the chapter, we use two sets of actual hormone concentration data sets, one for luteinizing hormone (LH) and one for growth hormone (GH, also called somatotropin). These are then examined using several analytical approaches. Software and documentation for many of these algorithms can be downloaded from Dr. Michael Johnson's site at the University of Virginia (see Internet Resources at the end of this chapter).

I. INTRODUCTION

The endocrine system consists of endocrine glands and the hormones they secrete. Hormones reach their target cells by traveling through the bloodstream. Figure 9-2 shows the location of the major endocrine glands of the human body.

The pituitary gland, often called the "master gland," is located at the base of the brain. It receives signals from the hypothalamus, a neurosecretory region of the brain. Figure 9-3 shows the positional relationship of the pituitary and hypothalamus. The hypothalamus enables communication between the nervous system and the endocrine system by producing releasing hormones and inhibiting hormones that act on the anterior pituitary. It also produces two hypothalamic hormones that are stored in, and released from, the posterior pituitary. Acting under this hypothalamic control, the anterior pituitary in turn produces hormones that themselves stimulate other endocrine glands.

The term *hormone* was originally applied to chemical substances secreted by endocrine glands and transported in the bloodstream to regulate the activity of *distant* target organs. This is the classical *endocrine* action. However, there are many other cell-signaling substances not produced by endocrine glands that have similar effects. Examples include the insulin-like growth factor type I (IGF-I) secreted by the liver; the histamines released by mast cells in response to injury, infection, or allergy; and the releasing hormones secreted by the hypothalamus. These other forms of chemical communication are called autocrine, paracrine, and neuroendocrine regulation, and are now generally grouped under the heading of endocrine signaling. *Autocrine* action occurs when a cell both secretes and has receptors for a regulatory

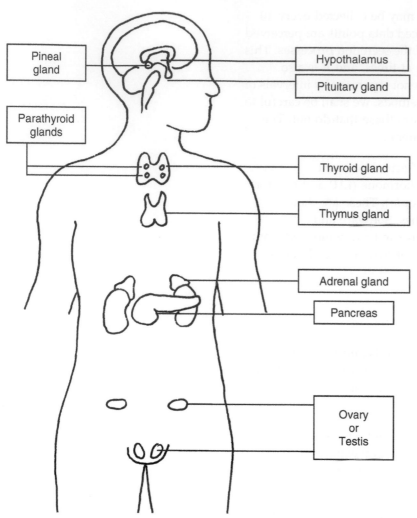

FIGURE 9-2.
Human endocrine glands. The parathyroid glands are located behind the thyroid gland, and the adrenal glands are located just above the kidneys.

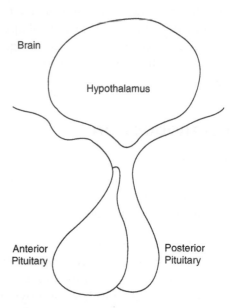

FIGURE 9-3.
Relationship between the hypothalamus (a region of the brain) and the pituitary gland. The anterior and posterior lobes of the pituitary have different functions, as noted in the text.

chemical. The secreted regulator enters the intercellular fluid and then binds the receptors, resulting in the regulation of the function of the *same* cell. *Paracrine* regulation involves release of a molecular signal that diffuses through the intercellular fluid and interacts with specific receptors on other *nearby* cells. *Neuroendocrine* regulation occurs when a neuroendocrine cell releases neurohormones into the bloodstream and these bind to target cell receptors.

Hormones are classified on the basis of their structure. *Peptide* or protein hormones (such as insulin, LH, or GH) are produced, like other proteins, on ribosomes. They are stored within the cells in secretory vesicles and are released by exocytosis in response to an appropriate stimulatory signal. They act by binding to specific receptors on the surface of target cells. In contrast, *steroid* hormones (such as mineralocorticoids, glucocorticoids, and sex steroids) are released immediately and not stored intracellularly. They are synthesized in response to an appropriate signal in different organs from a common precursor (cholesterol) and exert their action by diffusing through the plasma membrane of target cells and binding to intracellular receptors. The *fatty acid derivatives* include the prostaglandins, derived from the 20-carbon fatty acid arachidonic acid. Prostaglandins function through paracrine regulation and are the targets of the nonsteroidal, anti-inflammatory drugs, such as aspirin or ibuprofen. The *amines* (such as catecholamines and thyroid hormones) are amino acid derivatives, mostly synthesized from tyrosine residues, although melatonin is produced from tryptophan. Catecholamines (such as dopamine or norepinephrine) are the neurotransmitters of the autonomic nervous system, whereas the thyroid hormones affect almost every tissue, exerting growth, cardiovascular, and metabolic effects. Melatonin is an important regulator of circadian rhythms.

Once released, hormones are transported in the bloodstream, either free or bound to specialized carrier proteins. Even though only free hormones exert the effect, carrier proteins are important because they effectively modify the apparent kinetics of the active hormone. After reaching their target cells, hormones bind to specific receptors, which are protein molecules located on the membrane or inside the cell. This initiates a cascade of events culminating in a biological response specific to the target tissue.

We now provide some background about the mechanisms controlling the production of two pituitary hormones, LH and GH, which we shall be examining. In both cases, their production is controlled by neurohormones produced by the hypothalamus. We begin with LH.

Neurosecretory cells of the hypothalamus produce gonadotropin-releasing hormone (GnRH)—a short, 10–amino-acid peptide. The GnRH is transported through the portal veins to the anterior lobe of the pituitary, where the GnRH binds to its receptors on the pituitary cells. These cells respond by producing the gonadotropins LH and the follicle-stimulating hormone (FSH). LH and FSH enter the bloodstream from the pituitary, travel to the gonads, and subsequently exert their effects upon their target cells. Although they were identified and named in females, both LH and FSH also function in males. As in this chapter we shall be looking at the effects of these hormones on female fertility, however, we now focus on their activity in women.

In the ovary, FSH and LH stimulate the development of ovarian follicles, the production of the steroid hormones estrogen and progesterone, and ovulation. The ovaries regulate the production of FSH by releasing inhibin, which inhibits FSH production by the pituitary. The ovaries also exert control through the production of estrogen, which affects both the production of LH and FSH by the pituitary and the production of GnRH by the hypothalamus. After ovulation, LH supports the development of the empty ovarian follicle into the *corpus luteum* (yellow body), which secretes estrogen and progesterone to support the endometrium, and inhibin to block the production of FSH by the pituitary. The rising levels of progesterone and estrogen block the production of GnRH, LH, and FSH, preventing the induction of follicular development until the result of the previous ovulatory event is apparent. If fertilization and implantation occur, the *chorion* (one of the extraembryonic membranes of the new embryo) assumes responsibility for the maintenance of the endometrium. If not, the degeneration of the corpus luteum ensures that the endometrium is shed, and the cycle begins again. These events are summarized in Figure 9-4.

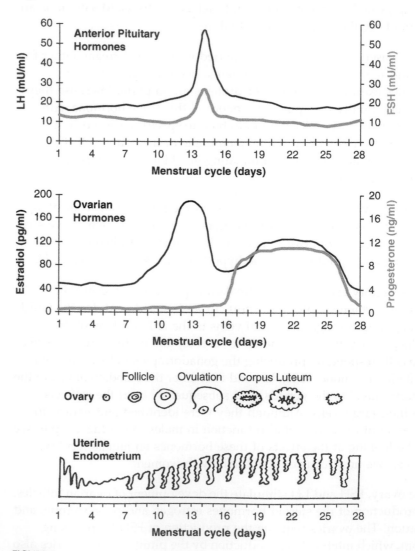

FIGURE 9-4.
Human menstrual cycle. The upper panels show the levels of the hormones LH, FSH, estrogen, and progesterone; and the lower panel shows the response of the ovarian follicle and uterine lining.

We turn now to GH. Like LH and FSH, GH is produced by the anterior pituitary in response to a hypothalamic releasing factor, the growth hormone–releasing hormone (GHRH). GHRH reaches the pituitary via the same circulatory system pathway taken by GnRH, where it binds its target cells and stimulates the release of GH into the bloodstream. The hypothalamus also produces an inhibitory substance, growth hormone–inhibiting hormone (GHIH, also called somatostatin), that decreases the production of GH by the pituitary. Generally speaking, if the levels of GH are low, the hypothalamus will produce GHRH. If the levels of GH are high, the hypothalamus will produce GHIH.

In summary, the endocrine system can be described as a complex of signaling mechanisms, directing and coordinating multiple functions in the organism. A remarkable feature of the system is its critical dependence upon the pattern of hormone release that encodes information necessary for the signaling mechanisms. *Although the average hormone levels are essential to the performance of the endocrine system, achieving these levels through secretion events with exact frequencies and amplitudes is of crucial importance.* For example, the same average concentration of a hormone can be achieved by a few secretion events with large amplitudes, by numerous frequent pulses of small amplitudes, or by an appropriate mix of large and small hormone releases. Although the average hormone concentration in all of these cases may be the same, the different profiles of the secretion events would represent different examples of endocrine signaling, only one of which would be functional in any given situation.

Serum levels of the pituitary hormones FSH and LH, as well as the ovarian hormones estrogen and progesterone, vary considerably over a normal 28-day menstrual cycle. Pituitary hormones, such as GH, prolactin, thyrotropin, adrenocorticotropic hormone, FSH, and LH, are secreted in a *pulsatile* manner (Veldhuis et al. [1987]), with their levels rising and falling multiple times per day because of bursts of secretion by the pituitary followed by periods of secretory inactivity. The number of secretory bursts per day varies with the hormone in question, the age, the gender, and the health of the individual. Figure 9-5 shows an example of the variation of blood serum LH levels over 24 hours in a healthy woman of reproductive age. LH levels rise and fall repeatedly each day, and this variation is vital to the performance of reproductive functions. Serum levels of GH for a healthy adult exhibit similar behavior over the course of a 24-hour period (see Figure 9-6).

The pulsatile nature of hormone release may easily be overlooked if hormone levels are measured on a timescale with nonoptimal resolution. Almost any introductory biology text would likely have a figure

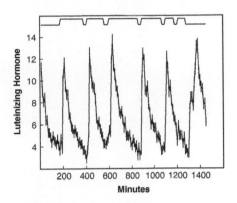

FIGURE 9-5.
A typical example of the variation of serum LH levels over 24 hours in a healthy woman of reproductive age sampled every 10 minutes. Individual data points are represented as vertical error bars corresponding to ± one standard error of the mean value (SEM), where the SEM is a function of the hormone concentration. Hormone levels rise and fall multiple times over the course of a single day. The SEMs were evaluated with Eqs. (9-1) and (9-2) introduced below, assuming an MDC = 1.0 and CV = 5.0%. The units for LH are mIU/ml. The upper sawtooth pattern represents peaks in the concentration time series as evaluated by the *CLUSTER* algorithm we shall describe later.

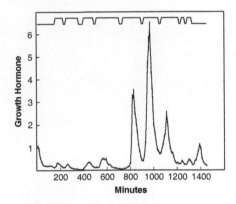

FIGURE 9-6.
A typical example of how serum growth hormone (GH) concentration changes over a 24-hour period in a normal healthy adult. The GH was measured every 10 minutes. The individual data points are represented with vertical lines of length corresponding to ± one SEM. Note that the SEM is a function of the hormone concentration. The SEMs were evaluated with Eqs. (9-1) and (9-2) introduced below, assuming an MDC = 0.0265 and CV = 8.23%. The units for GH are ng/ml. The levels of GH rise and fall multiple times over the course of a single day. The large pulses of GH occur during the sleep periods. The upper sawtooth pattern represents peaks in the concentration time series as evaluated by the *CLUSTER* algorithm we shall describe later.

similar to Figure 9-4. It shows the blood serum levels of FSH, LH, estrogen, and progesterone, and the events in the ovaries and uterine lining during a menstrual cycle. Unfortunately, Figure 9-4 gives the impression that hormone levels rise and fall smoothly on a timescale of days, which is incorrect. Instead, pituitary hormones are produced in a pulsatile manner; if they are not, the functions of the respective endocrine signaling pathways will be inhibited. For instance, it has been shown that the pulsatile nature of the GnRH signal is critical to its function of stimulating LH and FSH release. Administering the peptide as a nonvarying infusion not only fails to stimulate but, in fact, diminishes the response. If a woman's LH and FSH levels remained approximately steady over any given 24-hour period (as suggested by Figure 9-4), she would be infertile. Therefore, clinicians have used metering pumps to treat female infertility by administering GnRH in a normal pulsatile pattern in order to restore the signal and subsequently restore reproductive function.

A similar approach has been successful in cases of primary hypothalamic failure, such as Kallmann syndrome or hypothalamic amenorrhea, both of which are marked by a GnRH deficiency. The technique has also been used to *inhibit* gonadotropin secretion by means of long-acting, nonpulsatile GnRH analogues for the treatment of prostate cancer, premature puberty, or endometriosis. In either case, the pulsatile nature of the GnRH release is required for the pituitary to respond normally. However, in the latter three examples, the nonpulsatile GnRH will act to suppress the release of the gonadotropins LH and FSH by the pituitary. This results in growth suppression in prostate cancer, suppression of inappropriately early maturation in premature puberty, and suppression of symptoms in endometriosis. In summary, the GnRH–LH/FSH axis provides an impressive example of how obtaining specific quantitative knowledge about hormone secretion dynamics not only addresses important theoretical questions about the signaling mechanism of the endocrine system, but also results in successful therapeutic strategies. Children suffering from premature puberty have been restored to a normal developmental pattern, and previously infertile women have achieved normal pregnancies.

The techniques outlined above require objective quantitative methods for describing and comparing individual or multiple hormone secretion profiles. In particular, it is necessary to have (1) an objective definition of physiologically important properties of hormone concentration dynamics, (2) formal techniques for quantification of these aspects, and (3) standardized methods for comparison of hormone secretion data. A major challenge, however, is the high complexity of endocrine axes and the lack of sufficient direct experimental data. Direct measurement of some hormones of interest is either experimentally challenging or impossible, or unethical, especially in humans.[1] Scientists therefore face the problem of deriving the function and/or structure of an extremely complex biological system from a limited set of experimental data that frequently contain large uncertainties. The quantitative methods developed to support these efforts can be broadly divided into two types. *Statistical* methods are data-driven and provide an objective approach to deriving hormone secretion patterns from individual hormone time series. *Mathematical modeling* methods are based on physiological evidence and hypotheses regarding hormone secretion mechanisms. These two methods are not unrelated. Although some statistical methods are model-independent, other statistical methods frequently utilize mathematical models as part of the data analyses, and the results of a mathematical model are often validated through statistical data analyses.

In this chapter, we focus on the statistical approach. Mathematical modeling of hormone networks is discussed in the next chapter.

II. EXPERIMENTAL DESIGN, DATA COLLECTION, AND ERRORS OF MEASUREMENT

The primary objective when analyzing hormone data is to characterize the intrinsic pulsatile and nonpulsatile nature of the secretory process. These characteristics include the number, times, and masses of the pulsatile secretory events that increase the hormone concentration in the blood and the amount of basal (nonpulsatile) secretion. Before data collection is initiated, a carefully designed experimental protocol should be in place. Among other things, the protocol should specify the detailed mechanism of data collection, including whether repeated measurements will be used and specifying how often the measurements will be taken.

Because direct data collection at the endocrine glands where the hormones are secreted is often impossible, measurements of the hormone concentration in the blood are used to reconstruct the secretion patterns. This reconstruction is necessitated by the fact that the observed hormone concentrations in blood as a function of time are the result of a combination of hormone secretion into the blood and pharmacokinetic removal of the hormone from the blood. In addition, because the exact timing of hormone secretion events changes between closely matched subjects, and even when the same subject is resampled days or months later, considering repeated measurements is appropriate. Finally, an extremely important property of the data is the precision of the measured hormone concentrations, also known as *the*

---

1. Recall that hormones such as GnRH are secreted in tiny amounts by the hypothalamus, which is part of the brain. It is not feasible to directly measure the GnRH as it is secreted.

variance model of the data. The data points cannot all be measured with the same precision, and thus the analysis method must include weighting factors, as discussed in Chapter 8, to better describe the variable precision of the individual data points.

The data shown in Figures 9-5 and 9-6 were obtained by collecting blood samples from normal, healthy human volunteers every 10 minutes for a total of 144 data points in 24 hours. Each of the blood samples was assayed twice by a clinical laboratory. Hence, each of the data points in Figures 9-5 and 9-6 consists of four values:

1. The mean measured hormone concentration (Y_i);

2. The precision of the measured hormone concentration (SEM_i);

3. The time the blood sample was collected (X_i); and

4. The number of replicate assays.

Recall that the standard error of the mean (SEM) at each point is computed as the standard deviation of all readings at the data point (in this case, all readings at the specified time) divided by the square root of the number of repeated measurements. Given 15 or 20 repeated measurements on each data point, the values SEM_i are routinely used as estimates for the precision of measurements and to form the weighted sum of squared residuals (WSSR) with weights inversely proportional to SEM_i, as described in Chapter 8.

However, with only two or three replicates, as is the case of the data sets in Figures 9-5 and 9-6, the SEM of the replicates is far too inaccurate to be used as a realistic estimate of the precision of the hormone concentrations. Why were more repeated measurements not collected? For an optimal statistical analysis, it would also be better to collect many more data points than the 144 data points presented in Figures 9-5 and 9-6. Why were data points not collected over the course of multiple 24-hour periods?

The answers to these questions illustrate the competing factors researchers face when designing an experiment. On one hand, the more data points collected, the more accurate the statistical analyses will be. On the other hand, health and cost limitations may become a serious issue. For example, measuring each of the time points in Figures 9-5 and 9-6 requires a certain minimal volume of blood, and there are limits to the total amount of blood that can be safely drawn from a person. Each of these data points is also very expensive to collect and process; the typical cost of a clinical laboratory assay is approximately $50 per sample. Figure 9-5 has 144 data points, and each was assayed twice, so the cost for the assays alone was approximately $15,000. This does not include the time clinical staff spent collecting the blood samples, the cost of the hospital bed for the volunteer during the study, and payments to the volunteer. In addition, occasional experimental mishaps, such as failure to draw blood at a specified time, invalid assay readings, or contaminated samples, may occur. It would be unreasonable to assume that such mishaps require discarding all previously collected data points, especially for expensive data-collection protocols. Instead, statistical analyses that allow for missing values in the time-series data should be considered.

As noted, with only two or three replicates, the SEM cannot be used as an estimate of the precision of the hormone concentrations. As a rule of thumb, with less than 15 replicates a variance model must be created based upon the performance characteristics of the clinical laboratory assays. We present one such model next.

The minimal detectable concentration (MDC) is the lowest concentration that can be measured accurately. It is experimentally calculated as twice the SD of about 15 or 20 samples containing a hormone concentration of zero. Clinical laboratories will commonly report hormone levels that are less than the MDC as being too low to measure accurately. This creates serious problems for the proper analysis of hormone concentration time series. The algorithms require a numerical value for the concentration, not a "too low to measure." If these "too low to measure" values are replaced with 0.00, the analysis procedures are forced to find that basal (nonpulsatile) secretion does not exist. Yet, if these values are replaced with the MDC, the analysis procedures will incorrectly find a basal secretion yielding a concentration equal to the MDC. Also, if these values are treated as missing values, then valuable information (that the value is "too low to measure") is being neglected. Thus, the best treatment of "too low to measure" values, is to force the clinical laboratories to report the actual small value with a large experimental uncertainty.

To estimate the experimental uncertainties in such cases, a variance model must be created that accounts for certain performance characteristics of the clinical laboratory assays. One way to build such a model is to describe the way the experimental uncertainties of the assays change as the hormone concentration changes. Figure 9-7 depicts a typical variance model for hormone concentration assays expressed in terms of the coefficient of variation (CV). The CV is defined as the SD of the measurements divided by their mean value. A large value for the CV would indicate that the measured value might not represent the actual value accurately. This is more likely to happen at very low hormone concentrations below the MDC or at high hormone concentrations outside of the optimal range of the laboratory assay. In Figure 9-7, the CV increases to infinity at hormone concentrations approaching zero, decreases to a plateau in the optimal hormone

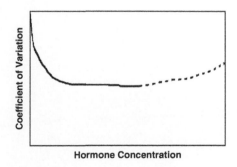

FIGURE 9-7.
A typical variance model, the coefficient of variation (CV), as a function of the hormone concentration. The CV is the standard error of the mean divided by the mean hormone concentration.

concentration range for the assay, and then increases when the concentration rises above the optimal range (dashed line).

When a concentration is significantly above the optimal range, the sample is usually diluted to fall within the optimal range and assayed again. Consequently, the dashed region of Figure 9-7 (and the associated experimental inaccuracies) can be dealt with by sample dilution. For low hormone concentrations, however, a mathematical model should be used to estimate the variance of the data.

One commonly used model for an empirical representation of the variance as a function of the hormone concentration is given in Eq. (9-1):

$$\text{Variance}([\text{Hormone}]) \approx \left[\left(\frac{MDC}{2} \right)^2 + \left(\frac{CV[\text{Hormone}]}{100} \right)^2 \right]. \qquad (9\text{-}1)$$

Here, CV is the percent coefficient of variation, and [Hormone] is the hormone concentration. Because the MDC and CV are routinely measured as a part of the quality control measurements performed by clinical laboratories and their values are well known, Eq. (9-1) can be used to provide a realistic estimate of the variance of the data. Once the variance is approximated, the SEM for the measured hormone concentration can be approximated by:

$$SEM \approx \sqrt{\frac{\text{Variance}([\text{Hormone}])}{N}}, \qquad (9\text{-}2)$$

where N is the number of replicates. Notice that Eq. (9-2) can now be used to provide a realistic estimate of the precision of the measured hormone concentration, even with a single measurement of the concentration.

To justify the choice of the variance model defined by Eqs. (9-1) and (9-2), recall that when more than 15 measurements are available for a data point, the SEM is calculated as:

$$SEM = \sqrt{\frac{\text{Variance}}{N}} = \frac{SD}{\sqrt{N}}.$$

In this expression, N is the number of measurements available for the data point, and SD is the standard deviation for those measurements. Thus, Eq. (9-2) estimates the SEM by using an approximation for the variance that cannot otherwise be accurately calculated because of the insufficient number of replicate measurements for each data point.

To substantiate Eq. (9-1), recall that at zero concentration, MDC is defined as MDC = 2(SD), where SD is the standard deviation at hormone concentration zero. Thus, at hormone concentration zero, the second term of the sum in Eq. (9-1) disappears, and we obtain

$$\text{Variance}([0]) = (\text{MDC}/2)^2 = (\text{SD})^2,$$

as expected. When the hormone concentration is not zero, the second term of the sum in Eq. (9-1) accounts for the additional variation at any given concentration that depends on the choice of the CV from Figure 9-7.

To summarize, in determining hormone pulsatility properties from experimental data, it is important to use analysis methods that can accommodate small numbers of data points with variable uncertainties and allow for missing values. We now discuss some of the computational procedures specifically developed for the analysis of hormone concentration time-series data. Although the topic of hormone pulsatility may seem rather specialized, many of the mathematical and statistical challenges are representative of a much broader class of quantitative problems.

We begin with some standard time-series analyses for detecting periodic behavior.

III. CLASSICAL METHODS FOR ANALYZING HORMONE CONCENTRATION TIME SERIES

A. Fourier and Power Spectrum Methods

Visual inspection of Figure 9-5 indicates the LH secretory events to be approximately equally spaced in time and of equal height. That is to say, the LH secretory profile in Figure 9-5 may appear to be *periodic*. To review the terminology, a function $G(t)$ is called periodic with period T, if for any value of t:

$$G(t) = G(t + T),$$

and T is the smallest positive constant for which the condition is satisfied.

The most common periodic functions are $G(t) = \sin(t)$ and $G(t) = \cos(t)$, and their period is $T = 2\pi$; that is, $\sin(t + 2\pi) = \sin(t)$ and $\cos(t + 2\pi) = \cos(t)$. Also, the variable t usually represents the time elapsed since a fixed moment in time.

EXERCISE 9-1

Show that each of the functions $h(t) = \sin(2\pi t)$ and $r(t) = \cos(2\pi t)$ has a period of $T = 1$.

EXERCISE 9-2

Show that the functions $h(t) = \sin(2\pi t/5)$ and $r(t) = \cos(2\pi t/5)$ have period $T = 5$.

To generalize the two previous exercises, if L is any real number, the functions $h(t) = \sin(2\pi t/L)$ and $r(t) = \cos(2\pi t/L)$ have period L. If L is the period of a function, its *frequency* (often denoted by ν) is defined by $\nu = 1/L$. The functions $h(t) = \sin(2\pi t/L)$ and $r(t) = \cos(2\pi t/L)$ have period L and frequency ν.

Often, the phenomenon in question may not be truly periodic but, instead, result from the compound effect of several factors, where each factor may itself be nearly periodic with different factors having different periods. For instance, to study the temperature of the Chesapeake Bay coastal water, we would have to consider daily and yearly cycles. We could expect that any measurable function of time $G(t)$ for the temperature will be a sum of (at least) two periodic functions with periods equal to 24 hours and 365 days, respectively: $G(t) = D(t) + Y(t)$. Furthermore, sun spot activity varies over an 11-year cycle; if its effects on temperature are to be taken into account, the temperature may be written as $G(t) = D(t) + Y(t) + S(t)$, with the function $S(t)$ having a period of 11 years.

When an observable function is a sum of periodic functions, or a sum of multiples of periodic functions, we say that the observable function is a linear combination of periodic components. In such cases, it is important that the periods or, equivalently, the frequencies of the different components, be identified from the observable function. This is accomplished by *Fourier transforming*. When applied to the function $G(t)$, the Fourier transform generates a function $f(\nu)$, called a *power spectrum function*, where ν denotes the frequency. For a given function $G(t)$, the power spectrum function gives a plot of the portion of a signal's power (energy per unit time) falling within given frequency bins (see Grafakos [2004] for the mathematical details).

The major practical problem is that the analytic form of the function $G(t)$ is rarely available. In most cases, we only have a set of discrete measurements of $G(t)$. Under such conditions, discrete approximations of the Fourier transform function are used as an approximation to the continuous Fourier transform. Various numerical methods are available for solving the problem. The most common of these methods is an algorithm developed by Tukey and Colley in 1965 called *Fast Fourier Transform* (FFT). Computer software systems, such as *MATLAB* and *BERKELEY MADONNA*, provide implementations of the algorithm. Our goal here is to describe how information about $f(\nu)$ can be used to identify the prevalent frequencies in the composition of the function $G(t)$.

Once the power spectrum function $f(v)$ is calculated, it provides information about the periodic components of the original function $G(t)$. Suppose, for example, a function $G(t)$ is exactly periodic with period L. More precisely, suppose we can write $G(t)$ as:

$$G(t) = a\cos(2\pi t/L) + b\sin(2\pi t/L),$$

where a and b are constants. The power spectrum function $f(v)$ will then have a peak at a frequency $v = 1/L$. This, of course, is only true if we have collected data over a period of time that allows periodic behaviors to have expressed themselves (i.e., the data should be collected over a sufficiently long time period T).

A sample graph for $f(v)$ is shown in Figure 9-8. Because the connection between the frequency v and the period L is immediate and given by $v = 1/L$, the graph of the power spectrum could be expressed as a function of v or as a function of L (panels A and B of Figure 9-8, respectively). In endocrinology, the period is more commonly used than the frequency.

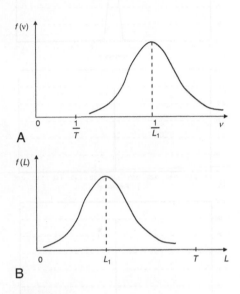

A

B

FIGURE 9-8.
The power spectrum as a function of the frequency (panel A) and the period (panel B).

Two things are important here:

1. The width of the peak depends on $1/T$, where T is the time interval over which the data are collected. Typically, collecting data over longer periods of time will make the peak clearer and better expressed.

2. For a fixed value T, the height of the peak is proportional to $a^2 + b^2$. Thus, increasing the magnitude of the coefficients a and b will make the peak higher.

Assume now that $G(t)$ is a sum of N sine and N cosine waves with periods $L_1, L_2, \ldots, L_N$. That is:

$$G(t) = a_1\cos\left(\frac{2\pi t}{L_1}\right) + b_1\sin\left(\frac{2\pi t}{L_1}\right) + \cdots + a_N\cos\left(\frac{2\pi t}{L_N}\right) + b_N\sin\left(\frac{2\pi t}{L_N}\right)$$

$$= \sum_{i=1}^{N} a_i\cos\left(\frac{2\pi t}{L_i}\right) + b_i\sin\left(\frac{2\pi t}{L_i}\right).$$

(9-3)

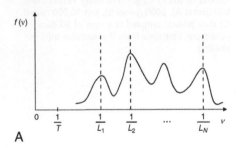

A

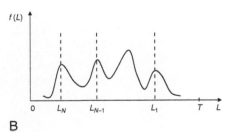

B

FIGURE 9-9.
The peaks in the power spectrum identify the prevalent frequencies (panel A) or periods (panel B).

In this case, the graphs of the power spectrum function are given in Figure 9-9. As before, the peaks are at the inverse periods $v_i = 1/L_i$ in panel A and at the periods L_i in panel B. The heights of the peaks are proportional to $a_i^2 + b_i^2$.

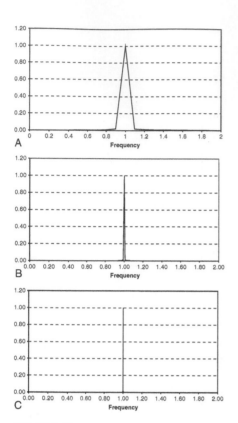

FIGURE 9-10.
Results of the FFT algorithm using, respectively, 500 (panel A), 5000 (panel B), and 50,000 (panel C) data points, sampled at a rate of 50 data points per unit time from the function $m(t) = \sin(2\pi t)$.

According to the *Fourier Theorem*, a broad class of time series can be represented as a sum of sine and/or cosine waves of the form given by Eq. (9-3) with periods L, $L/2$, $L/3$, and so on, where L is the maximal time range for the time series. One important prerequisite is that the time series does not show an overall trend over time. For example, the time series in Figures 9-5 and 9-6 do not appear to exhibit a pronounced upward or downward trend. Conversely, a time series containing height or weight data of a healthy child over several years will exhibit an upward trend. Time series that preserve their statistical characteristics such as mean and variance over time are called *stationary*.[2] Thus, a time series that exhibits a trend cannot be stationary.

The *power spectrum function* generated by the FFT is the sum of the squares of the amplitudes of these sine and/or cosine terms for each frequency (or period). The classical method for analysis of time-series data is to analyze the power spectrum of the data and identify the peaks, as illustrated in Figure 9-9. The peaks of the highest amplitude represent the prevalent frequencies/periods in the composition of the time series.

Consider, as an example, the function $m(t) = \sin(2\pi t)$ sampled 500 times over the interval [0,10] at equally spaced time instances. A FFT analysis should then identify the unique period $L = 1$, corresponding to a frequency $v = 1$, which is apparent in Figure 9-10(A). Sampling at the same sampling rate over a longer time interval (e.g., [0,100] and [0,1000] in Figures 9-10(B) and (C), respectively) sharpens the peak, more clearly identifying the dominant frequency, as expected.

The result of the FFT algorithm using, respectively, 500, 5000, and 50,000 data points, sampled at a rate of 50 data points per unit time, from the function $m(t) = \sqrt{2}\sin(2\pi t) + \sqrt{5}\cos(3\pi t) + \dfrac{1}{\sqrt{2}}\sin(7\pi t)$ is given in Figure 9-11. In general, if Y_i represents a discrete stationary time series of N data points at time values X_i ($i = 1,2,\ldots,N$), then Y_i can be exactly represented as the sum of sine and cosine waves[3] with a period L equal to the maximal time range (i.e., $T = 1440$ minutes for Figures 9-5 and 9-6) and where N is an odd number as:

$$Y_i = \frac{a_0}{2} + \sum_{n=1}^{(N-1)/2} \left(a_n \cos\left[\frac{2\pi n}{L} X_i\right] + b_n \sin\left[\frac{2\pi n}{L} X_i\right] \right). \qquad (9\text{-}4)$$

When N is an even number, the sum is from 1 to $(N-1)/2$ and either $a_{(N-1)/2} = 0$ or $b_{(N-1)/2} = 0$. Considering the power spectrum of such

---

2. The exact mathematical definition can be found in Box et al. (1994).
3. Equivalent representations involving only sine or only cosine functions with phase shift parameters are also valid. We will use these alternative forms in Chapter 11.

time series may sometimes effectively determine the periodic components.

Figures 9-12 and 9-13 present the power spectra of the data shown in Figures 9-5 and 9-6, respectively, presented as a function of the period of the sine or cosine components. The power spectra shown in Figures 9-12 and 9-13 are simply a plot of $a_n^2 + b_n^2$ as a function of L/n, for $n = 1, 2, \ldots, N$. There are several algorithms that can be used to calculate the unknown coefficients $a_0, a_n,$ and b_n ($n = 1, 2, \ldots$) and represent the data as sine and/or cosine series as shown in Eq. (9-4). As noted, the FFT algorithm is the most widely used. These figures were done with the FFT function of *MATLAB*, but many other software packages, including *BERKELEY MADONNA*, provide similar implementations.

The power spectrum of the LH example, Figure 9-12, indicates the data contain a periodic component of approximately 230 minutes (corresponding to $L = 1440$ and $n = 7$) corresponding to the visual impression from Figure 9-5. In contrast, the power spectrum of the GH example, Figure 9-13, gives no clear indication of a dominant frequency (or period). In addition, even in the more conclusive LH example, the calculated power spectra have not actually provided much information about the physiological mechanisms involved in hormone secretion into the blood or the kinetics of the elimination from the blood.

Summarizing, the FFT and power spectrum methods are limited to data sets of points equally spaced in time and having equal levels of experimental uncertainties. Also, it should be noted the power spectrum approach only evaluates the contributions of sine and cosine waves with integer harmonics. That is, only periods of L/n are considered, where L is the base period (1440 minutes for the present examples) and $n = 1, 2, 3, 4, \ldots$, and, thus, only variances carried by these frequencies will be explained by the FFT method.

B. Fractional Variance Methods

A different way to present these results is in terms of *fractional variance*, defined as the fraction of the variance explained by the model over the total variance of the data. Alternatively, one can consider the *fraction of remaining variance*, defined as the fraction of the variance that cannot be explained by the model over the total variance of the data. Both quantities are smaller than 1, and are often expressed as percentages. The two are easily related as:

$$\text{fractional remaining variance} = 1 - \text{fractional variance}.$$

Recall that we used a similar approach in Chapter 4 to quantify heritability, considering the fractional variance of a linear model with equal weights of the data points. In the current context, we want to

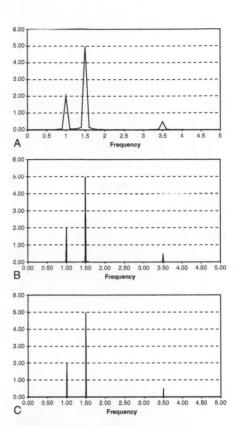

FIGURE 9-11.
Results of the FFT algorithm using, respectively, 500 (panel A), 5000 (panel B), and 50,000 (panel C) data points, sampled at a rate of 50 data points per unit time from the function $m(t) = \sqrt{2}\ \sin(2\pi t) + \sqrt{5}\ \cos(3\pi t) + \frac{1}{\sqrt{2}}\ \sin(7\pi t)$. As expected from the analytical form of $m(t)$, three dominant frequencies are identified at $v = 1$, $v = 3/2$, and $v = 7/2$, corresponding to the three periodic components with periods $L = 1$, $L = 2/3$, and $L = 2/7$ of $m(t)$. The amplitudes are proportional to the squares of the coefficients before the periodic components.

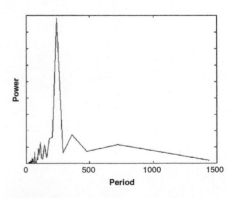

FIGURE 9-12.
The power spectrum of the LH data shown in Figure 9-5.

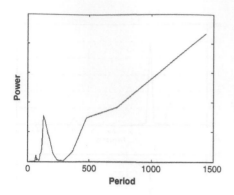

FIGURE 9-13.
The power spectrum of the GH data shown in Figure 9-6.

consider the fraction of the variance that can be described by a sine or cosine waveform of any particular period. The specific mathematical expression is given by Eq. (9-5). The denominator of this fraction is the weighted variance of the data minus the mean of the data. To compute this variance, we first perform a weighted least-squares fit of the data to a constant, as described in Chapter 8. Then, when the weighted least-squares value of a is found, the variance of the residuals is calculated as in Eq. (9-7). This gives the total variance of the data around a fixed level a, where the value of a has been chosen to minimize the weighted sum of squared residuals. The numerator of the fraction is the corresponding weighted variance-of-fit when the data are fit to the model given in Eq. (9-8)—a sum of a sine wave of period L, a cosine wave with period L, and a constant a_0. The estimates for a_0, a_L, and b_L are determined by a weighted linear least-squares procedure, and the variance explained by the model is calculated for those values as in Eq. (9-9).

$$\text{Fraction of remaining variance} = \frac{\text{Variance}_{\text{Equation}(9-9)}}{\text{Variance}_{\text{Equation}(9-7)}} \qquad (9\text{-}5)$$

$$Y_i \approx a \qquad (9\text{-}6)$$

$$\text{Variance} = \sum_i \left(\frac{Y_i - a}{SEM_i}\right)^2 = \sum_i R_i^2 \qquad (9\text{-}7)$$

$$Y_i \approx \frac{a_0}{2} + a_L \cos\left[\frac{2\pi}{L} X_i\right] + b_L \sin\left[\frac{2\pi}{L} X_i\right] \qquad (9\text{-}8)$$

$$\text{Variance} = \sum_i \left(\frac{Y_i - \dfrac{a_0}{2} - a_L \cos\left[\dfrac{2\pi}{L} X_i\right] - b_L \sin\left[\dfrac{2\pi}{L} X_i\right]}{SEM_i}\right)^2 = \sum_i R_i^2. \qquad (9\text{-}9)$$

The lower panel of Figure 9-14 presents a plot of the fraction of remaining variance after the LH data have been fit to a Fourier component, as in Eq. (9-8), as a function of the period L. In this example, there is a dominant Fourier component with a period of about 230 minutes that explains 50.2% of the variance of the data. This component is shown as a dashed line in the upper panel of Figure 9-14, along with the original LH data from Figure 9-5. Although the model explains half of the total variance, this Fourier component does not provide a perfect description of the experimental data because the physiology generates waveforms that are not a simple sum of a few sine/cosine

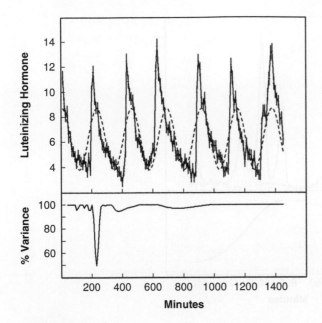

FIGURE 9-14.
The lower panel presents a plot of the percent variance remaining after the data from the LH example have been fit to a Fourier component, as in Eq. (9-6). A dominant periodic component with period of about 230 minutes explains 50.2% of the variance of the data and is shown as a dashed line in the upper panel along with the original LH data from Figure 9-5.

waves. The physiological waveform can, of course, be described by a series of many sine and/or cosine waves. However, it requires so many superimposed sine and/or cosine waves that our ability to interpret the results may be limited.

Figure 9-15 presents the analogous analysis of the GH example from Figure 9-6. In this example, a period of 1440 minutes only accounted for 22.9% of the variance; a period of 398 minutes accounted for 12.9%; and a period of 193 minutes accounted for 6.2%. It is clear this GH data cannot be adequately described by the sum of a few dominant Fourier components with periodicities of 193, 398, and 1440 minutes. It is interesting to note that this is an example of the multiple minima observed when fitting nonlinear equations which we described in Chapter 8.

C. Periodic Signal Averaging Methods

Figure 9-16 illustrates an alternate approach to the analysis of the luteinizing hormone example shown in Figure 9-5. This approach assumes the physiological signal occurs at a regular interval without assuming the waveform is a sine or cosine wave. The upper right panel of Figure 9-16 represents the percent of the variance remaining after an arbitrary waveform is subtracted from the data as a function of the period of the waveform. The period describing the most variance within

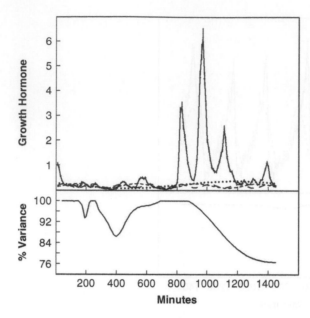

FIGURE 9-15.
The lower panel in Figure 9-13 plots the percent remaining variance, as a function of the period L, after the GH example has been fit to a Fourier component as in Eq. (9-6). The upper panel of Figure 9-13 presents the growth hormone data from Figure 9-6 and the weighted least-squares estimated Fourier components corresponding to periods of 1440 minutes (dotted line), 398 minutes (short dashed line), and 193 minutes (long dashed line).

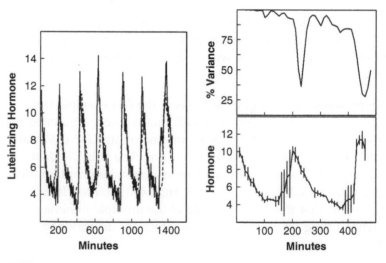

FIGURE 9-16.
Results from the periodic signal averaging method applied to the LH data in Figure 9-5.

the data is 460 minutes. With one data point sampled every 10 minutes, 460 minutes corresponds to 46 data points. The lower right panel presents a signal-averaged waveform that assumes a period of 46 data points. The first (from the left) point in the lower right panel is the average of the 1st, 47th, 93rd, and 139th data points. The second point in the lower right panel is the average of the 2nd, 48th, 94th, and 140th data points. The third is the average of the 3rd, 49th, 95th, and 141st data points, and so on. This process is repeated for a series of different assumed periods from three data points to $N/3$ data points, where N is the total number of data points. The repeated waveform shown in the lower right panel is clearly not a single sine or cosine wave. The upper right panel is the percent of the variance remaining after the waveform of period L is removed from the data plotted as a function of L. The dominant periodicities occur at periods of 230 and 460 minutes. The 460-minute waveform is essentially two cycles of the 230-minute waveform with slightly different amplitudes. The 230-minute waveform accommodates 63.7% of the variance, and the 460-minute periodic waveform describes 72% of the variance. Although this approach clearly provides a much better description of the data than does the Fourier approach shown in Figure 9-14, it is not perfect. Note that the alignment of the peaks of the data and the peaks of the periodic waveform is not exact. This method assumes the secretion events (i.e., the peaks) are equally spaced, whereas in reality they are not.

The corresponding analysis for the growth hormone example is not shown; however, as is evident from Figures 9-13 and 9-15, the GH is substantially less periodic, with dominant periods of 190 minutes and 400 minutes describing only 8.3% and 26.6% of the variance, respectively.

The analysis of the LH and GH data sets clearly indicates the functions describing the secretion of these hormones are not entirely, or even predominantly, periodic. This is a common feature of virtually every hormone concentration time series ever measured. Thus, Fourier and other algorithms which assume periodic events are not the best analytical methods for this type of data. In general, the objectives of hormone time series data analyses are to characterize the number, times, masses, and shapes of the pulsatile secretory events that increase the hormone concentration in the blood. The amount of basal (nonpulsatile) secretion is also of interest, as is the description of the time course of the removal of the hormones from the blood. The Fourier and signal-averaging approaches presented above cannot evaluate these characteristics well. Consequently, a group of statistically based algorithms, called *deconvolution methods*, have been developed to provide this information. Some of these methods do not require a constant measurement uncertainty, can accommodate missing values, and do not assume periodic secretion events.

D. *CLUSTER* Hormone Pulse Analysis Algorithm

One of the first statistically based algorithms was the *CLUSTER* algorithm (Veldhuis and Johnson [1986]; Urban et al. [1988]) depicted schematically in Figure 9-17. The *CLUSTER* algorithm functions as a

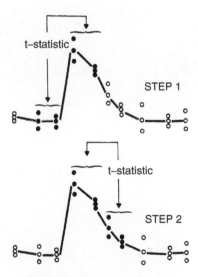

FIGURE 9-17.
Schematic illustration of the *CLUSTER* algorithm. (From Urban, R. J., Evans, W. S., Rogol, A. D., Kaiser, D. L., Johnson, M. L. and Veldhuis, J. D. [1988]. Contemporary aspects of discrete peak detection algorithms: I. The paradigm of the luteinizing hormone pulse signal in men. *Endocrine Reviews, 9(1)*, 3–37. © 1988 The Endocrine Society. Used by permission.)

two-step process. The first step looks for statistically significant increases within the data time series, while the second step looks for statistically significant decreases. Specifically, step 1 in Figure 9-17 is comparing a nadir (minimum) group size of 2, in this case data points 2 and 3 (each of which contains three replicates), with a peak (maximum) group size of 2, in this case data points 4 and 5. If this comparison indicates a statistically significant increase, as assessed by a grouped t-test, then the first data point in the peak group (in this case, data point 4) is marked as being a significant increase. This process is repeated with the group locations increased by one (i.e., points 3 and 4 are compared with points 5 and 6) until the end of the time series is reached. Every location corresponding to a statistically significant increase is recorded. Step 2 is identical to step 1, except that the grouped t-test is used to locate statistically significant decreases. Nadirs are then identified as significant decreases followed by significant increases, with peaks identified as the regions between the nadirs. One of the consequences of this definition of peaks and nadirs is that partial peaks at the beginning and the end of the hormone concentration time series are not identified. Neglecting these partial peaks is a design feature of the algorithm because the characteristics of a partial peak cannot be evaluated accurately.

The sawtooth pattern at the top of the LH and GH time series in Figures 9-5 and 9-6 is a diagrammatic depiction of the locations of the peaks and nadirs identified by the *CLUSTER* algorithm. The algorithm located six peaks in the LH time series and nine peaks in the GH time series. The *CLUSTER* algorithm provides a good illustration of the importance of using a correct variance model to evaluate the precision of the hormone concentrations. For example, the six peaks within the LH data in Figure 9-5 were based upon an MDC of 1 and a CV of 5%. If, however, we use MDC = 0.3 and CV = 3%, the *CLUSTER* algorithm will locate 12 statistically significant peaks in the LH time series. Clearly, the results obtained depend upon the assumed variance model for the data.

The *CLUSTER* algorithm provides some, but not all, of the desired characterizations of the hormone concentration time series. It provides information about the number, location, and size of the peaks in the data that meet its statistical criterion of significance. However, no information about the shape and size of the underlying secretion events and clearance mechanisms, which combine to create pulses, or any underlying basal secretion is provided by this method. Methods more powerful than *CLUSTER* in this regard are described next.

IV. DECONVOLUTION METHODS

Deconvolution methods are standard mathematical techniques widely used in science and engineering. A typical application is to remove the instrument response time from spectroscopic data (Jansson [1984]). In spectroscopy applications, the data are usually very accurate, with a much lower measurement uncertainty than the hormone concentration time series. In addition, spectroscopic applications will typically have many more data points and will not contain either outliers or missing values. Thus, the assumptions that are inherent in these methods may not be valid for hormone concentration time series data, where the experimental uncertainties (i.e., measurement errors) are substantially larger and variable and missing values and outliers are common.

A. Convolution Integral Model

This method is based on the assumption that the observed time dependence of the hormone concentration, $C(t)$, in the blood results from the coupling of two opposing physiological mechanisms—the rate of secretion, $S(t)$, into the blood and elimination, $E(t)$, from the blood, as shown in Figure 9-18. In Chapter 1, we explored a similar dependence in the context of designing optimal drug intake regimens. We examined the concentration of the drug in the bloodstream resulting from multiple doses administered at equal time intervals and discussed how physiological elimination affects the concentration.

Hormone concentrations in the blood are controlled by the same competing mechanisms. When the hormone is secreted by the endocrine glands, its concentration in the blood increases. Simultaneously, the pharmacokinetic processes eliminating the hormone from the blood are working to decrease its concentration. Thus, serum hormone concentration data cannot be used directly for assessing the hormone secretion. To obtain detailed information about the secretion events, the processes of basal secretion and the pharmacokinetic elimination must be decoupled. In other words, given the concentration function $C(t)$ and the elimination function $E(t)$, can the rate of secretion $S(t)$ be recreated?

Figure 9-19 provides a depiction of the coupling between secretion and elimination similar to Figure 1-25 of Chapter 1. The difference here is that we do not assume an instantaneous increase in concentration. The top panel of Figure 9-19 shows a typical secretion event (i.e., the rate of

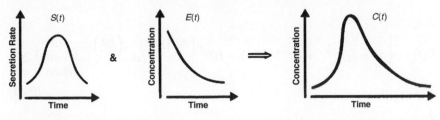

FIGURE 9-18.
Concentration results from the coupling of the secretion and the elimination. (From Veldhuis, J.D., Carlson, M.L. & Johnson, M.L. [1987]. The pituitary gland secretes in bursts: Appraising the nature of glandular secretory impulses by simultaneous multiple-parameter deconvolution of plasma hormone concentrations. *Proceedings of the National Academy of Sciences of the United States of America, 84,* 7686-7690.)

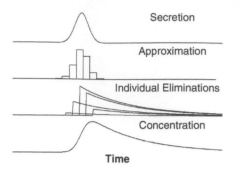

Secretion

Approximation

Individual Eliminations

Concentration

Time

FIGURE 9-19.
A graphical depiction of the coupling between secretion and elimination. (Figure 31.2 from Johnson, M.L., Straume, M. [1999]. Innovative quantitative neuroendocrine techniques. In Sex–steroid interactions with growth hormone [Serona Symposia]. pp. 318–326. New York: Springer-Verlag. © 1999 Springer-Verlag. With kind permission of Springer Science and Business Media. Reproduced with permission from BioSymposia, Inc. [formerly Serono Symposia USA].)

secretion into the serum as a function of time). The second panel approximates this secretion event with a series of rectangles. The third panel shows an elimination function for each of the rectangles in the panel above. For each rectangle of secretion, the concentration increases and subsequently decays according to the elimination function. The bottom panel of Figure 9-19 is the resulting total concentration as a function of time. It is the sum of all of the elimination time courses for the individual secretion rectangles. This panel shows the rapid increase in concentration followed by the slow decay typical of hormone concentration time-series data.

In Chapter 1, we assumed that drug elimination from the blood occurs at a rate proportional to the amount of the drug. This led to an exponential law for the decrease in the concentration (i.e., under this assumption $E(t) = e^{-kt}$, where $k = \ln(2)/HL$ and HL is the elimination half-life). The pharmacokinetic elimination of substances, including hormones, from the blood is a well-studied aspect of medical pharmacology. Because there are different factors contributing to the elimination of substances (each doing so in an exponential fashion), the most general form of the elimination function is given by the sum of exponential decays. The actual number of decays being summed can be large, because most substances are eliminated via multiple biochemical pathways. However, in the majority of cases, a single exponential decay is sufficient to describe the experimental data, and only in very rare cases are more than two exponential decays required.

The mathematical forms for the one- and two-exponential elimination functions are given in Eqs. (9-10) and (9-11), respectively. In Eq. (9-11), f_2 ($0 < f_2 < 1$) represents the fractional amplitude of the second elimination term, whereas HL_1 and HL_2 are the first and second elimination half-lives. These two equations are equal to zero when $t < 0$ so that the elimination does not occur before the secretion. A typical plot of these elimination functions is shown in the third panel in Figure 9-19.

$$E(t) = \begin{cases} e^{-kt} = e^{-\left(\frac{\ln 2}{HL}\right)t}, & \text{when } t \geq 0 \\ 0, & \text{when } t < 0 \end{cases} \tag{9-10}$$

$$E(t) = \begin{cases} (1-f_2)e^{-k_1 t} + f_2 e^{-k_2 t} = (1-f_2)e^{-\left(\frac{\ln 2}{HL_1}\right)t} + f_2 e^{-\left(\frac{\ln 2}{HL_2}\right)t}, & \text{when } t \geq 0. \\ 0, & \text{when } t < 0 \end{cases} \tag{9-11}$$

The stepwise rectangular approximation of the secretion rate may appear too inaccurate, but it can be improved by increasing the number of steps using more, narrower rectangles. For example, if one thinks of the amount of hormone being secreted in a small interval of time Δt after a fixed moment τ as being $S(\tau)\Delta t$, then we are in a situation like that in Chapter 1; namely, it is only a small inaccuracy to consider the amount of hormone $S(\tau)\Delta t$ to be totally and instantaneously delivered at time τ. Now, as in Chapter 1, if t is some time later than τ, the concentration of the hormone caused by this one secretion event is $S(\tau)\Delta t e^{-k(t-\tau)}$.

The result will become exact after passing to a limit $\Delta t \to 0$, where Δt is the width of the rectangles. Eqs. (9-12) and (9-13) give the resulting mathematical form of the combination of secretion and elimination in the form of a *convolution integral* for the concentration at time t.

$$C(t) = \int_0^t S(\tau)E(t-\tau)d\tau + C(0)E(t) \tag{9-12}$$

$$C(t) = \int_{-\infty}^t S(\tau)E(t-\tau)d\tau = S(t)*E(t). \tag{9-13}$$

We now formalize this idea and derive Eq. (9-12). Divide the interval $[0,t]$ into n equal subintervals of length $\Delta t = \dfrac{1}{n}$ and assume n instantaneous secretory events of magnitude $C_0, C_1, \ldots C_{n-1}$ have taken place at time instances $\tau_0 < \tau_1 < \cdots < \tau_{n-1} < t$, where τ_0 is in the interval $\left[0, \dfrac{t}{n}\right]$, τ_1 is in the interval $\left[\dfrac{t}{n}, 2\dfrac{t}{n}\right]$, and, in general, τ_m is in the interval $\left[m\dfrac{t}{n}, (m+1)\dfrac{t}{n}\right]$, $m = 0, 1, \cdots, n-1$ (see the horizontal axis in Figure 9-21). Assume also that the elimination function $E(t)$ is exponential, as in Eq. (9-10). At time t, because of hormone elimination, the residual amount from the secretion event at time τ_0 will be $C_0 e^{-k(t-\tau_0)}$. The secretion event at time τ_1 will contribute, by time t, a residual concentration of $C_1 e^{-k(t-\tau_1)}$, and so on. Thus, under the assumptions made, the hormone concentration in the blood at time t will be given by:

$$C(t) = C_0 e^{-k(t-\tau_0)} + C_1 e^{-k(t-\tau_1)} + \cdots + C_{n-1} e^{-k(t-\tau_{n-1})}. \tag{9-14}$$

Equation (9-14) is not quite exact because it assumed the secretion events at the specified moments are instantaneous. Graphing the function $C(t)$ from Eq. (9-14) will result in a function with jumps, similar to Figure 1-23 in Chapter 1 (the only difference is that the heights of the jumps, corresponding to the secretion amounts $C_0, C_1, \ldots C_n$ are now different). In reality, the hormone concentrations change in a continuous fashion, as the secretion events can never be instantaneous. As an approximation, however, a sharp increase in the hormone concentration can be considered the result of a secretion event with a secretion rate $S(t)$ that is constant over a very short interval of time and zero outside of this interval, as in Figure 9-20. The upper panel depicts such a secretion rate function $S(t)$ and the lower panel represents the corresponding

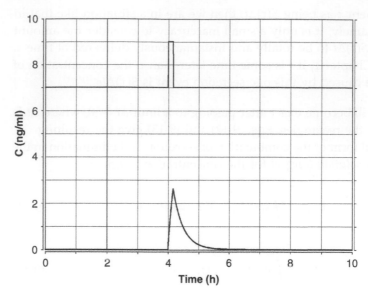

FIGURE 9-20.
Approximation of an instantaneous secretion event by infusion at a constant rate over a very short interval of time.

concentration function $C(t)$ as a result of this hormone secretion and exponential hormone elimination. The narrower the interval Δt is, the more pronounced the jump in the concentration function $C(t)$ will be, and the closer it will resemble an instantaneous secretion event.

Suppose now the secretion rate function $S(t)$ remains constant over each subinterval $\left[m\dfrac{t}{n}, (m+1)\dfrac{t}{n}\right]$, $m = 0, 1, \cdots, n-1$, and the value of the constant is equal to $S(\tau_m)$ (i.e., the secretion rate at time) τ_m (see Figure 9-21). Then, the amount secreted over the time interval $\left[0, \dfrac{1}{n}\right]$ is approximately

$$C_0 = S(\tau_0)\Delta t.$$

In the same way, the hormone secreted per unit volume during the time interval $\left[\dfrac{t}{n}, \dfrac{2t}{n}\right]$ can be approximated by:

$$C_1 = S(\tau_1)\Delta t,$$

and so on. The hormone secreted per unit volume during the time interval $\left[\dfrac{(n-1)t}{n}, t\right]$ can be approximated by:

$$C_{n-1} = S(\tau_{n-1})\Delta t.$$

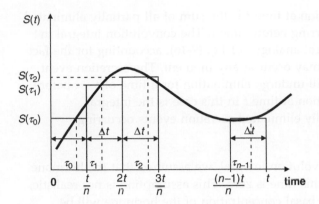

FIGURE 9-21.
Approximation of the secretion rate function $S(t)$. This approximation assumes that $S(t)$ remains constant over each interval $\left[\frac{(k-1)}{n}, \frac{kt}{n}\right]$.

Using Eq. (9-14), with the specific values above, we arrive at the following approximation for the concentration at time t:

$$C(t) \approx S(\tau_0) \Delta t\, e^{-k(t-\tau_0)} + S(\tau_1) \Delta t\, e^{-k(t-\tau_1)} + \ldots + S(\tau_{n-1}) \Delta t\, e^{-k(t-\tau_{n-1})},\ \text{i.e.}$$

$$C(t) \approx \sum_{m=0}^{n-1} S(\tau_m) e^{-k(t-\tau_m)} \Delta t. \qquad (9\text{-}15)$$

Note that Eq. (9-15) is based on two approximation assumptions. First, we assumed that the secretion events occurred instantaneously, and, second, we assumed that the rate of secretion $S(t)$ remained constant over each subinterval of length Δt. As $\Delta t \to 0$, both of these approximations will become more accurate, and thus Eq. (9-15) will give a better approximation for $C(t)$. Because the sum in Eq. (9-15) represents a Riemann sum for the function $S(\tau)e^{-(t-r)}$ over the interval $[0,t]$, taking the limit as $\Delta t \to 0$ in Eq. (9-15), gives:

$$C(t) = \lim_{\Delta t \to 0} \sum_{m=0}^{n-1} S(\tau_m) e^{-k(t-\tau_m)} \Delta t = \int_0^t S(\tau)e^{-k(t-\tau)}d\tau = \int_0^t S(\tau)E(t-\tau)d\tau,$$

where $E(t) = e^{-kt}$. This is the integral term of Eq. (9-12) for the most common case of exponential removal of the hormone from the bloodstream.

Identical derivations apply for any other form of the function $E(t)$, such as the sum of two exponential decays given in Eq. (9-11). For this more general case, Eq. (9-14) changes to:

$$C(t) = C_0 E(t - \tau_0) + C_1 E(t - \tau_1) + \cdots + C_n E(t - \tau_n). \qquad (9\text{-}16)$$

The first term of Eq. (9-16) represents the residual concentration at time t remaining from the secretion event at time τ_0. The second term is the residual concentration at time t from the secretion event at time τ_1, and so on. The concentration at time t is the sum of all partially eliminated secretion events occurring before time t. The convolution integral in Eq. (9-12) is the integral analogue of Eq. (9-16), accounting for the fact that secretion events may occur at any moment. The secretion event occurring at time t will undergo elimination for a time of $t - \tau$ before time t. The concentration at time t in this case is the integral over $[0,t]$ of all partially eliminated secretion events occurring before time t.

In calculating the convolution integral, we assumed that the hormone concentration before this time is zero. This assumption is not realistic, because at least some basal concentration of the hormone will be observed. The $C(0)$ in Eq. (9-12) represents the concentration at time $t = 0$, and the term $C(0)E(t)$ represents the residual of this concentration at time t. Alternatively, if the integration is over the interval $[-\infty,t]$ this extra term will be absorbed by the integral as in Eq. (9-13). The symbol * in Eq. (9-13) is the mathematical shorthand for a convolution integral.

In principle, Eqs. (9-12) and (9-13) can be used with a set of experimental data $C(t_1),C(t_2),\ldots$ and an assumed approximate value for the elimination half-life, HL, to determine the characteristics of the secretion rate as a function of time, $S(t)$. This process is known as deconvolution; that is, the inverse of a convolution.

B. Gold's Deconvolution Method

Gold's method is an example of one of the standard deconvolution techniques described above. Gold's method is an iterative approach for the solution of Eq. (9-13); that is, $C(t) = S(t) * E(t)$. As we saw in Chapter 8, an iterative method for $S(t)$ starts with an initial estimate and subsequently provides a better estimate based on the data. The process is repeated until the iterations do not change significantly from one cycle to the next. In Gold's method, the value of the secretion as a function of time after k iterations, $^kS(t)$, is expressed in terms of the previous iteration as:

$$^0S(t) = C(t)$$
$$^kS(t) = {}^{k-1}S(t)\frac{C(t)}{^{k-1}S(t) * E(t)}. \qquad (9\text{-}17)$$

In Eq. (9-17), the concentration as a function of time, $C(t)$, is used as the initial value for the secretion as a function of time, $^0S(t)$, to start the iteration. The function $E(t)$ is the exponential elimination function from Eq. (9-10).

Figure 9-22 presents the Gold's deconvolution of the GH data shown in Figure 9-6 assuming a 20-minute elimination half-life. It is clear there are at least three large secretion events and maybe as many as 20 small

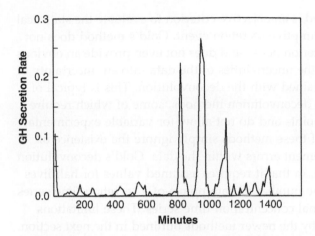

FIGURE 9-22.
Gold's deconvolution analysis of the growth hormone data shown in Figure 9-6. This calculation assumed a single elimination half-life of 20 minutes. The units of GH secretion are ng/ml/min.

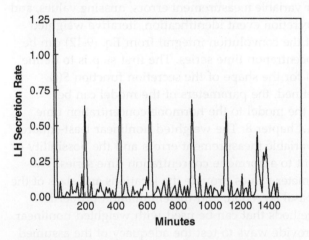

FIGURE 9-23.
Gold's method of deconvolution of the luteinizing hormone data shown in Figure 9-5. This calculation assumed a single elimination half-life of 50 minutes. The units of LH secretion are mIU/ml/min.

ones. The secretion appears to drop to nearly zero between some of the small secretion events, so it seems the contribution caused by basal secretion is either negligible or zero.

Figure 9-23 presents the Gold's deconvolution of the LH data shown in Figure 9-5 assuming a 50-minute elimination half-life. In this case, it appears there are seven or eight large secretion events. There also seem to be numerous small secretion events or perhaps even an elevated basal secretion. Because of the measurement errors within the data, an elevated constant basal secretion might look like a large number of very small secretion events.

Clearly, what is needed is an objective criterion to evaluate the statistical significance of a presumptive secretion event. Gold's method does not provide for such a criterion per se and does not even provide an obvious method to propagate the uncertainties of the data into an uncertainty of the secretion rates obtained with the deconvolution. This is typical of many of the standard deconvolution methods, some of which require equally spaced data points and do not allow for variable experimental uncertainties. Many of these methods simply ignore the existence of experimental measurement errors within the data. Gold's deconvolution method is also typical, in that it requires assumed values for half-lives and produces the same number of data points of a secretion time series as existed in the original concentration time series. These limitations are partially resolved by the newer methods outlined in the next section. More details about Gold's method can be found in Jansson (1984).

C. Multiparameter Deconvolution Methods

In order to account for variable measurement errors, missing values, and to perform accurate secretion event identification, iterative weighted least-squares fitting of the convolution integral from Eq. (9-12) can be used for hormone concentration time series. The first step is to define a mathematical model for the shape of the secretion function $S(t)$. Once this model is defined, the parameters of the model can be determined by fitting the model to the hormone concentration time series, as described in Chapter 8. The weighted nonlinear least-squares fitting addresses the variable measurement errors and the possibility of missing values inherent to a hormone concentration time series. The estimated model parameter values provide the locations and sizes of the secretion events, the basal secretion, and the elimination properties. The goodness-of-fit methods that can be used with weighted nonlinear least-squares fitting provide ways to test the adequacy of the assumed secretion model. Furthermore, the weighted nonlinear least-squares techniques also include methods to estimate the precision of the model parameters, which can be used to test the significance of individual secretion events. Such methods are usually referred to as multiparameter deconvolution methods.

The original multiparameter deconvolution method *DECONV* (Veldhuis et al. [1987]) operated under the assumption that the secretion events can be described as the sum of Gaussian curves occurring at different times, PP_k, and having different magnitudes, H_k, but all having the same width, *Secretion* SD, as is shown in Eq. (9-18).

$$S(t) = S_0 + \sum_k H_k e^{-\frac{1}{2}\left(\frac{t-PP_k}{Secretion\ SD}\right)^2}. \tag{9-18}$$

The positive constant S_0 in Eq. (9-18) is the basal secretion. One way to approximate the locations of the secretion events is to use statistical techniques to detect significant increases followed by significant decreases in the time series such as done by *CLUSTER*.

This algorithm suffers from three potential weaknesses. First, although it can easily be justified under the assumptions of the central limit theorem, some might find the assumption that the secretion event could be described by a Gaussian distribution questionable. Second, circa 1987 computer hardware was substantially slower than today's personal computers. Thus, the original multiparameter deconvolution algorithm and software did not include the most rigorous statistical tests for the existence of a secretion event, simply because the required computer resources were not available in 1987. Third, the user had to provide estimates, for the exact number of secretion events and their approximate locations and sizes. The *CLUSTER* method can be used to provide those estimates, but, as this method is just one out of many, the choice of initial values for the parameters will be somewhat subjective. Thus, multiparameter deconvolution alternatives that are more flexible with regard to assumptions and input information would be preferable.

One such algorithm, called *PULSE*, is a waveform-independent deconvolution method that is not based on the assumption that the secretion function is of the form given in Equation (9-18). In *PULSE*, the secretory pulses are assumed to have a general form that increases from a nadir to a peak and then decreases back to a nadir. It also eliminates the requirement to specify the number and approximate positions of the secretory events. Several other deconvolution techniques have been developed as described in Johnson and Veldhuis (1995) and Johnson et al. (2004) in an attempt to automate the original multiparameter deconvolution technique (e.g., *PULSE2* and *PULSE4*). The history of these automatic algorithms closely parallels the developments of available computer hardware. Faster computers mean that more computationally intensive statistical tests can be utilized.

Figures 9-24 and 9-25 present the results of applying of the *PULSE*, method to the LH and GH time series from Figures 9-5 and 9-6. The lower panel of these figures is the calculated secretion rate as a function of time, $S(t)$ in Eq. (9-17). The upper panel of these figures presents the calculated concentration as a function of time, $C(t)$, from Eq. (9-13). The original data points are represented as vertical error bars. The middle panel describes the corresponding residuals, R_i, for the values of the parameters minimizing the weighted least-squares norm in Eq. (9-19):

$$\text{Variance of Fit} = \sum_i \left(\frac{Y_i - C(t_i)}{SEM_i}\right)^2 = \sum_i R_i^2. \qquad (9\text{-}19)$$

Based on a visual inspection, it appears that the secretion patterns shown in Figures 9-24 and 9-25 provide a good description of the actual

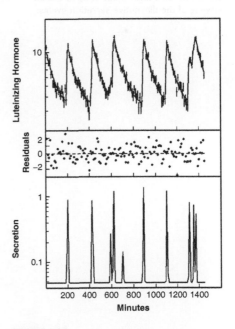

FIGURE 9-24.
Analysis of the luteinizing hormone data in Figure 9-5 by multiparameter deconvolution predicted 10 secretion events with $HL = 46.5$ minutes, $C(0) = 8.02$, $S_0 = 0.0478$, and *Secretion* $SD = 3.1$ minutes. The concentrations and secretion rates are expressed on a logarithmic scale to emphasize the small secretion events.

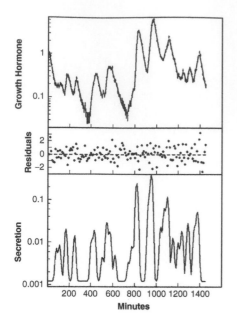

FIGURE 9-25.
Analysis of the growth hormone data in Figure 9-6 by multiparameter deconvolution predicted 24 secretion events with $HL = 20.4$ minutes, $C(0) = 1.42$, $S_0 = 0.00122$, and *Secretion* SD $= 9.7$ minutes. The concentrations and secretion rates are shown on a logarithmic scale to enable viewing of the diminutive secretion events.

experimental data. The residuals appear random, and there are no outliers.

Currently, an "automatic" multiparameter deconvolution technique is available (titled *AUTODECON*) that is maximally assumption-free and implements a rigorous statistical test for the existence of secretion events. In addition, it eliminates the subjective nature of the earlier algorithms by automatically inserting, and subsequently testing, the significance of presumed secretion events. This automatic algorithm is a combination of three modules; a *parameter fitting* module that performs weighted nonlinear least-squares parameter estimation, an *insertion* module that determines the location for and adds presumed secretion events, and the *triage* module that removes secretion events that are not statistically significant. We refer the reader to the references below for a comprehensive description.

Before we move on, we reiterate that in this chapter we focused on methods using hormone concentrations measured in the blood serum to determine whether the secretion events are periodic or pulsatile and to reconstruct the secretion levels from experimental data. Sophisticated mathematical methods, such as convolution–deconvolution techniques, are required in this case because of the ongoing physiological elimination of the hormone. An alternative approach would be to consider the factors impacting hormone secretion and model the network of interactions in order to obtain the secretion profile of the hormone in question. We consider such models in the next chapter.

REFERENCES

Box, G., Jenkins, G. M., & Reinsel, G. (1994). *Time series analysis: Forecasting and control* (3rd ed.). Upper Saddle River, NJ: Prentice Hall.

Grafakos, L. (2004). *Classical and modern Fourier analysis.* Upper Saddle River, NJ: Prentice Hall.

Guyton, A. C., & Hall, J. E. (2005). *Textbook of medical physiology* (11th ed.). New York: Saunders/Elsevier.

Jansson, P. A. (1984). *Deconvolution with applications in spectroscopy.* New York: Academic Press.

Johnson, M. L., & Veldhuis, J. D. (1995). Evolution of deconvolution analysis as a hormone pulse detection method. *Methods in Neuroscience, 28,* 1–24.

Johnson, M. L., Virostko, A., Veldhuis, J. D., & Evans, W. S. (2004). Deconvolution analysis as a hormone pulse-detection algorithm. In Johnson, M. L., & Brand, L. (Eds.), *Methods in Enzymology* (vol. 384, pp. 40–53). New York: Academic Press.

Urban, R. J., Evans, W. S., Rogol, A. D., Kaiser, D. L., Johnson, M. L., & Veldhuis, J. D. (1988). Contemporary aspects of discrete peak detection algorithms: I. The paradigm of the luteinizing hormone pulse signal in men. *Endocrine Reviews, 9,* 3–37.

Veldhuis, J. D., & Johnson, M. L. (1986). Cluster analysis: A simple, versatile, and robust algorithm for endocrine pulse detection. *American Journal of Physiology, 250*, E486–E493.

Veldhuis, J. D., Carlson, M. L., & Johnson, M. L. (1987). The pituitary gland secretes in bursts: Appraising the nature of glandular secretory impulses by simultaneous multiple-parameter deconvolution of plasma hormone concentrations. *Proceedings of the National Academy of Sciences of the United States of America, 84*, 7686–7690.

Willams, R. H., Larsen, P. R., Kronenberg, H. M., Melmed, S., Polonsky, K. S., Wilson, J. D., & Foster, D. W. (Eds.). (2002). *Williams textbook of endocrinology* (10th ed.). Philadelphia: W. B. Saunders Company.

INTERNET RESOURCES

http://mljohnson.pharm.virginia.edu/home.html
    Michael L. Johnson's page at the University of Virginia, from which the hormone pulsatility programs may be downloaded.
www.mathworks.com
    Home page for the suppliers of *MATLAB*.

FURTHER-READING

Johnson, M. L., & Straume, M. (1999). Innovative quantitative neuroendocrine techniques. In *Sex–steroid interactions with growth hormone* (Serona Symposia), pp. 318–326. New York: Springer-Verlag.

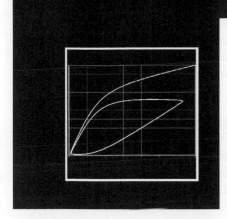

Chapter 10

ENDOCRINE NETWORK MODELING: FEEDBACK LOOPS AND HORMONE OSCILLATIONS

Life is a constant oscillation between the sharp horns of dilemmas.
Henry Louis Mencken (1880–1956)

In their article "What is a Biological Oscillator?" Friesen and Block wrote: "There can be little doubt that oscillations are an essential property of living systems. From primitive bacteria to the most sophisticated life forms, rhythmicity plays a vital role in providing for intercellular communication, locomotion, and behavioral regulation. Although the presence of biological rhythms has been recognized since antiquity, only recently has the origin of these rhythms been systematically addressed. At present, there are numerous descriptions of biochemical, biophysical, and physiological oscillations in the scientific literature . . . . Most recently, mathematical analysis has been applied to biological oscillators as well. . . ." (from Friesen and Block [1984], used with permission of the *American Journal of Physiology-Regulatory, Integrative, and Comparative Physiology*).

Since this was written, the importance of applying mathematical methods to examining the source, nature, mechanism, and stability of biologic oscillations has intensified considerably. In particular, the efforts to describe, explain, and predict oscillatory hormonal behavior are fundamentally interdisciplinary, with mathematics contributing its own arsenal of methods to the more traditional methods of biochemistry and physiology (see, for example, Farhy and Veldhuis [2005]; Farhy et al. [2002]; Wagner et al. [1998]; Keenan and Veldhuis [2001]; Farhy [2004]).

In Chapter 9, we described some statistical methods for examining the pulsatile nature of hormone release and quantifying the notion of secretion peaks. However, we did not discuss how the secretion events are regulated by the endocrine system. In this chapter, we construct and study mathematical models of hormone networks. The goal is to explore some of the endocrine mechanisms that control the secreting glands and cell groups to ensure precise hormone release with regard to amount, secretion times, and long-term secretion patterns. By these mechanisms, called *feedback mechanisms*, the body can sense that the concentration of a certain hormone has decreased and communicate to the secreting gland the amount of the additional hormone needed. The secretion rate will then be increased. This is an example of a

301

feedback mechanism, wherein the hormone regulates its own secretion. It is important to realize that such communications could rarely be considered instantaneous, and mathematical models often use *delay* factors to describe them.

One of the most biologically important and intriguing properties of hormones is the pulsatile pattern of their release, and this chapter will illustrate different conditions under which oscillations emerge in hormone concentration profiles. To approach the topic, we outline a formalism for mathematically modeling hormone systems and discuss the following questions:

1. What are the biological variables essential to the oscillator?

2. How do these essential variables interact?

3. Can these interactions lead to oscillations?

4. Under what conditions will oscillations be sustained?

I. INTRODUCTION

Oscillation can be described as a pattern in the dynamic plot of a measurable quantity (such as population size or hormone concentration in the bloodstream) that recurs with a relatively stable waveform and period. The important characteristics of an oscillation are the interpulse interval and the amplitude of the individual pulses (Figure 10-1, left panel). If the zenith–nadir difference in the amplitude of the oscillation is continuously decreasing, we have a case of damped oscillation (Figure 10-1, right panel). Note that the recurring pattern could be quite different from the well-known sine- or cosine-like waveforms (Figure 10-6). If a system contains more than one oscillating variable, another characteristic would be the phase relationship between these variables.

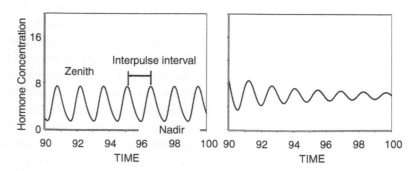

FIGURE 10-1.
Two oscillation patterns. Left panel: oscillations with constant amplitude; right panel: damped oscillations.

As we shall see, oscillations in endocrine physiology are often caused by (delayed) feedback loops. We begin with a heuristic explanation of this behavior.

Recall from Chapter 2 that if a system has a stable equilibrium state, its values stabilize around this state in the long run (i.e., the limit for $t \to \infty$ of the variable as a function of time is equal to the equilibrium state). An oscillatory system must include a restorative process that keeps the system close to its steady state, and oscillations can be sustained only if other factors, like inertia in physical systems, lead to overshooting the equilibrium value. For most cellular, endocrine, and neuronal oscillations, the critical factor that provides an overshoot is *delay*, which prevents the feedback restorative process from coming into full play until the equilibrium value has been passed (see Example 10-1).

Delays in biological systems can arise from many sources, and the debate about what causes delay and how best to model complex systems involving delay is far from over. The simplest situations involving delay are those with a certain time-offset or lag between an action triggered by one variable and the response to this action by a second variable. For example, suppose hormones A and B are involved in the control of a particular organismal function, and that hormone B turns off the production of hormone A. Suppose also that the inhibitory effect of Hormone B does not immediately follow an increase of hormone B in the bloodstream. The time elapsed between the increase in the concentration of B and the decrease in the concentration of A will be interpreted in the simulations below as an *explicit delay*. An explicit delay generally represents the amount of time necessary for a certain sequence of molecular and/or cellular events to occur. Explicit delays can vary greatly, depending upon the system at hand—the incubation period for an infectious disease represents one kind of explicit delay, and incubation periods can range from days to years. In this case, the delay results from the amount of time it takes for the pathogen to travel through the host's body and to multiply in the favored portion of the host's anatomy.

In other cases, the delay may not be explicit. In such cases, the delay would not formally reflect a certain period of time, but would result from a particular threshold value that must be met before the affected variable responds. In our hormone example (above), let us say that the level of hormone A will not be affected until the level of hormone B in the bloodstream reaches 300 pg/ml (picograms per milliliter). The delay then reflects the time necessary for the concentration of B to rise from its baseline levels and approach the threshold of 300 pg/ml. For another example, consider the Lotka–Volterra predator–prey model described by a coupled system with threshold values for the predator and prey populations (see Example 10-2). The population of the predator (owl) exerts control over the population of the prey (vole), and vice versa. In our work with this model, the thresholds were manifested as critical lines (the null clines) determined by the rate of change of the variables being equal to zero.

If a system involves more than one variable and the variables interact with one another, it is possible that one of them inhibits the growth of the other, which in turn stimulates the first one. This is, for example, the case in the Lotka–Volterra models. Such interaction is in essence a *(negative) feedback* and is often a factor that creates oscillations. When a single variable is considered, a self-inhibitory feedback is sometimes possible—in this case, we talk about *autofeedback* (see Example 10-1).

Whether a delay and/or a feedback will generate oscillatory behavior depends on the specific context of the problem and on the particular values of the system parameters. In this light, the presence of delays and/or feedbacks should be considered a factor that is *likely* to cause oscillations and should not be understood as a sufficient condition for oscillatory behavior in biological systems.

II. SYMBOLIC SCHEME REPRESENTATIONS OF THEORETICAL MODELS AND MODELING GOALS

Schematic diagrams are often used to show the most important components of a biological system and the connections between them. Standard symbols have been adopted to facilitate the display of information:

1. Rectangles (A, B, etc.) denote system variables (also referred to as *nodes*);

2. Lines (arrows) indicate specific relationships (also referred to as *conduits*) and are additionally marked with one of the symbols (+) or (−): a (+) indicates an excitatory action on the variable at which the line terminates, whereas a (−) denotes an inhibitory action; and

3. Triangles (D) on one or more of the lines indicate that a delay occurs from the change in the variable from which the line initiates until the corresponding effect is actually exerted.

For example, Figure 10-2 presents a schematic diagram of a network representing the interaction between two variables A and B. The line from A to B is marked with a (+), indicating an excitatory input. The line from B to A is marked with a (−) to indicate that B inhibits the

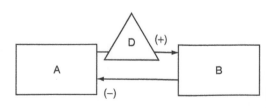

FIGURE 10-2.
Schematic diagram of a two-node network with feedbacks and delay.

growth of A. In addition, there is the presence of explicit delay for the stimulatory action of A upon B. This system contains one feedback, in which any one of the variables feeds back to suppress its own growth, and this feedback action is mediated by the other variable.

Example 10-1

Delayed Population Growth. Recall from Chapter 1 the classical logistic growth equation

$$\frac{dP}{dt} = r\left(1 - \frac{P(t)}{K}\right)P(t),$$

where K > 0 is the carrying capacity of the system, $r > 0$ is the inherent per capita growth rate for the population, and $P(t)$ is the size of the population at time t. It presents a model in which population size is limited by available resources. However, as we discussed, a limitation of this model is that it fails to take into consideration the time necessary for complex organisms to reach reproductive age. Thus, the diagram in Figure 10-3 illustrates a self-inhibitory effect (autofeedback) for the population size delayed by the delay time D necessary for each individual to reach reproductive age.

Notice that the origin of the line indicating excitatory input for the population size $P(t)$ is not shown. The input here is generated by the flow of natural resources that support the living organisms in the system. It also depends on the size of the population, and, de facto, P stimulates its own growth, providing an example of a self-stimulatory feedback mechanism that would force the system to explode if not restricted by the (negative) feedback. A differential equation corresponding to the diagram in Figure 10-3 is the classical delayed logistic growth equation:

$$\frac{dP}{dt} = r\left(1 - \frac{P(t-D)}{K}\right)P(t).$$

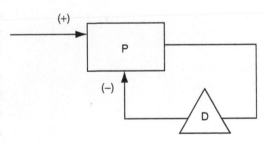

FIGURE 10-3.
Schematic representation of autofeedback with delay.

Example 10-2

A Predator–Prey Model. The diagram in Figure 10-4 schematically represents the Lotka–Volterra predator–prey model. Because the owls feed on the voles, the growth of the vole population causes growth in the owl population (excitatory input). Because the growth of the owl population causes the vole population to decline in size (inhibitory input), there is a negative feedback between the owls and the voles (see Chapter 2 for details).

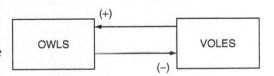

FIGURE 10-4.
Schematic presentation of the feedbacks in Lotka–Volerra's predator–prey model.

Example 10-3

The Growth Hormone Network. Growth hormone (GH), or somatotropin, is a major hormone regulating growth and metabolism. GH is secreted by the pituitary gland under the control of the following substances released by the hypothalamus: (1) The GH-releasing hormone (GHRH) that triggers the production and secretion of GH, and (2) somatostatin, or somatotropin release-inhibiting factor (SRIF), which is a GH secretion inhibitor. Numerous other substances could impact GH behavior. However, they are of secondary importance to the GH network and, for the sake of simplicity, will be omitted in the discussion that follows.

In a simplified view, GH secretion increases with increased hypothalamic GHRH secretion but is inhibited by SRIF (which acts as a suppressor for both the GHRH and GH). All three hormones are subject to exponential elimination with certain half-lives. Laboratory data in the adult male rat show that elevated concentrations of GH act by way of time-delayed feedback (D = 60–120 minutes) to stimulate SRIF release from the hypothalamus, which antagonizes GH release from the anterior pituitary gland and represses GHRH secretion. Schematically, this minimal GH network can be represented by the three-node diagram in Figure 10-5. The ellipses marked "elimination" signify that, in addition to the secretion rates of GH, GHRH, and SRIF controlled by the feedback mechanisms, the dynamic behavior of the system also depends upon the rates of continuous ongoing elimination for all three hormones.

Schematic representations (such as Examples 10-1 through 10-3) are based on significant prior knowledge of the functional connectivity of the system. Such knowledge is usually acquired through experimental work, data analysis, and, as we shall see below, mathematical modeling. Data analyses may also reveal certain specifics, such as periodicity, fluctuations, or time patterns. Once the schematic diagram is developed, it could be used as a basis for creating a dynamic model utilizing difference equations or differential equations. The model should be capable of reproducing and explaining key experimental observations.

Consider the schematic representation of the GH network in Figure 10-5, which is based on the physiological links between GH, GHRH, and SRIF described in Example 10-3. In this system, frequent measurements of GH concentrations in the bloodstream have unmasked complex patterns of gender-specific and developmentally regulated patterns of GH release in rats, sheep, and humans. In particular, GH secretion evolves as infrequent clusters of large pulses in adult male rodents, pubertal children, and young fasting or sleeping men and women, but unfolds as frequent, low-amplitude bursts in female rats and older,

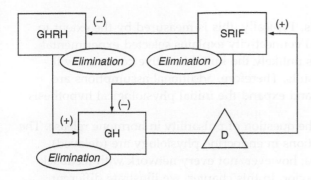

FIGURE 10-5.
Consensus GH network.

awake, or nutrient-replete humans. Figure 10-6 schematizes a typical GH secretion profile in the adult male rat (for real in vivo data see, for example, Lanzi and Tannenbaum [1992]). A model that describes the dynamics of GH concentration should reproduce the specifics of this pattern and suggest possible explanations for the mechanisms that are responsible for the observed dynamic behavior.

In this chapter, we discuss a general modeling approach to studying various mechanisms of hormone release control, similar to the GH network described above. The following three phases are fundamental to the modeling effort:

1. *Data analysis and exploration* of the specifics of hormone concentration time series, such as pulse detection, analysis of the variability and orderliness, determining the baseline secretion and half-life, and detecting the frequency of oscillations. As a result of the data analyses, selected experimental outcomes and hormone profile specifics are targeted for explanation in the modeling effort. Many aspects of this phase were described in Chapter 9.

2. *Formal network design* presenting an intuitive outline of the system's functional connectivity. This phase is based on analysis of available data from phase 1 and on the interactions between key system elements. The information is organized as a set of nodes and conduits in a hypothetical *formal endocrine network*.

3. *Dynamic modeling* of the formal network. The formal endocrine network from phase 2 is interpreted as a dynamic system and described with a set of coupled ordinary differential equations (ODEs). They give the time derivative of each network node and approximate all system positive and negative dose-responsive control links. The parameters in the ODEs must have a clear physiological meaning and are determined by comparing the model output with the phase 1 data.

The outcome of the modeling effort addresses the question of whether the design of the formal endocrine network is a good approximation of

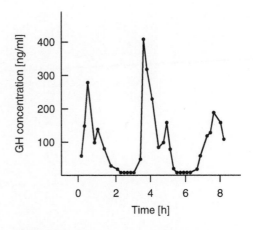

FIGURE 10-6.
Typical schematized adult male rat GH secretion profiles.

the actual hormone axis. Typically, this is measured by the extent to which the hypothesized connectivity explains selected experimental findings. However, it is unlikely the initial intuitive construct will provide satisfactory results. Therefore, additional assumptions are formulated that refine and expand the initial physiological hypothesis.

We specifically target the question of pulsatility in hormone release. The main sources of oscillations in endocrine physiology are (delayed) negative feedback loops; however, not every network with feedback generates periodic behavior. In this chapter, we illustrate different conditions under which oscillations emerge and perform quantitative analysis on various abstract endocrine networks, interpreted as dynamic systems. We shall be mainly concerned with phase 3 (above) and its relations to phases 1 and 2.

We begin by describing, through differential equations, an approximation of the evolution of the concentration of a single hormone secreted in the circulation under the control of one or more other regulators. We further simulate and analyze the interactions between system components (nodes) organized in different feedback networks. The main concepts are illustrated on two two-node models. System parameters are introduced on the basis of their physiologic meaning, and the effect of their modification is appraised. Oscillations caused by perturbations of systems with damped periodicity are distinguished from oscillations of systems with a genuine periodic behavior. In addition, we discuss the simulation of basic laboratory experimental techniques, point out some of their limitations, and suggest alternatives that reveal more network details.

In most of our examples, the underlying mathematical theory is not trivial. This is especially true for those models that explicitly include delays in the core system. Abstract mathematical details are generally avoided, and the focus is placed on numerical solutions and interpretations. As a rule, the simulated networks are abstract and do not correspond to a specific endocrine system. However, the constructs and the modeling techniques are fairly general and can be easily adapted to fit a particular physiology.

III. EVOLUTION AND CONTROL OF HORMONE CONCENTRATION

A. Rate of Change of Hormone Concentration

We begin by describing the quantitative approximation of the concentration dynamics of a single hormone secreted in its releasable pool. Recall from Chapter 9 that the rate of change of hormone concentration depends on two processes: Secretion and ongoing elimination. In Chapter 1, we discussed the equation $dC/dt = -\alpha C(t)$ describing the rate of change in the hormone concentration caused by the process of ongoing elimination. If the hormone is being secreted at the rate $S(t)$, the differential equation

$$\frac{dC}{dt} = -\alpha C(t) + S(t) \tag{10-1}$$

describes the change in the concentration. Here, as before, $C(t)$ is the hormone concentration in the corresponding pool; t is the time; $S(t)$ is the secretion rate; and the elimination is supposed to be proportional (with some rate elimination constant $\alpha > 0$) to the available concentration. Recall that the clearance constant $\alpha > 0$ and the half-life τ of the hormone in the blood are related through $\tau = \ln 2/\alpha$.

This model extends the model we considered in Chapter 1, Section IX. That model assumed instantaneous entry of drug into the bloodstream with every dose, and that doses were administered at equally spaced time intervals. Equation (10-1) extends this construct by allowing variable amount and continuous delivery with regard to time.

Also in Chapter 9, we discussed the following convolution integral as an alternative way to describe the processes of simultaneous secretion S and elimination E:

$$C(t) = \int_{-\infty}^{t} S(\tau)E(t - \tau)d\tau = (S * E)(t). \tag{10-2}$$

If $E(t) = e^{-\alpha t}$ [hence, $E(t - \tau) = E(t)E(-\tau)$] and we know the concentration at $t = 0$, $C_0 = C(0)$, Eq. (10-2) can also be given as:

$$C(t) = \int_{0}^{t} S(\tau)E(t - \tau)d\tau + C(0)E(t),$$

for positive values of t.

EXERCISE 10-1

Show that in the special case when $S(t) = S = const$ and $E(t) = e^{-\alpha t}$, the expression of the concentration function $C(t)$ becomes

$$C(t) = (C_0 - S/\alpha)e^{-\alpha t} + S/\alpha. \tag{10-3}$$

In Chapter 9, we derived the convolution representation (10-2) of $C(t)$ as the limit of a sequence of Riemann sums derived from discrete approximations. As the next exercise shows, when $E(t) = e^{-\alpha t}$, the convolution representation in Eq. (10-2) implies Eq. (10-1).

> **EXERCISE 10-2**
>
> Prove that if $C(t) = \displaystyle\int_{-\infty}^{t} S(z)e^{-\alpha(t-z)}dz$ then $\dfrac{dC(t)}{dt} = -\alpha C(t) + S(t)$.
>
> **Hint:** Follow the outline below:
>
> 1. Show that $C(t+h) = e^{-\alpha h}\displaystyle\int_{-\infty}^{t+h} S(z)e^{-\alpha(t-z)}dz$.
>
> 2. Show that $C(t+h) - C(t) = (e^{-\alpha h} - 1)\displaystyle\int_{-\infty}^{t} S(z)e^{-\alpha(t-z)}dz + e^{-\alpha h}\displaystyle\int_{t}^{t+h} S(z)e^{-\alpha(t-z)}dz$.
>
> 3. Show that $\dfrac{C(t+h) - C(t)}{h} = \dfrac{e^{-\alpha h} - 1}{h}C(t) + \dfrac{e^{-\alpha h}}{h}\displaystyle\int_{t}^{t+h} S(z)e^{-\alpha(t-z)}dz$.
>
> 4. Finally, to prove that $\dfrac{dC(t)}{dt} = -\alpha C(t) + S(t)$, use
>
> $$\lim_{h \to 0} \frac{1}{h}\int_{t}^{t+h} S(z)e^{-\alpha(t-z)}dz = S(t), \text{ and } \lim_{h \to 0}\frac{e^{-\alpha h} - 1}{h} = -\alpha.$$

Equation (10-1) can be implemented as a model describing the rate of change of hormone concentration in response to a specific pattern of hormone delivery/secretion. In the following two examples, the numerical simulations were performed with *BERKELEY MADONNA*.

Example 10-4

The plots in Figure 10-7 depict the simulated dynamics of a hormone in the circulation providing it is released endogenously (secreted internally) or administered exogenously (external delivery) as a bolus (left panel) or in a nonvarying fashion (right panel). In both simulations, the clearance constant is $\alpha = 3\text{h}^{-1}$ and the secretion (or infusion) rate is $20\,\text{ng/ml/h}$. The secretion is continuing for 10 minutes (left panel) or 180 minutes (right panel).

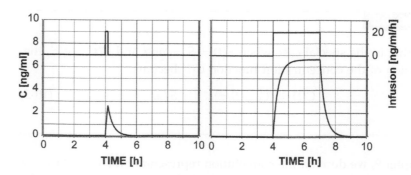

FIGURE 10-7.
Approximation of the raise and decay of a hormone administered as a (short-term) 10-minute (left) or (long-term) 3-hour constant infusion (right). The bottom line depicts the hormone concentration evolution, while the top curve illustrates the pattern of hormone delivery.

This example illustrates the expected concentration profiles of a hormone entering the bloodstream at one rate but for different periods of time. Note the difference in amplitude change of the hormone concentration as a function of the infusion/secretion length.

Example 10-5

This example illustrates the variation of hormone concentration in its releasable pool in response to a specific pattern of recurrent secretion events typical for some endocrine networks (e.g., the growth hormone or the luteinizing hormone axes). The more frequent delivery (left panel) causes incompleteness of the observable concentration pulses and a visible, gradual increase in peak amplitude. The simulations were carried out under the same conditions as in Example 4.

EXERCISE 10-3

Compare each of the graphs in Figure 10-8 and the simulated secretion events they represent with the graph in Figure 1-25, Chapter 1. What are the similarities and differences in the graphs and in the models they represent?

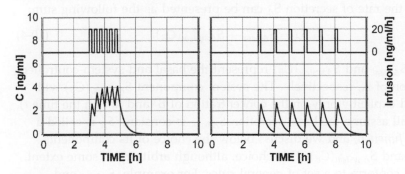

FIGURE 10-8.
Approximation of hormone concentration changes if the hormone is secreted in frequent (left) or infrequent (right) pulses. The bottom line is the hormone concentration, whereas the top curve marks the secretory pattern.

EXERCISE 10-4

Suppose the half-life of a hormone in the circulation is 12 min and at $t = 0$ the concentration of the hormone was equal to 0. At that moment, an intravenous infusion of the hormone at a constant rate was initiated. Three minutes later, the concentration of the hormone in the circulation was 500 ng/ml. Calculate the rate of exogenous infusion.

Hint: Use Eq. (10-3).

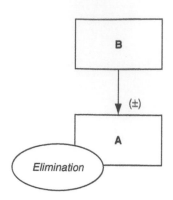

FIGURE 10-9.
Schematic presentation of a hormone network wherein the secretion of A is regulated by the concentration of hormone B (and also by elimination).

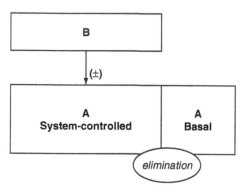

FIGURE 10-10.
Formal separation of the secretion of A into basal and system-controlled.

B. Secretion of One Hormone Controlled by the Concentration of Another

Up to this point, the events of secretion and elimination of a single hormone have been considered independent from any other hormone. In reality, there is considerable interaction between different hormones. We now consider the case when the secretion rate of hormone A is controlled by the concentration of hormone B. Figure 10-9 represents this situation.

The secretion of hormone A can be divided into two components, *basal* and *system-regulated* (Figure 10-10). The basal secretion represents the amount of hormone secreted independently of other system (model) components (e.g., hormone B, in our case). The system-regulated component corresponds to that part of the secretion exclusively related to other system hormones. Within this paradigm, if B is a stimulator, the basal secretion is what would remain if B were removed from the system. However, if B were an inhibitor, the basal secretion would be the release of A remaining after the action of B is applied to its full potency (e.g., by constant infusion of high, pharmacological doses of B).[1]

The above concept implies that the rate of change of the concentration C_A of hormone A from Eq. (10-1) can be written as:

$$\frac{dC_A}{dt} = -\alpha C_A(t) + S_A(t),$$

where the rate of secretion S_A can be presented as the following sum:

$$S_A = S_{A,basal} + S_{A,system}(C_B). \tag{10-4}$$

Here, $S_{A,basal}$ and $S_{A,system}(C_B)$ represent the basal and system-controlled secretion of A, respectively. The system component $S_{A,system}(C_B)$ will depend explicitly only on the concentration of B (and not on the time t). We shall assume the basal secretion $S_{A,basal}$ is constant. S_A is called a *control function*, and we discuss below the choice of its components $S_{A,basal}$ and $S_{A,system}(C_B)$. This choice, although arbitrary to some extent, should conform to a set of general rules. For example, $S_{A,basal}$ and $S_{A,system}(C_B)$ must be non-negative, because the secretion rate is always non-negative. Also, in most cases the system component of the control function will be monotone. The presence of hormone B can serve to either stimulate or inhibit the secretion of hormone A, but in either case we shall assume the effect is monotone. That is, if the effect of hormone B is to stimulate the secretion of A, then a higher concentration of B will cause a stronger stimulus (and higher secretion rate) of A. The function $S_{A,system}(C_B)$ will be monotone increasing as a function of C_B. Likewise, if the effect of B is to inhibit the production of A, the

---

1. The removal of hormone B could be performed by infusion of antibody to B, suppression of its secretion, or down-regulation or blockage of the receptors that mediate its action. For additional detail, see Section IV, part C. Note that this use of the term *basal secretion* is model-dependent and differs from the definition provided in Chapter 9.

stronger the presence of B, the lower the secretion of A, and the function $S_{A,system}(C_B)$ will be monotone decreasing. Monotone increase represents positive control, and monotone decrease represents negative control. Finally, the control function should be chosen to ensure that the concentration C_A will remain within its physiological range—between the minimal and maximal possible concentrations of hormone A.

From a physiological perspective, in order for a molecule of hormone B to exert its effect, it should bind to a specific receptor on the cell that produces and releases hormone A. As we saw in Chapter 7, hormone-receptor interactions obey laws of mass action, and the dissociation constant for the corresponding chemical reaction determines the affinity of the receptor for its ligand (hormone B). The higher the affinity of the receptor, the lower the hormone concentration required to elicit biological response. Sometimes, we say that changes in affinity (or in sensitivity) modulate the *potency* (power to produce the desired effect) of the hormone. On the other side, the *efficacy* (responsiveness; maximal effect that can be produced) of a hormone depends, among other things, on the number of receptors. This number may vary under different physiological conditions and affect the level of the response, but, generally, not the affinity. Therefore, it is desirable that the control function $S_{A,system}(C_B)$ explicitly embodies parameters corresponding to potency and efficacy.

These properties of $S_{A,system}(C_B)$ can be represented in a mathematical form by assuming that $S_{A,system}(C_B) = aF(C_B)$, where the parameter a represents the efficacy of hormone B, and F is a properly chosen, normalized (efficacy $= 1$) version of the control function. The desired criteria regarding $S_{A,system}(C_B)$ will now be satisfied if the following requirements are imposed on the normalized function:

1. $0 \leq F(C_B) \leq 1$;

2. F is monotone; and

3. F includes parameters that correspond to the potency of the action of B.

Among the various different functions satisfying the above properties, many authors choose to use the following nonlinear, sigmoid functions, known as *up- and down-regulatory Hill functions* to describe positive and negative hormone relationships:

$$F_{up(down)}(C) = \begin{cases} \dfrac{[C/T]^n}{[C/T]^n + 1} = \dfrac{C^n}{C^n + T^n} & (up) \\ \quad\quad or \\ \dfrac{1}{[C/T]^n + 1} = \dfrac{T^n}{C^n + T^n} & (down). \end{cases} \qquad (10\text{-}5)$$

As a justification for the name, consider the Hill equation given by Eq. (4-8) of Chapter 4, which was derived based on the presumed binding affinity of hemoglobin for oxygen molecules. Both functions are monotone and map the interval $[0, \infty]$ to $[0,1]$.

The parameter $T > 0$ in the definition of the functions (10-5) is called a *threshold*, and the power $n > 0$ is called a *Hill coefficient*. The threshold T in an up-regulatory Hill function is sometimes denoted by ED_{50} and is called *median effective dose*. Analogously, in a down-regulatory control function is referred to as ID_{50}, or *median inhibitory dose* (see Exercise 10-5). The parameters ED_{50} and ID_{50} approximate the potency of the regulatory hormone. The parameter n controls the slope of the Hill functions. For a fixed value of T, the larger the value of n, the steeper the slope. The Hill functions F_{up} and F_{down} are displayed in Figure 10-11 for $n = 3$ and $T = 10$. When $n = 1$ (the so-called Michaelis–Menten equation), the function has no inflection points, and its profile is a branch of a hyperbola.

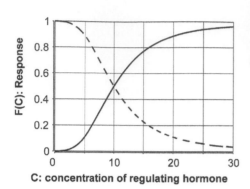

FIGURE 10-11.
Exemplary profiles of up-regulatory (black line) and down-regulatory (dotted line) Hill functions. In both examples and $n = 3$ and $T = 10$.

EXERCISE 10-5

Show that $F_{up} = 1 - F_{down}$, $F_{up}(T) = 1/2$, and $F_{down}(T) = 1/2$.

When Hill functions such as in Eqs. (10-5) are used, the mathematical form of the system component of the control function S_A from Eq. (10-4) can be defined as:

$$S_{A,system}(C_B) = aF_{up,(down)}(C_B). \tag{10-6}$$

The parameter a represents the efficacy of hormone B, the maximal effect B can produce on the secretion of A. With this choice of $S_{A,system}$, the control function S_A from Eq. (10-4) takes the form

$$S_A = S_{A,basal} + aF_{up,(down)}(C_B). \tag{10-7}$$

Figure 10-12 summarizes several examples, illustrating the changes in the control function in response to changes in the model parameters T, n, and a. We use the up-regulatory control function

$$S(C) = a\frac{(C/T)^n}{1 + (C/T)^n} = 20\frac{(C/10)^3}{1 + (C/10)^3}$$ as a reference (the described

changes refer to the following initial values of the parameters: $T = 10$, $n = 3$, and $a = 20$).

Figure 10-12 (left panel) represents changes in the response curve caused by varying the potency and/or efficacy (represented by the parameters T and a, respectively). Increasing T leads to decreasing the potency of the regulating hormone, because this means that higher concentrations of the regulatory hormone are necessary for reaching the median effective dose. In contrast, decreasing T would lead to increasing the potency of the regulating hormone (not shown). Changes in the value of the efficacy

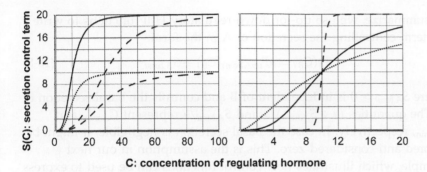

FIGURE 10-12.
Effect of manipulating the model. Left panel: Changes in the response curve caused by varying the potency and/or efficacy of the interaction. Solid line: control; upper dashed: decrease in potency of the regulating hormone (threefold increase in ED_{50}, that is, of the parameter T); dotted: twofold decrease in efficacy (that is, twofold increase in the value of a); lower dashed: combined twofold decrease in efficacy (that is, of the parameter a) and threefold increase in ED_{50} (that is, of the parameter T). Right panel: Alterations in the control response curve (black) associated with 10-fold increase (dashed) or twofold decrease (dotted) of the Hill coefficient n. Remark: Note the difference in the time scales in the left and right panels; the control function is the same in both plots.

(the parameter a) change the maximal value of the control function. The Hill coefficient n controls the slope of the control function. As n increases (values as large as 100 exist in biology; see Vrzheshch et al. [1994]; Mikawa et al. [1998]), the slope of the control function also increases as illustrated in Figure 10-12 (right panel). For large n, the control function acts as an on/off switch at the concentration value $C = T$. Plots similar to those in Figure 10-12 (left panel) are often seen in textbooks to illustrate the anticipated effect on percent of maximal response caused by decreased responsiveness and/or sensitivity.

The parameter a in Eq. (10-7) depends upon, and is determined from, the maximal possible attainable concentration of hormone A, $C_{A,max}$. The latter is the maximal physiologically possible endogenous concentration of A under a variety of conditions, including extremes such as responses to external high pharmacological stimulations. This maximal value may be known from experiments or hypothesized in case of mathematical simulations, in which case it could be considered a parameter of the model.

EXERCISE 10-6

Show that if a is the control coefficient from Eq. (10-7), then the quantities $(a + S_{A,basal})/\alpha$ and $S_{A,basal}/\alpha$ represent $C_{A,max}$ and $C_{A,min}$, respectively. Then, show that $a = \alpha C_{A,max} - S_{A,basal} = \alpha(C_{A,max} - C_{A,min})$.

Hint: Use Eq. (10-1), the fact that the maximal and minimal concentrations of A are achieved when $dC/dt = 0$, and the fact that $F_{up(down)}$ has maximal value 1 and minimal value 0.

To summarize, we use up- or down-regulatory Hill functions to write the term controlling the secretion of A in the form:

$$S_A(C_B) = aF_{up,(down)}(C_B) + S_{A,basal}, \qquad (10\text{-}8)$$

where $S_{A,basal} \geq 0$ is independent of B and controls the basal secretion of A. The quantities $(a + S_{A,basal})/\alpha$ and $S_{A,basal}/\alpha$ represent $C_{A,max}$ and $C_{A,min}$, respectively. In case the basal secretion is negligible, it might be ignored and considered zero. This is the assumption in our next example, which illustrates how control functions can be used to express the schematic diagrams describing the system in terms of coupled ordinary differential equations.

Example 10-6

Assuming that $S_{A,basal} = 0$ and $S_{B,basal} = 0$, write a system of differential equations describing the schematic hormone network in Figure 10-13:

<u>SOLUTION:</u>

We begin with the basic differential equations describing the rate of change of the concentrations C_A and C_B:

$$\frac{dC_A}{dt} = -\alpha_1 C_A + S_A \text{ and } \frac{dC_B}{dt} = -\alpha_2 C_B + S_B,$$

where $\alpha_1 > 0$ and $\alpha_2 > 0$ are the elimination constants of hormones A and B, and S_A and S_B are the respective control functions for the secretion rates. Because Figure 10-13 indicates that the increase of the concentration of A is inhibited by hormone B, we use a down-regulatory Hill function with parameters T_1 and n_1 to express

$$S_A(C_B) = a_1 F_{down}(C_B) = a_1 \frac{T_1^{n_1}}{(C_B)^{n_1} + T_1^{n_1}}.$$

As the increase of the concentration of B is stimulated by hormone A (as evident from the positive conduit indicated in Figure 10-13), we use an up-regulatory Hill function to express

$$S_B(C_A) = a_2 F_{up}(C_A) = \frac{(C_A)^{n_2}}{(C_A)^{n_2} + T_2^{n_2}}.$$

We need to account for the presence of delay in the way hormone A affects the secretion of hormone B. Because the delay D reflects the fact that secretion at time t is affected by the hormone action in a past moment, $t - D$, the control function S_B can be expressed as $S_B(t) = S_B(C_A(t - D))$. These considerations give the following system of differential equations representing the diagram from Figure 10-13:

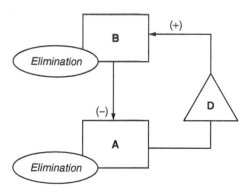

FIGURE 10-13.
Schematic hormone network for Example 10-6.

$$\frac{dC_A}{dt} = -\alpha_1 C_A + a_1 \frac{T_1^{n_1}}{(C_B)^{n_1} + T_1^{n_1}}$$

$$\frac{dC_B}{dt} = -\alpha_2 C_B + a_2 \frac{(C_A(t-D))^{n_2}}{(C_A(t-D))^{n_2} + T_2^{n_2}}.$$

The units in this model are

C_A, C_B, T_1, T_2	mass/volume(e.g., [ng/ml])
a_1, a_2	mass/volume/time(e.g., [ng/ml/h])
α_1, α_2	time^{-1}(e.g.,[h^{-1}])
D	time(e.g.,[h]).

In the sequel we avoid specifying the units, but the simulated profiles can be rescaled with ease to fit a particular physiology.

In this example, the assumptions $S_{A,basal} = 0$ and $S_{B,basal} = 0$ were made to emphasize the system interactions. As the basal secretion is independent from the hormone interactions, removing the assumption that the secretion rates $S_{A,basal}$ and $S_{B,basal}$ are negligible, leads to the following system of equations:

$$\frac{dC_A}{dt} = -\alpha_1 C_A + S_{A,basal} + a_1 \frac{T_1^{n_1}}{(C_B)^{n_1} + T_1^{n_1}}$$

$$\frac{dC_B}{dt} = -\alpha_2 C_B + S_{B,basal} + a_2 \frac{(C_A(t-D))^{n_2}}{(C_A(t-D))^{n_2} + T_2^{n_2}}.$$

Example 10-6 may be used to represent a possible dependence and interaction between two hormones. When hormone networks are considered, frequently more than one hormone controls the regulation of a specified hormone in the network, as shown, for example, on the formal diagram in Figure 10-5 depicting the consensus network of GH. Next, we consider how the ideas and mathematical formalism discussed in this section could be generalized and used to describe such multiple interactions.

C. Control of the Secretion of One Hormone by the Concentration of Multiple Hormones

---

Assume that two hormones instead of one control the secretion of A. We denote them by B and C, with corresponding time-varying concentrations $C_B(t)$ and $C_C(t)$. We shall still use the equation $\frac{dC_A}{dt} = -\alpha C_A(t) + S_A(t)$ to describe the rate of change in the concentration of A, but the control function S_A will now depend on the specific interaction between A, on the one hand, and B and C, on the other.

We use the notation $S_A = S_A(C_B, C_C)$ to express that the control function S_A depends (at any time t) on the concentrations of hormones B and C. The exact functional dependence is then determined by the specific physiological nature of the hormones' interaction. For example, if both B and C stimulate the secretion of A, and if they do so independently, we can use a control function of the form:

$$S_A(C_B, C_C) = a_B F_{up}(C_B) + a_C F_{up}(C_C) + S_{A,basal}. \qquad (10\text{-}9)$$

If B and C act simultaneously (e.g., the secretion of A requires the presence of both), the following control function may be appropriate:

$$S_A(C_B, C_C) = a F_{up}(C_B) F_{up}(C_C) + S_{A,basal}. \qquad (10\text{-}10)$$

On the other side, if the secretion of A is stimulated by B but is suppressed by C, the control function can be introduced as

$$S_A(C_B, C_C) = a F_{up}(C_B) F_{down}(C_C) + S_{A,basal}, \qquad (10\text{-}11)$$

if hormones B and C act simultaneously, or

$$S_A(C_B, C_C) = a_B F_{up}(C_B) + a_C F_{down}(C_C) + S_{A,basal}, \qquad (10\text{-}12)$$

if hormones B and C act independently. Note that Eq. (10-11) approximates a noncompetitive and simultaneous action of B and C. If B and C compete as they control the secretion of A, the secretion term can be described with a modified Hill function:

$$S_A(C_B, C_C) = a \frac{(C_B/T_B)^{n_B}}{(C_B/T_B)^{n_B} + (C_C/T_C)^{n_C} + 1} + S_{A,basal}. \qquad (10\text{-}13)$$

The latter form allows simulating competitiveness, understood as the capability of one hormone to overcome the effect of the other hormone, which cannot be achieved with the version utilized in Eq. (10-11). The latter is caused by the multiplicative form, wherein the value of $F_{down}(C_C)$ is an upper bound of the product $F_{up}(C_B) F_{down}(C_C)$ because both factors are less than 1.

Having outlined the mathematical formalism, we are now ready to investigate oscillating systems and the source of oscillations. As already noted, the mathematical theory is highly nontrivial. Thus, even in relatively simple cases of interactions between two hormones, a complete list of necessary and sufficient conditions for the system to oscillate may not be possible. In what follows, we present examples illustrating the different long-term behavior such systems may exhibit and compare the conditions under which such behaviors occur. We again begin with system networks involving only two hormones.

IV. OSCILLATIONS DRIVEN BY A SINGLE-SYSTEM FEEDBACK LOOP

In this section, we revisit models of formal networks similar to the model in Example 6, focusing on networks describing a feedback interaction between two hormones, A and B. The concentration of one, hormone A, regulates the secretion of another, hormone B, which in turn controls the release of A, as shown in Figure 10-14:

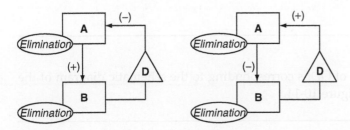

FIGURE 10-14.
Formal networks of two-node/one-feedback oscillators. The left panel depicts a network in which the main Hormone B is stimulated, while the other shows a model in which B is inhibited. D denotes a delay in the corresponding interconnection. In both networks, A and B are subject to elimination. (Reprinted from Farhy, L. S. [2004]. Modeling of oscillations in endocrine networks with feedback, *Methods in Enzymology, 384,* 54-81. Copyright 2004, with permission from Elsevier.)

Notice that in both constructs shown in Figure 10-14, any one of the hormones suppresses, indirectly, its own secretion. Therefore, these networks contain a negative feedback loop. Such networks are quite common in endocrinology. For example, in the male rat, GH stimulates the release of somatostatin with a lag of 60 to 120 minutes, and somatostatin, in turn, suppresses the release of GH.

As before, we assume the two hormones, A and B, are continuously secreted (driven by nonrhythmic excitatory inputs) in certain pool(s), such as the systemic circulation, where they are subject to elimination. The release of hormone B is regulated by hormone A. Hormone B itself exerts a delayed effect on the secretion of A. The interactions between A and B are assumed to be dose-responsive. For simplicity, we assume no delay in the action of A on B and no basal secretions.

The equations corresponding to the schemes in Figure 10-14 could be written following the pattern in Example 10-6. More specifically, the system of ODEs describing the schematic diagram in the left panel is:

$$\frac{dC_A}{dt} = -\alpha C_A(t) + a\frac{1}{(C_B(t-D_B)/T_B)^{n_B}+1}$$

$$\frac{dC_B}{dt} = -\beta C_B(t) + b\frac{(C_A(t)/T_A)^{n_A}}{(C_A(t)/T_A)^{n_A}+1}.$$

(10-14)

Because B inhibits the secretion of A, we used a down-regulatory Hill function in the first equation. This equation also accounts for the delay (via the parameter D_B) in the B → A system interaction depicted in the schematic diagram. In contrast, because A stimulates the secretion of B, an up-regulatory Hill function is used to represent the control function in the second Eq. (10-14).

We should note that, because of the presence of delay, solving these equations for $t \geq t_0$ requires preset initial conditions for C_B on the entire interval $[t_0 - D_B, t_0]$.

EXERCISE 10-7

Give the system of ODEs corresponding to the schematic diagram of the right panel in Figure 10-14.

A. Limit Cycles and Steady States

We now demonstrate that a nonzero delay and large nonlinearity in the control functions (sufficiently high Hill coefficients) can guarantee steady periodic behavior, because of the existence of a nontrivial limit cycle.

As the next exercise shows, the system defined by Eq. (10-14) has a unique equilibrium point (steady state). When there is no delay (that is, when $D_B = 0$), this equilibrium point is asymptotically stable and attracts all trajectories in the phase space (see Figure 10-15, left panel).

EXERCISE 10-8

For the model of Eq. (10-14) show that:

(a) The system has a unique equilibrium state.

(b) When $D_B = 0$, the equilibrium state of Eq. (10-14) is asymptotically stable.

Hint: Apply the theory presented in Chapter 2 by completing the following steps.

1. Represent the system defined by Eq. (8) in the form

$$x' = -\alpha x + F(y)$$
$$y' = -\beta y + G(x),$$

where $x = x(t)$ and $y = y(t)$, $\alpha, \beta, F, G > 0$, and F and G are monotonic decreasing and increasing, respectively.

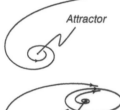

Attractor

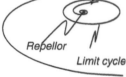

Repellor

Limit cycle

FIGURE 10-15.
Illustrative trajectories in the phase space (C_A, C_B) if the steady state is an attractor (top) or a repellor (bottom). In the latter case, a unique asymptotically stable periodic solution acts as limit cycle and attracts all system trajectories (except for the fixed point). (Reprinted from Farhy, L. S. [2004]. Modeling of oscillations in endocrine networks with feedback, *Methods in Enzymology, 384*, 54-81. Copyright 2004, with permission from Elseveri.)

2. Show that the system for the equilibrium state is

$$\alpha x = F(y)$$
$$\beta y = G(x)$$

3. Solve the above system for x and show that the resulting equation:

$$x = \frac{1}{\alpha} F\left(\frac{G(x)}{\beta}\right)$$

has only one positive solution. To do this, prove first that the function in the right-hand side of the above equation is monotonously decreasing. Do the same for y.

4. Investigate the stability by calculating the determinant and trace of the matrix

$$\begin{pmatrix} -\alpha & F'(y) \\ G'(x) & -\beta \end{pmatrix}.$$

Finalize your reasoning by demonstrating that in the equilibrium state the trace $-(\alpha + \beta) < 0$ and the determinant $\alpha\beta - F'(y)G'(x) > 0$.

The picture changes considerably in the presence of delays, because even a single nonzero delay (as in Eq. (10-14)) might change the properties of the steady state,[2] that may, for a certain range of delay values, become a repellor. In the latter case, there will exist a unique asymptotically stable periodic solution (which encircles the fixed point in the phase space) acting as a global limit cycle by attracting all trajectories, except the one originating from the fixed point (see the theorem of Poincaré–Bendixson in Chapter 2).

Although Poincaré–Bendixson's theorem gives a sufficient condition for the existence of a limit cycle, the verification of these conditions is often nontrivial, and we shall not focus on this question here. Instead, we examine the periodic solutions of two specific realizations of the networks shown in Figure 10-14. Each of these examples has a unique periodic solution and a unique repelling fixed point (Figure 10-15, right panel). We note that oscillations may be quite sensitive to changes in the model parameters and examine the system's response to external influences, such as changes to sensitivity, antibody infusion, and exogenous hormone infusion, expressed as appropriate modifications to the mathematical models.

---

2. The particular sensitivity analysis is nontrivial and is beyond the scope of this textbook. It consists of investigating the real part of eigenvalues, which are roots of equation containing a transcendental term, involving the delay. For more details, see Farhy and Veldhuis (2004).

Now consider a model described by the following core equations representing a possible formalization of the network where A stimulates the secretion of B (Figure 10-14, left panel):

$$\frac{dC_A}{dt} = -3C_A(t) + 20\frac{1}{[C_B(t-1)/2]^3 + 1}$$

$$\frac{dC_B}{dt} = -3C_B(t) + 40\frac{[C_A(t)/2]^3}{[C_A(t)/2]^3 + 1}.$$

(10-15)

The parameters are chosen to guarantee stable oscillations. The time plot and phase diagram of the (numerical) solution obtained with *BERKELEY MADONNA* are shown in Figure 10-16.

In this example, the parameter choice indicates[3] that we have surmised:

(a) The maximal attainable amplitude of C_B is 40/3.

(b) The maximal attainable amplitude of C_A is 20/3.

(c) ED_{50} for Hormone A is 2.

(d) ID_{50} for Hormone B is 2.

Even in this simple example, we have a variety of possibilities to model the interactions between A and B by considering changes to the parameters. It is important to understand in what way the parameter choice might affect the oscillatory properties of the system, and it is desirable to formulate a physiological explanation. For example, let's consider the following question: In what way will the model (10-15)

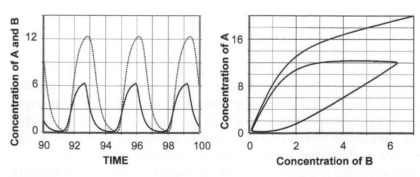

FIGURE 10-16.
Left panel: Dynamics of the concentration of A (solid) and B (dotted), for the model described by Eq. (10-15). Right panel: Corresponding phase diagram.

---

3. See Exercise 10-5 and the preceding discussion on the physiological meaning of the Hill function parameters.

respond to a fourfold decrease in feedback latency? That is, how will the model be affected by decreasing the delay time from D = 1 to D = 0.25?

To gain a better insight into this question, note that according to (a) above and Figure 10-16 (left panel), the peak response of B to endogenous A-drive is almost maximal. Therefore, an exogenous bolus of the stimulator A cannot bring forth dose-dependent release of B-secretion at levels notably higher than the typical endogenous B-concentration. This may be because the delay in the feedback is sufficiently long to provide enough time for the pulses to unfold. This is confirmed by the result produced by the model after 4-fold decrease in feedback latency. The output, presented in Figure 10-17, displays decreased peak amplitudes and elevated nadirs. The typical (endogenous) peak heights are lower than expected [(a) and (b) above], but the system is capable of responding to exogenous stimulation with higher pulses (Figure 10-17). Also, there is an increase in pulse frequency caused by the shorter feedback delay.

To provide an example that approximates the network where A inhibits the secretion of B (Figure 10-14, right panel), we use the following reference system of delayed ODEs:

$$\frac{dC_A}{dt} = -3C_A(t) + 20\frac{[C_B(t-1)/2]^3}{[C_B(t-1)/2]^3 + 1}$$

$$\frac{dC_B}{dt} = -3C_B(t) + 40\frac{1}{[C_A(t)/2]^3 + 1}. \tag{10-16}$$

Note the difference from Eq. (10-15): the up- (down)-regulatory function is replaced by a down- (up)-regulatory function. This system also has a stable periodic solution, a graph of which is presented in Figure 10-18:

We need to outline the framework that links the system parameters with experimental observations. This information is important

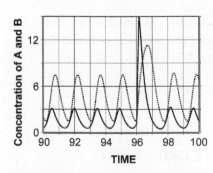
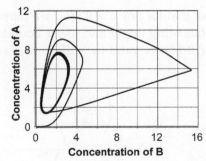

FIGURE 10-17.
Left panel: The effect of fourfold decrease in feedback latency [all other model parameters are the same as in Eq. (10-15)]. The black line illustrates the evolution of the stimulator A; the dotted line depicts hormone B. The plot also shows the model response to a brief bolus of hormone A introduced at t = 96:00. Right panel: Corresponding phase diagram.

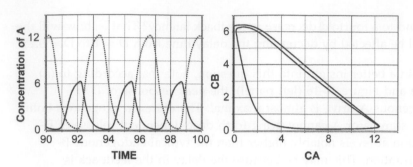

FIGURE 10-18.
Left panel: Evolution of the concentrations of hormone A (dotted) and hormone B (black), for the model described by Eq. (10-16). Right panel: Corresponding phase diagram.

when initial estimates for the parameters are made. We saw in Chapter 8 that providing good initial guesses for the values of the parameters may be critical to determining the correct values of the model parameters when nonlinear least-squares algorithms are applied to determine the best fit between data and model. Deriving dependencies between system parameters and experimental observations will also facilitate our discussion of changes in sensitivity.

B. Initial Parameter Estimates

Our purpose now is to derive simple conditions, broadly linking system parameters to experimental observations. Recall that in deriving the mathematical form of the control function S_A we found a relationship between the parameters of Eq. (10-8) and the maximal attainable hormone concentration (Exercise 10-6). Therefore, the elimination constants α and β and the coefficients a and b from Eq. (10-14) are linked with the maximal hormone concentrations in the following way: $C_{A,\max} = a/\alpha$ and $C_{B,\max} = b/\beta$. The following result shows that, after some time (depending on the initial conditions), the solutions will also be bounded away from zero.

EXERCISE 10-9

Prove that for any $\varepsilon > 0$ (and we may choose ε as small as we like) the following upper and lower bounds on the solution of the system Eq. (10-14) are valid for sufficiently large t:

$$0 < \frac{a}{\alpha}\frac{1}{\left(\dfrac{b}{\beta T_B}\right)^{n_B} + 1} - \varepsilon \le C_A(t) \le \frac{a}{\alpha} + \varepsilon$$

$$0 < \frac{b}{\beta}\frac{1}{\left(\dfrac{T_A}{\min C_A}\right)^{n_A} + 1} - \varepsilon \le C_B(t) \le b/\beta + \varepsilon. \tag{10-17}$$

Hint: Use the fact that if f and g are solutions to the differential equations $f' = -kf + F(t)$ and $g' = -kg + G(t)$ (with $k > 0$ and initial conditions $f(t_0) = g(t_0) > 0$) and $F(t) \geq G(t) \geq 0$, we have $f(t) \geq g(t)$ for all $t > t_0$.

Exercise 10-9 establishes that the evolving solution of the system (C_A, C_B) will approach the square $(0 \leq C_A \leq a/\alpha, 0 \leq C_B \leq b/\beta)$ even for those initial conditions that are outside of this square. On the other hand, if there is no external input in the system (no infusion of A or B), then $C_A < a/\alpha + \varepsilon$ after some time, and we get from Eq. (10-14) that the actual endogenous peak concentration of B will never reach b/β. In particular, with time its upper limit will approach

$$\frac{b}{\beta} \frac{1}{\left(\dfrac{T_A}{a/\alpha + \varepsilon}\right)^{n_A} + 1} \tag{10-18}$$

which is less than b/β. To get the above inequality, substitute the estimate for C_A in the term controlling the secretion in the second Eq. (10-14), and estimate the maximal concentration of B following the Hint in Exercise 10-9. We may work in a similar way to estimate the concentration of A using the second inequality in Eq. (10-11). (How?) Therefore, the solution of the unperturbed system (C_A, C_B) will be inside the square $(0 \leq C_A \leq a/\alpha, 0 \leq C_B \leq b/\beta)$ and the concentration of one hormone stimulated by an infusion of the other hormone will remain bounded in this square. (Why?) The latter justifies the previously used term *maximal attainable amplitude*.

All estimates may be further refined through a recurrent procedure inherent in the core system (Eq. (10-14)). For example, one can combine the two inequalities from Eq. (10-17) to get an explicit lower bound for C_B of

$$\frac{b}{\beta} \frac{1}{\left(\dfrac{\alpha T_A \left(\left(\dfrac{b}{\beta T_B}\right)^{n_B} + 1\right)^{n_A}}{a}\right) + 1} \leq C_B(t). \tag{10-19}$$

Accordingly, we can use this to write an explicit upper bound for C_A:

$$C_A \leq \frac{a}{\alpha} \frac{1}{\left(\dfrac{C_{B,min}}{T_B}\right)^{n_B} + 1} \leq \frac{a}{\alpha} \frac{1}{\left(\dfrac{M}{T_B}\right)^{n_B} + 1}, \text{ where } M = \frac{b}{\beta} \frac{1}{\left(\dfrac{\alpha T_A \left(\left(\dfrac{b}{\beta T_B}\right)^{n_B} + 1\right)}{a}\right)^{n_A} + 1}.$$

The inequalities derived above can assist in determining reasonable (initial) values for the model parameters. As we see next, they can also be used in examining changes in sensitivity.

C. Simulation of Feedback Experiments

The success of any modeling effort is measured by the model's potential to reproduce key feedback experiments. We now discuss ways for modeling the system reaction to certain experimental approaches, aimed at disclosing specific linkages within endocrine systems. We shall use reference systems, such as in Figure 10-14, to illustrate the modeling of three common lab techniques: administration of antibody to one of the nodes, sensitivity alterations, and infusion of one of the system nodes (hormones). We examine the corresponding model response and demonstrate that all of these conditions might disrupt the periodicity of the system.

1. Changes in sensitivity

Changes in the parameters of the Hill functions approximate alterations in system sensitivity. In the model from Eq. (10-15), this would correspond to changes in the threshold or in the Hill coefficient. Reducing (increasing) a threshold results in sensitivity increase (decrease). Changes in the Hill coefficient affect the slope of the control function. In general, increasing the Hill coefficient causes minor alterations in the frequency and does not disrupt the pulsatility of the hormones. In contrast, a decrease could effectively obliterate the oscillations by preventing the system from overshooting its steady state.

EXERCISE 10-10

Use Eq. (10-17) to show that pulses gradually shrink with:

(a) Decrease of T_A.

(b) Increase, but not decrease, of T_B.

Provide a heuristic explanation for the observed changes in model behavior.

Hint: Show that if $T_A \to 0$ then $C_B \to b/\beta$; and if $T_B \to \infty$ then $C_A \to a/\alpha$.

With appropriate computer software, one can study the specific effect of increases or decreases n_A, n_B, T_A, and T_B on the output. For the model given by Eq. (10-15), for example, investigating the numerical solutions indicates that:

1. Increases in n_B and/or n_A do not disrupt the pulsatility and yield minor increases in frequency and peak amplitude.

2. Decreases in n_B or n_A may alter the output, causing pulse shrinking and eventually loss of pulsatility.

3. Both decreases or increases of T_A can obliterate oscillations.

4. Increases, but not decreases, of T_B remove the pulsatility.

2. Antibody administration

Antibody molecules are made of proteins generated by the immune system. They are produced and circulated in the bloodstream in response to agents, usually cells or molecules, that the organism considers "not self." If these circulating antibodies come in contact with the agent, they are able to bind specifically to that foreign object—the target. This binding will result in several possible outcomes: The target may be inactivated; it may be now more easily destroyed by the immune system; or the target may now be unable to associate with the tissue. Accordingly, administration of a compound that acts like an antibody (Ab) to a certain substance, referred here as S, typically results in the de facto removal of S from the system. The rate of removal depends on the specific chemical reaction between Ab and S, and the process can be simulated by an increase of the elimination rate of S with a certain reaction-specific factor. The chemical reaction may change the single half-life pattern into a multiple half-life model (see Chapter 9). However, the single half-life approximation might still be sufficient in the simulations or used as a first step towards a more complex model.

To exemplify the concept, we simulated variable removal of the inhibitor A in the model described by Eq. (10-16) (see Figure 10-19). Three simulations were performed in which the coefficient clearance rate of A was increased gradually (90% increase was achieved in less than 3 hours) 4-fold (left), 8-fold (middle), or 16-fold (right) starting at $t = 88:00$.

Figure 10-19 exemplifies that increasing the elimination rate of a hormone could be used to simulate infusion of an antibody and almost a complete removal of one of the nodes. This may result in loss of periodicity (Figure 10-19, middle, right). The plot in Figure 10-19 (left panel) also captures a very interesting phenomenon predicted by the

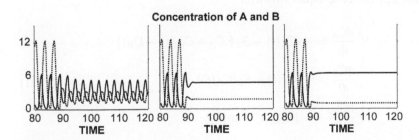

FIGURE 10-19.
Simulated variable infusion (starting at $t = 88:00$) of antibody to the inhibitor A (dotted line) in the reference model outlined in Eq. (10-16). The plots depict low (left panel), medium (middle panel), or almost complete (right panel) removal of A.

model: decrease in the peak amplitudes of B, even though an inhibitor is removed from the system.

3. Infusion of a system node

A typical approach to experimentally investigating the behavior of a particular endocrine network is to monitor the response to infusion of one of the system components. For example, if we want to explore the feedback exerted by GH on its own release, the experimental design would involve systemic administration of GH. If the infused hormone is indistinguishable from the endogenously secreted hormone, we have to estimate the amounts of hormone secreted internally by subtracting the model-predicted concentration of the infused hormone from the total measured hormone.

We start with an approximation of the concentration rate of change of a hormone that is simultaneously secreted and infused. The correct way to simulate infusion of a hormone, which is also a system node, would be to add an infusion term to the right-hand side of the corresponding ODE. This term should correspond to the infusion rate profile in the real experiment. Mathematically, it might be interpreted as change in the basal secretion. In terms of the model described by Eq. (10-14), if we are simulating infusion of hormone B, the corresponding equation becomes:

$$\frac{dC_B}{dt} = -\beta C_B(t) + S_B(C_A(t)) + \inf(t), \qquad (10\text{-}20)$$

where inf(t) is the infusion rate term. The solution of the above equation is the sum of both endogenous and exogenous concentrations of B. To follow the distinction explicitly, a new equation should be added to the system:

$$\frac{dC_{\inf}}{dt} = -\beta C_{\inf}(t) + \inf(t)$$

and $C_B(t)$ is replaced by $C_B(t) + C_{\inf}(t)$ in all model equations, except the one that describes the rate of change of the concentration of B. To sum up, the core equations are:

$$\frac{dC_A}{dt} = -\alpha C_A(t) + S_A\{[C_B + C_{\inf}](t - D_B)\}$$

$$\frac{dC_B}{dt} = -\beta C_B(t) + S_B(C_A(t)) \qquad (10\text{-}21)$$

$$\frac{dC_{\inf}}{dt} = -\beta C_{\inf}(t) + \inf(t).$$

The model defined by Eq. (10-21) is, in essence, a three-node/one-feedback construct, where exogenous B is the new node. The model output depicted in Figure 10-20 uses this system of differential equations to simulate exogenous infusion with the same assumptions used earlier for the simulations shown in Figure 10-17. Note that in Figure 10-17 we exploited Eq. (10-20), where the endogenous and the exogenous component were not separated. As expected, the model output for hormone A remains unchanged, and the sum of endogeneous and exogeneous B in Figure 10-20 equates to the profile shown in black in Figure 10-17, illustrating the equivalence of the two approaches. Another example is shown below in Section IV, Part E.

Changes in the profiles of the control functions can be used to model alterations in system sensitivity. The analysis shows that if a model has stable periodic behavior, the increase in one of the Hill coefficients would not change the system performance (see, for example, Glass and Kauffman [1973]). On the other hand, a decrease in the same parameter may transform the steady state from a repellor into an attractor and affect the periodic behavior. Changes in the action thresholds may also affect the periodicity. Exogenous hormone delivery can be simulated by a simple increase in the basal secretion, or by introducing a third node, if we would like to distinguish between the exogenous and endogenous components of one and the same substance, as we did in Eq. (10-21).

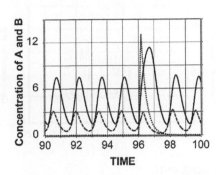

FIGURE 10-20.
Simulated bolus infusion (at $t = 96{:}00$) of the system hormone B (dashed line) in the model outlined in Eq. (10-15). The exogenous hormone is shown with a dotted line, whereas hormone A is plotted in black.

D. Oscillations Generated by a Perturbation

In the reference models in the previous section, the pulsatility was generated by a system having a unique periodic solution and a unique fixed repelling point. In this section, we demonstrate how oscillations appear as a result of disrupting a system that does not have a periodic solution and its fixed point is an asymptotically stable focus (Figure 10-15, left panel). We illustrate this concept with the following model:

$$\frac{dC_A}{dt} = -3C_A(t) + 60\frac{1}{\left[C_B(t - 0.25)/5\right]^2 + 1}$$

$$\frac{dC_B}{dt} = -3C_B(t) + 40\frac{\left[C_A(t)/4\right]^2}{\left[C_A(t)/4\right]^2 + 1}.$$

$(10\text{-}22)$

In this example, formalizing again the network in Figure 10-14 (left panel), the parameters are chosen so that there is no periodic solution [in contrast, for example, with Eq. (10-15)] and the unique fixed point attracts all trajectories in the phase space. Therefore, this system by itself cannot generate stable oscillations. However, if it is externally stimulated, it can be removed from its steady state, and oscillations will be generated. We can show this by simulating two brief (10-minute), unequal suppressions of the secretion of B at $t = 94$ and $t = 104$

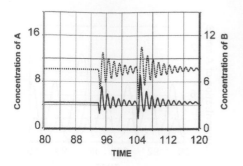

FIGURE 10-21.
Oscillations generated by perturbation of the system in Eq. (10-22). The plot depicts a brief (1/3 time units) suppression of the secretion of B (black line) at $t = 94$ and $t = 104$. The second perturbation was sevenfold higher.

(the second perturbation was sevenfold higher). The simulation was performed by decreasing the secretion rate of B (parameter b, which in this case is equal to 40) 3- and 21-fold, lasting 1/3 time units. Inhibition of B removes the trajectory in the phase space away from the fixed point in a dose-dependent manner, and the system gets enough energy to initiate a waning sequence of pulses, as shown in Figure 10-21. The frequency of the pulses is controlled by the coefficients of the core system Eq. (10-22), while the initial peak amplitude depends on the level of the perturbation.

In the above example, the perturbations were independent of the core system. In Section V, we show that delayed system feedback could evoke a similar effect, providing enough energy and generating oscillations in submodels with damped periodicity.

E. Identifying Nodes, Controlling the Oscillations

All of the system models considered so far were based on prior knowledge of the interaction between hormones, which was then utilized to create schematic diagrams describing the specific links of interaction (see, for example, Figures 10-5 and 10-14). We now examine possible approaches that would allow us to decide whether such interactions exist between hormones. We suggest experimental paradigms tailored to support or reject the hypothesis that two hormones, A and B, are interconnected in a specific oscillating networklike construct (like those in Figure 10-14).

We begin by describing a commonly encountered situation in which the results of mathematical simulations could provide valuable information for further experimental investigations. Consider, for example, a system in which B is the major oscillating hormone, its concentration in the bloodstream is readily assessable, and experimental data indicate that its release is controlled by another hormone, A. As frequently occurs, however, measuring hormone A directly may be experimentally difficult. For example, some human neuroregulators/ hormones, such as GHRH or gonadotropin-releasing hormone, are produced in the hypothalamus and control major pituitary peptides (such as GH and LH). Unfortunately, direct measurement of these hormones in the bloodstream is difficult, because they are secreted in small quantities and their concentration in the circulation is practically undetectable. When the concentration of A cannot be measured directly, the question of *whether a delayed feedback loop between A and B exists to drive the oscillations of B* cannot be answered directly, either. However, we can use mathematical models to facilitate the design of specific experiments exploring system connectivity. In this situation, the results of the real experiments are interpreted based on the outcome of the simulated experiments.

Assume that we would like to find evidence supporting the hypothesis that the inhibitor A is connected with B in a network similar to that shown in Figure 10-14 (right panel). We can try to monitor the system response to neutralization of the action of A (A-receptor blocker) or removal of A (antibody infusion) from the system. Because the model predicts gradual pulse shrinking toward the steady-state level (Figure 10-19), a similar experimental outcome would indicate system connectivity like that shown in Figure 10-14. As another example, we might be seeking support for the hypothesis that stimulator A is connected with B as shown in Figure 10-14 (left panel). In this case, administering a large, constant infusion of A should clamp the oscillations (by exceeding the action threshold, resulting in continuous response from the target organ). The latter concept is shown in Figure 10-22, which depicts a computer-generated prediction of the system response to infusion of hormone A [assuming that A stimulates B: Eq. (10-15)]. We simulated constant low (left panel) and high (right panel) infusion of A by increasing the basal A-secretion from zero to 3 or 6, starting at $t = 97$ [see Section IV, Part C.3 and Eq. (10-21)].

The model predicts gradual pulse shrinking toward the current steady-state level (the latter depends on the infusion rate). If exogenous A is sufficiently high (right panel), the pulses disappear. Similar outcomes observed in real experimental setting would suggest that A and B are connected as shown in Figure 10-14 (left panel).

This approach has limitations, however. These experiments cannot disclose whether A is actually involved in a feedback with B, or acts merely as a trigger to remove a certain subsystem from its steady state. For example, consider the two networks shown in Figure 10-23, and suppose only the concentrations of hormone B can be measured.

Assume that E stimulates B and its removal diminishes the secretion of B. Because endogenous E cannot be assessed, we have no direct means to

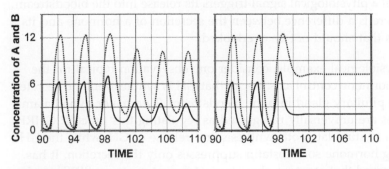

FIGURE 10-22.
System response to exogenous constant administration of the stimulating hormone A (black line). The plots show simulation of low (left panel) and high (right panel) infusion of A starting at $t = 97$.

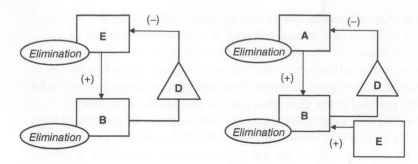

FIGURE 10-23.
Two hypothetical networks, in which hormone E stimulates the secretion of B. E is either involved in a delayed feedback (left panel), or perturbs the subsystem A-B (right panel). (Reprinted from Farhy, L. S. [2004]. Modeling of oscillations in endocrine networks with feedback, *Methods in Enzymology, 384*, 54–81. Copyright 2004, with permission from Elsevier.)

establish whether E is involved in a delayed feedback loop with B. Moreover, in both networks, constant high infusion of E (as proposed above) removes the pulsatility and elicits constant secretion of B. Therefore, a more sophisticated experiment is required to reveal whether E is indeed involved in a feedback loop with B (Figure 10-23, left panel) or acts by perturbing the A-B subsystem (Figure 10-23, right panel). A possible approach would be to block the endogenous secretion of E and then administer a single E bolus. The system response would be either a single spike of B secretion, if the network were that depicted on Figure 10-23 (left panel), or a waning train of several B pulses, if the network is the one shown on Figure 10-23 (right panel).

F. Separating Synthesis from Secretion

Although fluctuations of hormone concentration levels in the bloodstream reflect analogous secretion patterns, this does not necessarily mean that hormone synthesis follows the same patterns. For example, the hormone synthesis may be occurring at a low constant rate, with the amount of synthesized hormone accumulating in a separate pool until a physiological signal triggers its release into the bloodstream. Thus, there is a difference between the secretion of a hormone and its synthesis that should not be overlooked.

As a physiological example, consider one observation typical for the growth hormone control axis. Recall that GH is synthesized and secreted from the pituitary gland into the bloodstream. Two major hypothalamic hormones are universally recognized as regulating this process: GHRH is known to stimulate both synthesis and release of GH, while the GH-inhibiting hormone somatostatin suppresses only the secretion. It has been observed that constant, short-term (4 hours), systemic SRIF infusion initially suppresses GH release, but upon its withdrawal stimulates a large, reboundlike, GH release (Clark et al. [1988]). An intuitive explanation of this effect might be that because SRIF suppresses the secretion, but not the synthesis, during the 4 hours of systemic SRIF infusion, the releasable pool of GH has increased under persisting GHRH drive.

To model-test this prediction, it is appropriate to separate, on a network level, the hormone synthesis and storage from its release. This separation is important, because major network components affect these processes in different ways. Let us again consider the network from Figure 10-14 (left panel), in an attempt to explain a rebound release of B following a withdrawal of continuous infusion of a certain substance C. Assume that during the infusion of C the release of B was suppressed and we have evidence that C is not affecting the release of A. A possible explanation of the rebound phenomenon would be that C affects the release of B, but not its synthesis. However, because all conduits in the network are affected in this experiment, the intuitive reconstruction of all processes involved is not trivial.

One way to model this situation mathematically is to introduce a so-called storage pool, in which B is synthesized and held for release and another pool (e.g., the bloodstream) in which B is secreted. This adds a new equation to the model, describing the dynamics of the concentration of B in the storage pool. Denote this concentration by P_B. The following basic model assumptions would be appropriate:

1. The total concentration of B in the storage pool, P_B, is positively affected by the synthesis and negatively affected by the release.

2. The concentration P_B feedbacks on the synthesis of B and cannot exceed a certain absolute limit P_{max}.

3. The rate of release of B from the storage pool is stimulated by a high pool concentration, but might be inhibited by the concentration of B in the bloodstream.

4. B is subjected to elimination only after it is secreted.

Next, we construct the schematic diagram incorporating these assumptions. In the network from Figure 10-14 (left panel), in addition to A and B, there is a new substance, C, that inhibits the secretion (competing with A) but does not affect the synthesis and storage of B. This is shown in Figure 10-24.

Using Eq. (10-13) to approximate a "competitive" control function, we can describe the network with the following system of delayed ordinary differential equations:

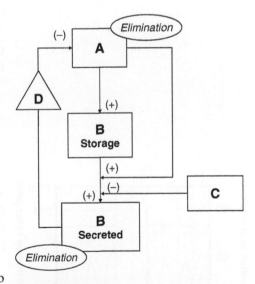

FIGURE 10-24.
Formal network that distinguishes between synthesis and release of hormone B. Hormone A stimulates the synthesis and secretion of B. A third hormone C suppresses the release of B, but not its synthesis. (Reprinted from Farhy, L. S. [2004]. Modeling of oscillations in endocrine networks with feedback, *Methods in Enzymology, 384,* 54–81. Copyright 2004, with permission from Elsevier.)

$$\frac{dC_A}{dt} = -\alpha C_A(t) + a \frac{1}{(C_B(t - D_B)/T_B)^{n_B} + 1}$$

$$\frac{dC_B}{dt} = -\beta C_B(t) + b \frac{(C_A(t)/T_{A,1})^{n_{A,1}}}{(C_A(t)/T_{A,1})^{n_{A,1}} + (C_C(t)/T_C)^{n_C} + 1} \frac{(P_B(t)/T_P)^{n_P}}{(P_B(t)/T_P)^{n_P} + 1}$$

$$\frac{dP_B}{dt} = c(P_{max} - P_B) \frac{(C_A(t)/T_{A,2})^{n_{A,2}}}{(C_A(t)/T_{A,2})^{n_{A,2}} + 1}$$

$$- b\theta \frac{(C_A(t)/T_{A,1})^{n_{A,1}}}{(C_A(t)/T_{A,1})^{n_{A,1}} + (C_C(t)/T_C)^{n_C} + 1} \frac{(P_B(t)/T_P)^{n_P}}{(P_B(t)/T_P)^{n_P} + 1}.$$

$$(10\text{-}23)$$

Here, for simplicity, we assumed that circulating levels of B do not feedback on the secretion. This corresponds to a model with a much higher concentration in the storage pool than in the circulation. The parameter c controls the rate of A-stimulated synthesis of B. The parameter θ represents the ratio between the volumes of the storage pool and the pool into which B is secreted. Typically, the second pool is larger and $\theta > 1$. It is assumed that the control functions that correspond to the A-driven synthesis and release are different with distinct thresholds $T_{A,1}$ and $T_{A,2}$ and corresponding Hill coefficients $n_{A,1}$ and $n_{A,2}$. The control, exerted on the secretion by the current concentrations of B in the storage pool, is presented by the up-regulatory function $\frac{(P_B(t)/T_P)^{n_P}}{(P_B(t)/T_P)^{n_P} + 1}$. The following values were assigned to the parameters in Eq. (10-23):

$$\alpha = 1; \quad \beta = 2; \quad \theta = 6; \quad a = 4; \quad b = 4000; \quad c = 2; \quad P_{max} = 900;$$

$$D_B = 2; \quad T_{A,1} = 4; \quad T_{A,2} = 3; \quad T_B = 40; \quad T_C = 10; \quad T_P = 500;$$

$$n_{A,1} = 2; \quad n_{A,2} = 2; \quad n_B = 3; \quad n_C = 2; \quad n_P = 2.$$

The infusion of C was approximated with a separate model equation (see Section IV, Part C.3) with an elimination constant equal to 5 and an infusion rate assumed to be a nonzero constant only during the time of infusion. The infusion is determined by

$$\inf(t) = \begin{cases} 0 & \text{if} & t < 60 \\ 3000 & \text{if} & 60 < t < 70 \\ 0 & \text{if} & t > 70 \end{cases}.$$

The model output is shown in Figure 10-25.

The plot depicts a reboundlike increase in the secretion of B following withdrawal of the inhibitor C. From a mechanistic point of view,

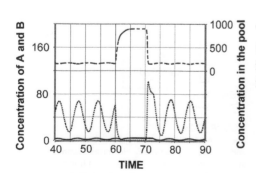

FIGURE 10-25.
Simulated rebound response following withdrawal of continuous C-infusion (timeline 60-70). Legend: A: black; B: dotted; B in the storage pool: dashed.

during the infusion of C the secretion of B is suppressed (but not the synthesis), and the concentration in the storage pool is increased. The concentration of A also increases, because low B levels cannot effectively block its release. Thus, the model explains the rebound jointly, by the augmented concentration in the storage pool and the increased secretion of A.

We would like to note that a network may incorporate a single feedback loop in a more complex way (e.g., via a combination of two or more nodes). Stability analysis of the steady state of such three-node networks with one feedback loop shows that these systems are capable of sustaining periodicity even without an explicit delay in the feedback loop, if the Hill coefficients are relatively high. The specific calculations can be found in Richelle (1977) and Thomas (1973).

In the case of two nodes and one negative feedback loop, the systems considered in this section always have only one fixed point (steady state), which is either a repellor or an attractor (Figure 10-15). In the first instance, the system has a unique limit cycle—a periodic solution, which attracts all trajectories in the phase space and thereby generates stable periodic behavior (Figure 10-16). In the second instance, the steady state is either a focus or a node and attracts all trajectories in the phase space. The construct displays damped periodic behavior only in the case of a focus. An external perturbation can initiate a waning train of pulses (Figure 10-1 , right panel; Figure 10-21) by removing the system from its steady state. Therefore, oscillations might be generated even by a system that does not have a periodic solution, and its fixed point is asymptotically stable. However, an external energy source should exist. The frequency of such oscillations is largely independent of the external perturbation (Figure 10-21).

V. NETWORKS WITH MULTIPLE FEEDBACK LOOPS

The available experimental data might suggest that the release of a particular hormone B is controlled by multiple mechanisms, with different periodicity in the timing of their action. This implies that probably more than one (delayed) feedback loop regulates the secretion of B and the formal endocrine network may include more than two nodes. In determining the elements to be included in the core construct, it is important to keep track of the length of the delays in the feedback action of all nodes of interest. For example, if the goal were to explain events recurring every one to three hours, the natural candidates to include in the formal network would be nodes involved in feedback relations with B with delays shorter than three hours. Long feedback delays cannot account for high-frequency events. In particular, if we hypothesize that a certain feedback is responsible for a train of pulses in the hormone concentration profile, the direct delay in this feedback must be shorter than the interpulse interval.

We now briefly discuss some features of abstract endocrine networks incorporating more than one delayed feedback loop. We shall consider networks with two (delayed) feedback loops, and where each loop accounts for its own oscillatory mechanism. Examples of two-feedback constructs are shown in Figure 10-26.

Note that each of the two three-node networks, shown in the middle panels of Figure 10-26, could be reduced to its corresponding two-node network from the top panels of Figure 10-26. For example, consider the three-node/two-feedback network shown in Figure 10-26, middle left panel. Assuming that both B and C can fully suppress the release of A (that is, when no basal secretion of A is assumed), we can describe the formal network by the system of delayed ODE:

$$\frac{dC_A}{dt} = -3C_A(t) + 10000 \frac{1}{[C_B(t-0.15)/100]^3 + 1} \frac{1}{[C_C(t)/70]^3 + 1}$$

$$\frac{dC_B}{dt} = -3C_B(t) + 4000 \frac{[C_A(t)/500]^2}{[C_A(t)/500]^2 + 1} \tag{10-24}$$

$$\frac{dC_C}{dt} = -8C_C(t) + 13200 \frac{[C_B(t-4)/200]^2}{[C_B(t-4)/200]^2 + 1}.$$

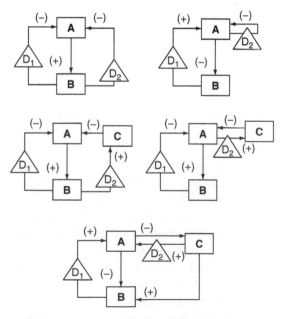

FIGURE 10-26.
Examples of hypothetical endocrine networks with more than one delayed feedback loop. In addition to the connections shown, all hormones are subject to elimination. (Reprinted from Farhy, L. S. [2004]. Modeling of oscillations in endocrine networks with feedback, *Methods in Enzymology, 384*, 54–81. Copyright 2004, with permission from Elsevier.)

This system is capable of generating recurring multiphase volleys by the mechanism described in Section IV, Part D, as illustrated in Figure 10-27. In particular, the A/B two-node subsystem, which does not have a periodic solution if C is taken out of the system, is perturbed by a delayed system loop via the third node C. This removes the system from its steady state and drives consecutive pulses during recurrent volleys.

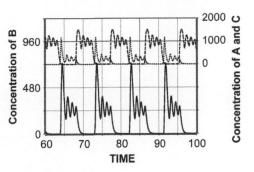

FIGURE 10-27.
Computer-generated output [concentration of A (dotted), B (black), and C (dashed)] of the core system Eq. (10-24).

The schematic diagram in the middle left panel of Figure 10-26, represented by Eq. (10-24), shows that hormone B reduces the secretion of A both directly and indirectly. The direct down-regulation occurs with a delay D_1 and corresponds to the left link of the diagram. The indirect effect of B is caused by up-regulating the secretion of a third hormone C, with a delay D_2. This third hormone then down-regulates the secretion of A. This observation shows that the sequence of nodes and conduits (B → C → A → B) is essentially a negative two-node delayed feedback loop: (B → A → B). Analogous model output can be achieved by reducing the three-node network to a two-node model with two feedbacks. Therefore, the system can be modeled by removing C from the system and introducing a correct delay in the conduit (B → A). The reduced network is the one shown in Figure 10-26, upper left panel (with, of course, D_2 different from the delay used to describe the B → C → A → B loop). A corresponding simplified system of delayed ODEs could be:

$$\frac{dC_A}{dt} = -3C_A(t) + 10000 \frac{1}{[C_B(t-0.15)/100]^3 + 1} \frac{1}{[C_B(t-4)/10]^3 + 1}$$

$$\frac{dC_B}{dt} = -3C_B(t) + 4000 \frac{[C_A(t)/500]^2}{[C_A(t)/500]^2 + 1}$$

$$(10\text{-}25)$$

and the model output (Figure 10-28) is similar to the hormone profiles shown in Figure 10-27.

Note that the schematic diagram in Figure 10-26 (upper left panel) corresponds to a situation where hormone B down-regulates the secretion of A through two different pathways. Therefore, additionally reducing the number of conduits in this diagram to obtain a network representation such as in Figure 10-14 (left panel) may not be possible. As a broad rule, decisions for reducing the number of nodes or links in the schematic diagrams should always be considered in the specific context of the particular physiology that is being investigated.

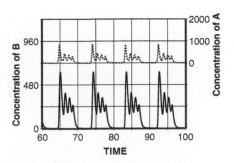

FIGURE 10-28.
Computer-generated output [concentration of A (dashed) and B (black)] of the core system Eq. (10-25).

Reducing the number of nodes and, therefore, the number of equations, from three to two decreases the number of parameters to be determined and the time needed for solving the equations numerically. This would be most important if multiple computer runs are required. Adding the third node to the formal network can only be justified if the goal is to simulate specific outcomes explicitly involving C; even then, the initial adjustment of the model would be significantly facilitated if C is added to the system only after validating the two-node construct. If the network is more complex, the attempt to reduce the number of nodes might not be possible. For example, the construct shown in Figure 10-26 (lower panel) cannot be transformed into a two-node model because of the high system interconnectivity.

In closing, we note that significant theoretical complications arise when networks have multiple steady states of different types. Methods such as Boolean formalization, described in Thomas (1973; 1983), could be used to analyze such systems. This method serves as an intermediary analysis between modeling phases 2 and 3 described in the first section of this chapter. The idea behind it is to describe complex systems in simpler terms that allow for preliminary finding of all stable and unstable steady states. Other complex endocrine networks with intertwined feedback loops are considered in Farhy and Veldhuis (2004; 2005), where their analysis strongly depends on the specific physiology.

REFERENCES

Clark, R. G., Carlsson, L. M.S, Rafferty, B., & Robinson, I. C. A. F. (1988). The rebound release of growth hormone (GH) following somatostatin infusion in rats involves hypothalamic GH-releasing factor release. *Journal of Endocrinology, 119,* 397–404.

Farhy, L. S., & Veldhuis, J. D. (2004). Putative growth hormone (GH) pulse renewal: Periventricular somatostatinergic control of an arcuate-nuclear somatostatin and GH-releasing hormone oscillator. *American Journal of Physiology-Regulatory, Integrative and Comparative Physiology, 286,* R1030–R1042.

Farhy, L. S., & Veldhuis, J. D. (2005). Deterministic construct of amplifying actions of ghrelin on pulsatile GH secretion. *American Journal of Physiology-Regulatory, Integrative and Comparative Physiology, 288,* R1649–R1663.

Farhy, L. S., Straume, M., Johnson, M. L., Kovatchev, B. P., & Veldhuis, J. D. (2002). Unequal autonegative feedback by GH models the sexual dimorphism in GH secretory dynamics. *American Journal of Physiology-Regulatory, Integrative and Comparative Physiology, 282,* R753–R764.

Farhy, L. S. (2004). Modeling of oscillations in endocrine networks with feedback. In Johnson, M. L., & Brand, L. (Eds.), *Methods in Enzymology* (vol. 384, pp. 54–81). New York: Academic Press.

Friesen, W. O., & Block, G. D. (1984). What is a biological oscillator? *American Journal of Physiology, 246,* R847–R853.

Glass, L., & Kauffman, S. A. (1973). The logical analysis of continuous non-linear biochemical control networks. *Journal of Theoretical Biology, 39,* 103–129.

Keenan, D. M., & Veldhuis, J. D. (2001). Disruption of the hypothalamic luteinizing hormone pulsing mechanism in aging men. *American Journal of Physiology-Regulatory, Integrative and Comparative Physiology, 281,* R1917.

Lanzi, R., & Tannenbaum, G. (1992). Time course and mechanism of growth hormone's negative feedback effect on its own spontaneous release. *Endocrinology, 130,* 780–788.

Mikawa, T., Masui, R., & Kuramitsu, S. (1998). RecA protein has extremely high cooperativity for substrate in its ATPase activity. *Journal of Biochemistry, 123,* 450–457.

Thomas, R. (1973). Boolean formalization of genetic control circuits. *Journal of Theoretical Biology, 42,* 563–585.

Thomas, R. (1983). Logical description, analysis and synthesis of biological and other networks comprising feedback loops. *Advances in Chemical Physics, 55,* 247–282.

Vrzheshch, P. V., Demina, O. V., Shram, S. I., & Varfolomeev, S. D. (1994). Supercooperativity in platelet aggregation: substituted pyridyl isoxazoles, a new class of supercooperativity platelet aggregation inhibitors. *FEBS Letters, 351,* 168–170.

Wagner, C., Caplan, S. R., & Tannenbaum, G. S. (1998). Genesis of the ultradian rhythm of GH secretion: a new model unifying experimental observations in rats. *American Journal of Physiology, 275,* E1046–E1054.

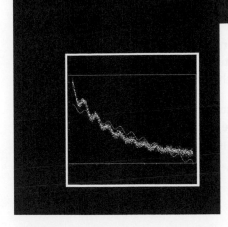

Chapter 11

DETECTING RHYTHMS IN CONFOUNDED DATA

The truth is rarely pure and never simple.

Oscar Wilde (1854–1900)

The rotation of the Earth produces the cycles of day and night that are of immense importance to all living organisms. Organisms that can anticipate these daily rhythms and adapt to them have a selective advantage. For example, the leaves of certain plants alternate between horizontal daytime and vertical nighttime positions, efficiently harvesting light during the day and protecting against radiative cooling at night. Many small, otherwise defenseless animals are largely nocturnal, sleeping during the day when sharp-eyed carnivores are active. Many desert-dwelling animals are also nocturnal, avoiding the daytime's intense heat and desiccating air. Most birds are diurnal, active and feeding during the day and returning to their roosts at night.

These daily biological cycles lasting approximately 24 hours are called *circadian rhythms*—from the Latin *circa* (about) and *dies* (day). A fascinating example of the importance of understanding circadian rhythms comes from the field of parasitology. Elephantiasis, a disease marked by the enormous swelling of parts of the body, is usually caused by the filarial worm *Wuchereria bancrofti*. This worm lodges in lymphatic vessels and hampers the return of lymph to the circulatory system, causing edema and swelling. The larvae of *W. bancrofti*, called microfilariae, exhibit a circadian rhythm with regard to their appearance in the bloodstream. In Africa, Asia, and parts of South America and the Caribbean, the microfilariae are present in large numbers during the night, but absent from the bloodstream during the day. The microfilariae of the South Pacific strain of *W. bancrofti*, on the other hand, are present in the bloodstream in the greatest numbers during the afternoon (Fontes et al. [2000]). Because the disease is diagnosed by observing microfilaria in a blood smear, these findings indicate the importance of knowing when to draw blood samples for testing.

Internal timekeeping mechanisms, often called biological clocks, are what give organisms the ability to anticipate such periodicities. By far, the best-studied of these biological clocks are those responsible for circadian rhythms. Circadian rhythms provide a survival advantage, and organisms with no circadian rhythms, or with malfunctioning circadian mechanisms, will be at a disadvantage. What happens to the mouse that ventures out of his home in the daytime? The hawk eats him for lunch!

In this chapter, we examine the mechanisms controlling the circadian rhythms in living organisms and some of the mathematical tools used to characterize and study rhythmic phenomena.

I. BIOLOGICAL CLOCKS

A. Introduction

Circadian rhythms were first described at the level of whole organisms. In 1729, the French astronomer Jean Jacques d'Ortous de Mairan (1678–1771) observed that the leaves of certain plants were perpendicular to the stem during the day, but parallel to the stem at night. This cycle continued even when the plant was placed in a dark closet. The French natural scientist Henri-Louis Duhamel du Monceau (1700–1782) was intrigued by de Mairan's observations and repeated the experiment in 1758 in a wine cellar, to confirm that the movements had not been caused by light leaking into de Mairan's closet. In 1832, the Swiss botanist Augustin Pyramus de Candolle (1778–1841) reported that the rhythms of the leaves of the sensitive plant, *Mimosa pudica*, persisted even when the plants were subjected to continuous artificial light. He also noted that, over time, the biological clocks of these plants ran faster than normal, with a period of 22 to 22.5 hours. Further experimentation allowed de Candolle to alter the clocks of the plants by changing the cycle of lighting, demonstrating that the plants could obtain cues from the exogenous light/dark cycle.

The existence of such circadian rhythms in animals has also been well-documented. For example, in 1914, J. S. Szymanski reported that goldfish swimming occurs with a daily rhythm. During the 1950s, Janet Harker used cockroaches, with their precisely timed nocturnal running activity, to physically locate the biological clock within these organisms. In the late 1950s, Patricia DeCoursey observed that flying squirrels' emergence from their dens exhibited closely controlled timing.

In the 1960s, Jurgen Aschoff of the Max Planck Institute conducted his so-called bunker studies, demonstrating the existence of human circadian rhythms. Participants in his studies were isolated from all external cues in an underground bunker, thus allowing their endogenous rhythms to be studied in a controlled manner. The volunteers remained in the underground apartment for one month, during which time their temperature, urine excretion, activity patterns, and performance on psychological tests were measured. The volunteers demonstrated undeniable circadian rhythms, with an average period of about 25 hours. Additional details about the studies described above can be found in the book *The Living Clocks* by R. R. Ward (1971).

Before proceeding, we should outline the defining characteristics of these circadian rhythms. First, they all have a period of approximately 24 hours. The period varies from species to species, and even among individuals within a species, and it may be slightly longer or slightly shorter than 24 hours. Second, the rhythms of an organism persist even with constant artificial light or a complete absence of light or of other external cues, and are thus shown to be endogenous to the organism. Third, circadian rhythms can be entrained by exposure to appropriate stimuli. Entrainment allows a continual resetting of the biological clock, ensuring that the organism will properly respond to changes. The changes may be gradual, such as changing day lengths that occur with the seasons, or more drastic, such as time zone changes during a long flight resulting in jet lag. Although the daily dark/light cycle is the principal entrainment stimulus, others, such as feeding and temperature, are known to play a role. Finally, in organisms whose cells are subject to changing operating temperatures, such as cold-blooded animals, insects, fungi, and plants, the circadian clock is temperature-compensated and continues to run at the same speed regardless of temperature.

B. Structure and Function of Biological Clocks in Mammals

In addition to their expression in whole organism behaviors, circadian rhythms have been demonstrated in organs and organ systems. They have been found to exist in the endocrine system, in the liver, pancreas, and digestive system, in muscle and adipose tissue, as well as in the circulatory and respiratory systems. Experimentation in vitro has shown that circadian rhythms may also be detected at the tissue, cellular, and molecular levels. Recent investigations have determined the molecular mechanisms responsible for these rhythms, and ongoing work continues to refine our understanding of these mechanisms.

It has become apparent that in many complex organisms the circadian clock is, in reality, multiple clocks. In fact, a hierarchy of timing mechanisms can be observed to be in operation. In mammals, the highest level of circadian rhythm control is exerted by the *suprachiasmatic nuclei* (SCN). The SCN are areas of the hypothalamus, found in the brain at the top of the brain stem. The SCN receive a neural signal from the retinas, and respond by producing signals controlling the action of a variety of other cells, tissues, and organs. In humans, the SCN controls the pineal gland, which produces a sleep-inducing hormone called melatonin. In the morning, when we first open our eyes, light hits the retinas and a nerve signal is carried to the SCN. The SCN then signals the pineal gland to turn off its production of melatonin. The daily light/dark cycle regulating the circadian clock is known as a *zeitgeber* (German for "time-giver").

Research has shown that a number of other mammalian cells, tissues, and organs have their own biological clocks that can operate in the absence of the SCN. Such clocks have been demonstrated to function in organs, such as the lung, liver, heart, skeletal muscle, and other parts of the brain, as well as in cultured fibroblasts (connective tissue cells). Yamazaki et al. (2000) used transgenic rats containing a reporter gene (one that makes a readily detectable protein) under the direction of circadian gene control regions to look for circadian rhythms in rats (see Section F later in this chapter). The particular gene arrangement was engineered so that the reporter protein would be produced in a rhythmic fashion in all tissues generating circadian rhythms. Rhythmicity was observed in the SCN, liver, lung, and skeletal muscle. Further research has demonstrated rhythmicity in other organs and tissues (Abe et al. [2002]). In mammals, the SCN has been found to act as the controller or pacemaker, keeping the other clocks coordinated through neural and hormonal mechanisms. Thus, a relationship between the body's many biological clocks exists, forming an internal temporal order (Richter et al. [2004]). For additional details, we refer the reader to the review articles by Richter et al. (2004) and Bell-Pedersen et al. (2005).

C. The Molecular Bases of Biological Clocks

The biological clocks responsible for circadian rhythms spring from multiple feedback mechanisms involving both positive and negative controls. Control of gene expression, protein–protein interactions, post-translational protein modification, nuclear transport, and protein degradation are all involved. In order to understand the control mechanism, we must first briefly review the flow of genetic information in the cell.

The cell's repository of genetic information is the deoxyribonucleic acid (DNA). Information stored in the DNA is copied into ribonucleic acid (RNA) and then used to direct the production of a protein. This idea, which is called the *central dogma* of molecular biology, is shown in Figure 11-1. DNA and RNA are polymers made up of subunits called nucleotides. A nucleotide consists of a sugar (ribose for RNA or deoxyribose for DNA), a phosphate group, and a nitrogenous base. Only four different nitrogenous bases are used for each nucleic acid. They are adenine (A), guanine (G), cytosine (C), and thymine (T) for DNA and adenine (A), guanine (G), cytosine (C), and uracil (U) for RNA. The structures of the four nucleotides used to make DNA are shown in Figure 11-2. The only parts that differ among the four structures are the nitrogenous bases. The genetic information in the nucleic acids is stored in the sequence of the nitrogenous bases along the chain.

Proteins are the functional elements in the cell. They are the enzymes that catalyze the cell's chemical reactions and are responsible for turning

DNA (contains genes)

Transcription (RNA synthesis)

mRNA (copy of gene)

Translation (protein synthesis)

Protein (cellular machinery)

FIGURE 11-1.
Information flow in a typical cell. Genes are sections of DNA that are transcribed into mRNA, which then goes to the ribosomes to guide the assembly of proteins.

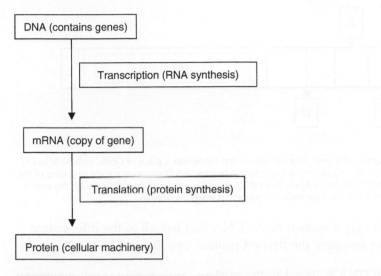

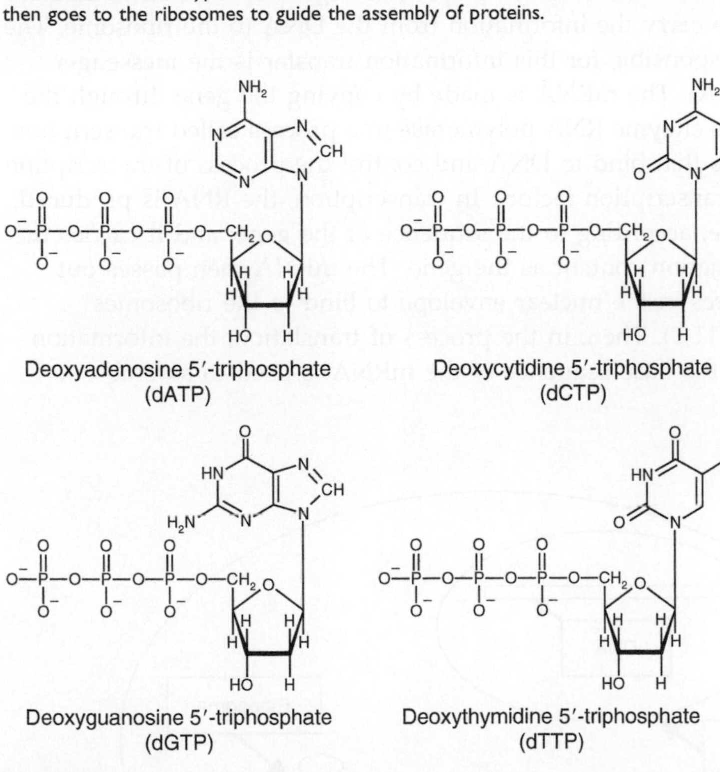

Deoxyadenosine 5′-triphosphate
(dATP)

Deoxycytidine 5′-triphosphate
(dCTP)

Deoxyguanosine 5′-triphosphate
(dGTP)

Deoxythymidine 5′-triphosphate
(dTTP)

FIGURE 11-2.
Structures of the four deoxynucleotides. Observe that two sections of each molecule shown—the phosphates (P- and O-containing groups) and the deoxyribose sugars (the pentagonal structures with O at the apex)—are identical. Only the nitrogenous bases (the one- or two-membered N-containing rings) differ.

genes on or off. They are responsible for cell structure, cell movement, and cell reproduction. They are also, as we shall see, responsible for circadian rhythms. In eukaryotic cells, the genes for all proteins (including the circadian timing proteins) are found in the nucleus.

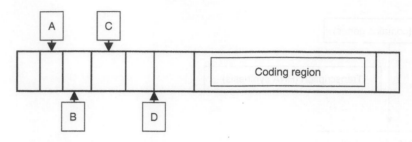

FIGURE 11-3.
A simplified diagram of a gene. The rectangular bar represents a piece of DNA, with A, B, and C representing sites for the binding of transcription factors, and D representing the beginning of the DNA to be transcribed into mRNA. The mRNA would continue to the right through the coding region, which contains the information necessary to make the protein.

A gene is simply a section of the DNA that has all of the information necessary to assemble the desired protein (see Figure 11-3).

Because the DNA is found in the nucleus, and protein synthesis occurs on ribosomes found in the cytoplasm (see Figure 11-4), an intermediary is needed to carry the information from the DNA to the ribosome. The molecule responsible for this information transfer is the messenger RNA (mRNA). The mRNA is made by copying the gene through the action of the enzyme RNA polymerase in a process called transcription. The proteins that bind to DNA and control the process of transcription are called transcription factors. In transcription, the RNA is produced, base by base, according to the sequence of the gene, and it carries the same information content as the gene. The mRNA then passes out through pores in the nuclear envelope to bind to the ribosomes (see Figure 11-5). Then, in the process of translation, the information encoded in the base sequence of the mRNA is used to assemble a protein.

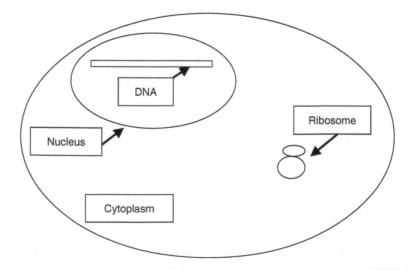

FIGURE 11-4.
A simplified diagram of a eukaryotic cell. Genes are made of DNA, found in the nucleus. Proteins are made on the ribosomes that are found in the cytoplasm.

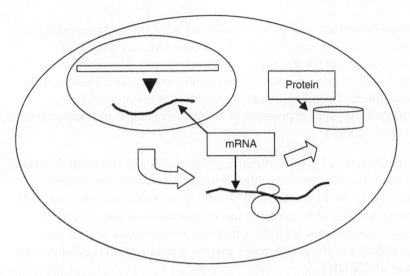

FIGURE 11-5.
Flow of genetic information in a cell. Information encoded in the DNA is copied into mRNA in the nucleus. The mRNA then leaves the nucleus and goes to the cytoplasm, where it binds to a ribosome. The information in the mRNA is used in the synthesis of a protein by the ribosome, and the protein can then perform its intended function.

Proteins may be altered, post-translationally, through the addition of chemical groups, such as sugars or lipids, or through the removal of part of the protein by cleavage with a protein-cutting enzyme. Proteins may be activated or inactivated by phosphorylation, the process of attaching phosphate groups. They may also be permanently inactivated by complete degradation. Many of these processes play a role in the circadian clock mechanisms.

D. Identification of Circadian Genes

Adult *Drosophila* (fruit flies) undergo eclosion (the process of emerging from their pupal cases) in the morning, when the high relative humidity allows their wings to inflate slowly and properly. In the 1960s, the timing of the process of eclosion in *Drosophila* was shown to vary with the genetic constitution of the flies. Under conditions of constant darkness (DD), different strains of flies gave different period lengths. The clear implication of this observation was that specific genes were responsible for the expression of this whole-organism behavior. In 1971, Konopka and Benzer reported the identification of long, short, and arrhythmic mutants of the *period* gene referred to, respectively, as *per*^{L}, *per*^{S}, and *per*^{0} (see Konopka and Benzer [1971]; Panda et al. [2002]).

Subsequently, cloning of the *period* gene permitted controlled studies that demonstrated that using genetic engineering to insert a wild-type *per* gene into *per*^{0} flies restored circadian rhythmicity. Also, increasing the expression of wild-type *per* was shown to result in shorter period length. Both *per* mRNA and protein were found to oscillate rhythmically in a manner consistent with the associated behavioral phenotype.

Wild-type *per* mRNA and protein oscillated with a near 24-hour rhythm, whereas *per^L* and *per^S* mRNAs and proteins oscillated with long (29-hour) and short (19-hour) periods, respectively. The *per^0* gene exhibited arrhythmic *per* mRNA and protein expression. These results provided the first demonstration of genetic expression patterning matching a behavioral expression of a whole-organism phenotype (see Panda et al. [2002].)

Anatomical studies have identified a set of neurons in the adult fly brain that is the activity-controlling site for behavioral circadian rhythms. This area of the fly brain seems to be analogous to the mammalian SCN. Additional studies of a variety of insect species have found similar molecular mechanisms, including proteins homologous to PER (the protein product of the *period* gene), involved in circadian regulation (see Panda et al. [2002]).

Continued research into the molecular mechanisms responsible for circadian rhythms revealed that *per* was not the only gene involved. Another circadian gene discovered in *Drosophila* is called *timeless* (*tim*). The timeless protein TIM forms a complex with the period protein PER that ultimately inhibits its own transcription (see following text and Figure 11-6). Additional genes have since been identified, and our understanding of the molecular mechanism for circadian control has

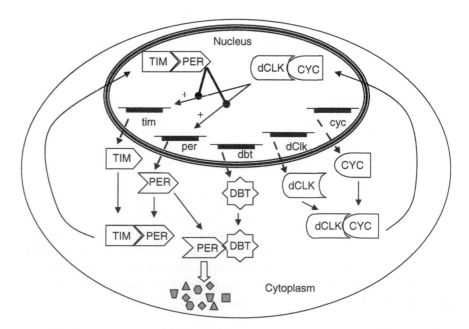

FIGURE 11-6.
Simplified schematic diagram of the *Drosophila* circadian clock mechanism. Abbreviations are as follows: *tim* and TIM are the *timeless* gene and protein, respectively; *per* and PER stand for *period* gene and protein, respectively; *dbt* and DBT for *double time*; *dClk* and dCLK for *clock*; *cyc* and CYC for *cycle*. The *doubletime* protein DBT phosphorylates free PER protein (i.e., any PER not bound to TIM) and facilitates its degradation. The dashed arrows represent transcription followed by translation. In the nucleus, the arrows indicated by a + indicate the elevation of transcription, and the arrows with the round heads indicate blocking of this activity.

been growing progressively clearer over the last decade. Studies of the molecular basis of circadian rhythms now use gene chip technology to determine what genes exhibit circadian or otherwise rhythmic temporal expression patterning at the level of systemwide phenomena. The use of gene chips will be discussed in the next chapter.

Both TIM and PER are made during the day, with the levels of TIM and PER proteins rising to a peak in early evening. TIM binds PER and the TIM/PER complex enters the nucleus and inhibits the action of another set of proteins, dCLK and CYC. These proteins are the products of the *dClock* (*dClk*) and *cycle* (*cyc*) genes, and they are transcription factors. In the absence of the TIM/PER complex, the dCLK and CYC proteins bind to each other and then bind to the controlling regions of the *per* and *tim* genes, turning on their expression so that *per* and *tim* mRNAs, and then proteins, are made. But when the TIM/PER complex enters the nucleus, it interferes with the ability of the dCLK/CYC complex to promote transcription of *tim* and *per* mRNA, so the transcription of *tim* and *per* mRNA stops. Both TIM and PER will be degraded by morning, the dCLK/CYC complex will bind to the *tim* and *per* genes and turn them on, and the amounts of *tim* and *per* mRNA, and then protein, will rise.

This cyclic behavior of mRNAs and proteins gives rise to the circadian rhythm of the fly. There are mammalian homologues of all of these genes, and their mRNAs and proteins undergo similar cyclic behaviors. Generally speaking, the more important a gene is, the more highly conserved it is. The extraordinary conservation of these genes across such diverse species indicates the powerful selective advantage conveyed by the circadian system.

The following works can be recommended for additional information on the identification of circadian genes: Panda et al. (2002), Reppert and Weaver (2002), Richter et al. (2004), and Bell-Pedersen et al. (2005).

E. Studying Circadian Phenomena with *Per-Luc* Bioluminescence

It is now well established that circadian timing control exists in effectively every cell of an organism. Current research is directed at determining how these numerous, widely distributed, rhythmic cells are orchestrated within the overall circadian mechanism. The challenge is to understand how the molecular and cellular circadian machinery produces the complex circadian behaviors manifested at the levels of tissues, organs, and whole organisms.

Circadian timing systems can be viewed as having a master clock in the brain and subsidiary clocks in other tissues. Experiments investigating communication between these levels are essential. In the SCN, this entails making observations of the electrical activities of individual neurons, as well as the collective electrical activities of

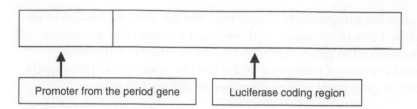

FIGURE 11-7.
Schematic diagram of a *per-luc* construct. The DNA from the controlling region of the *period* gene (the *period* promoter) has been fused to the DNA making up the coding region of the luciferase gene, producing a luciferase that is then under circadian control.

functionally coupled groups of neurons. But to probe the functional coupling between rhythms expressed in other non-SCN brain regions and peripheral tissues and organs, it is important to be able to study these non-SCN rhythms in vitro, as well as (eventually) in vivo.

Assessment of the regulatory patterning of the *period* circadian clock gene (*per*) became possible through the use of a reporter gene—in this case, a luciferase cDNA fused to the promoter region of *per* (see Figure 11-7). Luciferase is the enzyme responsible for bioluminescence—when luciferase acts upon its substrate, luciferin, light is produced. This recombinant reporter gene DNA, called a *per-luc* construct, was introduced into rat zygotes to produce what are called transgenic rats, carrying the *per-luc* DNA in every cell of their bodies. These rats are entrained to a circadian rhythm, and then the tissue to be studied is excised from the animal and put into tissue culture and supplemented with luciferin. Light emission from the cultured tissue then reveals when the *per* gene is being expressed, demonstrating the existence of a circadian rhythm in the tissue.

F. Technological Challenges in Analyzing Circadian Data

The bioluminescence time series data generated from *per-luc* experiments often show patterning in which average bioluminescence intensity is decreasing with time (drifting downward) as a result of the depletion of luciferin, a substrate necessary for light production by the luciferase enzyme (see Figure 11-8). In addition, as in Figure 11-8, a situation is often encountered in which the oscillatory magnitude changes with time. This represents a phenomenon referred to as *variance nonstationarity* (i.e., the variance exhibited in the time series changes as a function of time). In such cases, data normalization will be necessary to reveal the original rhythmic structure. Finally, further problems in the analysis of time series data come from the presence of noise that may often obscure the circadian rhythms.

Recall from Chapter 9 that most standard methods designed to determine periodic components in a time series, such as the fast Fourier transform (FFT) algorithm, require time series that do not have a trend. Variance nonstationarities and noise present additional

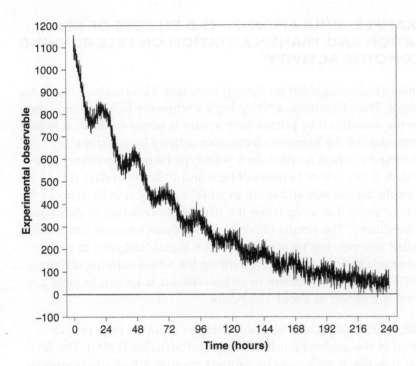

FIGURE 11-8.
A time series exhibiting a trend and variance nonstationarity.

problems. Thus, in many cases, some of the classical methods for analysis cannot be applied directly. To illustrate this last statement, we point to the work of Plautz et al. (1997), who reported that a secondary reproducible peak can be observed in the data, suggesting either an approximately half-circadian period component and/or the presence of two circadian rhythms out of phase with each other by approximately half a circadian period. Analysis by any method that assumes a single period estimate near a circadian range (i.e., in the vicinity of 24 hours) would be unable to accurately capture the dynamic behavior exhibited by this type of occurrence. This phenomenon is particularly pertinent in assessing the phase information of a rhythm from a noise-confounded time series. In such cases, algorithms that allow using appropriate filters in order to clean the data may be helpful.

In addition to the experiments using luciferase bioluminescence, numerous other experiments produce time series confounded with similar difficulties. Locomotor activity rhythms and neuronal firing patterns can also present data series possessing confounds that will challenge conventional attempts at analysis. Details can be found in Hurd et al. (1998), Sujino et al. (2003), Herzog et al. (1997), and Reppert and Weaver (2002).

For the rest of this chapter, we shall focus on techniques for analyzing confounded time series. We begin with an example designed to illustrate why such methods are important.

II. EXAMPLE: SIMULATION OF THE EFFECTS OF SCN ABLATION AND TRANSPLANTATION ON FREE-RUNNING LOCOMOTOR ACTIVITY

Like many rodents, golden (or Syrian) hamsters, *Mesocricetus auratus*, are nocturnal. Their locomotor activity begins when the lights go out, and one of the activities they pursue very avidly is wheel-running. It is easy to obtain data of the hamster's locomotor activity by electronically monitoring the wheel rotation. In a wild-type hamster, exposure to a light/dark (LD) cycle of 14 hours of light and 10 hours of dark (LD 14:10) will entrain the hamster and result in wheel-running activity at about the same time every day, even when the hamster is switched to dark/dark (DD) conditions. The results obtained under these constant conditions are called free-running because there is no signal (*zeitgeber*) to reset the hamster's biological clock. Measuring the wheel-running activity of a wild-type golden hamster in DD conditions is known to yield a free-running period of about 24.1 hours.

In 1988, Martin Ralph and Michael Menaker reported on a period mutation in the golden hamster (Ralph and Menaker [1988]). The first mutant hamster, a male, was recognized because it had a free-running period of 22.0 hours. Ralph and Menaker bred the mutant male with several normal females, and then subjected the F_1 animals to entrainment (LD 14:10) followed by constant dark to see how they behaved. Half of the F_1 animals had a free-running period averaging 22.3 hours. They were designated with the abbreviation T_s (for short *tau* or period). The other half had a period close to 24 hours. They were designated as T_n. When T_s animals were interbred, three types of offspring were produced: T_n with a period of 24 hours, T_s with a period of 22 hours, and a new super-short phenotype, T_{ss}, with a period of approximately 20 hours. These results are consistent with a mutation in a single gene, called *tau*, that is acting in a semidominant fashion (see Chapter 3). The normal hamster has two copies of the wild-type gene; the T_s hamster (with the 22-hour period) is a heterozygote with one normal and one mutant copy of *tau*; and the super-short mutant has two mutant copies. We shall refer to the homozygous super-short hamster as *tau^{ss}*.

The identity of the protein expressed by the *tau* gene is now known to be casein kinase I epsilon (CKIε). A kinase is an enzyme that places phosphate groups on proteins, in this case the mammalian proteins PER1 and PER2. It is similar in this way to the *Drosophila* gene double time (see Figure 11-6). Phosphorylation of a protein may alter its activity (increasing or decreasing it) or its cellular location or may mark it for destruction. In any event, the addition of such a large charged group is likely to be significant, and so it is not surprising that a change in the activity of CKIε because of mutation would have a radical effect on the period length.

For an example, we shall use simulated data of locomotor (wheel-running) activity of hamsters. We want to examine the relation between the SCN and the other biological clocks present in the organs, tissues, and cells of the animal.

The premise for our synthetic data sets is that the locomotor activity of a wild-type hamster and a *tau*^{ss} hamster has been observed over a period of time and the number of wheel rotations for every 6-minute interval recorded for both types. Next, the wild-type hamster has had its SCN destroyed, and SCN tissue from the *tau*^{ss} hamster has been transplanted into the wild-type hamster. After an appropriate recovery period, data are collected of the locomotor activity of the recipient wild-type hamster, providing a third data set. This premise is similar to a series of experiments reported by Ralph et al. (1990). Using the three data sets from this simulated experiment, we want to examine the effect induced on the circadian clock of the recipient animal. The following list presents a few possible scenarios that could occur as a result of the transplant:

1. Following the transplant, the recipient of the SCN exhibits rhythmic patterns that are essentially the same as those of the *tau*^{ss} donor. If this hypothesis could be corroborated from the data, this would present evidence that the SCN acts as a "master clock" and dominates the "local clocks" in the organs, tissues, and cells.

2. Following the transplant, the recipient of the SCN exhibits rhythmic patterns that are essentially the same as those of a wild-type hamster. If the data support this hypothesis, this would indicate that the rhythms of the local clocks have overridden the rhythms of the SCN.

3. Following the transplant, the recipient of the SCN exhibits new rhythmic patterns, different from those of both the wild-type and *tau*^{ss} hamsters. If the data supports this hypothesis, this may indicate a more complex relationship between the SCN and the local clocks, and the question of quantitatively characterizing this relationship becomes more interesting. For example, would the data support the hypothesis that after the transplant, the rhythmic patterns in the locomotor activity of the SCN recipient are a mix of those exhibited by the wild-type and the *tau*^{ss} hamsters before the transplant?

To answer these questions, we need to quantitatively characterize, as thoroughly as possible, the rhythmic behaviors in the three data sets and use these characterizations to develop and support our conclusions regarding these simulated experimental observations.

The data for this case study can be downloaded from our Web site (see Internet Resources at the end of this chapter), and we encourage the reader to do so and repeat some or all of the analyses presented below.

The file WT.NO represents 14 days of free-running locomotor activity of the wild-type hamster. The data were accumulated from an animal, under experimental conditions of DD, following a long-term regimen (2-week minimum) of 12 hours of light and 12 hours of dark (an LD 12:12 cycle).

The file TAUSS.NO represents 14 days of free-running locomotor activity of a *tau*^{ss} hamster. The data have been accumulated from an animal, under experimental conditions of DD, following a long-term exposure (two-week minimum) to 10 hours of light and 10 hours of dark (an LD 10:10 cycle).

The file TRNSPLNT.NO represents 14 days of free-running locomotor activity of the SCN transplant recipient.

We begin by plotting the data. The plots themselves make apparent some of the challenges in analyzing this simulated data. Clearly, all the plots exhibit a rhythmic pattern, with the pattern in the first data set being, perhaps, best expressed. Notice, however, that the rhythms are confounded. For the first two plots, the amplitudes change with time, and a shift that also grows with time is visible in the position of the peaks of the repeating patterns. In addition, the presence of noise makes it difficult to visually identify the exact location of the peaks. For the third plot, a rhythmic pattern is also clearly present, but it is more subtle. In order to address the questions raised above, we shall need to quantify the rhythms as accurately as possible and then analyze them appropriately.

This example illustrates some of the general analytical challenges presented by confounded time series, namely: (1) Mean and/or variance nonstationarities (i.e., time-dependent drifting and/or changes in oscillatory amplitude); (2) period and/or phase instability; and (3) noise. To overcome these challenges, the following question is important: What do we wish to learn from our data to be able to quantitatively characterize it? The typical analytical objectives of rhythmic analyses are to extract information about: (1) The period of expression; (2) the phase of expression; (3) the oscillatory amplitude of expression; and (4) the robustness of rhythmic expression (how strongly rhythmic the observed patterning is).

Even from the simple examples in Figure 11-9, it is clear that to answer these questions and the specific questions raised in the case study, we shall need tools that will allow us to work with confounded data. We would like (1) detrending strategies (to address mean nonstationarities, or drifting data); (2) strategies for data normalization (to address variance nonstationarities, or variable-dynamic-range data); and (3) analytical algorithms that, by design, attempt to accommodate the nonstationarities that may be present in uncorrected data series. Before proceeding with the example, we describe some tools for analyzing confounded time series. We begin by outlining some well-known fundamentals.

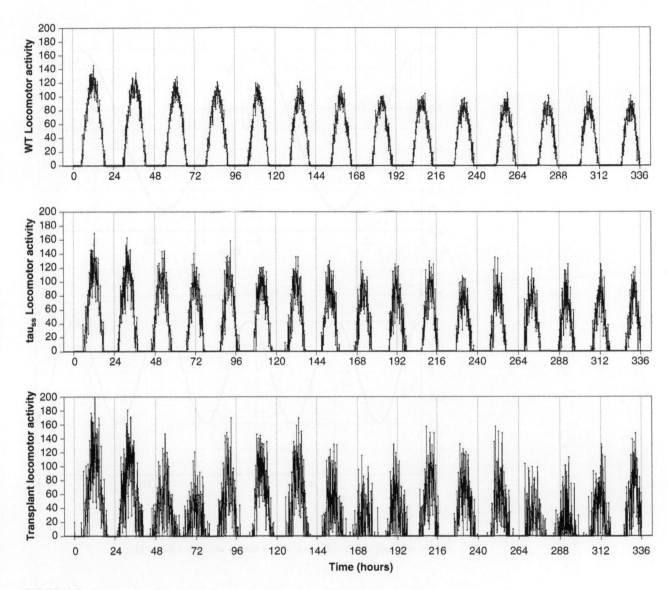

FIGURE 11-9.
Plots of the data from WT.NO (top panel), TAUSS.NO (middle panel), and TRNSPLNT.NO (bottom panel).

III. FUNDAMENTALS OF RHYTHMIC DATA AND TIME SERIES

A. Elementary Background

Recall that a purely periodic waveform might be of the form:

$$f(t) = \alpha \cos(\beta t + \gamma).$$

Each of the parameters α, β, and γ serves a particular function. The parameter α controls the amplitude of the wave. Figure 11-10(A) shows the graph of $\alpha\cos(t)$ where $\alpha > 0$. The parameter β controls the period (and therefore the frequency, because the frequency is the inverse of the

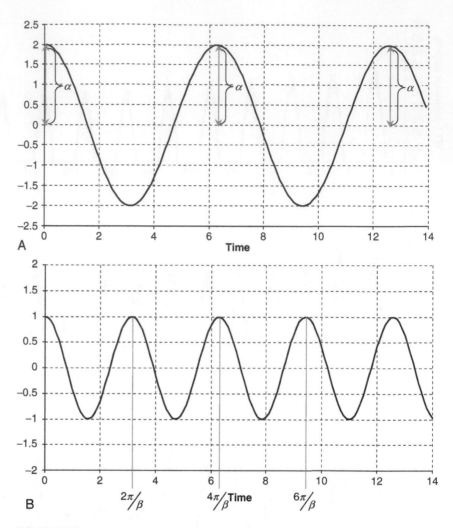

FIGURE 11-10.
Graphs of $\alpha\cos(t)$ with $\alpha > 0$ (panel A) and $\cos(\beta t)$ with $\beta > 1$ (panel B).

period). In Figure 11-10(B) we give the graph of $\cos(\beta t)$ with $\beta > 1$. Note that the larger β is, the higher the frequency and the shorter the period.

The parameter γ affects the phase shift of the graph. The graph is shifted to the left when $\gamma > 0$ and shifted to the right when $\gamma < 0$. It can also be interpreted as the time value at which the periodic or rhythmic pattern begins. The graph of $\cos(t + \pi/4)$ is shown in Figure 11-11(A), and the graph of $\cos(2t + \pi/4) = \cos(2(t + \pi/8))$ is shown in Figure 11-11(B).

A cosine-like pattern (or possibly a truncated cosine) is often found in circadian rhythm data, but there are also some variations. The amplitude might vary from one cycle to the next as in the graph shown in Figure 11-12(A), or the period may change by compressing (or expanding) as in the graphs shown in Figure 11-12(B).

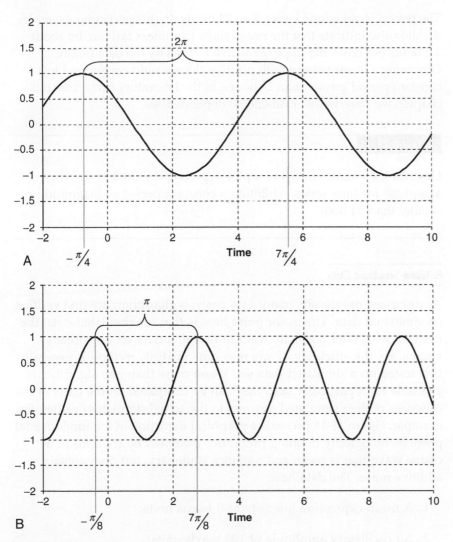

FIGURE 11-11.
Graphs of cos(t + π/4) (panel A) and cos(2t + π/4) = cos(2(t + π/8)) (panel B).

One way to identify whether the period changes and how it changes is by the phase shift. This is particularly useful when an estimate is available for the length of the cycle, such as 24 hours for circadian rhythms. Consider the graph shown in Figure 11-13(A), which is periodic with an exact period of 24 hours. In Figure 11-13(B), we have plotted the phase shift for each cycle. In the context of circadian rhythms, this can be interpreted as the time of day at which the maximum value (also called *acrophase*) occurs. If there is no change in the period, the maximum value will always occur at the same time every day, and the plot will be composed of values arranged in a horizontal line [see Figure 11-13(B)]. Now consider the graph in Figure 11-13(C), where the phase is shifting by a constant amount when considered relative to a 24-hour period. Figure 11-13(D) shows the time at which the maximal value occurs for each day. As the line has positive slope, this means that the maximal value occurs later and later each day.

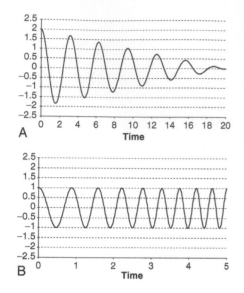

FIGURE 11-12.
A cosine-like rhythmic pattern with decreasing
amplitude (panel A) and shrinking period (panel B).

The period is, therefore, larger than 24 hours. In this particular example, the data also indicate that the phase shifts in a linear fashion, by about 2.6 hours per day—the slope of the line formed by the data points. Observing a linear pattern with positive slope would correspond to a constant period greater than 24 hours. In the laboratory project for this chapter, we observe other functional dependencies.

EXERCISE 11-1

Describe the characteristics of a plot depicting the time of daily acrophase for time series exhibiting a constant period of magnitude smaller than 24 hours.

B. Using Simulated Data

As discussed previously, many data analysis algorithms are first verified on simulated data. The major point here is that for simulated data, the answers we would normally want to find in a data set are already known. So, the efficiency of any new data analysis method is generally first tested on a simulated data set. When more than one algorithm is available for a particular task, they can be compared on the basis of the closeness of their generated answers to the actual simulation values. For example, Figure 11-14 presents a graphical depiction of the fundamental phenomenologically defining properties exhibited by rhythms of a cosine wave that is mean and variance stationary, but does contain additive noise. The data have:

1. A mean expression intensity of 0 y-axis units;

2. An oscillatory amplitude of 100 y-axis units;

3. A period of oscillation of 24 hours;

4. A phase reference point, in this case the time of acrophase, at 0 hours; and

5. Gaussian-distributed random noise added, such that the standard deviation of the noise is 10 y-axis units.

The situation becomes more challenging when the data contain a trend and the variance of the data changes with time. Figure 11-15 introduces mean and variance nonstationarities on top of a rhythm similar to that presented in Figure 11-14. The data set in Figure 11-15 represents a cosine wave possessing mean and variance nonstationarities exhibiting:

1. A mean expression intensity that is time-dependent, such that at time zero, the mean expression intensity is 1000 y-axis units, but it decays in magnitude in exponential manner to a final

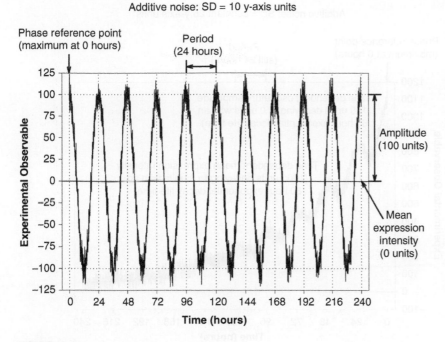

Additive noise: SD = 10 y-axis units

Phase reference point
(maximum at 0 hours)

Period
(24 hours)

Amplitude
(100 units)

Mean
expression
intensity
(0 units)

FIGURE 11-14.
Mean and variance stationary noisy cosine wave.

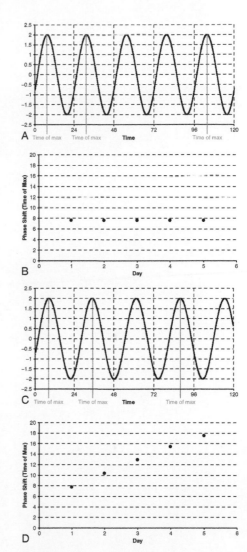

FIGURE 11-13.
An example of period shifting. For the graph in panel A, the period does not change, and the acrophase (the point of daily maximum) occurs at the same time each day (panel B). For the graph in panel C, the period is larger than 24 hours, and the daily maximum is delayed by about 2.6 hours from day to day (panel D).

value of 0 y-axis units with an 80-hour exponential decay lifetime [i.e., 80 hours is the time for the mean value to decay to $1/e$ of its magnitude 80 hours prior ($1/e = \sim 1/2.72 = \sim 0.368$)];

2. An oscillatory amplitude that is also time-dependent, such that at time zero, the oscillatory amplitude is 100 y-axis units, but it too decays in magnitude in an exponential manner to a final value of 0 y-axis units, again, with an 80-hour exponential decay lifetime;

3. A period of oscillation of 24 hours;

4. A phase reference point (again, the maximum) at 0 hours; and

5. Gaussian-distributed random noise added, such that the standard deviation of the noise is 25 y-axis units.

We use the simulated data from Figures 11-14 and 11-15 as test data for several of the analyses discussed below. Subsequently, we apply those analyses to quantitatively address the questions posed in the Example presented in Section II.

IV. DATA PREPROCESSING STRATEGIES

If mean and/or variance nonstationarities present formidable confounds, then data preprocessing strategies may be needed before

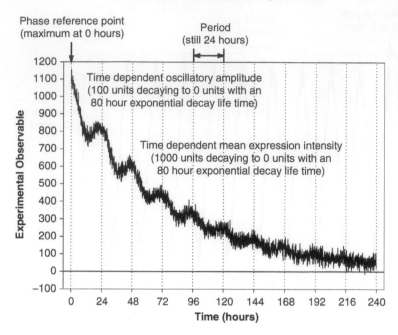

FIGURE 11-15.
Mean and variance nonstationary noisy cosine wave.

performing data analysis. Various algorithms exist for noise and trend removal. We do not attempt to present them here in detail but, rather, introduce them and concentrate on discussing their differences and similarities.

A. Data Filtering

Data filtering is generally applied to remedy the presence of noise. However, as with any preprocessing of data, data filtering may either lead to information loss ("leaky filters") or to altering the data by introducing, for example, a phase change in the data. The filtering software used for the examples in this chapter is called *ARFILTER*. Its implementation uses a forward–backward linear exponential (i.e., first-order) autoregressive filtering strategy, as reported in Orr and Hoffman (1974), and we refer the reader to this article for the mathematical description and details. It should be stressed, however, that a particularly attractive feature of *ARFILTER* is that it results in *zero phase change* of the output. Different algorithms implementing similar noise reduction techniques for time series can be found in several commercially available software packages. For example, using the exponential smoothing option available in *MATLAB* would result in similar output, and similar options are available in *SPSS Trends* and *SAS*.

If the input time series for *ARFILTER* is $x(t)$, the algorithm produces an output series $u(t)$ that represents the low-frequency component(s) of the original data series. The filtering procedure uses a single parameter ρ, the value of which is optimized so that the residuals, defined by $[x(t) - u(t)]$, are least-squares minimized, as described in Chapter 8. The detrended residuals $[x(t) - u(t)]$ then represent the residual high-frequency component(s) that were filtered out by this forward–backward autoregressive filtering process. It is important to note that, depending on the situation, either $u(t)$ or $[x(t) - u(t)]$ may be used in subsequent analysis: $u(t)$ if the objective was to remove excessive high-frequency noise from the original data series, and $[x(t) - u(t)]$ if the objective was to remove confounding, low-frequency trending from the original data series. Figures 11-16 and 11-17 illustrate the use of *ARFILTER* with the noisy simulated data sets from Figures 11-14 and 11-15. The time series $u(t)$ and $[x(t) - u(t)]$ are presented to the right of the original time series $x(t)$. An important

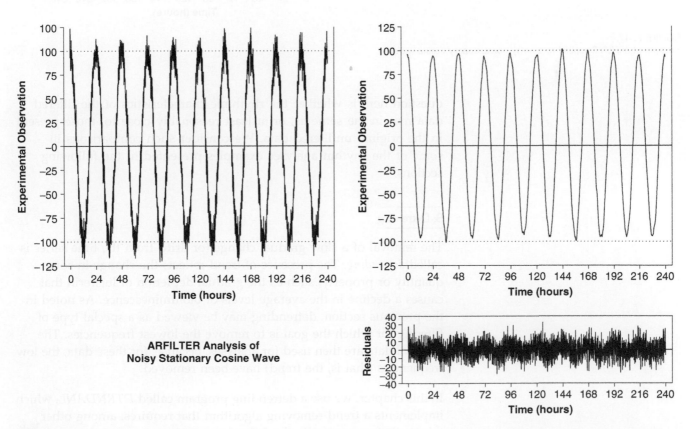

FIGURE 11-16.
Example of *ARFILTER* applied to a noisy stationary cosine wave.

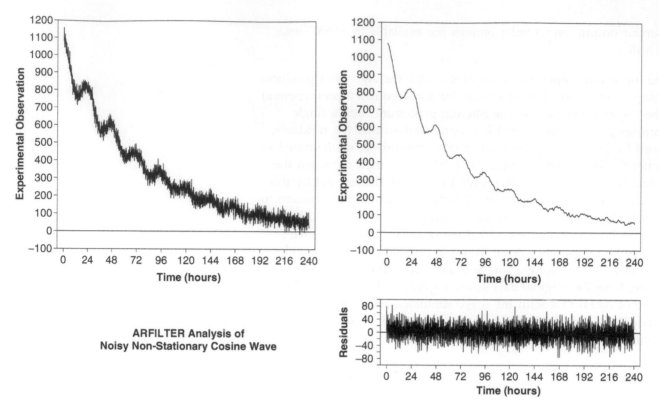

FIGURE 11-17.
Example of *ARFILTER* applied to a noisy nonstationary cosine wave.

question here is whether the rhythmic characteristics of the filtered data remain the same as, or at least reasonably close to, those present in the original unfiltered data. We consider such comparisons in some of the rhythm analyses examples presented in the following sections.

B. Detrending

The removal of a slow gradual change (or drift) from the time series is called *detrending*. The presence of trend reflects the change in some quantity or property, such as the gradual depletion of luciferin that causes a decline in the average levels of bioluminescence. As noted in the previous section, detrending may be viewed as a special type of filtering for which the goal is to remove the lowest frequencies. The residual data are then used for analyses, because, for these data, the low frequencies (that is, the trend) have been removed.

In this chapter, we use a detrending program called *DTRNDANL*, which implements a trend-removing algorithm that requires, among other things, that users specify the following inputs:

1. The value of a filter window (FWINDOW) to apply; and

2. Whether the detrended data are to be presented in original y-value space or in terms of standard normal deviates (SND) space, with the latter being used to address variance nonstationarities.

The *DTRNDANL* software also allows users to specify one of several ways of reporting the pointwise uncertainties with the detrended data, such as uniform weighting, the arithmetic standard error of the mean (SEM) of calculated detrended values, and others.

Here, again, we should point out that many commercially available software packages provide detrending options. *MINITAB* and *Microsoft Excel*, for example, allow for trend removal of known shapes, such as linear, exponential, S-shaped, and others. The functional form of the trend, however, is often not obvious.

The *DTRNDANL* algorithm does not assume any specific functional form for the trend. It begins by considering a data sequence of length FWINDOW beginning with the first data point. A linear regression detrending is performed on this subseries data sequence. Subsequently, the values of the detrended subsequence are stored in their original units, or one additional step is performed on the detrended subseries values: each value is divided by the standard deviation of the subseries prior to storing. In the latter case, we say that the result is stored in *standard normal deviate space* (SND-space). The algorithm then repeats this process by starting its second pass with the second point in the original time series, its third pass with the third point, and so on, until terminating the detrending process when the filter window requests a subseries analysis that extends beyond the last point of the original time series data. All values stored at x-value locations are then averaged to produce the final, detrended time series sequence of values. When the method is applied, the averaging of the sequential linearly detrended subseries acts to remove the trend from the data. Further, if the data are divided by the SD, then variance nonstationarity is also reduced.

The selection of an appropriate value for a filter window is critical for successful application of this algorithm. For example, to detrend circadian rhythms data that are recorded in units of hours, a value for the filter window of 24 hours would be an appropriate choice (assuming that the dominant rhythm exhibited by the data has a period near 24 hours). Figures 11-18 and 11-19 visualize the output of *DTRNDANAL* with input provided by the data sets from Figures 11-14 and 11-15, respectively.

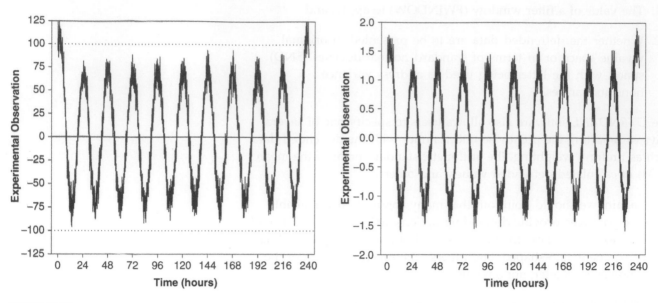

FIGURE 11-18.
An example of *DTRNDANL* applied to the noisy stationary cosine wave originally presented in Figure 11-14, again employing a fixed-period sliding window detrend that assumes an approximately 24-hour intrinsic periodicity. The panel on the left is the result produced by leaving the data series being detrended in original data space, whereas the panel on the right is the result produced by processing and converting the data series being detrended to SND-space.

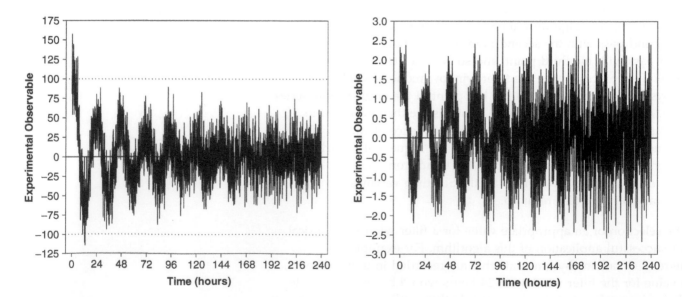

FIGURE 11-19.
An example of *DTRNDANL* applied to the noisy nonstationary cosine wave originally presented in Figure 11-15. *DTRNDANL* employs a fixed-period sliding window detrending algorithm that will perform trend removal assuming good/reasonable prior knowledge of intrinsic periodicity, such as approximately 24 hours in the case of circadian rhythmicity. The panel on the left is the result produced by leaving the data series being detrended in original data space, whereas the panel on the right is the result produced by processing and converting the data series being detrended to SND-space.

V. METHODS FOR RHYTHM ANALYSIS AND ANALYTICAL STRATEGIES

Our objective in this chapter is to present methods for detecting rhythms in confounded data, whether stationary original data series or preprocessed nonstationary data that have been made (more closely) stationary. Presented here are a few analytical strategies that have found common usage for these purposes.

A. Model-Dependent Algorithms

Model-dependent algorithms assume a specific analytic form for the rhythmic wave and then employ nonlinear least-squares (NLLS) techniques, as described in Chapter 8, for estimating the values of the model parameters from the data. We consider two such algorithms, out of several in the literature.

1. Cosin2nl

This is an algorithm to assess the period, phase, and amplitude of a one-component cosine function with a linear trend of the form:

$$y(t) = c_0 + c_1 t + \alpha \cos \left[\frac{2\pi(t + \phi)}{\tau} \right],$$

where $y(t)$ is the time series on which analysis is being performed, c_0 is a constant offset term, c_1 is a slope of the linear trend, t is time, and α, ϕ, and τ are the amplitude, phase, and period, respectively, of the cosine function. The parameters of this function are then estimated by nonlinear least-squares minimization of the Gauss–Newton type, as described in Chapter 8. The procedure allows for nonlinear asymmetric joint confidence limits for all parameters to be calculated, if desired, at any user-specified confidence probability level. The details of the procedure can be found in Straume et al. (1991).

An amplitude term significantly different from zero indicates a statistically significant rhythm at the specified level of confidence probability. If the confidence limits of the amplitude term encompass zero, however, the rhythm is not statistically significant at the specified level of confidence probability. As expected from a NLLS algorithm, *COSIN2NL* requires user-specified initialization in the form of initial guesses for the values of the parameters of the cosine model. Thus, it is not a fully objective analytical strategy because it may be susceptible to the influence of user-introduced bias.

Figures 11-20 and 11-21 illustrate the results of applying this algorithm to the data sets presented in Figures 11-14 and 11-15. The estimates for the model parameters, together with their 95% confidence intervals, are also presented. The specific details appear in the figure legends.

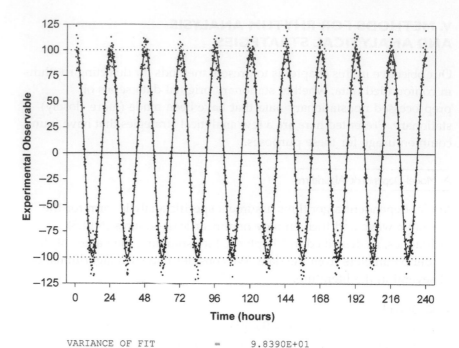

```
VARIANCE OF FIT          =      9.8390E+01
         WITH 2396 DEGREES OF FREEDOM
SUM OF RESIDUALS SQUARED =      2.3574E+05
SQUARE ROOT OF VARIANCE  =      9.9192E+00

FITTED PARAMETER VALUES AND CONFIDENCE LIMITS
Amp  =  9.963258E+01 +/-  9.650230E-01
      (  9.866756E+01 ->  1.005976E+02)
Per  =  2.399660E+01 +/-  1.062584E-02
      (  2.398598E+01 ->  2.400723E+01)
Phi  = -2.112440E-02 +/-  6.268264E-02
      ( -8.380704E-02 ->  4.155825E-02)
DC   =  5.486899E-02 +/-  9.708050E-01
      ( -8.979580E-01 ->  1.043652E+00)
Slop = -3.297019E-05 +/-  6.891264E-03
      ( -7.051850E-03 ->  6.730677E-03)
```

FIGURE 11-20.
COSIN2NL analysis of the noisy stationary cosine wave originally presented in Figure 11-14. In this instance, *COSIN2NL* performed a single-component cosine analysis in which five parameters were nonlinear least squares (NLLS) estimated (Amp, Per, Phi, DC, and Slop; where DC refers to the estimated value at time zero on the x-axis). NLLS was performed by a modified Gauss–Newton method. The convergence criterion was set to a fractional change in variance of 10^{-9}. Approximate nonlinear asymmetric joint confidence limits were calculated at 95% confidence (in which lower and upper parameter confidence limits are estimated independently; the values reported in parentheses; the values reported following +/− are one-half the difference between the estimated upper and lower confidence limits). The results of this analysis thus indicate that (1) the estimated oscillatory amplitude is 99.63 ± 0.97 y-axis units; (2) the estimated period is 23.997 ± 0.011 hours; (3) the estimated phase is −0.021 ± 0.063 hours; (4) the estimated value at time zero (DC) is 0.05 ± 0.97 y-axis units; and (5) the estimated slope is 0.0000 ± 0.0069 y-axis units per hour.

Notice that the *COSIN2NL* algorithm performs very well when the data set from Figure 11-14 is used as input—the amplitude, period, phase, and slope of the model-assumed linear trend are determined quite precisely (compare with the known characteristics of the simulated data

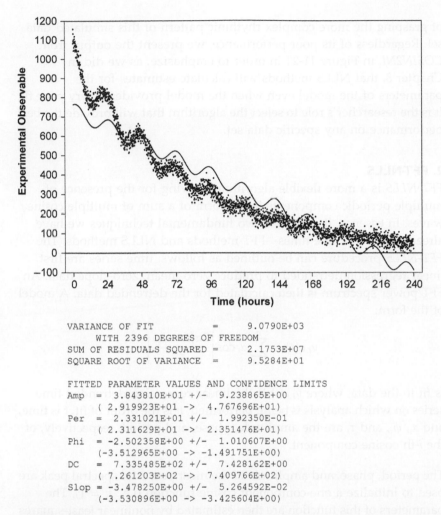

```
VARIANCE OF FIT          =       9.0790E+03
      WITH 2396 DEGREES OF FREEDOM
SUM OF RESIDUALS SQUARED =       2.1753E+07
SQUARE ROOT OF VARIANCE  =       9.5284E+01

FITTED PARAMETER VALUES AND CONFIDENCE LIMITS
Amp  =  3.843810E+01 +/-   9.238865E+00
     ( 2.919923E+01 ->   4.767696E+01)
Per  =  2.331021E+01 +/-   1.992350E-01
     ( 2.311629E+01 ->   2.351476E+01)
Phi  = -2.502358E+00 +/-   1.010607E+00
     (-3.512965E+00 ->  -1.491751E+00)
DC   =  7.335485E+02 +/-   7.428162E+00
     ( 7.261203E+02 ->   7.409766E+02)
Slop = -3.478250E+00 +/-   5.264592E-02
     (-3.530896E+00 ->  -3.425604E+00)
```

FIGURE 11-21.
COSIN2NL analysis of the noisy nonstationary cosine wave originally presented in Figure 11-15. As in Figure 11-20, *COSIN2NL* performed a single-component cosine analysis in which five parameters were nonlinear least squares (NLLS) estimated (Amp, Per, Phi, DC, and Slop; where DC refers to the estimated value at time zero on the x-axis). NLLS was performed by a modified Gauss–Newton method. The convergence criterion was set to a fractional change in variance of 10^{-9}. Approximate nonlinear asymmetric joint confidence limits were calculated at 95% confidence (in which lower and upper parameter confidence limits are estimated independently; the values reported in parentheses; the values reported following +/− are one-half the difference between the estimated upper and lower confidence limits). The results of this analysis thus indicate that (1) the estimated oscillatory amplitude is 38.4 ± 9.2 y-axis units; (2) the estimated period is 23.31 ± 0.20 hours; (3) the estimated phase is −2.5 ± 1.0 hours; (4) the estimated value at time zero (DC) is 733.5 ± 7.4 y-axis units; and (5) the estimated slope is −3.478 ± 0.053 y-axis units per hour.

set from Figure 11-14). Notice, however, that the model does not accommodate the data from Figure 11-15 particularly well (see Figure 11-21). Indeed, assuming a linear trend is not realistic for the data set in Figure 11-15. Also, as the model assumes a single cosine wave; it is not capable of capturing the decreasing oscillatory amplitudes that are apparent in Figure 11-15. As a result, *COSIN2NL* is incapable of grasping the more complex rhythmic pattern of this simulated data set. Regardless of its poor performance, we present the output from *COSIN2NL* in Figure 11-21 in order to emphasize, as we did in Chapter 8, that NLLS methods will calculate estimates for the parameters of the model even when the model provides a very poor fit. It is the researcher's role to select the algorithm that will provide the best performance on any specific data set.

2. FFT-NLLS

FFT-NLLS is a more flexible algorithm allowing for the presence of multiple periodic components in the form of a sum of multiple cosine waves. In essence, it combines two fundamental techniques we have already used multiple times—FFT methods and NLLS methods. The *FFT-NLLS* procedure can be outlined as follows: time series are first linear regression detrended to produce zero-mean, zero-slope data. An FFT power spectrum is then calculated for the detrended data. A model of the form:

$$y_{LR}(t) = \sum_{i=1}^{n} \alpha_i \cos\left[\frac{2\pi(t + \phi_i)}{\tau_i}\right]$$

is fit to the data, where $y_{LR}(t)$ is the linear regression detrended time series on which analysis is being performed, n is the order of fit, t is time, and α_i, ϕ_i, and τ_i are the amplitude, phase, and period, respectively, of the i-th cosine component.

The period, phase, and amplitude of the most powerful spectral peak are used to initialize a one-component cosine function (i.e., $n = 1$). The parameters of this function are then estimated by nonlinear least-squares minimization as in *COSIN2NL*. Upon convergence, approximate nonlinear asymmetric joint confidence limits are estimated for all parameters (period, phase, amplitude, and constant offset) at 95% confidence probability. If the amplitude is significantly different from zero, then the procedure is repeated at the next higher order. The two most powerful FFT spectral peaks are then used to initialize a two-component cosine function (i.e., $n = 2$) that is subsequently NLLS minimized to the linear regression detrended data, and confidence limits are again evaluated. This process is repeated iteratively until at least one cosine component is identified with an amplitude that is not statistically significant.[1]

The statistical significance of each derived rhythmic component is assessed by way of the relative amplitude error (RAE), defined as the

_____

1. Other possible scenarios for terminating the procedure along with more detailed description of the *FFT-NLLS* procedure can be found in Plautz et al. (1997).

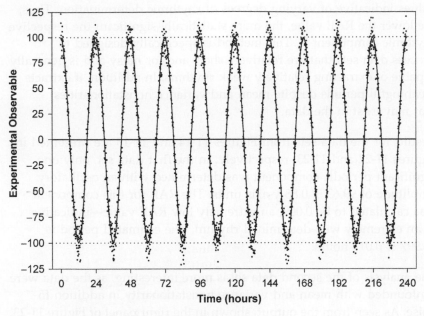

```
        Variance of fit =   9.83490E+01
             Error code =          0
        Std dev of fit  =   9.91711E+00
             NDF         =       2397
        Conv criterion  =   1.00000E-06
        Conf Prob       =   0.9500
          DC Ampl = -5.61565E-02 +/-   6.22836E-01
                    [-6.78992E-01 to   5.66679E-01]
          Ampl ( 1) =   9.96325E+01 +/-   8.80296E-01
                    [ 9.87522E+01 to   1.00513E+02]
          Per  ( 1) =   2.39965E+01 +/-   1.13200E-02
                    [ 2.39851E+01 to   2.40078E+01]
          Phase ( 1) = -2.20003E-02 +/-   6.52157E-02
                    [-8.72160E-02 to   4.32154E-02]
```

FIGURE 11-22.
FFT-NLLS analysis of the noisy stationary cosine wave originally presented in Figure 11-14. In this instance, the method terminated execution after requiring only a single-component cosine analysis in which three parameters of periodic rhythms expressed in the data were found: the amplitude, the period, and the phase (Ampl, Per, and Phase). NLLS was performed by a modified Gauss–Newton method. The convergence criterion was set to a fractional change in variance of 10^{-6}. Nonlinear asymmetric joint confidence limits were calculated at 95% confidence (in which lower and upper parameter confidence limits are estimated independently; the values reported in brackets; the values reported following +/− are one-half the difference between the estimated upper and lower confidence limits). The results of this analysis indicate that (1) the estimated oscillatory amplitude is 99.63 ± 0.88 y-axis units (producing a RAE of 0.009, not shown on printout); (2) the estimated period is 23.997 ± 0.011 hours; and (3) the estimated phase is −0.022 ± 0.065 hours. The extremely low RAE value is indicative of an extremely well-determined rhythm.

following ratio: in the numerator, the amplitude error (one-half the difference between the upper and lower 95% amplitude confidence limits) to, in the denominator, the most probable derived amplitude magnitude. Theoretically, this metric will range from 0.0 to 1.0; 0.0 indicates a rhythmic component known to infinite precision (i.e., zero error); 1.0 indicates a rhythm that is not statistically significant (i.e., error equal to the most probable amplitude magnitude); and intermediate values indicative of varying degrees of rhythmic determination. Thus, the lower the RAE value, the more statistically significant the respective rhythmic component is. This method is specifically designed to process data sets that are relatively short and/or noisy and is generally capable of extracting relatively weak rhythms. In addition, it extracts meaningful periods despite mean and variance nonstationarities that may exist in the data.

Results for the data sets from Figures 11-14 and 11-15 are presented in Figures 11-22 and 11-23, respectively. In the first data set, only one significant periodic component was determined with an oscillatory amplitude of 99.63 ± 0.88 y-axis units. The RAE for this component was calculated to be 0.009, an extremely low RAE value—indicative of an extremely well-determined rhythm. The estimated period is 23.997 ± 0.011 hours.

The analysis of the second data set is more interesting, as the data were confounded with mean and variance nonstationarity in addition to noise. As seen from the output, shown in the right panel of Figure 11-23, four periodic components were identified with the respective average estimates for the periods and nonsymmetric 95% confidence intervals shown in Table 11-1.

The RAE associated with the periodic component of 23.4 is the smallest, with a value of 0.183 (not shown on printout). In addition, we point out the first period listed in the first column appears to be the effect of an attempt to fit the data trend as a periodic component.

B. A Model-Independent Algorithm: *PHASEREF*

The last method we introduce for assessing period, oscillatory amplitude, and phase information from a rhythmic data series is referred to as *PHASEREF*. This is a maximally assumption-free strategy in which no model form for any rhythms is assumed. However, the interpretation of results does require the assumption that there exists in the data series being analyzed one dominant, primary rhythmic component, the period of which is (approximately) known a priori. *PHASEREF* is a modification of a method presented by Meerlo et al. (1997). It requires the user to provide two period values with which to calculate two sets of smoothed running average values of the data series to be analyzed.

One smoothing filter should have a period value close to that expected for the dominant rhythm being expressed in the data series. For a circadian time series, this value would be approximately 24 hours. The result of smoothing the data with this period filter is an appropriately smoothed baseline series in which the dominant-period rhythm is nearly completely removed. Only long-term trends

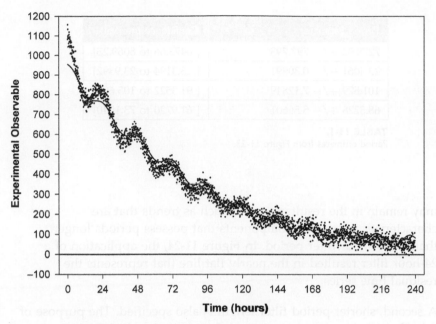

```
    Variance of fit =   1.53920E+03
        Error code =        0
    Std dev of fit =   3.92326E+01
            NDF     =     2388
    Conv criterion =   1.00000E-06
        Conf Prob  =   0.9500
     DC Ampl   =   5.07169E+04 +/-  1.14171E+04
              [ 3.88735E+04 to  6.17077E+04]
     Ampl ( 1) =   5.08134E+04 +/-  1.14174E+04
              [ 3.89694E+04 to  6.18042E+04]
      Per ( 1) =   7.27832E+03 +/-  7.97793E+02
              [ 6.47366E+03 to  8.06925E+03]
    Phase ( 1) =   3.51944E+03 +/-  3.98592E+02
              [ 3.11776E+03 to  3.91495E+03]
     Ampl ( 3) =   2.93980E+01 +/-  5.37680E+00
              [ 2.47165E+01 to  3.54701E+01]
      Per ( 3) =   2.34061E+01 +/-  3.09914E-01
              [ 2.33194E+01 to  2.39392E+01]
    Phase ( 3) =  -1.91805E+00 +/-  6.77985E-01
              [-2.61967E+00 to -1.26370E+00]
     Ampl ( 2) =   7.23998E+00 +/-  5.95002E+00
              [ 3.65280E+00 to  1.55528E+01]
      Per ( 2) =   1.01879E+02 +/-  7.12849E+00
              [ 9.13522E+01 to  1.05609E+02]
    Phase ( 2) =   5.79012E-02 +/-  1.23911E+01
              [-1.83886E+01 to  6.39350E+00]
     Ampl ( 4) =   5.96442E+00 +/-  5.40078E+00
              [ 1.20917E+00 to  1.20107E+01]
      Per ( 4) =   6.83228E+01 +/-  5.56601E+00
              [ 6.19720E+01 to  7.31040E+01]
    Phase ( 4) =  -3.31417E+00 +/-  1.18759E+01
              [-2.08387E+01 to  2.91315E+00]
```

FIGURE 11-23.
FFT-NLLS analysis of the noisy nonstationary cosine wave originally presented in Figure 11-15.
In this instance, *FFT-NLLS* identified four rhythmic components that were considered statistically
significant at 95% confidence. The results of this analysis produce, as a circadian rhythm estimate,
a rhythmic component exhibiting (1) an estimated oscillatory amplitude of 29.4 ± 5.4 y-axis units
(RAE of 0.183); (2) the estimated period is 23.41 ± 0.31 hours; and (3) the estimated phase is −1.92
± 0.68 hours.

Period Estimate	Confidence Interval
7278.32 +/− 797.793	[6473.66 to 8069.25]
23.4061 +/− 0.30991	[23.3194 to 23.9392]
101.879 +/− 7.12849	[91.3522 to 105.609]
68.3228 +/− 5.56601	[61.9720 to 73.1040]

TABLE 11-1.
Period estimates from Figure 11-23.

may remain in the resultant series, such as trends that are characteristic of spectral components that possess periods longer than the 24-hour filter period. In Figure 11-24, the application of a 24-hour filter resulted in the nearly flat line that represents the residual time series.

A second, shorter-period filter length is also specified. The purpose of this filter is to generate a smoothed data series in which high-frequency noise is filtered out but in which the presumed longer-period dominant

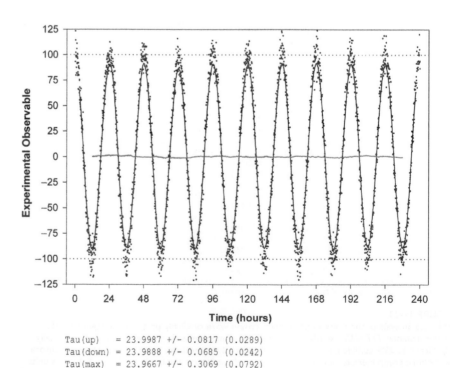

```
Tau(up)   = 23.9987 +/- 0.0817 (0.0289)
Tau(down) = 23.9888 +/- 0.0685 (0.0242)
Tau(max)  = 23.9667 +/- 0.3069 (0.0792)
```

FIGURE 11-24.
PHASEREF analysis of the noisy stationary cosine wave originally presented in Figure 11-14. Values of 24 hours and 6 hours are used for the running average filters. The phase reference points are derived as the times of upward and downward crosses, and times of maxima/minima. The mean period values (+/− SD, with SEM values in parentheses) are presented next to the graph for (1) consecutive up-crosses; (2) consecutive down-crosses; and (3) consecutive times of maxima and minima considered jointly in the average.

rhythm is retained, together with any longer-period trends. Next, the times of occurrence of up-crossings and down-crossings of the short-period filtered curve with regard to the baseline curve are computed (see Figure 11-24). The times of maximum and minimum differences between the two curves are also calculated from the difference of the short-period smoothed series and the baseline series. Period estimates for the dominant rhythm can now be obtained from successive differences between up-crosses, maxima, down-crosses, and minima, respectively.

A typical implementation of this analytical strategy might entail calculating a 6-hour and a 24-hour running average of the original data series. The crossings of these two smoothed lines provide the rising and falling phase markers for each cycle. The maximum differences between the smoothed curves for each cycle calculated between the peak and the trough allow calculating the amplitude of each cycle. The time of peak provides a third phase marker (assuming that acrophase is to be used as the phase reference marker of record). More details can be found in Abe et al. (2002).

Although maximally assumption-free, attempts to apply this method directly may meet with technical difficulties. For example, the period estimates presented in Figure 11-25 differ significantly from both those obtained for the same data sets through the use of *FFT-NLLS* and the actual simulated value of 24 hours. The tabulated summary of the algorithm output presented in Figure 11-26 provides an explanation. Whereas the expected estimates for TAU(up), TAU(down), and TAU (max) are in the vicinity of 24 hours, numerous instances appear in which values for period estimates are considerably shorter than 24 hours. This is a consequence of noise confounds creeping in, beginning at about 100 hours of x-axis time and manifesting consistently beyond about 200 hours of x-axis time. In such cases, preprocessing of the data through filtering may be beneficial. We present such examples in the next section.

VI. PREPROCESSING BEFORE ANALYSIS

We note that, in general, preprocessing of the data may introduce bias into the analytical results. In some cases, however, preprocessing may be necessary if the presence of significant confounds hinders the direct analysis of the time series. Pros and cons should always be carefully weighed before the use of preprocessing techniques. The next two exercises illustrate this point.

A. *ARFILTER* Followed by Rhythms Analysis

Examples are provided here of analyses of the noisy, nonstationary time series introduced in Figure 11-15, except this time preprocessed

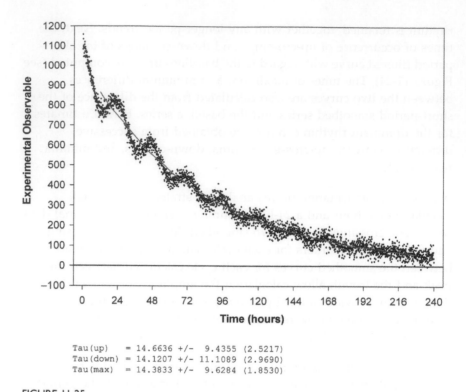

Tau(up)   = 14.6636 +/-   9.4355 (2.5217)
Tau(down) = 14.1207 +/- 11.1089 (2.9690)
Tau(max) = 14.3833 +/- 9.6284 (1.8530)

FIGURE 11-25.
PHASEREF analysis of the noisy nonstationary cosine wave originally presented in Figure 11-15, again calculated with 6-hour and 24-hour filters. However, whereas clean period estimates were obtained from the analysis shown in Figure 11-24, the present case shows sufficient noise is present so as to confound a clean assessment of period, arising primarily at the long-time end of the data series (the details of which are presented in Figure 11-26).

to remove noise prior to analysis. Although this is not a good idea in general because removing noise structure prior to analysis will typically introduce (sometimes considerable) bias into analytical results, it may prove beneficial to an analysis like that performed by *PHASEREF*.

EXERCISE 11-2

Figures 11-27 through 11-30 present the outputs of the procedures *COSIN2NL*, *FFT-NLLS*, and *PHASEREF* performed on time series from Figure 11-15 after preprocessing by *ARFILTER*. Compare the results with those presented in Figures 11-21, 11-23, 11-25, and 11-26 showing the outputs of the same procedures performed on the original time series. Comment on the differences and whether or not, in your opinion, using filtering was beneficial and/or necessary in each case.

XOVER(up)	XOVER(down)	AMPL(max)	TIME(max)	ALPHA	RHO	TAU(up)	TAU(down)	TAU(max)
1.801500E+01								
	3.003500E+01	6.147595E+01	2.475000E+01	1.202000E+01				
4.241500E+01		-5.964130E+01	3.660000E+01		1.238000E+01	2.440000E+01		
	5.350500E+01	5.369855E+01	4.765000E+01	1.109000E+01			2.347000E+01	2.290000E+01
6.593500E+01		-4.243335E+01	6.015000E+01		1.243000E+01	2.352000E+01		2.355000E+01
	7.809500E+01	2.992264E+01	7.305000E+01	1.216000E+01			2.459000E+01	2.540000E+01
8.997500E+01		-3.398944E+01	8.305000E+01		1.188000E+01	2.404000E+01		2.290000E+01
	1.013450E+02	2.957397E+01	9.535000E+01	1.137000E+01			2.325000E+01	2.230000E+01
1.013550E+02		-1.528931E-02	1.013500E+02		1.000214E-02	1.138000E+01		1.830000E+01
	1.015650E+02	4.588318E-01	1.014500E+02	2.099991E-01			2.200012E-01	6.099998E+00
1.145950E+02		-2.159099E+01	1.069500E+02		1.303000E+01	1.324000E+01		5.599998E+00
	1.256750E+02	2.053612E+01	1.213500E+02	1.108000E+01			2.411000E+01	1.990000E+00
1.388550E+02		-1.934850E+01	1.306500E+02		1.317999E+01	2.425999E+01		2.370000E+01
	1.498650E+02	1.715137E+01	1.452500E+02	1.101001E+01			2.419000E+01	2.390000E+01
1.611250E+02		-1.900436E+01	1.543500E+02		1.125999E+01	2.227000E+01		2.370001E+01
	1.723950E+02	1.754298E+01	1.657500E+02	1.127000E+01			2.253000E+01	2.050000E+01
1.862750E+02		-1.420439E+01	1.797500E+02		1.387999E+01	2.514999E+01		2.539999E+01
	1.974150E+02	1.120005E+01	1.905500E+02	1.114000E+01			2.501999E+01	2.480000E+01
1.983450E+02		-2.171257E+00	1.980500E+02		9.300079E-01	1.207001E+01		1.830000E+01
	1.984350E+02	3.676605E-02	1.983500E+02	8.999634E-02			1.020004E+00	7.800003E+00
2.098350E+02		-7.790092E+00	2.008500E+02		1.140001E+01	1.149001E+01		2.800003E+00
	2.171150E+02	9.428452E+00	2.135500E+02	7.279999E+00			1.868001E+01	1.520000E+01
2.216850E+02		-5.838020E+00	2.195500E+02		4.569992E+00	1.184999E+01		1.870000E+01
	2.218350E+02	4.202080E-01	2.217500E+02	1.500092E-01			4.720001E+00	8.199997E+00
2.219850E+02		-6.716156E-02	2.218500E+02		1.499939E-01	3.000031E-01		2.300003E+00
	2.220550E+02	4.871750E-02	2.220500E+02	6.999207E-02			2.199860E-01	3.000031E-01
2.222750E+02		-6.159210E-01	2.221500E+02		2.200012E-01	2.899933E-01		2.999878E-01
	2.225150E+02	5.733643E-01	2.223500E+02	2.400055E-01			4.600067E-01	3.000031E-01
2.233050E+02		-1.382374E+00	2.231500E+02		7.899933E-01	1.029999E+00		1.000000E+00
	2.277250E+02	4.844704E+00	2.265500E+02	4.420013E+00			5.210007E+00	4.199997E+00

FIGURE 11-26.
Verbose, tabulated summary of the *PHASEREF* analytical session presented in Figure 11-25. The columns of interest are:
XOVER(up)—the times at which the short-period smoothed series crosses baseline upward.
XOVER(down)—the times at which the short-period smoothed series crosses the baseline downward.
TIME(max)—the times of extremes, which may be maxima or minima.
TAU(up)—times between up-crossings (estimated as the difference between the lines in the first column). For example, 24.4 = 42.415 − 18.015, and so on.
TAU(down)—times between down-crossings (estimated as the difference between the lines in the second column). For example, 23.47 = 53.505 − 30.035, and so on.
TAU(max)—contains the respective times between maxima and times between minima. For example, 22.9 = 47.65 − 24.75 and 23.55 = 60.15 − 36.6.

B. *DTRNDANL* Followed by Rhythms Analysis

Examples are provided here of analyses of the noisy, nonstationary time series introduced in Figure 11-15. This time, the series has been preprocessed to remove trend prior to analysis. This type of preprocessing may be necessary before the use of spectral analysis type methods on time series with aggressive trends, because they are valid only under the assumptions that the time series being analyzed is trend-free.

EXERCISE 11-3

Figures 11-31 through 11-34 present the outputs of the procedures *COSIN2NL*, *FFT-NLLS*, and *PHASEREF* performed on the detrended

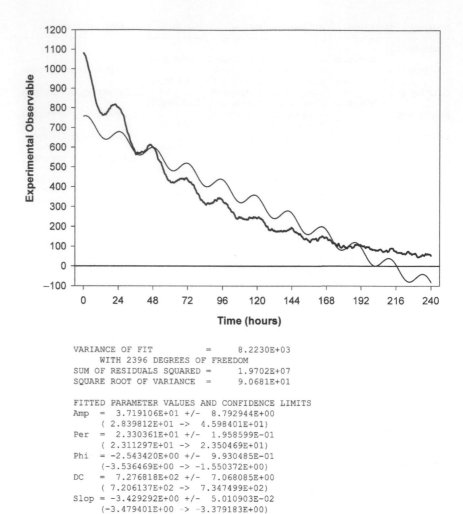

```
VARIANCE OF FIT          =      8.2230E+03
     WITH 2396 DEGREES OF FREEDOM
SUM OF RESIDUALS SQUARED =      1.9702E+07
SQUARE ROOT OF VARIANCE  =      9.0681E+01

FITTED PARAMETER VALUES AND CONFIDENCE LIMITS
Amp  =   3.719106E+01 +/-   8.792944E+00
     ( 2.839812E+01 ->   4.598401E+01)
Per  =   2.330361E+01 +/-   1.958599E-01
     ( 2.311297E+01 ->   2.350469E+01)
Phi  =  -2.543420E+00 +/-   9.930485E-01
     (-3.536469E+00 ->  -1.550372E+00)
DC   =   7.276818E+02 +/-   7.068085E+00
     ( 7.206137E+02 ->   7.347499E+02)
Slop =  -3.429292E+00 +/-   5.010903E-02
     (-3.479401E+00 ->  -3.379183E+00)
```

FIGURE 11-27.
COSIN2NL analysis of the noisy nonstationary data series originally presented in Figure 11-15, in which preprocessing by *ARFILTER* was performed prior to rhythms analysis.

(using *DTRNDANAL*) time series from Figure 11-15. Compare these results with those presented in Figures 11-21, 11-23, 11-25, and 11-26, which give the outputs of the same procedures for the original time series. Comment on whether, in your opinion, using preprocessing was beneficial and/or necessary in each case.

VII. EXAMPLE ANALYSIS: SIMULATION OF THE EFFECTS OF SCN ABLATION AND TRANSPLANTATION ON FREE-RUNNING LOCOMOTOR ACTIVITY

We now come back to the case study described in Section II. We have two major goals. The first is to characterize, as thoroughly as

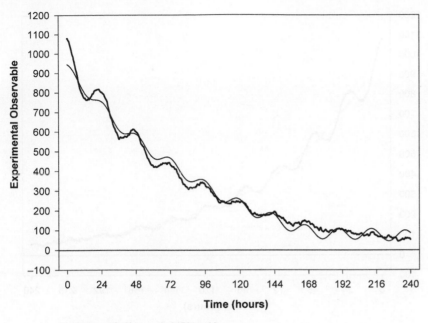

```
      Variance of fit =  8.94701E+02
           Error code =     0
      Std dev of fit =  2.99116E+01
               NDF     =   2388
      Conv criterion  =  1.00000E-06
          Conf Prob   = 0.9500
       DC Ampl  =  5.19081E+04 +/-  1.09720E+04
              [ 4.01079E+04 to  6.20519E+04]
      Ampl ( 1) =  5.20033E+04 +/-  1.09721E+04
              [ 4.02028E+04 to  6.21469E+04]
       Per ( 1) =  7.40877E+03 +/-  7.69920E+02
              [ 6.58932E+03 to  8.12916E+03]
     Phase ( 1) =  3.58464E+03 +/-  3.84854E+02
              [ 3.17516E+03 to  3.94487E+03]
      Ampl ( 3) =  2.82494E+01 +/-  4.11893E+00
              [ 2.47775E+01 to  3.30154E+01]
       Per ( 3) =  2.34058E+01 +/-  2.83412E-01
              [ 2.33486E+01 to  2.39154E+01]
     Phase ( 3) = -1.93907E+00 +/-  5.37706E-01
              [-2.49976E+00 to -1.42435E+00]
      Ampl ( 2) =  7.09566E+00 +/-  4.64823E+00
              [ 4.51094E+00 to  1.38074E+01]
       Per ( 2) =  1.02121E+02 +/-  6.15426E+00
              [ 9.20261E+01 to  1.04335E+02]
     Phase ( 2) =  1.91086E-01 +/-  9.90647E+00
              [-1.52590E+01 to  4.55390E+00]
      Ampl ( 4) =  5.80006E+00 +/-  4.13437E+00
              [ 2.32895E+00 to  1.05977E+01]
       Per ( 4) =  6.82809E+01 +/-  3.33319E+00
              [ 6.44606E+01 to  7.11270E+01]
     Phase ( 4) = -3.49172E+00 +/-  8.18569E+00
              [-1.54624E+01 to  9.09025E-01]
```

FIGURE 11-28.
FFT-NLLS analysis of the noisy nonstationary data series originally presented in Figure 11-15, in which preprocessing by *ARFILTER* was performed prior to rhythms analysis.

possible, the rhythmic behavior of the pretransplant locomotor data from the wild-type hamster (Figure 11-9, top panel) and the *tau^ss* hamster (depicted in the middle panel of Figure 11-9. The second is to examine and characterize the rhythmic behavior of

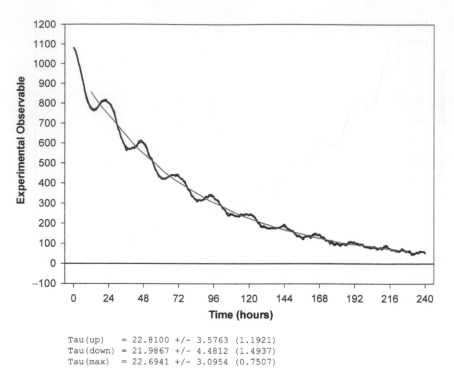

```
Tau(up)   = 22.8100 +/- 3.5763 (1.1921)
Tau(down) = 21.9867 +/- 4.4812 (1.4937)
Tau(max)  = 22.6941 +/- 3.0954 (0.7507)
```

FIGURE 11-29.
PHASEREF analysis of the noisy nonstationary data series originally presented in Figure 11-15, in which preprocessing by *ARFILTER* was performed prior to rhythms analysis.

XOVER(up)	XOVER(down)	AMPL(max)	TIME(max)	ALPHA	RHO	TAU(up)	TAU(down)	TAU(max)
1.795500E+01								
	3.000500E+01	5.852899E+01	2.430000E+01	1.205000E+01				
4.242500E+01		-5.671198E+01	3.610000E+01		1.242000E+01	2.447000E+01		
	5.360500E+01	4.967926E+01	4.755000E+01	1.118000E+01			2.360000E+01	2.325000E+01
6.598500E+01		-4.058145E+01	5.970000E+01		1.238000E+01	2.356000E+01		2.360000E+01
	7.814500E+01	2.760773E+01	7.315000E+01	1.216000E+01			2.454000E+01	2.560000E+01
8.999500E+01		-3.163821E+01	8.355000E+01		1.185001E+01	2.401000E+01		2.385000E+01
	1.013950E+02	2.708063E+01	9.535000E+01	1.139999E+01			2.325000E+01	2.220000E+01
1.146050E+02		-1.982954E+01	1.070500E+02		1.321001E+01	2.461000E+01		2.350000E+01
	1.256250E+02	1.881995E+01	1.211500E+02	1.102000E+01			2.423000E+01	2.580000E+01
1.388450E+02		-1.670316E+01	1.306500E+02		1.322000E+01	2.424000E+01		2.359999E+01
	1.496650E+02	1.586432E+01	1.445500E+02	1.081999E+01			2.403999E+01	2.340000E+01
1.612150E+02		-1.696100E+01	1.547500E+02		1.155000E+01	2.237000E+01		2.410001E+01
	1.722350E+02	1.535486E+01	1.661500E+02	1.102000E+01			2.257001E+01	2.159999E+01
1.862450E+02		-1.131422E+01	1.799500E+02		1.400999E+01	2.503000E+01		2.520000E+01
	1.975650E+02	9.363792E+00	1.905500E+02	1.132001E+01			2.533000E+01	2.440001E+01
2.097450E+02		-6.061455E+00	2.013500E+02		1.217999E+01	2.350000E+01		2.140001E+01
	2.169650E+02	6.859566E+00	2.136500E+02	7.220001E+00			1.939999E+01	2.309999E+01
2.232450E+02		-4.011070E+00	2.197500E+02		6.279999E+00	1.350000E+01		1.839999E+01
	2.278850E+02	3.197918E+00	2.264500E+02	4.639999E+00			1.092000E+01	1.280000E+01

FIGURE 11-30.
Verbose, tabulated summary of the *PHASEREF* analytical session presented in Figure 11-29. Whereas the expected estimates for TAU(up), TAU(down), and TAU(max) are to be in the vicinity of 24 hours, a few instances in which values for period estimates are considerably shorter than 24 hours arise as a consequence of noise confounds beginning to creep in, although this time, not until about 217 hours of x-axis time (compare the unfiltered analysis shown in Figure 11-26).

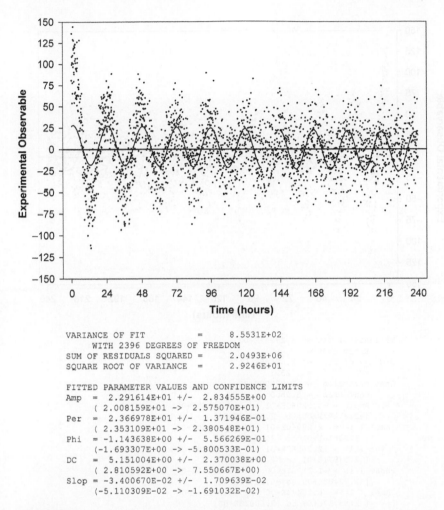

```
VARIANCE OF FIT           =      8.5531E+02
     WITH 2396 DEGREES OF FREEDOM
SUM OF RESIDUALS SQUARED =      2.0493E+06
SQUARE ROOT OF VARIANCE  =      2.9246E+01

FITTED PARAMETER VALUES AND CONFIDENCE LIMITS
Amp  =   2.291614E+01 +/-  2.834555E+00
       ( 2.008159E+01 ->  2.575070E+01)
Per  =   2.366978E+01 +/-  1.371946E-01
       ( 2.353109E+01 ->  2.380548E+01)
Phi  =  -1.143638E+00 +/-  5.566269E-01
       (-1.693307E+00 -> -5.800533E-01)
DC   =   5.151004E+00 +/-  2.370038E+00
       ( 2.810592E+00 ->  7.550667E+00)
Slop =  -3.400670E-02 +/-  1.709639E-02
       (-5.110309E-02 -> -1.691032E-02)
```

FIGURE 11-31.
COSIN2NL analysis of the noisy nonstationary data series originally presented in Figure 11-15, in which detrending by *DTRNDANL* was performed prior to rhythms analysis.

the wild-type hamster's post-transplant locomotor data (bottom panel, Figure 11-9) and the effect of SCN transplantation on the functioning of its circadian clock. The approach that we describe here is just one of many possible ways to analyze the data, and we encourage the reader to examine alternative strategies.

The data files (WT.NO, TAUSS.NO, and TRNSPLNT.NO) and the outputs of numerous analyses utilizing some or combinations of some of the algorithms we describe in this chapter can be downloaded from our Web site (see Internet Resources at the end of this chapter). To begin with, we use *FFT-NLLS* to identify the periodic components. A summary of the output for each of the files is presented in Figure 11-35.

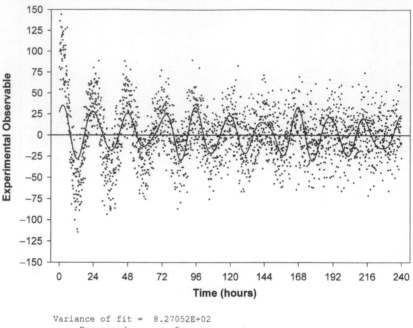

```
Variance of fit =   8.27052E+02
       Error code =        0
 Std dev of fit  =   2.87585E+01
             NDF  =      2391
Conv criterion   =   1.00000E-06
      Conf Prob  = 0.9500
   DC Ampl  = -3.52063E-01 +/-   2.52091E+00
           [-2.87297E+00 to  2.16885E+00]
  Ampl ( 1) =   2.28960E+01 +/-   3.53500E+00
           [ 1.93610E+01 to  2.64310E+01]
   Per ( 1) =   2.36563E+01 +/-   1.59310E-01
           [ 2.34971E+01 to  2.38158E+01]
 Phase ( 1) = -1.21761E+00 +/-   6.05645E-01
           [-1.82408E+00 to -6.12795E-01]
  Ampl ( 2) =   6.10444E+00 +/-   3.57133E+00
           [ 2.53757E+00 to  9.68023E+00]
   Per ( 2) =   1.83330E+01 +/-   2.90947E-01
           [ 1.80420E+01 to  1.86239E+01]
 Phase ( 2) = -4.25990E+00 +/-   1.75813E+00
           [-5.98599E+00 to -2.46973E+00]
  Ampl ( 3) =   4.79961E+00 +/-   3.57176E+00
           [ 1.22786E+00 to  8.37137E+00]
   Per ( 3) =   1.49053E+01 +/-   2.58201E-01
           [ 1.46439E+01 to  1.51603E+01]
 Phase ( 3) = -3.50373E+00 +/-   1.80036E+00
           [-5.30032E+00 to -1.69960E+00]
```

FIGURE 11-32.
FFT-NLLS analysis of the noisy nonstationary data series originally presented in Figure 11-15, in which detrending by *DTRNDANL* was performed prior to rhythms analysis.

Using the RAE as a criterion for significance, a period of 24.5 hours is identified for the wild-type hamster with RAE = 0.021. The 12.25-hour "half-period" is caused by an attempt at shape accommodation of the low-activity "flat-spots" in the data series. The third component has a RAE of 0.458, much greater than 0.021, and we can thus safely conclude that the rhythm with 24.5-hour

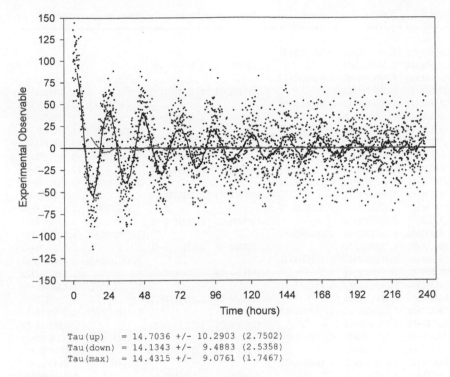

```
Tau(up)   = 14.7036 +/- 10.2903 (2.7502)
Tau(down) = 14.1343 +/-  9.4883 (2.5358)
Tau(max)  = 14.4315 +/-  9.0761 (1.7467)
```

FIGURE 11-33.
PHASEREF analysis of the noisy nonstationary data series originally presented in Figure 11-15, in which detrending by *DTRNDANL* was performed before rhythms analysis.

period is the most robust. Similar considerations lead to identifying the primary period for the *tau^{ss}* hamster to be 20 hours (with a half-period effect observed there, as well). For the post-transplant data, four periodic components are identified with average periods of 20 hours, 24.5 hours, 9.98 hours, and 104.9 hours, with the 20-hour and the 24.5-hour components representing the most dominant spectral contributions (i.e., largest absolute amplitudes and smallest RAE values). Thus, further analyses are needed for TRNSPLNT.NO in order to identify the most robust periodic components.

Next, because of the variance nonstationarities in the data, it may be prudent to consider applying *DTRNDANL* with output in SND-space. If *PHASEREF* will be involved, filtering to remove the high frequency noise may also be useful. The output is presented in Figure 11-36.

The use of *PHASERF* with the data from Figure 11-36 may now be employed to determine additional estimates for the rhythms. The

XOVER(up)	XOVER(down)	AMPL(max)	TIME(max)	ALPHA	RHO	TAU(up)	TAU(down)	TAU(max)
1.843500E+01								
	3.003500E+01	4.680085E+01	2.475000E+01	1.160000E+01				
4.203500E+01		-4.013690E+01	3.660000E+01		1.200000E+01	2.360000E+01		
	5.328500E+01	4.168316E+01	4.765000E+01	1.125000E+01			2.325000E+01	2.290000E+01
6.536500E+01		-2.980821E+01	5.855000E+01		1.208000E+01	2.333000E+01		2.195000E+01
	7.810500E+01	2.206183E+01	7.305000E+01	1.274001E+01			2.482000E+01	2.540000E+01
8.986500E+01		-2.452057E+01	8.305000E+01		1.175999E+01	2.450000E+01		2.450000E+01
	1.012150E+02	2.266267E+01	9.535000E+01	1.135000E+01			2.310999E+01	2.230000E+01
1.144050E+02		-1.517857E+01	1.045500E+02		1.319000E+01	2.454000E+01		2.150000E+01
	1.255150E+02	1.545501E+01	1.213500E+02	1.111000E+01			2.430000E+01	2.600000E+01
1.387650E+02		-1.390635E+01	1.306500E+02		1.325000E+01	2.436000E+01		2.609999E+01
	1.498650E+02	1.354610E+01	1.452500E+02	1.110001E+01			2.435001E+01	2.390000E+01
1.608450E+02		-1.499517E+01	1.543500E+02		1.098000E+01	2.208000E+01		2.370001E+01
	1.609250E+02	2.500519E-02	1.608500E+02	8.000183E-02			1.106000E+01	1.560001E+01
1.610750E+02		-2.908133E-01	1.610500E+02		1.499939E-01	2.299957E-01		6.699997E+00
	1.717150E+02	1.406143E+01	1.657500E+02	1.064000E+01			1.078999E+01	4.899994E+00
1.853950E+02		-9.638412E+00	1.797500E+02		1.368001E+01	2.432001E+01		1.870000E+01
	1.961350E+02	9.234318E+00	1.905500E+02	1.073999E+01			2.442000E+01	2.480000E+01
1.961650E+02		-1.311388E-01	1.961500E+02		2.999878E-02	1.076999E+01		1.639999E+01
	1.970050E+02	1.904894E+00	1.965500E+02	8.400116E-01			8.700104E-01	6.000000E+00
2.068750E+02		-7.307346E+00	2.008500E+02		9.869995E+00	1.071001E+01		4.700012E+00
	2.071450E+02	3.679020E-01	2.070500E+02	2.700043E-01			1.014000E+01	1.050000E+01
2.097750E+02		-3.851472E+00	2.087500E+02		2.629990E+00	2.899994E+00		7.899994E+00
	2.164750E+02	8.621487E+00	2.135500E+02	6.700012E+00			9.330002E+00	6.500000E+00
2.234750E+02		-6.468973E+00	2.195500E+02		7.000000E+00	1.370001E+01		1.080000E+01
	2.238050E+02	1.014919E+00	2.236500E+02	3.299866E-01			7.329987E+00	1.009999E+01
2.239350E+02		-4.329326E-01	2.238500E+02		1.300049E-01	4.599915E-01		4.300003E+00
	2.242350E+02	7.028205E-01	2.240500E+02	3.000031E-01			4.300079E-01	4.000092E-01
2.242850E+02		-1.605172E-01	2.242500E+02		5.000305E-02	3.500061E-01		3.999939E-01
	2.279150E+02	4.983418E+00	2.267500E+02	3.629990E+00			3.679993E+00	2.699997E+00

FIGURE 11-34.
Verbose, tabulated summary of the *PHASEREF* analytical session presented in Figure 11-33. Whereas the expected estimates for TAU (up), TAU(down), and TAU(max) are to be in the vicinity of 24 hours, numerous instances in which values for period estimates are considerably shorter than 24 hours arise as a consequence of noise confounds beginning to creep in at about 160 hours of x-axis time.

estimates from the *FFT-NLLS* results can be used for initializing the *PHASEREF* procedures.

Instead of presenting the verbose output from *PHASEREF*, we give a plot in Figure 11-37 of acrophase values with a fitted least-squares regression line.

For the wild-type hamster, the acrophase is delayed by about 0.5 hours from day to day, an observation reiterated by the estimated slope of the regression line (slope = 0.50215 +/− 0.01139). For the *tau*^{ss} hamster, the acrophase occurs earlier and earlier each day, by about a 4-hour difference (slope = −4.001286 +/−0.014826). These estimates, consistent with previous results from *FFT-NLLS*, allow us to conclude that the wild-type and *tau*^{ss} hamsters exhibit circadian rhythms with periods of 24.5 hours and 20.0 hours, respectively. The amplitudes and

FFT-NLLS of WT.NO

File Name	Amplitude	Period	Phase	Rel-Amp	Rel-Per	Rel-Phi
wt.no	4.796E+01	2.450E+01	-1.199E+01	0.021	0.000	0.003
	(9.882E-01)	(1.080E-02)	(8.043E-02)			
	2.040E+01	1.225E+01	3.448E-01	0.048	0.001	0.013
	(9.886E-01)	(1.041E-02)	(1.638E-01)			
	2.158E+00	2.155E+01	7.079E+00	0.458	0.008	0.081
	(9.894E-01)	(1.827E-01)	(1.743E+00)			

FFT-NLLS of TAUSS.NO

File Name	Amplitude	Period	Phase	Rel-Amp	Rel-Per	Rel-Phi
tauss.no	4.783E+01	2.000E+01	8.032E+00	0.029	0.001	0.007
	(1.402E+00)	(1.354E-02)	(1.318E-01)			
	1.904E+01	1.000E+01	-1.982E+00	0.074	0.001	0.017
	(1.406E+00)	(8.692E-03)	(1.692E-01)			

FFT-NLLS of TRANSPLNT.NO

File Name	Amplitude	Period	Phase	Rel-Amp	Rel-Per	Rel-Phi
trnsplnt.no	3.574E+01	2.000E+01	7.950E+00	0.078	0.001	0.014
	(2.773E+00)	(2.808E-02)	(2.726E-01)			
	1.316E+01	2.453E+01	-1.147E+01	0.212	0.006	0.048
	(2.789E+00)	(1.455E-01)	(1.169E+00)			
	1.026E+01	9.982E+00	-2.232E+00	0.272	0.004	0.084
	(2.791E+00)	(4.256E-02)	(8.373E-01)			
	8.214E+00	1.049E+02	-1.753E+01	0.333	0.046	0.098
	(2.738E+00)	(4.857E+00)	(1.033E+01)			

FIGURE 11-35.
FFT-NLLS analysis of the original data.

phase shifts for the rhythm are estimated as part of the output in Figure 11-35.

The decreasing patterns in acrophase times for the transplant data also indicate a shorter than 24-hour period, but the linear model may not be the best fit, because there appears to be a pattern to the residuals. This may be indicative of a more complex rhythmic component structure, composed of at least two periodic components. Thus, additional analyses are needed in order to determine the composition of the rhythm. In Figure 11-38, we present the output from the *FFT-NLLS* procedure on the detrended transplant data, as well as on the *ARFILTER*ed original data TRNSPLNT.NO. The detrended data are of particular interest, because the large estimated period of 104.9 hours in Figure 11-35 may be related to an attempt to fit for the trend. Observe that the large period estimate is no longer present for the detrended data (top panel of

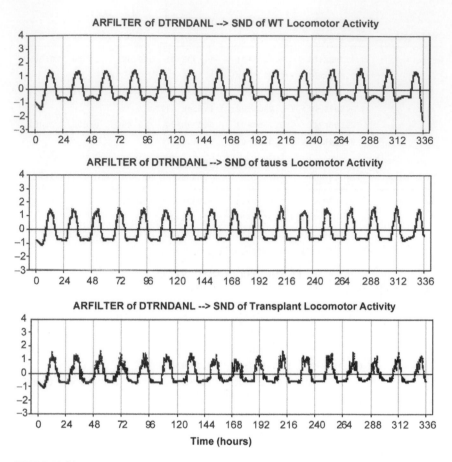

FIGURE 11-36.
ARFILTER output of for the detrended data (plotted in SND-space).

Figure 11-38), but the component with period 24.5 hours is also gone. In contrast, using the *ARFILTER*ed data shows both the 20-hour component and the 24.5-hour component (and also the ~10-hour period).

Judging by the combined information that we have presented so far, it is possible to hypothesize that the transplant hamster exhibits rhythms that are a mix of the 24.5-hour and 20.0-hour rhythms characteristic of the wild-type and the *tau*^{ss} hamsters. If this can be further corroborated, it would indicate that both the rhythms of the *tau*^{ss} SCN and that of the original wild-type are evident. The extent to which these rhythms participate in the rhythmic formation of the post-transplant data can be assessed by the ratio of the spectral peak magnitudes of the periodic components, measured in this case by the ratio of the respective

```
FITTED PARAMETER VALUES AND CONFIDENCE LIMITS
DC   = 1.193061E+01 +/-  8.383989E-02 ( 1.184677E+01 ->  1.201445E+01)
Slop = 5.021500E-01 +/-  1.139110E-02 ( 4.907589E-01 ->  5.135411E-01)
```

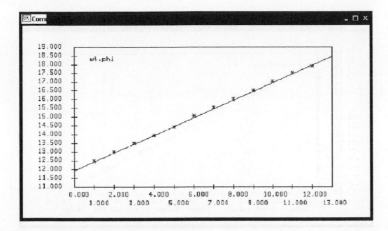

A

```
FITTED PARAMETER VALUES AND CONFIDENCE LIMITS
DC   =  1.197362E+01 +/-  1.348000E-01 ( 1.183882E+01 ->  1.210842E+01)
Slop = -4.001286E+00 +/-  1.482594E-02 (-4.016111E+00 -> -3.986459E+00)
```

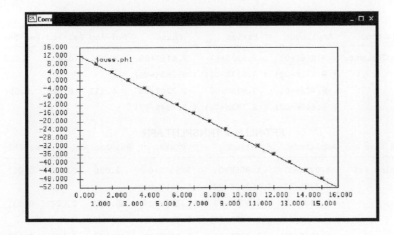

B

FIGURE 11-37.
Plots of the daily times of maximum obtained by using *PHASEREF.* The phase shifts are evident from
the nonzero slopes of the fitted least-squares regression lines.

(Continued)

amplitudes. From the data in Figures 11-35 and 11-38, we can
estimate this ratio to be approximately from 3:1 to 5.5:1. Further
analyses will be necessary in order to obtain a more accurate
estimate (the known relative contributions in this simulated example
are 75% 20-hour rhythm and 25% 24.5-hour rhythm). Outputs from
such additional procedures using the data from this example can
be downloaded from our Web site. We encourage the reader to consider
them and continue the exploration.

```
FITTED PARAMETER VALUES AND CONFIDENCE LIMITS
DC   =  1.233070E+01 +/-  7.228050E-01 ( 1.160789E+01 -> 1.305350E+01)
Slop = -4.041382E+00 +/-  7.950711E-02 (-4.120889E+00 -> -3.961875E+00)
```

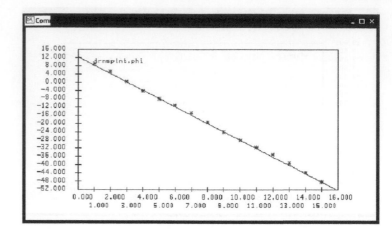

C

FIGURE 11-37 Cont'd.

FFT-NLLS of TRNSPLNT.DTR

File Name	Amplitude	Period	Phase	Rel-Amp	Rel-Per	Rel-Phi
trnsplnt.dtr	9.016E-01	1.993E+01	7.446E+00	0.071	0.002	0.016
	(6.432E-02)	(3.337E-02)	(3.263E-01)			
	2.772E-01	9.975E+00	-2.320E+00	0.233	0.003	0.054
	(6.448E-02)	(2.783E-02)	(5.404E-01)			

FFT-NLLS of TRNSPLNT.AFR

File Name	Amplitude	Period	Phase	Rel-Amp	Rel-Per	Rel-Phi
trnsplnt.arf	8.029E-01	1.995E+01	7.607E+00	0.030	0.001	0.007
	(2.376E-02)	(1.322E-02)	(1.308E-01)			
	2.448E-01	9.975E+00	-2.323E+00	0.097	0.002	0.031
	(2.381E-02)	(1.562E-02)	(3.066E-01)			
	1.433E-01	2.425E+01	1.103E+01	0.166	0.005	0.037
	(2.383E-02)	(1.095E-01)	(8.988E-01)			

FIGURE 11-38.
Results from *FFT-NLLS* from the *ARFILTER*ed data.

REFERENCES

Abe, M., Herzog, E. D., Yamazaki, S., Straume, M., Tei, H., Sakaki, Y., Menaker, M., & Block, G. D. (2002). Circadian rhythms in isolated brain regions. *Journal of Neuroscience, 22,* 350–356.

Bell-Pedersen, D., Cassone, V. M., Earnest, D. J., Golden, S. S., Hardin, P. E., Thomas, T. L., & Zoran, M. J. (2005). Circadian rhythms from multiple oscillators: Lessons from diverse organisms. *Nature Reviews Genetics, 6,* 544–566.

Fontes, G., Rocha, E. M. M., Brito, A. C., Fireman, F. A. T., & Antunes, C. M. F. (2000). The microfilarial periodicity of *Wuchereria bancrofti* in north-eastern Brazil. *Annals of Tropical Medicine and Parasitology, 94,* 373–379.

Herzog, E. D., Geusz, M. E., Khalsa, S. B. S., Straume, M., & Block, G. D. (1997). Circadian rhythms in mouse suprachiasmatic nucleus explants on multimicroelectrode plates. *Brain Research, 757,* 285–290.

Hurd, M. W., DeBruyne, J., Straume, M., & Cahill, G. M. (1998). Circadian rhythms of locomotor activity in zebrafish. *Physiology & Behavior, 65,* 465–472.

Konopka, R. J., & Benzer, S. (1971). Clock mutants of *Drosophila melanogaster*. *Proceedings of the National Academy of Sciences of the United States of America, 68,* 2112–2116.

Meerlo, P., van den Hoofdakker, R. H., Koolhaus, J. M., & Daan, S. (1997). Stress-induced changes in circadian rhythms of body temperature and activity in rats are not caused by pacemaker changes. *Journal of Biological Rhythms, 12,* 80–92.

Orr, W. C., & Hoffman, H. J. (1974). A 90-min cardiac biorhythm: Methodology and data analysis using modified periodograms and complex demodulation. *IEEE Transactions on Biomedical Engineering, 21,* 130–143.

Panda, S., Hogenesch, J. B., & Kay, S. A. (2002). Circadian rhythms from flies to human. *Nature, 417,* 329–335.

Plautz, J. D., Straume, M., Stanewsky, R., Jamison, C. F., Brandes, C., Dowse, H. B., Hall, J. C., & Kay, S. A. (1997). Quantitative analysis of *Drosophila period* gene transcription in living animals. *Journal of Biological Rhythms, 12,* 204–217.

Ralph, M.R, & Menaker, M. (1988). A mutation of the circadian system in golden hamsters. *Science, 241,* 1225–1227.

Ralph, M.R, Foster, R. G., Davis, F. D., & Menaker, M. (1990). Transplanted suprachiasmatic nucleus determines circadian period. *Science, 247,* 975–978.

Reppert, S. M., & Weaver, D. R. (2002). Coordination of circadian timing in mammals. *Nature, 418,* 935–941.

Richter, H. G., Torres-Farfan, C., Rojas-Garcia, P. P., Campino, C., Torrealba, F., & Seron-Ferre, M. (2004). The circadian timing system: Making sense of day/night gene expression. *Biology Research, 37,* 11–28.

Straume, M., Frasier-Cadoret, S. G., & Johnson, M. L. (1991). Least squares analysis of fluorescence data. In Lakowicz, J. R., (Ed.), *Topics in Fluorescence Spectroscopy, Volume 2: Principles* pp. 117–240. New York: Plenum Press.

Sujino, M., Masumoto, K., Yamaguchi, S., van der Horst, G. T. J., Okamura, H., & Inouye, S. T. (2003). Suprachiasmatic nucleus grafts restore circadian behavioral rhythms of genetically arrhythmic mice. *Current Biology, 13,* 664–668.

Ward, R. R. (1971). *The living clocks.* New York: Alfred A. Knopf.

Yamazaki, S., Numano, R., Abe, M., Hida, A., Takahashi, R., Ueda, M., Block, G. D., Sakaki, Y., Menaker, M., & Tei, H. (2000). Resetting central and peripheral circadian oscillators in transgenic rats. *Science, 288,* 682–685.

INTERNET RESOURCES

www.biomath.sbc.edu/data.html
Sweet Briar College Biomathematics web site with simulated data for the hamster SCN transplant example and output from analyses.

FURTHER READINGS

Hamada, T., LeSauter, J., Venuti, J. M., & Silver, R. (2001). Expression of period genes: Rhythmic and nonrhythmic compartments of the suprachiasmatic nucleus pacemaker. *Journal of Neuroscience, 21,* 7742–7750.

Johnson, M. L., & Frasier, S. G. (1985). Nonlinear least-squares analysis. In Hirs, C. H. W., & Timasheff, S. N. (Eds.), *Methods in Enzymology* (vol. 117, pp. 301–342). New York: Academic Press.

Matsuno, H., Tanaka, Y., Aoshima, H., Doi, A., Matsui, M., & Miyano, S. (2003). Biopathways representation and simulation on hybrid functional Petrinet. *Silico Biology, 3,* 32.

Reppert, S. M., & Weaver, D. R. (2001). Molecular analysis of mammalian circadian rhythms. *Annual Review of Physiology, 63,* 647–676.

Stokkan, H., Yamazaki, S., Tei, H., Sakaki, Y., & Menaker, M. (2001). Entrainment of the circadian clock in the liver by feeding. *Science, 291,* 490–493.

Tukey, J. W. (1963). An introduction to the frequency analysis of time series. In Thompson, J. R., & Brillinger, D. R., (Eds.), *Mathematics 596, Spring 1963.* Princeton, NJ: Princeton University.

Welsh, D. K., Logothetis, D. E., Meister, M., & Reppert, S. M. (1995). Individual neurons dissociated from rat suprachiasmatic nucleus express independently phased circadian firing rhythms. *Neuron, 14,* 697–706.

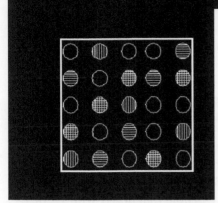

Chapter 12

USING MICROARRAYS TO STUDY GENE EXPRESSION PATTERNS

Fabricating and Using Microarrays

Analysis of Microarray Data

Microarrays and Circadian Rhythms

Progress has been much more general than retrogression.
Charles Robert Darwin (1809–1882)

The *per-luc* reporter gene constructs described in the last chapter are an excellent tool for identifying cells, tissues, and organs with their own circadian rhythms. However, many genes other than the *period* gene are expressed in a circadian manner. Some of these genes, such as *clock* and *timeless*, have already been identified and studied, but there are undoubtedly others currently unidentified. To understand circadian functioning in greater depth, we would like to be able to identify and characterize all genes that are under circadian control.

As we have seen, genes and proteins work together in living organisms in an intricate network to ensure proper functioning at the system level. Traditional methods that are designed to study single genes are incapable of capturing these complex dynamics, and new tools are needed to investigate these interactions. The best tool currently available for simultaneously examining the expression of multiple genes is the *microarray*, also called the DNA array, the DNA chip, or the gene chip. Microarrays are now widely used in biological and biomedical research, as it becomes increasingly clear that to understand biological function we must discover and understand patterns of gene expression. Microarrays allow us to analyze the expression of all of the thousands of genes of an organism. This also allows us to identify genes expressed in a rhythmic fashion.

Microarray technology provides a means to examine the effect of gene regulation on physiologic functions, to study the genetic mechanisms of certain diseases, and to assess the effectiveness of new therapies. Applications include studying embryonic development in order to prevent and treat birth defects, investigating pancreatic function to improve treatment for diabetes, and examining cancer cell behavior for the purpose of designing new approaches to treatment.

For example, cancer is not a single disease. There are many different kinds of cancer, arising in different organs and tissues through the accumulated mutation of multiple genes. Traditionally, cancer diagnoses have relied on microscopic examination of cell appearance, which may be incapable of distinguishing among cancer types. More recently, cancers have been characterized by the level of expression of particular genes. It has become apparent that the best way to distinguish many different kinds of cancer is by examining their patterns of gene expression. Knowing these patterns would allow an accurate diagnosis and an immediate selection of appropriate therapies, as well as facilitating the development of better treatments.

The power of microarray technology can be seen in the case of diffuse large B-cell lymphoma (DLBCL).[1] In the United States alone, more than 20,000 people are diagnosed with DLBCL each year. The disease progresses rapidly, and although many people diagnosed with DLBCL were treated successfully, more than half of the patients were not. Microscopic examination of cells from the disease revealed no basis for predicting which cases would result in a successful outcome and which would not. In 2000, Alizadeh et al. reported on a microarray-based experiment that distinguished, on the basis of gene expression patterns, two different types of DLBCL (Alizadeh et al. [2000]). One type, which they called "germinal center B-like DLBCL," was associated with much better long-term survival rates than the other type, which they called "activated B-like DLBCL." Further studies identified additional gene expression profiles and allowed even better discrimination among DLBCL types. The ability to accurately diagnose DLBCL has led to more accurate prognoses and the opportunity for improved selection among the available therapeutic options. More importantly, knowledge of the gene expression profiles of DLBCL will allow the development of targeted treatments, some of which are already in clinical trials (see, for example, Abramson and Shipp [2005]).

The use of microarrays in the last decade has provided the opportunity to compare expression patterns among thousands of different genes under different conditions and to determine those with similar expression patterns. When two or more genes have similar expression profiles, they may be regulated by the same factors (co-regulated) and may be functionally related. Identifying such groups of genes provides clusters that warrant further detailed examination with regard to their expression control mechanisms.

Another potential use of microarray technology is the study of *cancer chronotherapy*, an approach that takes into account the fact that some cancer therapies work better when administered at certain times of the day than at others (see Mormont and Levi [2003]). Comparing the circadian patterns of cancer cell gene expression with such clinical observations should allow identification of genes that are important to the success or failure of the treatment. Characterization of these genes should allow better understanding of these timing phenomena, which

---

1. DLBCL is a common form of non-Hodgkin's lymphoma. Lymphomas are cancers arising from lymphocytes, the cells of the immune system.

should then lead to the design of improved chemotherapeutic agents and regimens.

Nowadays, the amount of microarray-based literature is vast, and the methods for analyzing gene chip data are becoming increasingly more technical. As microarrays of increasing gene density are made, we need better methods for processing and organizing huge amounts of data for easy access and analyses. Microarray research has been supported by innovative methods from the new field of *bioinformatics*, the discipline that studies the application of computer technology to the management of biological information. Emerging at the interface of biology, mathematics, and computer science, bioinformatics allows data organization, manipulation, and analysis at a scale that would otherwise be impossible.

We begin this chapter with a brief review of the technologies used to manufacture microarrays and some of the ways in which they are being utilized. We then describe some methods for data processing and analyses designed to quantify the results from microarray experiments. We present some of the mathematical background pertinent to designing clustering methods and then turn to some analytical tools developed to study circadian gene expression.

I. FABRICATING AND USING MICROARRAYS

A. Producing Labeled cDNA

Recall from Chapter 11 that the ultimate result of gene expression is the production of proteins, which are the *translation* products of messenger RNAs (mRNAs). Gene expression occurs via RNA synthesis, called *transcription*, the process in which one of the two strands of the DNA is used to direct the production of an RNA molecule, as diagrammed in Figure 12-1. The nitrogenous bases of RNA (A, C, G, and U) hydrogen-bond to the nitrogenous bases of the DNA, with A binding to U and C binding to G.

The product of the transcription of protein coding genes is mRNA, so we can isolate mRNA from the cells of interest and use the amount of mRNA as a proxy for the amount of protein produced by the cell. Because mRNA is an extremely fragile, short-lived molecule, vulnerable to seemingly ubiquitous ribonuclease enzymes, it would be difficult to use mRNA in microarray-based experiments. However, the mRNA can be converted into a more durable form called *complementary DNA* (*cDNA*). The cDNA is made from the mRNA and is complementary to it. This means that the cDNA can be used in place of the mRNA to investigate gene expression.

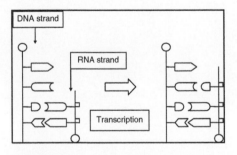

FIGURE 12-1.
Graphic representation of transcription (RNA synthesis). The DNA strand at left is being copied during the production of RNA. The RNA is here distinguished from the DNA by the small boxes attached to the diagram, which represent the additional hydroxyl groups that differentiate the ribose sugar of RNA from the deoxyribose sugar of DNA.

The DNA double helix is held together through hydrogen bonding of A to T and C to G, as is shown in Figure 12-2. Because these hydrogen bonds are weak, it is possible to separate the two strands of a DNA molecule into two single-stranded DNA molecules by manipulating the pH, the salt concentration, or the temperature of the DNA solution. The converse is also true: A cDNA molecule can bind to a complementary DNA sequence under the appropriate conditions.

Suppose that we wish to examine the differences in gene expression between normal liver cells and cancerous liver cells. We begin the process of making cDNA by isolating total RNA from the cells of interest. We disrupt the cell membrane and inactivate and remove all of the proteins that may damage or contaminate the RNA. Total RNA includes ribosomal RNA (rRNA), transfer RNA (tRNA), and small nuclear RNA (snRNA), as well as mRNA. Because mRNA is a small

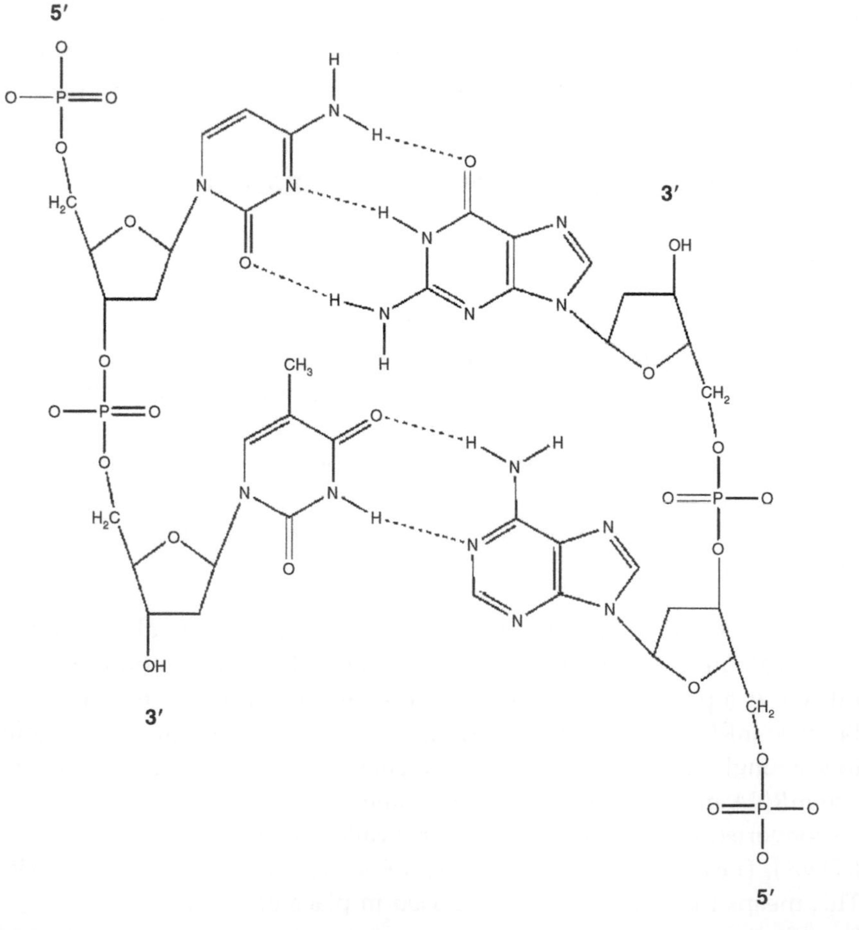

FIGURE 12-2.
Hydrogen bonding between nitrogenous bases. Note the two bonds (represented by dotted lines) between A and T and the three bonds between G and C.

proportion of the total RNA in a cell, it may be necessary to purify the mRNA away from the rest of the RNA population.

The method for doing this uses a post-transcriptional modification of eukaryotic mRNA that has, at the 3' end, a long string of "A" nucleotides forming the *poly(A) tail* (see Figure 12-3). The addition of the poly(A) tail occurs after the RNA has been made (thus the term post-transcriptional), but before it is used for translation. A string of T nucleotides, also known as an oligo(dT) molecule, would be complementary to a poly(A) tail, and oligo(dT) molecules attached to a solid support can therefore be used to capture the mRNA in a total RNA sample. This is usually done by putting the solid support with the bound oligo(dT) into a tube or cylinder, called a column, and allowing the total cellular RNA mixture to pass through the column. The mRNA will stick to the oligo(dT) and the remaining RNAs will flow out of the column. The column is washed with buffer and the purified mRNA is then collected from the column by altering the salt concentration of the wash buffer. Alternatively, because of the poly(A) tail, one can go directly to the next step without isolating mRNA.

We convert mRNA to cDNA through a process called *reverse transcription*. This process, diagrammed in Figure 12-3, utilizes a retroviral enzyme properly called an *RNA-dependent DNA polymerase,*

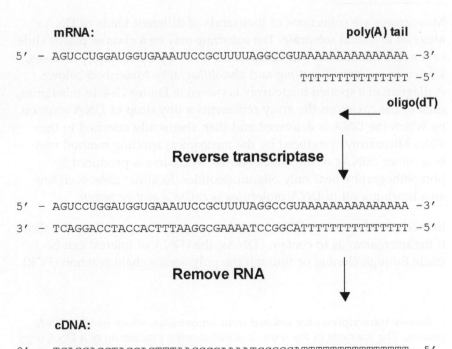

mRNA: **poly(A) tail**

5' – AGUCCUGGAUGGUGAAAUUCCGCUUUUAGGCCGUAAAAAAAAAAAAAAAA –3'

                                    TTTTTTTTTTTTTTTTT –5'

                                                    oligo(dT)

Reverse transcriptase

5' – AGUCCUGGAUGGUGAAAUUCCGCUUUUAGGCCGUAAAAAAAAAAAAAAAA –3'

3' – TCAGGACCTACCACTTTAAGGCGAAAATCCGGCATTTTTTTTTTTTTTTT –5'

Remove RNA

cDNA:

3' – TCAGGACCTACCACTTTAAGGCGAAAATCCGGCATTTTTTTTTTTTTTTT –5'

FIGURE 12-3.
Reverse transcription of a small mRNA. The reverse transcriptase enzyme is responsible for the extension of the oligo(dT). Note that the new strand is DNA. The mRNA may be removed enzymatically or through treatment with a strong base.

and colloquially called *"reverse transcriptase."*[2] This method takes advantage of the poly(A) tail on the RNA by using oligo(dT) molecules as primers to begin the reverse transcription process. The reverse transcriptase enzyme incorporates the correct deoxyribonucleotide building blocks into the growing chain, making a DNA copy of the mRNA (note that the DNA uses T nucleotides instead of the U nucleotides of the mRNA molecule).

Next, the cDNA needs to be labeled to allow us to distinguish between normal and cancer cell cDNAs. Two different techniques can be used to accomplish this: *direct enzymatic incorporation* and *chemical coupling*. In direct enzymatic labeling, the most common technique, a fluorescently labeled nucleotide is introduced into the cDNA as it is being made. When two different kinds of cell mRNAs are used, fluorescent dyes are used with one emission wavelength for the normal cell cDNA and another emission wavelength for the cancer cell cDNA. Two commonly used dyes are Cy3 and Cy5, which emit green and red light, respectively. This method allows the simultaneous use of both cDNAs to probe a single microarray. In chemical coupling, a modified nucleotide is introduced during cDNA synthesis. This modified nucleotide is then labeled through a second step, a chemical reaction with the fluorescent dye.

B. Making the Microarray

Microarrays are collections of thousands of different kinds of DNA attached to a solid substrate. The substrate may be a glass or plastic slide or a nylon membrane. Microarrays are made by one of two processes, known as *mechanical spotting* and *photolithography* (described below). A diagram of a spotted microarray is shown in Figure 12-4. In this figure, each of the circles on the array represents a tiny drop of DNA solution, by which the DNA is delivered and then chemically attached to the slide. Microarrays produced by the mechanical spotting method may bear either cDNAs or oligonucleotides.[3] Microarrays produced by photolithography bear only oligonucleotides. In either case, each tiny area bearing a set of DNA molecules is called a *spot* or *feature*.

In mechanical spotting, DNA is made first and then placed on the slide. If the microarray is to contain cDNAs, the cDNA of interest can be made through cloning or through the polymerase chain reaction (PCR).

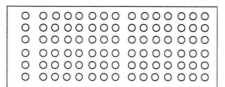

FIGURE 12-4.
Graphic representation of a spotted microarray. Each spot or feature contains cDNA or oligonucleotides.

2. Reverse transcriptases are isolated from retroviruses, which have an RNA genome. The life cycle of the retrovirus includes the production of a DNA copy of their RNA genome using reverse transcriptase and then the integration of the DNA copy into the host cell's own DNA.
3. Oligonucleotides are relatively short polymers made of several nucleotides. Sequences can be selected or built to represent specific genes of interest, as described in the next section.

If the cDNAs are to be made by cloning, recombinant plasmids bearing the cDNAs are introduced into bacteria, which will then make many copies of the cDNAs. If the cDNAs are to be amplified via PCR, the replicative power of the *Thermus aquaticus* DNA polymerase is used to make the copies. If the microarray is to contain oligonucleotides, a DNA synthesizer is programmed to produce DNAs with the desired sequences.

Once the DNAs are produced, they may be spotted onto the solid support by inkjet-style printer heads or by pins dipped into the DNA and then touched to the surface of the support. Robots move the inkjets or pins, so that the placement of features is precise and reproducible. The use of robots allows high-density placement of tiny features. Each of the thousands of spots on a single slide may represent a different gene.

In photolithography, on the other hand, the oligonucleotides are synthesized in place in a process which is akin to that used to produce integrated circuits. Light is used to activate one step in the process, and masks are used to block the light for those portions of the chip that are not to participate in the reaction. The process employed by Affymetrix, Inc. to produce its widely used GeneChip® is shown in Figure 12-5. The first step in the manufacturing process requires that a quartz wafer be covered with a light-sensitive protecting group. This group binds to all available sites on the wafer and prevents nucleotides

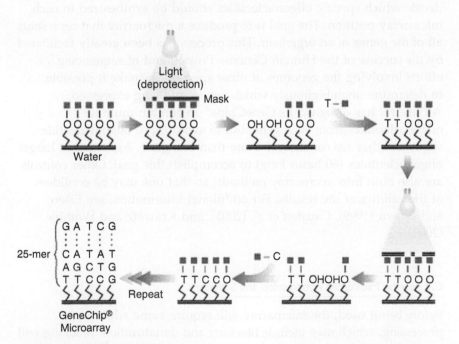

FIGURE 12-5.
Affymetrix, Inc. uses a unique combination of photolithography and combinatorial chemistry to manufacture GeneChip® Arrays. (Image courtesy of Affymetrix, Inc.)

from binding nonspecifically to the wafer. A carefully designed opaque mask is placed over the wafer, and a light is turned on, deprotecting the exposed parts of the wafer. The deprotected spots now have free hydroxyl groups on their surfaces, whereas the areas under the mask do not. The first nucleotide in the sequence can now be added, and it will bind to the free hydroxyl groups. Each nucleotide to be added will also carry a light-sensitive protecting group. Excess nucleotide will be removed; a new mask will be applied; and the light will shine again, deprotecting only those areas not covered by the new mask. Now the second nucleotide to be added will be able to bind to the free hydroxyl groups, some of which will be on the wafer, and some of which will be on the first nucleotide. The process will be repeated again and again, until each spot on the wafer bears an oligonucleotide that is 25 nucleotide units long. This process allows extremely high feature density, as is shown in graphic form in Figure 12-6(A). Figure 12-6(B) shows a finished GeneChip.®

The technologies for microarray production are constantly evolving. NimbleGen Systems, Inc., for example, uses a technique called maskless photolithography, utilizing tiny movable mirrors to direct the light to specific spots, rather than masks to block it. Oligonucleotides may also be synthesized in situ through the use of inkjet printers, with the printer head delivering the required nucleotides in the desired sequence. This is the method used by Agilent Technologies, Inc.

Regardless of the arraying technology used, the manufacturer must decide which specific oligonucleotides should be synthesized in each microarray position. The goal is to produce a microarray that represents all of the genes of an organism. This process has been greatly facilitated by the success of the Human Genome Project, and of sequencing efforts involving the genomes of other species. To make it possible to determine unambiguously which genes are being expressed, Affymetrix has designed its GeneChips® to contain multiple oligonucleotides from each gene, but to avoid those oligonucleotide sequences that are common to more than one gene. Agilent uses longer oligonucleotides (60 bases long) to accomplish this goal. Other controls are also built into microarray methods, so that one may be confident of the validity of the results. For additional information, see Eisen and Brown (1999), Causten et al. (2003), and Krawetz and Womble (2003).

FIGURE 12-6.
Panel A: Graphic representation of a GeneChip® microarray, showing the high density of features possible with photolithography. Each tiny square represents a different oligonucleotide. Panel B: Affymetrix GeneChip® probe array. Image courtesy of Affymetrix, Inc.

C. Using the Microarray: Hybridization and Scanning

Before being used, the microarray will require some additional processing, which may include blocking and denaturation. Blocking will inactivate any free reactive groups that might bind the labeled cDNAs nonspecifically, reducing background fluorescence. Microarrays with spotted cDNAs should also be heated to denature the cDNAs and thus render them single-stranded and available to hybridize with the labeled cDNAs.[4]

The next step in using a microarray is *hybridization*. In our example, the labeled cDNAs from the normal and cancerous liver cells are mixed together in an appropriate buffer and placed on the microarray. The cDNAs will then bind to any features that carry complementary DNA. Some features will bind only normal cell cDNA; some features will bind only cancer cell DNA; some features will bind both normal and cancer cell cDNAs; and some features will bind neither cDNA. Following the hybridization step, the microarray is washed to remove any unbound cDNAs. Figure 12-7(A) represents a microarray after hybridization and washing. Note that at this point in an actual microarray experiment, there will be no visible difference between the hybridized microarray and a nonhybridized array. The hybridized microarray must be scanned to determine which features have bound which cDNAs. Figure 12-7(B) shows an image from a scanned GeneChip.® This chip has been hybridized and washed, and the light intensity in each spot is proportional to the amount of labeled cDNA binding at that spot.

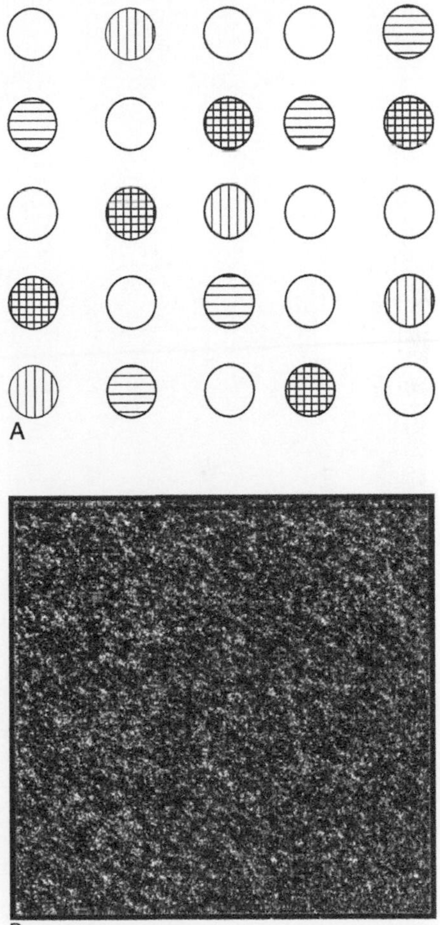

The process of *scanning* involves the use of a light source, usually a laser, that will excite the fluorescent label on the cDNAs. Each fluor molecule has a characteristic excitation wavelength and a characteristic emission wavelength. The excitation beam will excite electrons in the fluors to a higher energy state, and then the fluors release some of their excess energy in the form of visible light as the electrons decay back to their original energy states. A detector is used to measure the emitted light. The greater the amount of labeled cDNA bound to a feature, the greater the amount of light emitted when the microarray is scanned. The emitted light from both fluors will be measured and recorded as image files. Laser scanning confocal microscopy can be used for this process, and a number of different types of scanners are commercially available.

II. ANALYSIS OF MICROARRAY DATA

A. Filtering and Normalization

Once the image files are acquired, the images are converted into quantitative data measuring the intensity of fluorescence at each spot. The first step is to determine where the spots are, how much of the fluorescence at each spot is caused by the hybridization, and how much may be caused by nonspecific background fluorescence. Determining

FIGURE 12-7.
Panel A: Graphic representation of microarray following hybridization and washing. Each circle represents a gene. The circles filled with vertical lines represent hybridization with the cDNA from one cell type; those filled with horizontal lines represent hybridization with the cDNA from the other cell type; and those filled with grids represent hybridization with cDNA from both cell types. Empty circles represent features which did not hybridize with either cDNA; panel B: Actual output from an Affymetrix GeneChip®. Data from an experiment showing the expression of thousands of genes on a single GeneChip®. (Image courtesy of Affymetrix, Inc).

___
4. Oligonucleotides are already single-stranded and do not need to be denatured.

the location of the spots is facilitated by the fact that the spots are arranged in a regular array, using either a rectangular grid or a tighter "orange packing" layout. Image-processing software is used to define the location of the spots and to quantify the amount of fluorescence for each spot. The quantification involves measuring the intensity of the individual pixels that make up the image of the spot and a separate measurement of the pixels that make up the background. The background fluorescence is subtracted from the spot fluorescence, and the amount of expression is recorded. When the goal is to estimate the difference between normal and cancer cells, as in our example, the ratios of red fluorescence (cancer cells) divided by green fluorescence (normal cells) may be recorded as well. If we were to assign values to the spots shown in Figure 12-7(A), we might obtain a result such as that shown in Table 12-1. Spots with no fluorescence are omitted.

Unfortunately, using a ratio means that a gene that has a twofold increase in expression in the cancer cells, such as gene D, will give a ratio of 2, but a gene that has a twofold decrease in expression in the cancer cells, such as gene G, will give a ratio of 0.5. In fact, all of the decreased expression will be squeezed between 0 and 1, whereas the increases in expression, such as the 10-fold increase in genes A and H, can have extremely high numbers. One solution to this problem is to transform the ratios to logarithms. This is usually done using base 2, as is shown in Table 12-2.

Spot Position			Fluorescence[*]		Ratio
Row	Column	Gene "Name"	Red	Green	Red/Green
1	2	A	1,000	100	10
1	5	B	120	1,200	0.1
2	1	C	100	1,000	0.1
2	3	D	1,500	750	2
2	4	E	150	1,500	0.1
2	5	F	1,800	1,200	1.5
3	2	G	600	1,200	0.5
3	3	H	800	80	10
4	1	I	1,500	1,500	1
4	3	J	50	500	0.1
4	5	K	1,800	360	5
5	1	L	900	90	10
5	2	M	180	900	0.2
5	4	N	1,000	1,500	0.67

TABLE 12-1.
Synthetic data for Figure 12-7(A).

[*]Fluorescence is rarely reported in absolute units. Since the conditions vary widely among experiments because of equipment differences, the results are given on an intensity scale relative to each experiment.

| Spot Position | | Fluorescence | | | Ratio | Log |
Row	Column	Gene "Name"	Red	Green	Red/ Green	(Base 2)
1	2	A	1,000	100	10	3.32
1	5	B	120	1,200	0.1	–3.32
2	1	C	100	1,000	0.1	–3.32
2	3	D	1,500	750	2	1.00
2	4	E	150	1,500	0.1	–3.32
2	5	F	1,800	1,200	1.5	0.58
3	2	G	600	1,200	0.5	–1.00
3	3	H	800	80	10	3.32
4	1	I	1,500	1,500	1	0.00
4	3	J	50	500	0.1	–3.32
4	5	K	1,800	360	5	2.32
5	1	L	900	90	10	3.32
5	2	M	180	900	0.2	–2.32
5	4	N	1,000	1,500	0.67	–0.58

TABLE 12-2.
Synthetic data for Figure 12-7(A), with logarithmic transformation of red/green ratios.

EXERCISE 12-1

Compare the ratio and logarithmic data from Table 12-2. Characterize each as a decrease or an increase in expression. Why is using the log base 2 a better situation than using the ratios?

Note that in Tables 12-1 and 12-2 we did not list the features from Figure 12-7(A) that had shown no fluorescence at all. This procedure is called *filtering*. In a real microarray experiment, we might also want to remove all of the features that had very low values. This type of filtering process would reduce the size of the data set (thus increasing processing speed) and remove the lowest-quality data because any spots with intensity near the background level are likely to be measured with questionable accuracy. Genes with missing values in replicate measurements may also be filtered. Finally, depending on the experiment, it may be important to only consider genes that changed expression by a given amount, such as a factor of two. As with any preprocessing of data, filtering may result in loss of information. When carefully used, however, it increases processing speed and accuracy without a significant risk of eliminating any important genes.

Through image processing, background correction, and filtering, the information from the microarray that was initially stored as an image is converted to a table of values for each gene present on the microarray.

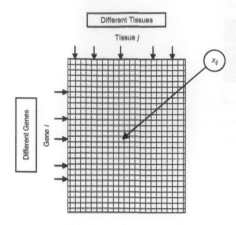

FIGURE 12-8.
Gene expression matrix for a comparative tumor gene expression study. Each row represents the expression values for a different gene, and each column represents the values for a different tumor sample.

Tables such as these are called *gene expression matrices*. For a typical gene expression experiment, the matrices would include the values from a number of arrays, representing different patients, treatments, or cell types, as depicted in Figure 12-8. Mathematically, the information can be stored as a matrix

$$x = \begin{pmatrix} x_{11} & x_{12} & \dots & x_{1n} \\ x_{21} & x_{22} & \dots & x_{2n} \\ \vdots & & & \\ x_{m1} & x_{m2} & \dots & x_{mn} \end{pmatrix} \qquad (12\text{-}1)$$

of size $m \times n$ where m is the number of genes and n is the number of different tissues, with x_{ij} denoting the value assigned to the i-th gene in the j-th sample (Figure 12-8).

Thus far, we have only focused on the conceptual side of the hybridization experiment, leaving the experimental details aside and making some implicit assumptions. For example, we have assumed that equal amounts of mRNA were obtained from both cell types. We have also assumed that the cancer cell cDNA, incorporating Cy5, was labeled to the same degree as the normal cell cDNA, incorporating Cy3. Finally, we have assumed that the two fluorescently labeled cDNAs are detected with equal efficiency. Experimentally, there are numerous reasons that may cause these assumptions to not be true, resulting in systematic bias and providing sources for systematic variance in the gene expression levels across experiments. Thus, compensatory techniques are necessary to remove bias and make the experimental results comparable.

One such technique, called *normalization,* allows the results to be adjusted to compensate for a systemic problem (bias) in the data caused by technical variations. For instance, this technique can be used to compare data from different arrays or different color channels. Normalization procedures require a set of genes to be used as a basis for comparison. The procedures may use the set of all genes on the array and measure an aggregate characteristic, such as total fluorescence intensity. Alternatively, normalization may look at a subset of the genes in the experiment. These may be housekeeping genes, which should be expressed equally in all of the cell types under study. The experiment may also include artificially introduced controls (such as bacterial genes introduced into a mammalian expression assay) which may be used as a normalization set.

Another type of normalization, pertinent to the clustering techniques examined below, is gene normalization across tissue samples. This is done to adjust for different scales of expressions. Assume for example, that the gene expressions of five genes, denoted by A, C, D, E, and F, have been measured in four different tissue types and the results plotted in Figure 12-9(A). Notice that genes A and C are co-regulated across

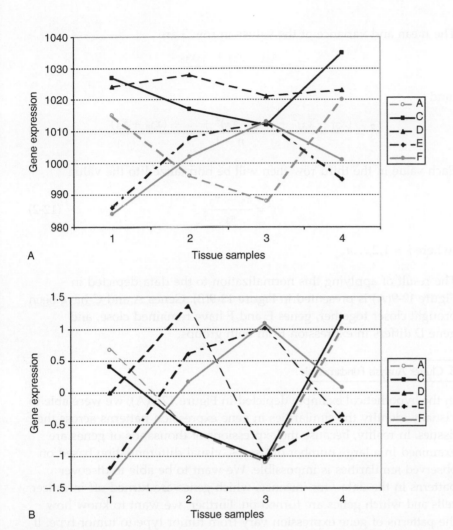

FIGURE 12-9.
Effects of normalization. The data in panel A have been normalized, and the results plotted in panel B.

samples, although gene C is expressed at a relatively higher level. Genes E and F are also co-regulated, whereas gene D does not seem to change expression much across samples. It would be desirable to group genes based on similar expression patterns, so A and C should be grouped together, although in the current situation C might be grouped with D because of their similar average values across samples. In such cases, normalizing the gene expressions in the expression matrix across the samples would be appropriate.

This can be done by calculating the mean value and standard deviation for each row of the matrix X, and then normalizing each entry of the row by subtracting the mean value and dividing by the standard deviation. For the entries in row 3, for example, the calculations are as follows:

The mean and variance of the values in row 3 are

$$\overline{x}_3 = \frac{x_{31} + x_{32} + \ldots x_{3n}}{n}$$

and

$$s_3^2 = \frac{(x_{31} - \overline{x}_3)^2 + (x_{32} - \overline{x}_3)^2 + \ldots + (x_{3n} - \overline{x}_3)^2}{n - 1}$$

Each value in the third row then will be normalized to the value

$$y_{3j} = \frac{x_{3j} - \overline{x}_3}{s_3}, \tag{12-2}$$

where $j = 1,2,\ldots.n$.

The result of applying this normalization to the data depicted in Figure 12-9(A) is presented in Figure 12-9(B). Genes A and C have been brought closer together, genes E and F have remained close, and gene D differs in expression from both groups.

B. Cluster Analysis Fundamentals

In the hypothetical example depicted in Figure 12-9(A), we were able to visually identify the similarities in gene expression patterns across the tissues. In reality, because the expressions of thousands of genes are examined in a large number of tissues, visual differentiation based on observed similarities is impossible. We want to be able to discover patterns in the data—for instance, which genes are turned off in cancer cells and which genes are turned on. Further, we want to know how the patterns of gene expression vary from tumor type to tumor type. If there are patterns of gene expression that are common to certain tumor types, this may be indicative of common functionality. The fact that some genes are expressed in similar patterns does not necessarily mean that their gene products interact with each other, but they might. If we do not know which genes are co-expressed, we cannot study them to determine whether they are interacting. Clearly, we need a quantitative method that will allow us to detect these patterns reliably, so that we can find the "needles" of specific gene information in this "haystack" of data regarding thousands of genes.

The methods available to classify co-expressed genes into groups can be broadly divided into two categories called supervised and unsupervised learning. In *supervised learning*, the genes are divided into a fixed number of predefined groups. These could be defined qualitatively, for example as "diseased" or "normal," or be quantitatively defined by their number. In *unsupervised learning*, the genes are grouped into categories based on similarities in their expression profiles. The computational method used to perform the partition into groups is generally referred to as *cluster analysis*.

Cluster methods can be further divided into several types. *Divisive clustering* begins by considering all genes as a single group, which is then partitioned into subgroups in a way that maximizes the difference between them. In contrast, *agglomerative clustering* begins by grouping the two genes with the most similar expression patterns, and then treating them as a single entity in the succeeding steps. It then groups the next most similar pair or adds the grouped pair to another gene if there is no other more similar pair. In both cases, it is necessary to provide a strict mathematical measure for dissimilarity, and we turn to this question next.

Mathematically, the *dissimilarity measure*[5] between gene expression profiles can be defined as a function of the respective rows of the data matrix X from Eq. (12-1) that quantitatively determines how different the gene expressions are. Using $x_i = (x_{i1}, x_{i2}, \ldots x_{in})$ and $x_k = (x_{k1}, x_{k2}, \ldots x_{kn})$ to denote the ith and k^{th} rows of X, we denote the dissimilarity measure between them by $d_{i,k} = d(x_i, x_k)$. A variety of choices exists for the specific functional form of $d(x_i, x_k)$, but it must satisfy the following distance axioms:

1. $d(x_i, x_k) \geq 0$ for any two vectors x_i, x_k; that is, the distance should be always positive;

2. $d(x_i, x_k) = d(x_k, x_i)$; that is, the distance should be symmetric; and

3. $d(x_i, x_k) \leq d(x_i, x_s) + d(x_s, x_k)$ for any vector x_s; that is, the distance should satisfy the triangle inequality.

By far, the most commonly used dissimilarity measure is based on the *Euclidean distance* defined as

$$d_E(x_i, x_k) = \sqrt{\sum_{j=1}^{n} (x_{ij} - x_{kj})^2}. \tag{12-3}$$

Other commonly used distance measures are the *Pearson correlation distance*, given by

$$d_P(x_i, x_k) = \frac{1}{n-1} \sum_{j=1}^{n} \left(\frac{x_{ij} - \bar{x}_i}{s_i}\right) \left(\frac{x_{kj} - \bar{x}_k}{s_k}\right); \tag{12-4}$$

the *Manhattan* or *block distance*, defined as

$$d_B(x_i, x_k) = \sum_{j=1}^{n} |x_{ij} - x_{kj}|; \tag{12-5}$$

and the *Chebyshev distance*:

$$d_c(x_i, x_k) = \max_j |x_{ij} - x_{kj}|. \tag{12-6}$$

---

5. Dissimilarity measures are also sometimes called distances.

Numerous other more complex distance measures are being used in research studies, and no single standard has emerged yet. As different dissimilarity measures will often change the clustering of the microarray data, comparison of results across different measures should not be made.

EXERCISE 12-2

Show that $d_E(x_i, x_k)$ from Eq. (12-3) satisfies the distance axioms 1 through 3.

Hint: First prove the result for $n = 2$.

EXERCISE 12-3

Show that $d_B(x_i, x_k)$ from Eq. (12-5) satisfies the distance axioms 1 through 3.

EXERCISE 12-4

Show that $d_C(x_i, x_k)$ from Eq. (12-6) satisfies the distance axioms 1 through 3.

Gene "Name"	Tissue Types	
	1	2
A	1.5	−0.4
B	1.4	−0.5
C	1.1	−0.3
D	−1.2	0.5
E	−1.4	0.8
F	−1.6	0.2

TABLE 12-3.
Gene expression values for a set of hypothetical tumor sampling.

C. Cluster Analysis Methods

1. Hierarchical clustering

Consider the set of gene expression values shown in Table 12-3. For this illustration, we consider the gene expression values measured for only two hypothetical tumor samples. In this case, $n = 2$, and the expression values can be depicted as points on a two-dimensional coordinate system. We can also label the rows of the data matrix X by the gene names A, B, C, ..., instead of $x_1, x_2, x_3, \ldots$. Now, the Euclidean distance from Eq. (3) is the usual geometric distance between points in the plane. For our example,

$$d(A, C) = \sqrt{(1.5 - 1.1)^2 + (-0.4 - (-0.3))^2} = \sqrt{0.17} = 0.412.$$

See Figure 12-10.

After the distances between any two rows of the data matrix X are computed, it is convenient to store the data again as a matrix, called a *proximity matrix*. The proximity matrix for the data in Table 12-3 is:

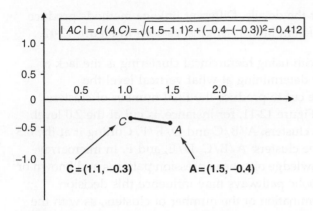

FIGURE 12-10.
In the case of two tissue samples, the distance from Eq. (12-3) is the geometric distance between points in the plane. The distance between the gene expressions A and C is depicted.

$$
D = \begin{array}{c} \\ A \\ B \\ C \\ D \\ E \\ F \end{array}
\begin{pmatrix}
\overset{A}{0.000} & \overset{B}{0.141} & \overset{C}{0.412} & \overset{D}{2.846} & \overset{E}{3.138} & \overset{F}{3.158} \\
0.141 & 0.000 & 0.361 & 2.786 & 3.087 & 3.081 \\
0.412 & 0.361 & 0.000 & 2.435 & 2.731 & 2.746 \\
2.846 & 2.786 & 2.435 & 0.000 & 0.361 & 0.500 \\
3.138 & 3.087 & 2.731 & 0.361 & 0.000 & 0.632 \\
3.158 & 3.081 & 2.746 & 0.500 & 0.632 & 0.000
\end{pmatrix}. \qquad (12\text{-}7)
$$

The clustering process begins with finding the smallest value in the proximity matrix and merging the respective genes into a cluster. For this proximity matrix, the smallest distance is 0.141, and thus the first cluster will contain A and B. In the next step, the proximity matrix is updated as follows: Genes A and B are replaced by the midpoint between them, and the distances from the other genes to this midpoint are calculated, resulting in a matrix with fewer rows and columns. The process continues until all genes are merged into a single cluster. For our example, after the initial cluster A/B is formed, D and E will be merged together. The A/B group would then be clustered with C. The D/E group would then be combined with F. Finally, we link the A/B/C cluster to the D/E/F cluster, as we have no more genes to link. A map of this clustering, called a *dendrogram*, is shown in Figure 12-11. The lengths of the dendrogram branches denote the distances at which the clusters are merged.

The clustering method described here is known as *average linkage*, as each cluster was represented by the midpoint between the newly merged genes, and this midpoint was then used for updating the proximity matrix. In *complete linkage*, on the other hand, the distances between each gene in the new cluster and the genes in the other clusters are calculated, and the largest distance is used in the proximity matrix. In *single linkage*, the smallest of these distances is used as the distance between the clusters. Several other linkage methods exist, such as the *centroid method* and *Word* linkage, and we refer the reader to Amaratunga

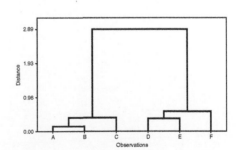

FIGURE 12-11.
Dendrogram with average linkage and Euclidean distance for the gene expression matrix in Table 12-3. Genes A and B have the most similar expression patterns, followed by genes D and E. The combined expression pattern of D and E is similar to that of F, and the combined expression pattern of A and B is similar to that of C. Finally, the A/B/C cluster is linked to the D/E/F cluster.

and Cabrera (2004) for the details. Different linkage methods produce dendrograms with different branch lengths, as seen in Figure 12-12.

A common problem with using hierarchical clustering is the lack of standardized rules for determining at what vertical level the dendrogram should be cut to produce the final number of clusters. If the dendrogram in Figure 12-11, for instance, is cut at the 2.0 level, there will be only two clusters: A/B/C and D/E/F. Cutting it at the 0.5 level will produce three clusters: A/B/C, D/E, and F. In microarray applications, prior knowledge of gene expression patterns and known or putative genetic metabolic pathways may influence this decision. Nonetheless, the determination of the number of clusters, as with the choice of dissimilarity measure or linkage method, remains somewhat arbitrary. We refer the reader to Everitt et al. (2001), where some guidance is provided.

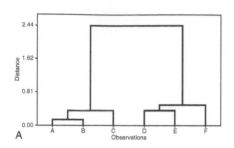

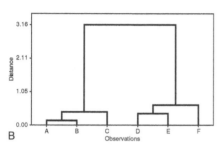

FIGURE 12-12.
Comparison of dendrograms from hierarchical complete linkage (A) and single linkage (B) clustering. Euclidean distance was used in both cases. Notice the differences in the vertical scale and compare with the result for average linkage in Figure 12-11.

2. K-Means clustering

K-Means clustering is a supervised learning method designed to group the data into a fixed number, k, of clusters. This number may be determined by the design of the experiment or based on prior knowledge. In many cases, however, the method is used to partition the data into different numbers of groups until a satisfactory result is achieved. As in the other methods, the objective is to determine the most appropriate grouping of the data.

The process is iterative and begins by selecting k genes at random to form k groups. At each step, the distances between any of the genes to the centers of the k groups are calculated, with genes then assigned to the group with the closest centers. The original k genes serve as initial centers of the k groups. In the following steps, the new centers are calculated as the averages of the genes assigned to the group. The process continues until no gene is reallocated to a new cluster. At each iteration, cluster statistics can be computed to assess the strength of the clusters, and some standard statistical packages, such as *MINITAB* and *SPSS*, give their values as part of the output. The *average intercluster distance* is defined as

$$D_{AV} = \frac{1}{N} \sum_{i=1}^{k} \sum_{m=1}^{n_i} d(y_{im}, \overline{y}_i), \tag{12-8}$$

where y_{im} is the m-th member of the i-th cluster, $\overline{y}_i$ is the center of the i-th cluster, and n_i is the number of members in the i-th cluster. The *total within cluster sum of squares* is computed by first finding the sums of squared residuals from the cluster centers and then adding them across all clusters. In mathematical form,

$$S = \sum_{i=1}^{k} \sum_{m=1}^{n_i} \sum_{j=1}^{n} \left(y_{imj} - \overline{y}_{ij}\right)^2, \tag{12-9}$$

where k is the number of clusters, n_i is the number of members in the i-th cluster, y_{imj} is the j-th coordinate of the m-th member of the i-th cluster, and $\bar{y}_{ij}$ is the j-th coordinate of the center of the i-th cluster. In general, smaller values of D_{AV} and S indicate more strongly expressed clustering within groups.

A variation of the classical K-Means algorithm uses the quantity from Eq. (12-8) to reassign data points to clusters in ways that decrease the value of S. The algorithm terminates when two different iterations return essentially the same value for S within a small tolerance. Another variation applies the K-Means method multiple times with different randomly selected initial cluster centers from the data. The final clustering is chosen to have the smallest within-cluster sum of squares, S.

Clustering procedures are included in a variety of commercial and noncommercial computer software packages. We encourage you to use Michael Eisen's *Cluster* and *Tree View* programs (Eisen et al. [1998]) to examine the diffuse large B-cell lymphoma (DLBCL) data (available on his Web site; see Internet Resources at the end of this chapter). This should allow you to duplicate the result reported in Alizadeh et al. (2000), identifying two types of B-cell lymphoma: germinal center B-like DLBCL and activated B-like DLBCL. In Figure 12-13, we reproduce one of the diagrams reported in the paper that includes both row and column clustering.

VI. MICROARRAYS AND CIRCADIAN RHYTHMS

A. Introduction

In Chapter 11, we discussed some of the genetic mechanisms of circadian rhythms that are expressed at the level of the whole organism and in diverse organs, tissues, and cells. Some of the genes involved have been identified, but we would like to discover what other genes are expressed in a circadian manner. With microarray technologies, we no longer need to confine our molecular investigations to the *period* gene or the other known genes, such as *tim*, *clock*, or *cycle*. Now we can look at the gene expression patterning of many hundreds or thousands of genes simultaneously and explore in greater detail an organism's system of temporal regulation and control of circadian patterns.

Many gene expression studies[6] have contributed to our current understanding of the molecular basis of circadian clocks and how the oscillators in different tissues and organs may be coordinated. The eventual goals of such studies are to develop an accurate systems-level

---

6. Reviewed in Reppert and Weaver (2002), Richter et al. (2004), and Bell-Pedersen et al. (2005).

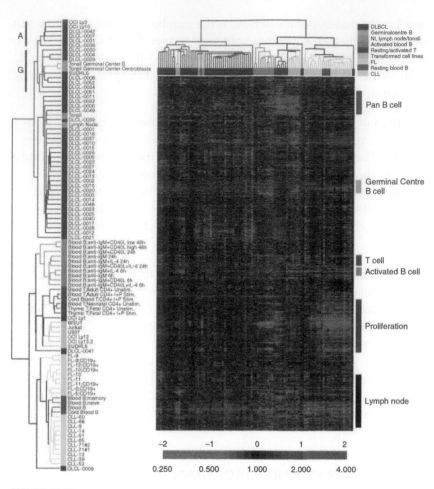

FIGURE 12-13.
A microarray graph showing the results of with row and column hierarchical clustering. The column clustering is obtained by applying the clustering methods to the transposed data matrix X^T. (Figure 13 from Alizadeh et al. [2000]. © 2000 Macmillan Publishers Ltd. *Nature, 403,* 503–511. Reprinted by permission.)

understanding of the interacting regulatory networks that underlie molecular, behavioral, and physiological circadian control. Microarrays are proving to be central to this research (see Duffield [2003]), and their use will undoubtedly continue to be essential.

Although we are beginning to understand some of the intracellular molecular mechanisms of circadian systems, the manner in which they give rise to cyclic temporal patterning in behavior and physiology is still largely unknown. A major challenge will be to experimentally unravel the details of the mechanistic control of circadian biology, especially with regard to environmental stimuli. By exploring circadian control and regulation, we are also gaining a greater understanding of how information is communicated across and between hierarchical biological levels: molecular, cellular, tissue, organ, organ system, and whole organism. Although it is biological questions that drive specific aims of inquiry, quantitative analytical solution strategies to assist in the process are emerging as critically important.

We now discuss the computational challenges of assessing the presence of circadian rhythms in time series microarray data.

B. Computational Challenges

If we wish to study a circadian phenomenon, we need to collect data as frequently as possible and for as long as possible. To detect genes expressed in a circadian manner, a series of mRNA samples must be collected over time. Challenges present at the experimental level, however, always limit the number of data points that can be obtained for the analysis of any time series.

Consider, for example, a study designed to examine circadian gene expression patterns in mouse liver or pancreas, with experiments performed in triplicate and with data points being collected every 4 hours for 48 hours. We would need to purchase and entrain 39 mice (3 mice per 13 time points) under dark/light conditions over an appropriately long period of time, before they are transferred into constant dark at the beginning of the experiment. Three mice will be sacrificed at the beginning of the experiment, and every 4 hours thereafter, and their organs of interest excised. We would then need to extract mRNA from each of the organs from the 39 mice, and each mRNA sample then needs to be converted to labeled cDNA and then be hybridized to a microarray. Scanning and analyzing each array will then be needed to obtain gene expression data. Thus, there are two major factors limiting the number of data points that can be feasibly obtained in each time series. First, in most cases the labor and expense involved in collecting each data point would preclude using intervals shorter than four hours apart. Second, because of the general dampening of the circadian rhythm under conditions of constant darkness (see Ceriani et al. [2002]), the time window for the entire experiment will be likely limited to about 48 hours.

The challenges for circadian analysis of gene chip–derived time series are thus considerable, as data sets presented for analysis are typically characterized by (1) extremely sparse determination (often only 13 points at a 4-hour sampling frequency for 48 hours); (2) extremely high dimensionality (on the order of 10^4 gene IDs per microarray in current Affymetrix implementations); and (3) low replicate numbers (thus limiting pointwise reliability, primarily because of the considerable financial costs of multiple chips per experimental time point). The sparse number of data points for each gene expression time series renders the use of many conventional methods for rhythm analysis inappropriate because such methods typically require much larger samples to generate statistically significant results. Instead, idiosyncratic algorithms specifically designed to statistically address such limitations have been developed and applied to microarray time-series analysis. Note that it is imperative that such algorithms possess high levels of automation and efficiency, given the huge number of genes on the microarray that are being examined for circadian behavior.

C. Statistical Assessment of Daily Rhythms in Microarray Data

The *COSOPT* algorithm described in this section has been used successfully in several studies for the analysis of microarray data in *Arabidopsis, Drosophila,* and mammalian systems (see Edwards et al. [2006]; Ceriani et al. [2002]; and Panda et al. [2002]). More methods for statistical assessment of circadian rhythms in gene expression patterning can be found in Straume (2004).

COSOPT accommodates variable weighting of individual time points, such as standard errors of the mean (SEMs) from replicate measurements or errors derived from preprocessing. *COSOPT* utilizes user-provided estimates of the circadian period entered as the value and range of the assumed period. Test periods are then calculated, uniformly spaced in the assumed range. The computational process begins by importing the time series on which an arithmetic linear regression detrending is performed. The mean and SD of the detrended time series are then calculated.

For each test period τ, 101 test cosine basis functions
$y_b(t) = \cos\left(\dfrac{2\pi(t+\varphi)}{\tau}\right)$ of unit amplitude are considered, varying over a range of phase values φ between $\left[-\dfrac{\tau}{2}, \dfrac{\tau}{2}\right]$. The number of cosine functions is chosen to allow that phase be considered in increments of 1% of each test period. Next, for each test cosine basis function $y_b(t)$, *COSOPT* calculates the least-squares optimized linear correspondence between the linear-regression–detrended data, $y_{lr}(t)$, according to the model

$$y_{lr}(t) = \alpha + \beta y_b(t).$$

The optimization is across all values of t, in terms of the parameters, α and β. The quality of optimization possible by each test cosine basis function is quantitatively characterized by the sum of squared residuals χ^2 between $y_{lr}(t)$ and the model given by $\alpha + \beta y_b(t)$.

The values of χ^2 are used to identify the optimal phase with the smallest value for χ^2 providing the optimal correspondence between $y_{lr}(t)$ and $y_b(t)$ (see Figure 12-14). The values for α and β, least-squares fit for the optimal phase value, now represent the optimized measures of the average expression and magnitude of the oscillatory amplitude expressed by $y_{lr}(t)$ (as modeled by a cosine wave of the corresponding period).

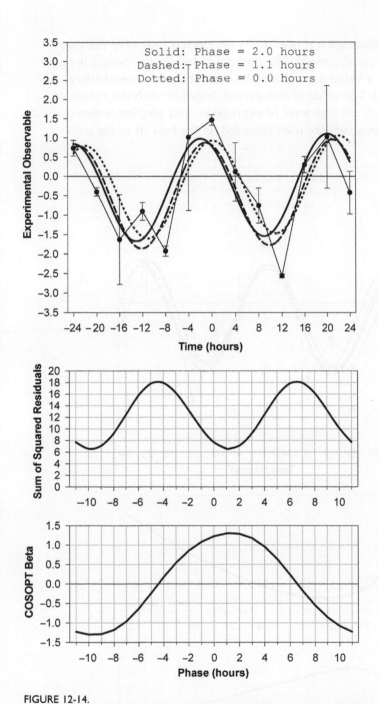

FIGURE 12-14.
Phase optimization for a presumed 22-hour circadian period. Three of the 101 cosine waves with a different phase ranging from −11 to 11 hours are depicted in the top panel, together with the 13 data points measured over 48 hours at equal 4-hour intervals. The optimal phase is identified as that minimizing the sum of squared residuals χ^2. The middle panel presents a plot of the χ^2 values for the 101 phase values tested. The optimal phase for this period is then obtained as one of the values minimizing χ^2. In this case, we obtain Phase = 1.1 hours. The corresponding β values are shown in the bottom panel. For this example, the optimal phase is 1.1 hours, and the corresponding amplitude β is 1.31.

A similar optimization procedure is performed next over the user-provided range of periods. The optimal value for the period is again identified as a value minimizing the sum of squared residuals χ^2 (see Figure 12-15). The value of this period, together with the values of optimal phase, average level of expression, and amplitude determined on the previous step are then recorded as the best fit to the data.

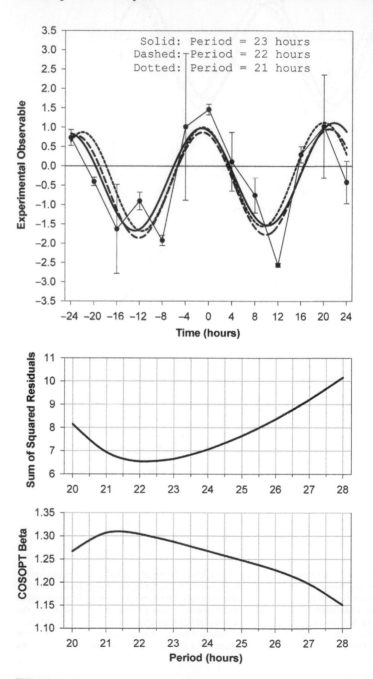

FIGURE 12-15.
Period optimization over a user-defined range of 20 to 28 hours for the period. The minimum of the sum of squared residuals χ^2 is achieved at period $\tau = 22$ hours (middle panel). The respective values of average expression and amplitude are $\alpha = -0.3055$ and $\beta = 1.31$ (bottom panel).

To assess statistically the probability that a significant rhythm is present in $y_{lr}(t)$ (in relation to, or as modeled by, a cosine functional form of the corresponding period and optimal phase), *COSOPT* employs empirical resampling methods applied directly in terms of the parameter β at each test period and corresponding optimal phase. One thousand Monte Carlo cycles are carried out in which surrogate realizations of $y_{lr}(t)$ are generated by both (1) randomly shuffling the temporal sequence of the original data; and (2) adding pseudo–Gaussian-distributed noise to each surrogate point in proportion to the corresponding value of point uncertainty (e.g., replicate SEM). In this way, both the influence of temporal patterning and the magnitude of pointwise experimental uncertainty are specifically accounted for in the surrogate realizations. Then, as with the original $y_{lr}(t)$ sequence, optimal values of α and β are determined, along with a corresponding χ^2, and retained in memory for each surrogate at each test period/optimal phase.

For each test period/optimal phase, the mean and standard deviation of the surrogate β values are then calculated. These values, in relation to the β value obtained for the original $y_{lr}(t)$ series, are then used to calculate a one-sided significance probability p based on a normality assumption (which is, in fact, satisfied by the distribution of β values obtained from the 1000 randomized surrogates). This probability is then multiple measures corrected (MMC) for the number of original data points comprising the time series to obtain the probability $p(\text{MMC}) = 1 - (1 - p)^N, N = 13$, which provides a more conservative assessment of significance. A summary of the analytical session is then produced for each time series, composed of entries for only those test periods that correspond to χ^2 minima.

In order to assess the performance of *COSOPT*, simulated data sets were prepared to approximate previously encountered gene chip profiles from experimental examinations of expression time series (Harmer et al. [2000]; Panda et al. [2002]; Ceriani et al. [2002]). One thousand surrogate data sets were prepared at each condition considered (see below), in which time series possessed 13 data points, representing 48 hours of observation obtained at 4-hour sampling intervals. All time series were surrogate realizations of a 24-hour-period cosine wave ranging in representational time from -24 hours to $+24$ hours, at which acrophase (the time of maximum) occurred at time zero. All data sets were composed of N(0,1) noise, to which 24-hour cosine profiles were added to produce data with signal-to-noise ratios of either 0, 1, or 2. This was achieved by adding nothing, or a unit-amplitude cosine wave, or an amplitude-2 cosine wave, respectively.

At each signal-to-noise ratio, replicate sampling also was varied, ranging from 1, 2, 3, 4, or 5 replicate observations being averaged per data point. A final variable considered in the analyses was whether or not replicate pointwise uncertainties were explicitly considered. This variable allows for employing either (1) pointwise SEMs in variable weighting for statistical considerations, or (2) no weighting, assuming, in effect, each data point to be known with infinite precision (i.e., with associated pointwise SEMs of zero).

Tables 12-4 through 12-8 summarize the results obtained from simulations with each of the aforementioned analytical strategies as a function of simulation condition. As expected, higher signal-to-noise ratios and/or higher numbers of replicates yield more accurate and more statistically significant results. In the tables, values reported are means from analyses of the 1000 surrogate time series for each condition tested. Associated standard deviations appear in parentheses immediately below the mean values. The column headings have the following meanings:

- $\pm$ *SEM* refers to whether (Yes) or not (No) individual pointwise replicate uncertainties were considered during statistical analysis;

- *S/N* denotes signal-to-noise ratio;

- % refers to percentage of files identified as circadianly rhythmic (i.e., with periods such that 20 hrs $< \tau <$ 28 hrs);

- *MeanExpLev* denotes mean expression level of the identified time series (theoretically zero); values in parentheses are SDs;

- *Period* denotes the period of the identified time series (theoretically 24);

- *Phase* denotes time of acrophase of the identified time series (theoretically zero);

- *Ampl* denotes oscillatory amplitude of the identified time series (theoretically either 2, 1, or 0);

- *Prob* denotes uncorrected significance probability of the identified time series;

- *Prob(MMC)* denotes multiple measures corrected significance probability of the identified time series. It is considered circadianly rhythmic if Prob(MMC) < 0.05; and

- N/D means the value was not determined.

In conclusion, despite the considerable challenges for circadian analysis of gene chip-derived time series, the *COSOPT* algorithm appears to perform admirably. It extracts reliable estimates of period, phase, and oscillatory amplitude. It requires no user initialization, is stable, and appears to produce unbiased parameter value estimates. It is readily

±SEM	S/N	%	MeanExpLev	Period	Phase	Ampl	Prob	Prob (MMC)
Yes	2	100.0	0.150 (0.127)	24.019 (0.708)	−0.001 (0.341)	2.015 (0.166)	6.57e−4 (2.23e−4)	8.51e−3 (2.87e−3)
No	2	100.0	0.150 (0.127)	24.019 (0.708)	−0.001 (0.341)	2.015 (0.166)	4.10e−4 (1.44e−4)	5.32e−3 (1.86e−3)
Yes	1	66.7	0.079 (0.128)	24.027 (1.365)	0.025 (0.667)	1.110 (0.131)	2.29e−3 (8.16e−4)	2.94e−2 (1.03e−2)
No	1	96.6	0.074 (0.127)	24.056 (1.383)	−0.004 (0.677)	1.045 (0.159)	1.09e−3 (6.77e−4)	1.41e−2 (8.69e−3)
Yes	0	0.0	—	—	—	—	—	—
No	0	2.2	0.040 (0.137)	23.514 (2.585)	−0.289 (7.797)	0.463 (0.109)	2.91e−3 (8.24e−4)	3.72e−3 (1.03e−2)

TABLE 12-4.
COSOPT performance assessment with $N = 5$ replicates.

±SEM	S/N	%	MeanExpLev	Period	Phase	Ampl	Prob	Prob (MMC)
Yes	2	99.9	0.150 (0.141)	23.986 (0.741)	0.017 (0.377)	2.030 (0.195)	7.82e−4 (2.90e−4)	1.01e−2 (3.74e−3)
No	2	100.0	0.150 (0.141)	23.986 (0.741)	0.017 (0.377)	2.030 (0.196)	4.53e−4 (1.91e−4)	5.88e−3 (2.47e−3)
Yes	1	50.5	0.081 (0.137)	23.905 (1.285)	0.042 (0.690)	1.175 (0.151)	2.49e−3 (8.07e−4)	3.19e−2 (1.02e−2)
No	1	92.6	0.073 (0.141)	23.933 (1.405)	0.043 (0.751)	1.074 (0.180)	1.26e−3 (7.60e−4)	1.62e−2 (9.73e−3)
Yes	0	0.0	—	—	—	—	—	—
No	0	2.3	−0.003 (0.125)	22.970 (2.363)	−1.789 (7.025)	0.567 (0.135)	2.45e−3 (9.22e−4)	3.14e−2 (1.16e−2)

TABLE 12-5.
COSOPT performance assessment with $N = 4$ replicates.

±SEM	S/N	%	MeanExpLev	Period	Phase	Ampl	Prob	Prob (MMC)
Yes	2	99.8	0.151 (0.159)	24.037 (0.865)	0.017 (0.467)	2.033 (0.221)	1.08e−3 (4.83e−4)	1.40e−2 (6.20e−3)
No	2	100.0	0.151 (0.159)	24.035 (0.865)	0.016 (0.467)	2.032 (0.222)	5.35e−4 (2.35e−4)	6.93e−3 (3.04e−3)
Yes	1	30.0	0.081 (0.152)	24.099 (1.457)	0.040 (0.844)	1.264 (0.164)	2.73e−3 (7.88e−4)	3.49e−2 (9.91e−3)
No	1	83.1	0.078 (0.157)	24.024 (1.574)	0.047 (0.919)	1.113 (0.195)	1.50e−3 (8.48e−4)	1.93e−2 (1.08e−2)
Yes	0	0.0	—	—	—	—	—	—
No	0	3.4	0.027 (0.164)	23.018 (2.365)	0.928 (5.959)	0.668 (0.148)	2.79e−3 (8.77e−4)	3.57e−2 (1.10e−2)

TABLE 12-6.
COSOPT performance assessment with $N = 3$ replicates.

An Invitation to Biomathematics **Chapter Twelve**

±SEM	S/N	%	MeanExpLev	Period	Phase	Ampl	Prob	Prob (MMC)
Yes	2	94.8	0.157 (0.191)	24.016 (1.068)	−0.008 (0.561)	2.070 (0.254)	1.59e−3 (7.72e−4)	2.05e−2 (9.85e−3)
No	2	99.7	0.157 (0.191)	24.007 (1.080)	−0.008 (0.561)	2.052 (0.264)	7.10e−4 (3.66e−4)	9.19e−3 (4.72e−3)
Yes	1	12.2	0.073 (0.220)	24.014 (1.588)	−0.205 (0.938)	1.427 (0.188)	2.67e−3 (8.09e−4)	3.42e−2 (1.02e−2)
No	1	64.2	0.076 (0.197)	23.923 (1.762)	−0.045 (1.100)	1.205 (0.220)	1.83e−3 (9.63e−4)	2.35e−2 (1.22e−2)
Yes	0	0.0	—	—	—	—	—	—
No	0	2.5	−0.047 (0.192)	23.516 (2.174)	−1.049 (7.219)	0.704 (0.148)	2.43e−3 (8.63e−4)	3.11e−2 (1.09e−2)

TABLE 12-7.
COSOPT performance assessment with $N = 2$ replicates.

±SEM	S/N	%	MeanExpLev	Period	Phase	Ampl	Prob	Prob (MMC)
Yes	2	N/D	N/D	N/D	N/D	N/D	N/D	N/D
No	2	92.0	0.155 (0.278)	24.058 (1.494)	0.048 (0.753)	2.149 (0.354)	1.28e−3 (7.94e−4)	1.65e−2 (1.02e−2)
Yes	1	N/D	N/D	N/D	N/D	N/D	N/D	N/D
No	1	33.7	0.098 (0.286)	23.935 (2.054)	0.142 (1.296)	1.490 (0.280)	2.12e−3 (9.43e−4)	2.72e−2 (1.20e−2)
Yes	0	N/D	N/D	N/D	N/D	N/D	N/D	N/D
No	0	1.5	−0.024 (0.377)	23.313 (2.307)	1.258 (6.676)	1.035 (0.234)	2.49e−3 (8.15e−4)	3.19e−2 (1.03e−2)

TABLE 12-8.
COSOPT performance assessment with no replicates.

amenable to completely automated implementation and is sufficiently rapid to complete analysis of 20,000 or more GeneIDs in only a few hours. This allows the identification of genes not previously known to be under circadian control, and that identification is the first step toward building a better understanding of circadian phenomena.

REFERENCES

Abramson, J. S., & Shipp, M. A. (2005). Advances in the biology and therapy of diffuse large B-cell lymphoma: Moving toward a molecularly targeted approach. *Blood, 106,* 1164–1174.

Alizadeh, A. A., Eisen, M. B., Davis, R. E., Ma, C., Lossos, I. S., Rosenwald, A., Boldrick, J. C., Sabet, H., Tran, T., & Yu, X. (2000). Distinct types of diffuse large B-cell lymphoma identified by gene expression profiling. *Nature, 403,* 503–511.

Amaratunga, D., & Cabrera, J. (2004). *Exploration and analysis of DNA microarray and protein array data.* New York: John Wiley & Sons.

Bell-Pedersen, D., Cassone, V. M., Earnest, D. J., Golden, S. S., Hardin, P. E., Thomas, T. L., & Zoran, M. J. (2005). Circadian rhythms from multiple oscillators: Lessons from diverse organisms. *Nature Reviews Genetics, 6,* 544–566.

Causton, H. D., Quackenbush, J., & Brazma, A. (2003). *Microarray gene expression data analysis: A beginner's guide.* Malden, MA: Blackwell Publishing.

Ceriani, M. F., Hogenesch, J. B., Yanovsky, M., Panda, S., Straume, M., & Kay, S. A. (2002). Genome-wide expression analysis in Drosophila reveals genes controlling circadian behavior. *Journal of Neuroscience, 22,* 9305–9319.

Duffield, G. E. (2003). DNA microarray analyses of circadian timing: The genomic basis of biological time. *Journal of Neuroendocrinology, 15,* 991–1002.

Edwards, K. D., Anderson, P. E., Hall, A., Salathia, N. S., Locke, J. C. W., Lynn, J. R., Straume, M., Smith, J. Q., & Millar, A. J. (2006). Flowering locus C mediates natural variation in the high-temperature response of the *Arabadopsis* circadian clock. *The Plant Cell, 18,* 639–650.

Eisen, M. B., & Brown, P. O. (1999). DNA arrays for analysis of gene expression. In Weissman, S. M., (Ed.), *Methods in Enzymology* (vol. 303, pp. 179–205). New York: Academic Press.

Eisen, M. B., Spellman, P. T., Brown, P. O., & Botstein, D. (1998). Cluster analysis and display of genome-wide expression patterns. *Proceedings of the National Academy of Sciences of the United States of America, 95,* 14863–14868.

Everit, B., Landau, S., & Leese, M. (2001). *Cluster analysis.* London: Oxford University Press.

Harmer, S. L., Hogenesch, J. B., Straume, M., Chang, H. S., Han, B., Zhu, T., Wang, X., Kreps, J. A., & Kay, S. A. (2000). Orchestrated transcription of key pathways in Arabidopsis by the circadian clock. *Science, 290,* 2110–2113.

Krawetz, S. A., & Womble, D. D. (2003). *Introduction to bioinformatics: A theoretical and practical approach.* Totowa, NJ: Humana Press.

Mormont, M. C., & Levi, F. (2003). Cancer chronotherapy: Principles, applications, and perspectives. *Cancer, 97,* 155–169.

Panda, S., Antoch, M. P., Miller, B. H., Su, A. I., Schook, A. B., Straume, M., Schultz, P. G., Kay, S. A., Takahashi, J. S., & Hogenesch, J. B. (2002). Coordinated transcription of key pathways in the mouse by the circadian clock. *Cell, 109,* 307–320.

Reppert, S. M., & Weaver, D. R. (2002). Coordination of circadian timing in mammals. *Nature, 418,* 935–941.

Richter, H. G., Torres-Farfan, C., Rojas-Garcia, P. P., Campino, C., Torrealba, F., & Seron-Ferre, M. (2004). The circadian timing system: Making sense of day/night gene expression. *Biology Research, 37,* 11–28.

Straume, M. (2004). DNA microarray time series analysis: Automated statistical assessment of circadian rhythms in gene expression patterning. In Brand, L., & Johnson, M. L., (Eds.), *Methods in Enzymology* (vol. 383, pp. 149–166). New York: Academic Press.

INTERNET RESOURCES

http://rana.lbl.gov/EisenSoftware.htm
Michael Eisen's site, with *Cluster* and *Tree View* programs and diffuse large B-cell lymphoma (DLBCL) data available for download.

http://www.affymetrix.com
Home page of Affymetrix, Inc., with links to information about their product designs, manufacturing processes, microarray hybridization, and description of their GeneChip® and other products.

FURTHER READING

Bendat, J. S., & Piersol, A. G. (2000). *Random data: Analysis and measurement procedures* (3rd ed.). New York: John Wiley & Sons.

Butte, A. (2002). The use and analysis of microarray data. *Nature Reviews Drug Discovery, 1*, 951–960.

Young, M. W., & Kay, S. A. (2001). Time zones: A comparative genetics of circadian clocks. *Nature Reviews Genetics, 2*, 702–715.

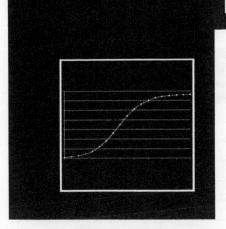

Solutions

Chapter 1

Exercise 1-1.

a) $P_n - P_{n-1} = kP_{n-1}$ can be rewritten as
$P_n = P_{n-1} + kP_{n-1} = (1+k)P_{n-1}$.

b) Because a) applies for any value of $n = 1, 2, 3, \ldots$, applying it repeatedly leads to the chain of equalities:

$$P_n = (1+k)P_{n-1} = (1+k)(1+k)P_{n-2} = (1+k)^2 P_{n-2}$$
$$= (1+k)^3 P_{n-3} = \ldots (1+k)^n P_0.$$

Exercise 1-3.

Parts a) and b):

The tables below are analogous to Tables 1-2 and 1-3 in the text.

Time [hours] n	Biomass P_n	Change in Biomass $P_n - P_{n-1}$	$k = (P_n - P_{n-1})/ P_{n-1}$	Predicted Biomass for $k = 0.606$
0	9.6			9.6
1	18.3	8.7	0.906	15.4
2	29	10.7	0.585	24.8
3	47.2	18.2	0.628	39.8
4	71.1	23.9	0.506	63.9
5	119.1	48	0.675	102.6
6	174.6	55.5	0.466	164.7
7	257.3	82.7	0.474	264.5

TABLE 1-1.
Estimation of k from biomass data using the discrete model from Eq. (1-1). The average value for k is $k = 0.606$. (Note that without $k = 0.906$, the average is $k = 0.556$.)

c) The graph is shown in Figure 1-1. The model does not appear capable of remaining accurate; because the exponential function is unbounded, it predicts unlimited growth.

Time t [hours]	Biomass $P(t)$	$r = \ln(P(t+1)) - \ln(P(t))$	Predicted Biomass for $r = 0.470$	Relative Error $[\%] = \frac{\|\text{Predicted} - \text{Actual}\|}{\text{Actual}} 100$
0	9.6		9.6	0.000
1	18.3	0.645	15.4	16.066
2	29	0.460	24.6	15.256
3	47.2	0.487	39.3	16.692
4	71.1	0.410	62.9	11.514
5	119.1	0.516	100.7	15.482
6	174.6	0.383	161.1	7.756
7	257.3	0.388	257.7	0.152

TABLE 1-2.
Estimation of r from biomass data using the continuous model from Eq. (1-7). The average value for r is $r = 0.470$.

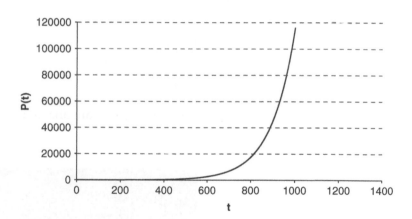

FIGURE 1-1.
Graph of Eq. (1-4), with $r = 0.470$, depicting the solution of the continuous model defined by Eq. (1-2).

Exercise 1-5.

a) When $P(t) < K$, the population will be increasing, because
$$\frac{dP}{dt} = a\left(1 - \frac{P(t)}{K}\right)P(t) > 0.$$

b) When $P(t) = K$, the population remains unchanged, because
$$\frac{dP}{dt} = a\left(1 - \frac{P(t)}{K}\right)P(t) = 0.$$

c) When $P(t) > K$, the population will be decreasing, because
$$\frac{dP}{dt} = a\left(1 - \frac{P(t)}{K}\right)P(t) < 0.$$

Exercise 1-7.

If there were a time when $P(t) = K$, the value of $P(t)$ would remain equal to K for all times thereafter (since $\dfrac{dP}{dt} = 0$ for all times thereafter).

Exercise 1-9.

a) The graph of T versus $\dfrac{dT}{dt}$ is presented in Figure 1-3.

b) The equilibrium states are where $\dfrac{dT}{dt} = 0$; that is, where the graph in Figure 1-3 crosses the horizontal axis. The equilibrium states $T = -15$ and $T = 22$ are stable. More specifically, if $T < -15$, then $\dfrac{dT}{dt} > 0$, and T will increase toward -15, and if $T > -15$, but less than 5, then $\dfrac{dT}{dt} < 0$, and T will decrease toward -15.

c) (i) If $T(t_0) = 23$, then $\dfrac{dT}{dt} < 0$, so T will decrease;

   (ii) If $T(t_0) = 18$, then $\dfrac{dT}{dt} > 0$, so T will increase;

   (iii) If $T(t_0) = -16$, then $\dfrac{dT}{dt} > 0$, so T will increase;

   (iv) If $T(t_0) = 4$, then $\dfrac{dT}{dt} < 0$, so T will decrease.

d) The time trajectories are shown in Figure 1-4.

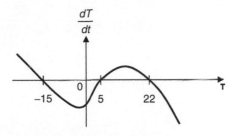

FIGURE 1-3.
Graph of T versus $\dfrac{dT}{dt}$ for Exercise 1-9, part a.

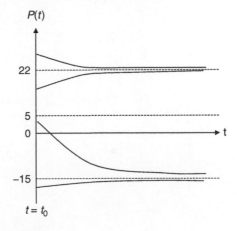

FIGURE 1-4.
The time trajectories of the temperature $T(t)$ for the 4 cases in part c of Exercise 1-9.

Exercise 1-11.

From Eq. (1-25) in the text, the equilibrium states are found by setting the change $P_{n+1} - P_n$ equal to zero for all values of n = 0,1,2, … This is equivalent to having $P_n = P$, for all n = 0,1,2, …. Thus, setting P_{n+1} and P_n equal to P in Eq. (1-25), we obtain $a\left(1 - \dfrac{P}{K}\right)P = 0$.

Solving for P yields the equilibrium states $P = 0$ and $P = K$.

Exercise 1-13.

One serious limitation is that the model does not provide means for taking into account that only individuals who have reached a certain level of maturity can reproduce. The model from Eq. (1-28) bases the net per capita rate of growth r on the population size of D units of time earlier. However, to determine the population's rate of change $\dfrac{dP}{dt}$, this per capita rate of growth is applied to the *current* population size $P(t)$.

This, among other things, implies that even the babies in the current population will be contributing to the rate of population change at time t, which is unrealistic.

Exercise I-15.

Rewrite $\dfrac{dC}{dt} = -rC$ as $\dfrac{dC}{C} = -rdt$. Integrating both sides gives

$\displaystyle\int \dfrac{dC}{C} = -r\int dt$ and thus $\ln|C| = \ln C = -rt + k$, where k is an arbitrary constant. Therefore, $C(t) = e^{-rt+k} = e^k e^{-rk}$. Substituting $t = 0$, we obtain $e^k = C(0)$, which shows that the solution of Eq. (1-29) is given by $C(t) = C(0)e^{-rt}$.

Exercise I-17.

Given that the half-life $\tau = 2.5$ hours, we can compute the elimination rate constant $r = \dfrac{\ln(2)}{\tau} = \dfrac{\ln(2)}{2.5} = 0.2773$. Next, we need to find t such that $C(t) = C(0)e^{-(0.2773)t} = (0.01)C(0)$, or $e^{-(0.2773)t} = (0.01)$. Solving for t, we obtain $-0.2773t = \ln(0.01) = -4.6053$, or $t \approx 16.6$ hours.

Exercise I-19.

Because $R_n = C([e^{-Tr}]^n + [e^{-Tr}]^{n-1} + [e^{-Tr}]^{n-2} + \ldots + e^{-Tr})$ and

$$R_{n+1} = C([e^{-Tr}]^{n+1} + [e^{-Tr}]^n + [e^{-Tr}]^{n-1} + [e^{-Tr}]^{n-2} + \ldots + e^{-Tr})$$
$$= R_n + C[e^{-Tr}]^{n+1}, \text{ the sequence } \{R_n\} \text{ is increasing. Thus}$$

$R_n < R = \lim\limits_{n\to\infty} R_n$. Physiologically, R represents the level of saturation of the drug concentration observed when equal doses are given at equal time intervals.

Exercise I-21.

Rewrite $MEC = \dfrac{MTC - MEC}{e^{Tr} - 1}$ as $e^{Tr} - 1 = \dfrac{MTC - MEC}{MEC} = \dfrac{MTC}{MEC} - 1$,

where from $e^{Tr} = \dfrac{MTC}{MEC}$. Taking logarithms from both sides gives

$Tr = \ln\dfrac{MTC}{MEC}$ and $T = \dfrac{1}{r}\ln\dfrac{MTC}{MEC}$.

Exercise I-23.

If we assume that a pharmaceutical company follows the objectives 1 through 4 outlined in the exercise, we may hypothesize that the initial dose C_0 is approximately MTC. Further, as we assume that the regular doses after the initial dose are determined as $C \approx MTC - MEF = C_0 - MEF$, we can compute $MEF \approx C_0 - C$. Furthermore, we can hypothesize that since the time between the doses is given, the elimination constant can also be estimated from the available information. Namely, solving the expression $T = \frac{1}{r} \ln \frac{MTC}{MEC}$ derived in Exercise 21 for r one obtains $r = \frac{1}{T} \ln \frac{MTC}{MEC} = \frac{1}{T} \ln \frac{C_0}{C - C_0}$.

Exercise 1-25.

Enter the model in *BERKELEY MADONNA* as below and run it. Then follow the instructions for importing a data set and comparing the results.

STARTTIME = 0

STOPTIME=20

DT = 0.02

d/dt(P) = a* (1-P/K)*P

init P = 9.6

K=660

a = 0.608

Chapter 2

Exercise 2-1.

Taking in mind that $I + S = N$, we write
$$\frac{dI}{dt} = \alpha IS - \beta I = \alpha I(N - I) - \beta I = \alpha I\left(N - \frac{\beta}{\alpha} - I\right) < 0, \text{ if } N - \frac{\beta}{\alpha} < 0.$$
This means that $I(t)$ is decreasing and there is no epidemic.

If $N - \frac{\beta}{\alpha} = 0, \frac{dI}{dt} = \alpha I\left(N - \frac{\beta}{\alpha} - I\right) = -\alpha I^2 < 0$, and the result is the same.

Exercise 2-3.

Some of the limitations of the model are:

a) Once recovered, individuals are considered susceptibles right away. This will not be true for diseases that confer even temporary immunity.

b) The assumption that the disease is spread via direct contact has severe limitations. For instance, the model would not be appropriate for infections that are airborne.

c) The model assumes a constant recovery rate and infection rate, regardless of age or geographic or socioeconomic factors.

d) The model assumes that infectives and susceptibles remain uniformly mixed with time, which may not always be true.

Exercise 2-5.

Additional factors may include the mobility of infectives and susceptibles and how the amount of contact affects the likelihood of infection.

Exercise 2-7.

The result appears plausible for some diseases. If the infection has a low infection rate and high removal rate, it will be hard to transmit and easy to cure. Under such assumptions, it would be reasonable to expect that the infection will be eliminated long before the entire population is infected.

Exercise 2-9.

a) Upon infection, an individual moves to the new group E where they need to remain for exactly D units of time. After having spent time D in this state, an individual becomes infectious and is, therefore, moved to the group of infectives I.

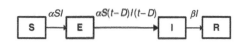

FIGURE 2-1.
Flowchart illustrating the rates of transition between the groups for the model in Exercise 2-9, a.

The block diagram for this case is given in Figure 2-1 with the "in-flow" and "out-flow" rates labeling the arrows. The mathematical form of those rates is justified below.

In comparison with the SIR model, the rate of infecting susceptibles at time t (that is, the rate of flow from S to E at time t) is the same: $\alpha S(t)I(t)$. Because each individual will be spending exactly time D in E, the rate of flow from E to I at time t will be exactly the rate of flow from S to E at time $t - D$; that is $\alpha S(t - D)I(t - D)$. The rate of flow from I into R is the same as that in the SIR model. Because a fraction β of the infectives at recovering per unit time at any given time t, the rate of flow from E into R is $\beta I(t)$. Recall that $1/\beta$ gives the average time over which an infected person will remain infectious.

Thus, based on the block diagram in Figure 2-1 and the explanation provided above, the equations are:

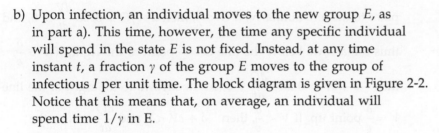

FIGURE 2-2.
Flowchart illustrating the rates of transition between the groups for the model in Exercise 2-9, b.

$$\frac{dS}{dt} = -\alpha I(t)S(t)$$

$$\frac{dE}{dt} = \alpha I(t)S(t) - \alpha I(t-D)S(t-D)$$

$$\frac{dI}{dt} = \alpha I(t-D)S(t-D) - \beta I(t)$$

$$\frac{dR}{dt} = \beta I(t)$$

Notice that $\dfrac{dS}{dt} + \dfrac{dE}{dt} + \dfrac{dI}{dt} + \dfrac{dR}{dt} = 0$.

b) Upon infection, an individual moves to the new group E, as in part a). This time, however, the time any specific individual will spend in the state E is not fixed. Instead, at any time instant t, a fraction γ of the group E moves to the group of infectious I per unit time. The block diagram is given in Figure 2-2. Notice that this means that, on average, an individual will spend time $1/\gamma$ in E.

The equations are:

$$\frac{dS}{dt} = -\alpha I(t)S(t)$$

$$\frac{dE}{dt} = \alpha I(t)S(t) - \gamma E(t)$$

$$\frac{dI}{dt} = \gamma E(t) - \beta I(t)$$

$$\frac{dR}{dt} = \beta I(t)$$

Notice that, again, $\dfrac{dS}{dt} + \dfrac{dE}{dt} + \dfrac{dI}{dt} + \dfrac{dR}{dt} = 0$.

Exercise 2-11.

a) The null clines for V are computed by setting $\dfrac{dV}{dt} = (\alpha - \gamma O)V = 0$ and the null clines for O by setting $\dfrac{dO}{dt} = (-\delta + \varepsilon V)O = 0$. The first of these equations yields $V = 0$ or $O = \dfrac{\alpha}{\gamma}$. The second gives the null clines for O: $O = 0$ or $V = \dfrac{\delta}{\varepsilon}$.

b) The equilibrium states are determined by solving the system of equations

$$\frac{dV}{dt} = (\alpha - \gamma O)V = 0$$

$$\frac{dO}{dt} = (-\delta + \varepsilon V)O = 0$$

The equilibrium states are $V = 0, O = 0$ and $V = \frac{\delta}{\varepsilon}, O = \frac{\alpha}{\gamma}$.

c) The phase diagram is presented in Figure 2-3. If $O > \frac{\alpha}{\gamma}$, then $\alpha - \gamma O < 0$, so $\frac{dV}{dt} < 0$, and thus the arrows above the line $O = \frac{\alpha}{\gamma}$ point to the left. Similarly, if $O < \frac{\alpha}{\gamma}$, then $\alpha - \gamma O > 0$, so $\frac{dV}{dt} > 0$, and thus the arrows below the line $O = \frac{\alpha}{\gamma}$ point to the right. If $V > \frac{\delta}{\varepsilon}$, then $-\delta + \varepsilon V > 0$ and $\frac{dO}{dt} > 0$, so the arrows to the right of the line $V = \frac{\delta}{\varepsilon}$ point up. If $V < \frac{\delta}{\varepsilon}$, then $-\delta + \varepsilon V < 0$ and $\frac{dO}{dt} < 0$, so the arrows to the left of the line $V = \frac{\delta}{\varepsilon}$ point down.

d) Neutrally stable.

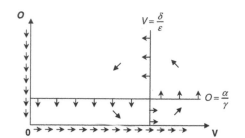

FIGURE 2-3.
Phase diagram for Exercise 2-11, C.

Exercise 2-13.

When $O = V, \frac{dO}{dt} = \delta O \left(1 - \frac{O}{V}\right) = 0$ and

$$\frac{dV}{dt} = \frac{2}{3}V - \frac{V^2}{6} - \frac{V^2}{1+V} = -\frac{V^2 + 3V - 4}{6(1+V)}.$$ Because

$$-(V^2 + 3V - 4) = -(V+4)(V-1), \ -(V^2 + 3V - 4) \begin{cases} >0 \ for \ V<1 \\ <0 \ for \ V>1 \end{cases},$$

and thus $\frac{dV}{dt} > 0$ for $V < 1$ and $\frac{dV}{dt} < 0$ for $V > 1$. Therefore, along the $O = V$ null cline, the directional fields are horizontal (because $\frac{dO}{dt} = 0$ and there is no change for O), pointing right for $V < 1$ (where $\frac{dV}{dt} > 0$ and, therefore, V increases), and pointing left for $V < 1$ (where $\frac{dV}{dt} < 0$ and, therefore, V decreases).

When $O = \frac{2}{3}\left(1 - \frac{V}{4}\right)(1 + V)$, $\frac{dV}{dt} = 0$, and $\frac{dO}{dt}\begin{cases} > 0, \text{when } O < V \\ < 0, \text{when } O > V \end{cases}$. Note

that the points below the line $O = V$ correspond to $O < V$ while those above the line $O = V$ correspond to $O > V$. Thus, for the points on the parabola $\frac{2}{3}\left(1 - \frac{V}{4}\right)(1 + V)$ that are below the line $O = V$, there is no change for V and O is increasing. This corresponds to vertical directions of change, with the arrows pointing upward. For the points above the line $O = V$, there is no change for V and O is decreasing. This corresponds to vertical directions of change, with the arrows pointing downward.

Exercise 2-15.

a) The term $r(N, P)$ represents the per capita rate of growth for the population N, as a function of the sizes N and P of the two competing populations. In the same way as in deriving the logistic growth equation for a single population, the model assumes that the per capita net rate of growth $r(N, P)$ varies with the change of population sizes. Because two populations are competing for the same resource, $r(N, P)$ should be decreasing as P increases. The reason for this behavior is as follows. If $P = 0$, we already know from Exercise 2-14 that the population size N follows a logistic equation. When $P \neq 0$, the per capita net rate of growth for N will also have to account for the presence of the other population of size $P(t)$ that is competing for the same resource.

b) This part is similar, referring to the net per capita rate of growth $k(N, P)$ for the population P.

c) In part a) we noted that when $P(t) \neq 0$, the per capita net rate of growth $r(N, P)$ for N will also have to account for the presence of the population P competing for the same resource. Also recall that in the logistic Eq. (1-12), the term N/K can be considered as representing the size of the population as a fraction of its carrying capacity. For Eq. (1-12), the rate of growth slows down as the fraction N/K approaches 1.

In the same way, the fraction $(N + bP)/K$ can be considered to measure the combined size of the two populations as a fraction of the carrying capacity K of the population N. The parameter b is the "conversion factor" between the species. To make this clearer, imagine, for example, that we refer to a population N that is growing in a flask where each organism needs a unit volume to survive. If the flask has a volume of 1000, it will not have room for more than 1000 population N organisms. Because the population P, however, is growing in the same flask, their sizes will also have to be taken into consideration. This is a population of different species, and an organism of this type may need more or less volume to survive when compared with that needed for a species of type N. Assume that each P species needs ½ unit of volume. Then, two species of type P will fill the same volume as one species of type N. In this case, we will use $b = ½$ in the model. If, on the other hand, each P species needs 5 units of volume, one organism of type P will compare with 5 species of type N in terms of volume. In this case, we will use $b = 5$ in the model.

Therefore, the factor $N + bP$ represents the combined sizes of the N and P populations, appropriately scaled to be measured in the same units as the carrying capacity K. The fraction $(N + bP)/K$ can then be considered as representing the size of the combined (N and P) populations as a fraction of the carrying capacity K.

For the second of Eqs (2-15) the scaling is done the other way in order to compare the combined sizes of the N and P populations measured in the same units as the carrying capacity M of the population P. The parameter g is the "conversion factor" from species of type N to species of type P in order to consider the combined sizes of the two populations as a fraction of the carrying capacity M for the population P.

Exercise 2-17.

Denote

$$\frac{dN}{dt} = r(N,P)N = a\left(1 - \frac{N+bP}{K}\right)N = f(N,P)$$

$$\frac{dP}{dt} = k(N,P)P = c\left(1 - \frac{P+gN}{M}\right)P = g(N,P).$$

$$\frac{\partial f}{\partial N} = a - \frac{2a}{K}N - \frac{ab}{K}P \qquad \frac{\partial f}{\partial N}(0,0) = a$$

$$\frac{\partial f}{\partial P} = -\frac{ab}{K}N \qquad \frac{\partial f}{\partial P}(0,0) = 0$$

Then,

$$\frac{\partial g}{\partial N} = -\frac{cg}{M}P \qquad \frac{\partial g}{\partial N}(0,0) = 0$$

$$\frac{\partial g}{\partial P} = c - \frac{2c}{M}P - \frac{cg}{M}N \qquad \frac{\partial g}{\partial P}(0,0) = c$$

Thus, the stability of the point $(0, 0)$ is determined by the matrix $J = \begin{pmatrix} a & 0 \\ 0 & c \end{pmatrix}$, and because $\det(J) = ac > 0$ and $\operatorname{trace}(J) = a + c > 0$, the equilibrium point $(0, 0)$ is a repeller, for any values of $a > 0$ and $c > 0$.

Chapter 3

Exercise 3-1.

a) The "**A**" site must have at least one dominant allele if the dominant phenotype is observed. The probability for this to occur is $^3/_4$ because there are 4 possibilities (**AA**, **Aa**, **aA**, and **aa**), of which three have at least one **A** allele. In a similar way, the possibilities for the "**B**" site are **BB**, **Bb**, **bB**, and **bb**) of which only one (**bb**) has the recessive phenotype. Because the "**A**" and "**B**" sites are independent, the probability that the **A** phenotype is dominant and the **B** phenotype is recessive is $(^3/_4)(¼) = 3/16$.

b) In this case, we need to multiply the probability that the **A** phenotype is recessive (¼) and the probability that the **B** phenotype is dominant ($^3/_4$). Thus, the probability in question is $(¼)(^3/_4) = 3/16$.

c) In this case, we need to multiply the probabilities that both the **A** and **B** phenotypes will be recessive, which gives $(¼)(¼) = 1/16$.

d) From the 16 possibilities given in the Punnett square in Figure 3-8, the combinations that correspond to the dominant **A** phenotype and recessive **B** phenotype are **AAbb**, **Aabb**, and **aAbb**. Thus, the probability for this is 3/16, the same as determined in part a). The combinations that correspond to the recessive **A** phenotype and dominant **B** phenotype are **aaBB**, **aaBb**, and **aabB**, reflecting the probability we found in part b). There is only one combination among the 16 that corresponds to recessive **A** and **B** phenotypes: **aabb**. Thus, the probability for this is 1/16, as we computed in part c). Finally, there are 9 combinations that correspond to dominant **A** and **B** phenotypes, reflecting a probability of 9/16. The probabilities of 9/16, 3/16, 3/16, and 1/16 correspond exactly to the 9:3:3:1 ratio of the phenotypes determined from the Punnett square.

Exercise 3-3.

The genotypes corresponding to each of the blood types A, B, AB, and O, and the blood type frequencies are presented in the table below:

Blood Types	Genotypes	Blood Type Frequencies
A	$I^A I^A$ or $I^A i$	$p^2 + 2pr$
B	$I^B I^B$ or $I^B i$	$q^2 + 2qr$
AB	$I^A I^B$	$2pq$
O	ii	r^2

For example, the blood type frequency for type A is $p^2 + 2pr$ because type A may have two genotypes: $I^A I^A$ or $I^A i$. The frequency of the $I^A I^A$ genotype is p^2, because the frequency of the I^A is p and the two sites are independent. The frequency of the $I^A i$ genotype is $2pr$ because the frequency of i is r, and this allele can appear either on the maternal of the paternal chromosome (that is, because $I^A i$ and $i I^A$ represent the same genotype).

Exercise 3-5.

$\binom{4}{k} = \dfrac{4!}{k!(4-k)!}$. Substituting subsequently k =0, 1, 2, 3, and 4 gives the desired values. For example, $\binom{4}{2} = \dfrac{4!}{2!(4-2)!} = \dfrac{4!}{2!2!} = \dfrac{1\cdot2\cdot3\cdot4}{(1\cdot2)(1\cdot2)} = 6.$ In the same way, $\binom{4}{0} = 1, \binom{4}{1} = 4, \binom{4}{3} = 4, \binom{4}{4} = 1$ as in Table 3-2.

Chapter 4

Exercise 4-1.

HYPOTHESES

The following two hypotheses may be formulated. Hypothesis 1 can be tested using the Student t test. The testing of hypothesis 2 requires tests not covered in the text, but we have included it here for readers with broader statistical backgrounds.

(1) Magnetism influences the height of germinated seedlings;

(2) Magnetism influences seed germination.

Study design and grouping variable: First, in order to proceed with statistical comparisons, we introduce a grouping variable MAGNET=1 for the bowl placed on a magnet and MAGNET=0 for the other bowl. Because the two bowls are independent, we have an independent group design with two groups of seeds identified by the grouping variable MAGNET. All further statistical tests will contrast MAGNET=0 versus. MAGNET=1. Because none of the hypotheses is directional, two-tail significance levels will be used.

Hypothesis 1 requires us to judge whether the germinated seeds grew taller under the influence of a magnet. Thus, for this analysis we need to select only the seeds that have germinated and to compare their average heights (variable HEIGHT). Because height is a continuous variable that tends to be normally distributed, we use the independent-group t test to compare HEIGHT across the groups defined by MAGNET. The results

	Test of Hypothesis 1		Test of Hypothesis 2	
	Average (St. Error) Height of Germinated Seeds	t test	Number (%) of Seeds Germinated	Mann-Whitney Nonparametric Comparison
MAGNET=0	24.5 (0.34)	$t = -4.57$ $p < .001$	43 (86%)	$Z = -1.32$ (two-tail) $p = 0.18$
MAGNET=1	26.9 (0.41)		47 (94%)	

TABLE 4-1.
Results for Exercise 4-1.

of such a comparison are presented in Table 4-1. The t test shows that the difference in the germination rate is *highly significant* ($p < .001$).

Figure 4-1 presents graphically the results from the height comparison of germinated plants, including average height on/off magnet and the standard error of the mean for each group. Because the standard errors do not overlap, we can visually conclude that the two groups of seeds have grown differently.

Hypothesis 2 requires us to judge whether the *count* of germinated seeds is greater under the influence of a magnet. Thus, a nonparametric test should be used. In order to execute such a test, we first recode all data into a new binary variable: GERM=1 if the seed germinated and GERM=0 if the seed did not germinate. Then we apply a Mann–Whitney test comparing the variable GERM across the groups defined by MAGNET. The results of such a comparison are presented in Table 4-1 below. The Mann–Whitney test shows that the difference in the germination rate is *not significant* ($p = .18$).

Future Research: Because the nonparametric test did not find a significant difference in the seed germination rate, but the significance level is generally low, further investigation with a larger sample size would determine whether magnetism has any influence on the rate of seed germination.

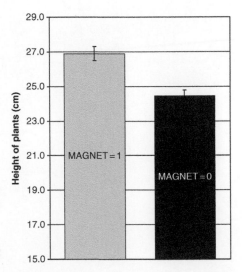

FIGURE 4-1.
Average height of lentil plants (cm) at 2 weeks after planting.

Chapter 5

Exercise 5-1.

a) Because 1 mmol/L = 18 mg/dl, the target range of [3.9, 10] mmol/L corresponds to [70, 180] mg/dl. The entire range of [1.1, 33.3] mmol/L corresponds to [20, 600] mg/dl.

b) The conditions

$$f(33.3, \alpha, \beta) = -f(1.1, \alpha, \beta) \quad \text{and}$$

$$f(10, \alpha, \beta) = -f(3.9, \alpha, \beta)$$

correspond to the following system of equations for α, $\beta > 0$.

$$[(\ln(33.3))^{\alpha} - \beta] = -[(\ln(1.1))^{\alpha} - \beta]$$

$$[(\ln(10))^{\alpha} - \beta] = -[(\ln(3.9))^{\alpha} - \beta]$$

This can be rewritten as

$$(\ln(33.3))^{\alpha} + (\ln(1.1))^{\alpha} = 2\beta \qquad (5\text{-}1)$$

$$(\ln(10))^{\alpha} + (\ln(3.9))^{\alpha} = 2\beta,$$

where from, the following equation for $\alpha > 0$ is obtained:

$$(\ln(33.3))^{\alpha} + (\ln(1.1))^{\alpha} - (\ln(10))^{\alpha} + (\ln(3.9))^{\alpha} = 0.$$

This is a nonlinear equation that cannot be solved directly, but using Derive, MATLAB, or any other system or a scientific calculator that can estimate the solution numerically, we obtain $\alpha = 1.0329$. Substituting this value in Eq. (5-1), we obtain

$$\beta = \frac{1}{2}[(\ln(33.3))^{1.0329} + (\ln(1.1))^{1.0329}] = 1.8707.$$

c) The value of γ is determined from the condition

$$\gamma \cdot [(\ln(33.3))^{1.0329} - 1.8707] = \sqrt{10}, \text{ or}$$

$$\gamma = \frac{\sqrt{10}}{[(\ln(33.3))^{1.0329} - 1.8707]} = 1.7740.$$

Exercise 5-3.
....

The LBGI and HBGI are most sensitive to the addition of extremely low and extremely high BG measurements, respectively. In general, measurements in the hypoglycemic range will cause the LBGI to increase and measurements in the hyperglycemic range will cause the HBGI to increase. It should be noted that because both the LBGI and HBGI are averaging over the whole data set of BG measurements, the indices are relatively stable with regard to extreme low or high BG values.

Exercise 5-5.
....

The standard deviation provides a measure for how spread the data is around the average value. The results from Table 5-2(A) in the text confirm that in T1DM, the BG deviates from the desired average (which would, ideally, be in the target range) significantly more than in T2DM. However, because the standard deviation does not provide any indication for the direction of the deviations from the average, it cannot be effective as a measure for hypoglycemia or hyperglycemia.

Chapter 6

Exercise 6-1.

The mean value of a data set does not indicate in any way how large or small the variability is. The following two data sets, for example, both have mean value equal to 3: A = {3, 3, 3, 3, 3} and B = {1, 2, 3, 4, 5}. The set B, however, exhibits more variability.

Exercise 6-3.

Similarities: (1) As in the definition for rl and rh given by Eq. (5-5) in Chapter 5 of the text, the functions $rl(x)$ and $rh(x)$ describe the risk for deceleration or acceleration, respectively, associated with each measurement's deviation from a preferred target value. (2) In both cases, the risk increases with the square of the distance of the measurement from a preferred target value. (3) In both cases, the risks for deviations to the left or the right of the target value are computed separately.

Differences: (1) The functions $rd(x)$ and $ra(x)$ are computed directly from the set of RR measurements while the functions $rl(x)$ and $rh(x)$ are computed for the data set of transformed BG measurements (via the transformation $f(BG)$ from Eq. (5-4) of the text. (2) For the functions $rd(x)$ and $ra(x)$ the risk is measured as the square of the distance from the median m of the data set of RR values. For the functions rl and rh from Chapter 5 the risk is measured as the square of the distance from the value 0—the center of the transformed, via the transformation $f(BG)$, euglycemic BG range.

Exercise 6-5.

The solution of this problem includes three steps: (1) Compute the median of RR intervals, which in this case is 666.67 milliseconds; (2) compute $ra(x)$ and $rd(x)$ for each RR interval, and (3) compute R_1 and R_2 as the average of $ra(x)$ and $rd(x)$. The results of the computations are given in Table 6-2.

Exercise 6-7.

There are four different 2-term subsequences of S2: {1,1}, {0,1}, {1,0}, and {0,0}.

RR Interval (milli-seconds)	$rd(x)$	$ra(x)$
666.7	0.00	0.00
666.7	0.00	0.00
659.3	0.00	54.76
666.7	0.00	0.00
689.7	529.00	0.00
652.2	0.00	210.25
666.7	0.00	0.00
681.8	228.01	0.00
666.7	0.00	0.00
645.2	0.00	462.25
	$R_2 = 75.70$	$R_1 = 72.73$

TABLE 6-2.
Results for Exercise 6-5.

Exercise 6-9.

Both sequences have standard deviation SD = 0.5, and, thus, the tolerance value for both sequences is $t = r \cdot SD = (0.2)(0.5) = 0.1$. Therefore, two sequences are "matches" only if they are identical.

For the sequence S1: 0, 0, 1, 0, 0, 1, 0, 0, 1, all subsequences of length 2 (beginning at up to N-m = 9–2 = 7) in the series are: 00, 01, 10, 00, 01, 10, 00. The total number of template matches of length $m = 2$ is B = 2 + 1 + 1 + 2 + 1 + 1 + 2 = 10. Further, all subsequences of length m + 1 = 3 are: 001, 010, 100, 001, 010, 100, 001 and the total number of template matches of length 3 is A = 2 + 1 + 1 + 2 + 1 + 1 + 2 = 10. Therefore, SampEn = $-\ln(A/B) = -\ln(10/10) = 0$.

For the sequence S2: 1, 0, 0, 0, 1, 0, 1, 0, 0, all subsequences of length 2 (beginning at up to N-m = 7) in the series are: 10, 00, 00, 01, 10, 01, 10, and all subsequences of length 3 are: 100, 000, 001, 010, 101, 010, 100. Thus, B = 2 + 1 + 1 + 1 + 2 + 1 + 2 = 10 and A = 1 + 0 + 0 + 1 + 0 + 1 + 1 = 4, and SampEn = $-\ln(A/B) = -\ln(4/10) = 0.9163$.

The SDs for the two sequences are the same because the sequences have exactly the same numbers of 0s and 1s.

Chapter 7

Exercise 7-1.

a) The rates of change in the concentrations of A, B, and C for the balanced equation $mA + nB \leftrightarrow C$, will be given by the differential equations:

$$\frac{d[C]}{dt} = k_1[A]^m[B]^n - k_2[C]$$

$$\frac{d[A]}{dt} = k_2[C] - k_1[A]^m[B]^n, \tag{7-1}$$

$$\frac{d[B]}{dt} = k_2[C] - k_1[A]^m[B]^n$$

where k_1 and k_2 are the respective reaction rate constants for the reactions $mA + nB \rightarrow C$ and $C \rightarrow mA + nB$.

In equilibrium, the rates of change of the concentrations are zero; that is, $\frac{d[C]}{dt} = 0, \frac{d[A]}{dt} = 0$, and $\frac{d[B]}{dt} = 0$. From Eqs. (7-1), we obtain $k_1[A]^m[B]^n - k_2[C] = 0$, which yields the law of mass action

$$[C] = \frac{k_1}{k_2}[A]^m[B]^n = K_a[A]^m[B]^n \text{ with } K_a = \frac{k_1}{k_2}.$$

(b) Because m molecules of type A and n molecules of type B are necessary to produce one molecule of type C, the association rate in the rate of change $\frac{d[C]}{dt}$ in Eq. (7-1) is now proportional to the number of ways that m molecules of type A can "meet" n molecules of type B. Mathematically, this is equivalent to counting the number of ways in which m molecules of type A and n molecules of type B can be chosen independently, given the concentrations [A] and [B]. Because each of the type A molecules can be chosen in numbers of ways proportional to [A], the number of ways to independently choose m type-A molecules is proportional to $[A]^m$. The justification of this claim is as follows: If N denotes the total number of type-A molecules, the number of ways to choose m molecules without replacement is $N(N-1)\ldots(N-m+1)$. Because N is much bigger than m, the following approximation will be quite accurate: $(N-1) \approx N$, $(N-2) \approx N \ldots (N-m+1) \approx N$. Thus, $N(N-1)\ldots(N-m+1) \approx N^m$. Because N is proportional to [A], the last approximation shows that the number of ways to choose m type-A molecules is proportional to $[A]^m$. In the same way, the number of ways to independently choose n type-B molecules is proportional to $[B]^n$. Since we assume that type A and type B molecules are uniformly mixed at all times and chosen independently from one another, the number of ways to choose m molecules of type A and n molecules of type B will be proportional to $[A]^m[B]^n$. The coefficient of proportionality is accounted for in the constant k_1, which also reflects the likelihood that a molecule of type C will be produced when m molecules of type A "meet" n molecules of type B.

Exercise 7-3.

Graphs for Eq. (7-9) for different values of n and $k = 0.5$ are shown in Figure 7-1(A). Graphs for Eq. (7-9) for n = 2 and different values for k are shown in Figure 7-1(B).

a) As expected, when k is kept fixed, the curve becomes steeper and reaches saturation levels faster as n increases.
b) When n is kept fixed, the curve becomes steeper and reaches saturation levels faster as k increases.

Exercise 7-5.

a) For $\Xi_4 = 1 + K_{41}[O_2] + K_{42}[O_2]^2 + K_{43}[O_2]^3 + K_{44}[O_2]^4$,
we calculate $\frac{\partial \Xi_4}{\partial [O_2]} = K_{41} + 2K_{42}[O_2] + 3K_{43}[O_2]^2 + 4K_{44}[O_2]^3.$

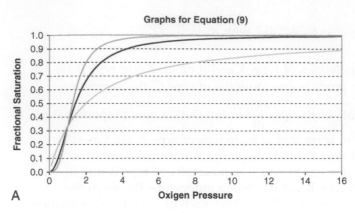

FIGURE 7-1(A).
Graphs for Equations (7-9) for $k = 0.5$ and values for $n = 1$ (light gray), $n = 2$ (black) and $n = 3$ (dark gray).

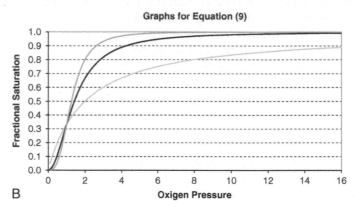

FIGURE 7-1(B).
Graphs for Equations (7-9) for $n = 2$ and values for $k = 0.1$ (light gray), $k = 0.3$ (black) and $k = 0.5$ (dark gray).

Thus, $[O_2] \cdot \dfrac{\partial \Xi_4}{\partial [O_2]} = K_{41}[O_2] + 2K_{42}[O_2]^2 + 3K_{43}[O_2]^3 + 4K_{44}[O_2]^4$

and subsequently $\overline{Y}_4 = \dfrac{\overline{N}}{4} = \dfrac{1}{4} \dfrac{[O_2]}{\Xi_4} \dfrac{\partial \Xi_4}{\partial [O_2]}$

$$= \frac{1}{4} \frac{K_{41}[O_2] + 2K_{42}[O_2]^2 + 3K_{43}[O_2]^3 + 4K_{44}[O_2]^4}{1 + K_{41}[O_2] + K_{42}[O_2]^2 + K_{43}[O_2]^3 + K_{44}[O_2]^4}, \text{ verifying Eq. (7-22).}$$

b) For $\Xi_2 = 1 + K_{21}[O_2] + K_{22}[O_2]^2$, the calculations are similar.

$\dfrac{\partial \Xi_2}{\partial [O_2]} = K_{21} + 2K_{22}[O_2]$ and therefore $[O_2] \cdot \dfrac{\partial \Xi_2}{\partial [O_2]} = K_{21} + 2K_{22}[O_2]$

$= K_{21}[O_2] + 2K_{22}[O_2]^2$. Thus $\overline{Y}_2 = \dfrac{\overline{N}}{2} = \dfrac{1}{2} \dfrac{[O_2]}{\Xi_2} \dfrac{\partial \Xi_2}{\partial [O_2]}$

$$= \frac{1}{2} \frac{K_{21}[O_2] + 2K_{22}[O_2]^2}{1 + K_{21}[O_2] + K_{22}[O_2]^2}.$$

Exercise 7-7.

Using that the binding polynomial is the sum of the concentrations of all binding species present in solution, we obtain (using Eqs. (7-11) in the text) that

$$\Xi_4 = [Hb_4] + [Hb_4O_2] + [Hb_4(O_2)_2] + [Hb_4(O_2)_3] + [Hb_4(O_2)_4]$$

$$= [Hb_4] + K_{41}[Hb_4][O_2] + K_{42}[Hb_4][(O_2)]^2 + K_{43}[Hb_4][(O_2)]^3$$

$$+ K_{44}[Hb_4][(O_2)]^4,$$

where K_{4i}, $i = 1,2,3,4$, are the equilibrium constants from Eqs. (7-12). If we now take, as in the derivation of Eqs. (7-27) in the text, the concentration of the reference state to be the unoxygenated tetrameric hemoglobin concentration and express hemoglobin concentration as a fraction of the unoxygenated hemoglobin concentration, we obtain (because in these units $[\alpha\beta] = 1$) that

$$\Xi_4 = 1 + K_{41}[O_2] + K_{42}[(O_2)]^2 + K_{43}[(O_2)]^3 + K_{44}[(O_2)]^4.$$

Chapter 8

Exercise 8-1.

We need to show that the solution (a,b) of the system of equations

$$a \sum_i X_i^2 + b \sum_i X_i - \sum_i X_i Y_i = 0$$

$$a \sum_i X_i + nb - \sum_i Y_i = 0,$$

is of the form given in Eq. (8-5) in the text. Expressing a from the first equation and substitution in the second gives

$$a = \frac{-b \sum_i X_i + \sum_i X_i Y_i}{\sum_i X_i^2}$$

$$\frac{-b \left(\sum_i X_i\right)^2 + \left(\sum_i X_i Y_i\right)\left(\sum_i X_i\right)}{\sum_i X_i^2} + nb - \sum_i Y_i = 0,$$

or

$$-b\left(\sum_i X_i\right)^2 + \left(\sum_i X_i Y_i\right)\left(\sum_i X_i\right) + nb\left(\sum_i X_i^2\right)$$

$$-\left(\sum_i X_i^2\right)\left(\sum_i Y_i\right) = 0,$$

which can also be written as

$$[n\left(\sum_i X_i^2\right) - \left(\sum_i X_i\right)^2]b + \left(\sum_i X_i Y_i\right)\left(\sum_i X_i\right)$$

$$-\left(\sum_i X_i^2\right)\left(\sum_i Y_i\right) = 0,$$

and $b = \dfrac{\left(\sum_i X_i^2\right)\left(\sum_i Y_i\right) - \left(\sum_i X_i Y_i\right)\left(\sum_i X_i\right)}{n\left(\sum_i X_i^2\right) - \left(\sum_i X_i\right)^2}$, establishing the

expression for b in Eqs. (8-5).

Substituting now this value in the second of Eqs. (8-4) gives

$$a\sum_i X_i + n\frac{\left(\sum_i X_i^2\right)\left(\sum_i Y_i\right) - \left(\sum_i X_i Y_i\right)\left(\sum_i X_i\right)}{n\left(\sum_i X_i^2\right) - \left(\sum_i X_i\right)^2} - \sum_i Y_i = 0,$$

or

$$a\sum_i X_i + \frac{n\left(\sum_i X_i^2\right)\left(\sum_i Y_i\right) - n\left(\sum_i X_i Y_i\right)\left(\sum_i X_i\right) - n\left(\sum_i X_i^2\right)\sum_i Y_i + \left(\sum_i X_i\right)^2\sum_i Y_i}{n\left(\sum_i X_i^2\right) - \left(\sum_i X_i\right)^2} = 0,$$

and therefore $a = \dfrac{n\left(\sum_i X_i Y_i\right)\left(\sum_i X_i\right) - \left(\sum_i X_i\right)^2\sum_i Y_i}{\left(\sum_i X_i\right)\left(n\left(\sum_i X_i^2\right) - \left(\sum_i X_i\right)^2\right)} =$

$\dfrac{n\left(\sum_i X_i Y_i\right) - \left(\sum_i X_i\right)\sum_i Y_i}{n\left(\sum_i X_i^2\right) - \left(\sum_i X_i\right)^2}$, establishing the expression for a in

Eqs. (8-5).

Exercise 8-3.

We use Eqs. (8-5) from the text. From Table 8-1, compute

$$\sum_i X_i Y_i = (0.25)(1.3) + (0.5)(2.7) + (0.75)(3.3) + (1.0)(5.1) = 9.25,$$

$$\sum_i X_i = 0.25 + 0.5 + 0.75 + 1.0 = 2.5$$

$$\sum_i Y_i = 1.3 + 2.7 + 3.3 + 5.1 = 12.4$$

$$\sum_i X_i^2 = (0.25)^2 + (0.5)^2 + (0.75)^2 + (1.0)^2 = 1.875.$$

Substitution into Eqs. (8-5) yields:

$$a = \frac{4(9.25) - (2.5)(12.4)}{4(1.875) - (2.5)^2} = \frac{6}{1.25} = 4.8,$$

$$b = \frac{(1.875)(12.4) - (9.25)(2.5)}{4(1.875) - (2.5)^2} = \frac{0.125}{1.25} = 0.1.$$

Exercise 8-5.

$$P = \begin{bmatrix} \dfrac{\partial G(guesses, X_1)}{\partial\, guesss_1} & \dfrac{\partial G(guesses, X_1)}{\partial\, guesss_2} & \cdots \\[2ex] \dfrac{\partial G(guesses, X_2)}{\partial\, guesss_1} & \dfrac{\partial G(guesses, X_2)}{\partial\, guesss_2} & \cdots \\[2ex] \cdots & \cdots & \cdots \\[2ex] \dfrac{\partial G(guesses, X_n)}{\partial\, guesss_1} & \dfrac{\partial G(guesses, X_n)}{\partial\, guesss_2} & \cdots \end{bmatrix}$$

$$Y^* = \begin{bmatrix} Y_1 - G(guesses, X_1) \\ Y_2 - G(guesses, X_2) \\ \cdots \\ Y_n - G(guesses, X_n) \end{bmatrix}, \text{ and } \varepsilon = \begin{bmatrix} answer_1 - guess_1 \\ answer_2 - guess_2 \\ \cdots \end{bmatrix}.$$

The matrix P has as many columns as the number of parameters in the model. The number of parameters is also equal to the number of rows for the matrix ε.

Exercise 8-7.

When d is fixed (and thus not a parameter to be estimated from the data), the model has the form $Y = G(a, b, c; X) = a\sin(2\pi X/d) + b\cos(2\pi X/d) + c$. Then $\dfrac{\partial G(a, b, c; X_i)}{\partial a} = \sin(2\pi X_i/d)$, $\dfrac{\partial G(a, b, c; X_i)}{\partial b} = \cos(2\pi X_i/d)$, and

$\dfrac{\partial G(a,b,c;X_i)}{\partial c} = 1$. Because none of the first-order derivatives depends on the parameters a, b, and c, all second-order derivatives will be equal to zero. Therefore, this is a linear model.

Exercise 8-9.

We need to show that the values of the parameters determined by the Gauss-Newton method are exactly the values of the parameters for which all first-order derivatives of the weighted sum of squared residuals function WSSR with respect to the model parameters are equal to zero. In other words, we need to show that

$\dfrac{\partial WSSR}{\partial\,parameter_j} = 0$, where the function WSSR is defined by

$$WSSR = \sum_i \left(\frac{Y_i - G(parameters; X_i)}{SEM_i}\right)^2.$$

A direct computation shows that

$$\frac{\partial WSSR}{\partial\,parameter_j} = -2\sum_i \frac{1}{SEM_i^2}\left[Y_i - G(parameters; X_i)\right]\frac{\partial G(parameters; X_i)}{\partial\,parameter_j},$$

and therefore the elements of the matrix $P^T Y^*$ in the Gauss–Newton least squares method are proportional to the partial derivatives of the function WSSR. Because the Gauss–Newton method terminates when it finds the solution $\varepsilon = 0$ where $\varepsilon = (P^T P)^{-1}(P^T Y^*)$, this means that for those values of the parameter either $(P^T P)^{-1} = 0$ or $(P^T Y^*) = 0$. However, $(P^T P)^{-1}$ cannot be zero, because it is invertible. This means that when $\varepsilon = 0$, $(P^T Y^*) = 0$. We have already shown that the elements of $(P^T Y^*)$ are proportional to the partial derivatives of the function WSSR with respect to the model parameter, establishing that $\dfrac{\partial WSSR}{\partial\,parameter_j} = 0$ at the values of the parameters for which $\varepsilon = 0$. Therefore, the Gauss–Newton algorithm determines the weighted least squares estimates for the model parameters.

Chapter 9

Exercise 9-1.

Notice that $h(t+1) = \sin(2\pi(t+1)) = \sin(2\pi t + 2\pi) = \sin(2\pi t)$, the last equality following from the fact that the function $\sin(x)$ is periodic with period 2π. This shows that $h(t+T) = h(t)$ for $T = 1$. Next, the value for $T > 0$ is the smallest, because 2π is the smallest positive value for which $\sin(2\pi t + 2\pi) = \sin(2\pi)$ (because 2π is the period of the sine function.) In the same way, $r(t + 1) = \cos(2\pi(t + 1)) = \cos(2\pi t + 2\pi) = \cos(2\pi t) = r(t)$, which shows that $r(t)$ is periodic with period $T = 1$.

Chapter 10

Exercise 10-1.

With $S(t) = S = const$ and $E(t) = e^{-\alpha t}$, the integral in Eq. (10-2) in the text becomes

$$C(t) = \int_0^t Se^{-\alpha(t-\tau)}d\tau + C(0)e^{-\alpha t} = C_0e^{-\alpha t} + Se^{-\alpha t}\int_0^t e^{\alpha\tau}d\tau$$

$$= C_0e^{-\alpha t} + \frac{S}{\alpha}e^{-\alpha t}(e^{\alpha\tau})\Big|_{\tau=0}^{\tau=t}$$

$$= C_0e^{-\alpha t} + \frac{S}{\alpha}e^{-\alpha t}(e^{\alpha t} - 1) = C_0e^{-\alpha t} + \frac{S}{\alpha} - \frac{S}{\alpha}e^{-\alpha t}$$

$$= \left(C_0 - \frac{S}{\alpha}\right)e^{-\alpha t} + \frac{S}{\alpha},$$

verifying the claim.

Exercise 10-3.

Similarities: In both figures, the concentration exhibits steep increases at certain moments because of drug intake (in Figure 1-23) or hormone secretion (in Figure 10-8). When the hormone secretion or drug intake events are occurring frequently, the concentration peaks form an increasing sequence that appears to plateau with time. When the hormone secretion or drug intake events are occurring infrequently, the concentration decays to approximately baseline levels after each peak, and there is no apparent trend of increase in the peak levels of the concentration.

Differences: The model depicted in Figure 1-23 assumed instantaneous increase in the concentration following drug intake, which was a simplifying assumption that facilitated the model development, but never occurs in reality. In Figure 1-23, this reflects the vertical jumps in the concentration function that are associated with the administration of each dose. The model depicted in Figure 10-8, on the other hand, does not assume instantaneous hormone secretion and subsequent instantaneous increases of the hormone concentration. Instead, the secretion is assumed to occur at a constant rate over short intervals of time that corresponds to steep (but not instantaneous!) increases of the hormone concentration over these time intervals.

Exercise 10-5.

By definition, $F_{down}(C) = \dfrac{T^n}{C^n + T^n}$, $F_{up}(C) = \dfrac{C^n}{C^n + T^n}$ and therefore

$$1 - F_{down}(C) = 1 - \frac{T^n}{C^n + T^n} = \frac{C^n + T^n - T^n}{C^n + T^n} = \frac{C^n}{C^n + T^n} = F_{up}(C). \text{ Further,}$$

$$F_{up}(T) = \frac{T^n}{T^n + T^n} = \frac{T^n}{2T^n} = \frac{1}{2}. \text{ In the same way,}$$

$$F_{down}(T) = \frac{T^n}{T^n + T^n} = \frac{T^n}{2T^n} = \frac{1}{2}.$$

Exercise 10-7.

The following ODEs describe the network dynamics.

$$\frac{dC_A}{dt} = -\alpha C_A(t) + a F_{up}(C_B(t - D))$$

$$\frac{dC_B}{dt} = -\beta C_B(t) + b F_{down}(C_A(t)).$$

Using the definitions for $F_{up(down)}$ from Eq. (10-5), the equations become:

$$\frac{dC_A}{dt} = -\alpha C_A(t) + a \frac{(C_B(t - D_B))^{n_B}}{(C_B(t - D_B))^{n_B} + (T_B)^{n_B}} = -\alpha C_A(t) + a \frac{(C_B(t - D_B)/T_B)^{n_B}}{(C_B(t - D_B)/T_B)^{n_B} + 1}$$

$$\frac{dC_B}{dt} = -\beta C_B(t) + b \frac{(T_A)^{n_A}}{(C_A(t))^{n_A} + (T_A)^{n_A}} = -\beta C_B(t) + b \frac{1}{(C_A(t)/T_A)^{n_A} + 1}.$$

Exercise 10-9.

Let $\varepsilon > 0$ be arbitrary.

First, we compare the solution $C_A(t)$ of Eqs. (10-14) to the solution of the equation:

$$\frac{dG}{dt} = -\alpha G(t) + a.1$$

$$G(0) = C_A(0)$$

Assuming for simplicity that $t_0 = 0$ and applying the hint (because the control function is $\dfrac{1}{(C_B(t - D_B)/T_B)^{n_B} + 1} \leq 1$) we get that $C_A(t) \leq G(t)$ for all $t > 0$. On the other hand, we can solve the above problem for $G(t) : G(t) = [C_A(0) - a/\alpha]e^{-\alpha t} + a/\alpha$. Because $\lim e^{-\alpha t} = 0$ for $t \to \infty$, for any $\varepsilon_1 > 0$ we can find a positive constant T_1 such that for all $t \geq T_1$ we have $C_A(t) \leq G(t) \leq a/\alpha + \varepsilon_1$.

In the same way, for the right-hand side of the second inequality in Eqs. (10-17), one can show that there exists another positive constant T_2 such that for all $t \geq T_2$ we have $C_B(t) \leq b/\beta + \varepsilon_1$. Therefore, if we choose $T_3 = \max(T_1, T_2)$, both inequalities are satisfied for $t \geq T_3 : C_A(t) \leq G(t) \leq a/\alpha + \varepsilon_1$ and $C_B(t) \leq b/\beta + \varepsilon_1$.

Next, we compare $C_A(t)$ and the solution of the problem

$$\frac{dF}{dt} = -\alpha F(t) + a\frac{1}{([b/\beta + \varepsilon_1]/T_B)^{n_B} + 1}.$$

$$F(T_3) = C_A(T_3).$$

Because for $t \geq T_3$, $\dfrac{1}{([b/\beta + \varepsilon_1]/T_B)^{n_B} + 1} \leq \dfrac{1}{(C_B(t - D_B)/T_B)^{n_B} + 1}$, the hint gives us that $F(t) \leq C_A(t)$ for $t \geq T_3$. Setting now

$a_1 = a\dfrac{1}{([b/\beta + \varepsilon_1]/T_B)^{n_B} + 1}$ and solving for F, we obtain that:

$C_A(t) \geq [C_A(T_1) - a_1/\alpha]e^{-\alpha t} + a_1/\alpha$ for $t \geq T_3$. Because $\lim e^{-\alpha t} = 0$ for $t \to \infty$, for any $\varepsilon > 0$ we can find a positive constant $T_4 > T_3$ such that for all $t \geq T_4$ we have $C_A(t) > a_1/\alpha - \varepsilon/2$.

Further, because $a_1 = a\dfrac{1}{([b/\beta + \varepsilon_1]/T_B)^{n_B} + 1}$ approaches $a\dfrac{1}{(b/[\beta T_B])^{n_B} + 1}$ as ε_1 approaches zero, we can choose $\varepsilon_1 > 0$ so small that

$a_1/\alpha > \dfrac{a}{\alpha}\dfrac{1}{(b/[\beta T_B])^{n_B} + 1} - \varepsilon/2$, which establishes that

$C_A(t) > a_1/\alpha - \varepsilon/2 > \dfrac{a}{\alpha}\dfrac{1}{(b/[\beta T_B])^{n_B} + 1} - \varepsilon/2 - \varepsilon/2 = \dfrac{a}{\alpha}\dfrac{1}{(b/[\beta T_B])^{n_B} + 1} - \varepsilon,$

and completes the proof of the first of the inequalities in Eqs. (10-17).

To complete the left-hand side of the second inequality we apply the hint in a similar manner as above and we use the fact that

$$\frac{(C_A(t)/T_A)^{n_A}}{(C_A(t)/T_A)^{n_A} + 1} = \frac{1}{(T_A/C_A(t))^{n_A} + 1} > \frac{1}{(T_A/\min C_A)^{n_A} + 1}.$$

Chapter 11

Exercise 11-1.
....

With a constant period of less than 24 hours, the acrophase will be occurring earlier and earlier each day. Thus, the plot depicting the time of daily acrophase will be comprised of data points forming a line with a negative slope. As an illustration, the plot in Figure 11-1 depicts the daily times of acrophase for a data series exhibiting a constant period of 20 hours.

Exercise 11-3.
....

COSIN2NL: The results for the detrended time series are close to those for the original time series. The period is estimated to be 23.3 ± 0.199 hours for the original series and 23.7 ± 0.137 for the detrended time series. Thus, detrending does not appear to be either necessary or beneficial for this analysis.

FFT-NNLS: For this analysis, the results for the detrended time series are substantially different when compared to those for the original time series. For the original series, a period of 7278 ± 798 was identified (Figure 11-23) which can be explained as an attempt to fit for the trend, and is no longer present after the detrending. The circadian component for the detrended series is estimated at 23.66 ± 0.16 hours (which is close to 23.41 ± 0.31 hours for the original one). In addition, FFT-NLLS for the detrended series identified two periodic components: 18.33 ± 0.29 hours and 14.91 ± 0.26 hours (Figure 11-32). It appears that analyzing the detrended series may be worthwhile, especially for eliminating the large-period estimates present in the results for the original time series.

PHASEREF: For this analysis, the results for the detrended time series are close to those for the original time series. In both cases (Figure 11-25

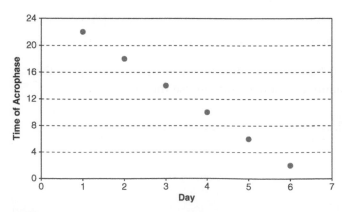

FIGURE 11-1.
A plot depicting the daily time of acrophase for a time series with a period of 20 hours.

and Figure 11-33), the PHASEREF algorithm fails to properly identify the simulated value of 24 hours for the period because of noise confounds that are still present after the detrending. This algorithm is not substantially influenced by the presence of or the lack of a trend and, thus, the detrending is not beneficial or necessary in this case.

Chapter 12

Exercise 12-1.

Positive values in the Log column correspond to an increase in expression, while negative values correspond to a decrease. The characterization as increase/decrease for each spot position is given in Table 12-1 below.

Spot position		Gene	Ratio		Expression
Row	Column	"name"	Red/Green	Log	increase/decrease
1	2	A	10	3.32	Increase
1	5	B	0.1	−3.32	Decrease
2	1	C	0.1	−3.32	Decrease
2	3	D	2	1.00	Increase
2	4	E	0.1	−3.32	Decrease
2	5	F	1.5	0.58	Increase
3	2	G	0.5	−1.00	Decrease
3	3	H	10	3.32	Increase
4	1	I	1	0.00	No Change
4	3	J	0.1	−3.32	Decrease
4	5	K	5	2.32	Increase
5	1	L	10	3.32	Increase
5	2	M	0.2	−2.32	Decrease
5	4	N	0.67	−0.58	Decrease

TABLE 12-1.
Data for gene expression increase/decrease related to Exercise 12-1.

Using the logarithm of the ratio instead of the ratio itself has the advantage that the increases and decreases of the same magnitude give values of the same magnitude that differ only by their sign. For example, a ten-fold increase in expression in the cancer cells, such as gene A, will give a logarithm of the ratio of 3.32, and a gene that has a ten-fold decrease in expression in the cancer cells, such as gene B, will give a logarithm of the ratio of −3.32. In contrast, a ten-fold increase in expression in the cancer cells for gene A, will give a ratio of 10, but a ten-fold decrease in expression in the cancer cells for gene B, will give a ratio of 0.1.

In other words, when the ratio is used, the range of values indicating an increase in gene expression will be $(1, \infty)$ while the range of values for the ratios that indicate a decrease will only be $(0,1)$. This is not optimal, because the whole range of decrease in expression is being squeezed in the interval $(0,1)$. The use of a logarithm of the ratio alleviates this problem: the range of values indicating increase is now $(0, \infty)$ while the range of values indicating a decrease in expression is $(-\infty, 0)$. This symmetry allows for an easy interpretation of the magnitude of increase/decrease. More specifically, values of the logarithm that have the same magnitude but different signs correspond to expression increases (positive values) and decreases (negative value) of equal magnitudes.

Exercise 12-3.

Let $x_i = (x_{i1}, x_{i2}, \ldots, x_{in})$ and $x_k = (x_{k1}, x_{k2}, \ldots, x_{kn})$ be the i-the and k-th row of the gene expression matrix. We want to show that the dissimilarity measure

$$d(x_i, x_k) = \sum_{j=1}^{n} |x_{ij} - x_{kj}|$$

satisfies conditions (1) through (3).

(1) Because the absolute value is always non-negative and the dissimilarity measure is defined as a sum of absolute values, $d(x_i, x_k) \geq 0$ is satisfied.

(2) $d(x_i, x_k) = d(x_k, x_i)$ is obvious, because for any two numbers a and b $|a - b| = |b - a|$, and thus

$$d(x_i, x_k) = \sum_{j=1}^{n} |x_{ij} - x_{kj}| = \sum_{j=1}^{n} |x_{kj} - x_{ij}| = d(x_k, x_i).$$

(3) To prove that $d(x_i, x_k) \leq d(x_i, x_s) + d(x_s, x_k)$, we will use the well-known triangle inequality for real numbers, which states that for any real numbers a, and b, $|a| + |b| \geq |a + b|$.

Now,

$$d(x_i, x_s) + d(x_s, x_k) = \sum_{j=1}^{n} |x_{ij} - x_{sj}| + \sum_{j=1}^{n} |x_{sj} - x_{kj}|$$

$$= \sum_{j=1}^{n} |x_{ij} - x_{sj}| + |x_{sj} - x_{kj}| \geq \sum_{j=1}^{n} |x_{ij} - x_{kj}| = d(x_i, x_k),$$

establishing that $d(x_i, x_k) \leq d(x_i, x_s) + d(x_s, x_k)$. The last inequality in the chain of inequalities above follows from the triangle inequality because

$$|x_{ij} - x_{sj}| + |x_{sj} - x_{kj}| \geq |(x_{ij} - x_{sj}) + (x_{sj} - x_{kj})| = |x_{ij} - x_{kj}|.$$

INDEX

A

AAOG. *See* American Academy of Obstetrics and Gynecology
AAP. *See* American Academy of Pediatrics
Ackers model, hemoglobin-oxygen binding, 224–226
Acrophase, 443
Adair equations, hemoglobin-oxygen binding, 221–222
Adenosine triphosphate (ATP), 212
AIRFILTER software, 360–362, 362f, 373, 376f, 377f, 378f, 384f
Alleles, 100, 103
  dominant, 428
Amensalism, 53
American Academy of Obstetrics and Gynecology (AAOG), 185
American Academy of Pediatrics (AAP), 182
Analysis, analytical strategies, 365–373
ApEn. *See* Approximate entropy
Approximate entropy (ApEn), 204
Association equilibrium constant, 214
Asymptotic standard errors, data fitting, procedure objectives, 259–260
ATP. *See* Adenosine triphosphate

B

Beltrami, Edward, 89
BERKELEY MADONNA software, 280, 310, 322
  discrete models solutions and, 47, 49
  dynamical systems, 45–49, 47f, 48f, 49f
  numerical solutions and, 47f, 49f
Best fit, 6, 7f. *See also* Data fitting
BG. *See* Blood glucose
Binding polynomials, 225
  fractional saturation functions derivation, 226–231
Binding reactions, 213–217. *See also* Cooperative binding, blood oxygen transport; Ligand binding, data fitting
  cooperative binding, blood oxygen transport, 213–217
  drug action, drug-receptor complex, 213–214
  fractional saturation, 216–217, 217f, 219–221, 226–230
  mass action law, 214

Binomial Theorem, 122
Bioinformatics, 391
Biological clocks, 341–351. *See also* Circadian rhythms; Rhythmic data
  entrainment, 343
  molecular basis, 344–347
  structure, function in mammals, 343–344
Bioluminescence studies, circadian rhythms, 349–350, 350t
Biomass, 11, 419t, 420t
Blood glucose (BG), 151, 159f. *See also* Risk analysis, blood glucose data
  group comparisons, 169–172, 170f
  high BG index, validation, 175, 432
  low BG as hypoglycemia predictor, validation, 172–175, 173f, 174f, 432
  measurement scale symmetrization, 159–163
  ranges, 154–155, 154f
  risk function, 163–165, 164f, 165f
  risk indices, 165–166
  scale conversion, 160f
  scale transformation, 162f, 163f
  self-monitoring, 157–158
  T1DM patient and fluctuations of, 155f
  T1DM *vs.* T2DM, 168f, 169f, 170f, 432
Blood oxygen transport. *See* Cooperative binding, blood oxygen transport
Blood type, 429
  frequency, 429
Bohr effect, hemoglobin-oxygen binding, 218–219
Bonferroni corrections, 171
Boveri, Theodor, 101

C

Carnivory, 53
Carrying capacity, 12–14, 16, 427
CDC. *See* Centers for Disease Control
cDNA. *See* Complimentary DNA
Centers for Disease Control and Prevention (CDC), 185
Central Limit Theorem, 125
Chaos theory, 30
Chi-square distribution, 133, 133f
Cholera, 56, 57t
Chromosome Theory of Inheritance (Boveri, Sutton), 101

Chromosomes, heredity, 102–107, 102f, 104f, 105f
Circadian rhythms, 341–343. *See also* Rhythmic data
  circadian gene identification, 347–349, 348f
  computational challenges, 409–410
  data analysis, 350–351, 351f
  Drosophila, 347–348, 348f
  microarray technology and, 407–416
  per-luc bioluminescence studies, 349–350, 350t
  SCN and, 343–344
  TIM/PER protein complex, 348–349, 348f
  variance nonstationary, 350
  zeitgeher, 343
Clinical blood glucose optimization, 153–154
Cluster analysis, 402–407, 404t, 405f, 408f
CLUSTER hormone pulse algorithm, 287–288, 287f
Codominance, 119
Coefficient of variation (CV), 277
Commensalism, 53
Competition, exploitive, resource *vs.* interference, 53
Competitive interaction model, 91–94, 92f, 94f
Complex hereditary patterns, 118–120
Complimentary DNA (cDNA), 391–394
Continuous mathematics, 9
Continuous population growth model, 8–11
Continuous traits, 120, 121–127
Convolution integral model, 289–294, 289f, 290f, 292f, 293f
Cooperative binding, blood oxygen transport
  binding reactions, 213–217
  hemoglobin-oxygen binding, mathematical models, 213–226
  introduction, 211–213
Cosin2nl algorithm, 365–368, 366f, 367f, 376f, 379f
COSOPT algorithm, 410–416, 411f, 412f, 415–416f
CV. *See* Coefficient of variation
Cytology, 101

447